Angiogenesis in Health, Disease and Malignancy

Ali Salajegheh

Angiogenesis in Health, Disease and Malignancy

Springer

Ali Salajegheh
School of Medicine
Griffith University
Gold Coast, Queensland, Australia

ISBN 978-3-319-28138-4 ISBN 978-3-319-28140-7 (eBook)
DOI 10.1007/978-3-319-28140-7

Library of Congress Control Number: 2016933859

Printed on acid-free paper

This Springer imprint is published by Springer Nature
The registered company is Springer International Publishing AG Switzerland.

I would like to dedicate this book to my parents, my sister and brother, and my family for their unconditional love and support and to the love of my life for being my soul mate, my partner in life, and my best friend. I owe a great debt of gratitude to my teachers, mentors, and elders for their inspiring guidance and care, and I would like to thank everybody especially all my students who were important to the prosperous insight of this book, as well as expressing my apology that I could not mention them personally one by one.

Contents

Chapter 1
Introduction to Angiogenesis in Normal Physiology, Disease and Malignancy

Abstract Vasculogenesis is the de novo formation of blood vessels by mesodermal progenitors undergoing differentiation to endothelial cells. New vasculature formation from pre-existing vasculature occurs through the physiological remodelling process known as angiogenesis. Angiogenesis is implicated in the proliferation and growth of both physiologically normal and neoplastic tissues, through the establishment of vascular supply, essential for delivering growth requirements such as oxygen and nutrients. Many inhibitory and promoter genes regulate this process, however the role of specific genes in the shift from normal angiogenesis to tumour initiation is complex and thus poorly understood. This book investigates the process of angiogenesis in malignancies. The functional role of key regulatory factors will be examined in the context of normal healthy condition. Then the association of these factors to disease and malignancy, cancer proliferation and progression will be discussed. New insights into the role of angiogenesis and the therapeutic inhibition of its regulators will be investigated due to the great potential for exploitation in the development of a novel treatment for cancer.

Keywords Angiogenesis • Normal physiology • Disease • Malignancy

1.1 Introduction to Angiogenesis

There are two mechanisms for the formation of blood vessels in the body; angiogenesis and vasculogenesis. Both are essential in formation of the vascular network in our bodies.

Vasculogenesis is the formation of blood vessels from angioblasts. It is a dynamic process that involves cell–cell and cell–extracellular matrix interactions directed by growth factors and morphogens (Carmeliet 2005). This process includes differentiation of mesodermal stem cells into angioblasts, and then the differentiation of angioblasts into endothelial cells under the command of growth factors. It occurs inside the embryo and it initiates the first vascular system of our body.

Angiogenesis is the term used to describe the process of growth or formation of new blood vessels from pre-existing blood vessels (Folkman and Shing 1992). This process begins in utero and prolongs throughout life. Angiogenesis is an essential

A. Salajegheh, *Angiogenesis in Health, Disease and Malignancy*,
DOI 10.1007/978-3-319-28140-7_1

process in normal physiology in the body for it has major roles in healing, reproduction, growth of tissues and also diseases (Felmeden et al. 2003). The role of angiogenesis and its contribution to disease has been a topic of discussion for the last few decades (Folkman 1996). Metabolically active tissues in the body are always in close proximity to bloody capillaries for essential processes such as exchange of nutrients and metabolites must occur (Adair and Montani 2010). It is responsible for both remodelling and expansion of the existing network formed by vasculogenesis. Angiogenesis exists as a very important process in the body due to the fact that all tissues require oxygen. Oxygen is delivered through capillaries which are formed through angiogenesis. The control of angiogenesis in the body is dependent on a specific tissue's metabolic needs. The reason angiogenesis has gained much publicity in the world of health is due to its influence on both many diseases and cancer. Malignant growth is dependent on the formation of these new blood vessels, that supply nutrition to the tumour site, and provide a way for certain cells and survival factors to enter and leave circulation (Folkman 2006; Seaman et al. 2007). Chemical factors such as angiogenic growth factors, angiogenesis inhibitors and other genetic factors can cause abnormalities in the process of angiogenesis (Seaman et al. 2007). This lack of control results in either too much or too less angiogenesis occurring (Albini et al. 2012). The stimulation of angiogenesis is now considered therapeutic approach in terms of treating peripheral arterial disease, ischemic heart disease, and in wound care, however, in cases of ophthalmic conditions, cancer, rheumatoid arthritis, and other diseases, limitation of angiogenesis seems a better option (Folkman and Shing 1992).

In general, angiogenesis occurs in a series of steps that are initiated by a certain cell not having enough oxygen. This cell will release proangiogenic growth factors which attract inflammatory and endothelial cells that then proliferate and grow. Endothelial cells in the target location then secrete metalloproteases which digest the blood vessel walls allowing these cells to go towards the angiogenic stimulus. The process is completed through anastomosis which joins the capillaries coming from the arterioles and venules, establishing proper blood flow in the body (Carmeliet 2005). For angiogenesis to occur in the body, its positive regulation through growth factors must overcome the factors that inhibit angiogenesis. Angiogenesis is a complex process composed of many steps and a combination of factors such as cells, soluble factors and extracellular matrix (ECM) components (Liekens et al. 2001). Embryonic mesenchymal cells differentiate into endothelial cells to develop into blood vessels (Felmeden et al. 2003). There are growth factors that have been described to be associated with the positive regulation of angiogenesis; these include vascular endothelial growth factor (VEGF), transforming growth factors (TGF-beta), fibroblast growth factors (FGF), epidermal growth factor (EGF). On the other hand, there are also growth factors that have been shown to have little effect on proliferation of endothelial cells in angiogenesis (Carmeliet 2005). There are two types of angiogenesis that occur in adults: these are sprouting angiogenesis and intussusceptive angiogenesis.

Sprouting angiogenesis essentially refers to branching of existing blood vessels via enzymatic degradation of capillary basement membrane, proliferation and migration of endothelial cells towards angiogenic stimulus through the extracellular matrix, formation of a tube with lumen, fusion of vessels to form network and finally stabilization of vessels by recruitment of pericytes (Adair and Montani 2010). Angiogenic stimuli induce the endothelial cells of existing vessels to proliferate and increase in permeability. Next, the extracellular matrix of the proximal tissue gets degraded to give space for the new blood vessel. The endothelial cells continue sprouting and proliferating into the provisional extracellular space, and form the lumen of the developing capillary. In the final stages of angiogenesis stabilization occurs, where a mature vascular basement membrane covers and pericytes surround the new vessel to protect and support it (Hong et al. 2014; Senger and Davis 2011).

In terms of sprouting angiogenesis, the first growth factor was discovered in the 1980s which was fibroblast growth factor. FGF triggers all the main steps in the angiogenesis process and is produced by many cells, among which are macrophages and tumour cells. Despite the fact that FGF is not released on a regular basis, after its secretion into the extracellular matrix the angiogenesis cascade is initiated. Various research points to the fact that vascular endothelial growth factor (VEGF) is a very important part of the initiation of angiogenesis as well. Essentially, it activates the proliferation of endothelial cells and is also a powerful initiator of vasodilation in blood vessels. Furthermore, it is a stimulator of metalloproteinases and plasminogens (Hillen and Griffioen 2007).

To go into more depth on the process, we can use the role of VEGF as an example. VEGF provides an angiogenic stimulus that helps the endothelial tip cell sprout through the extra-cellular matrix thus forming the foundation for a new capillary to develop. The tip cells are aided by processes on them called filodopia which release proteolytic enzymes that establish a path through the ECM. The sprouting of the tip cells with their filodopia is also heavily dependent on the VEGF gradient as the filodopia are covered in many VEGF receptors. Following the tip cells, are endothelial stalk cells which cause the elongation of the sprout. These stalk cells are what forms the capillary lumen. The uniting of two tip cells at a centre of VEGF release forms one continuous lumen through which blood can perfuse. The VEGF stimulus dies down after the tissues in that area are sufficiently oxygenated (Carmeliet 2005).

Intussusceptive angiogenesis, in comparison, is a much newer discovery which was first observed in the lungs (Burri and Djonov 2002). It is a much faster and efficient form of angiogenesis as there is no endothelial cell proliferation and it is merely the splitting of a blood vessel into two by reorganization of the existing endothelial cells so that the cells extend into the lumen (Adair and Montani 2010; Carmeliet 2005). This ultimately causes an increase in the overall complexity of the network and occurs throughout life like sprouting angiogenesis.

1.2 Angiogenesis in Normal Physiology

During embryonic development the primary capillary plexus is developed by vasculogenesis. Vasculogenesis differs from angiogenesis in that the formation of new vasculature does not happen from pre-existing vessels but rather from endothelial cell precursors, called angioblasts. The primary capillary plexus is then remodelled by the process of angiogenesis (Papetti and Herman 2002). The growth of newly developed cells is angiogenesis-dependent; all cells require nutrition, oxygen and removal of wastes to be able to function (Folkman 1996). The growth of the vascular network that is the cardiovascular system is one of the earliest events of angiogenesis occurring during embryonic development (DeWitt 2005). After birth the blood vessels remain quiescent however it may also arise during adulthood such as seen in reproduction. During the menstrual cycle the development of the endometrium occurs resulting in highly-developed vascularity. The endometrium eventually sheds causing breakages in the surrounding blood vessels, which then undergo angiogenesis in order to repair the endometrial tissue (Smith 2001; Folkman 2006). Another instance where angiogenesis may occur later in life, and is of quite significance is in the regeneration of damaged tissues during the wound repair and healing process. During the process of wound healing, sprouting angiogenesis takes place resulting in endothelial projections that enter the blood clot over the wound (Tonnesen et al. 2000). The organised vascularity assists in the deposition of a collagen matrix within the granulation tissue and as it matures into a scar, the dense network of blood vessels eventually decreases (Tonnesen et al. 2000). Angiogenesis should be also enhanced in numerous nonrelated diseases such as in ischemic tissues or cardiac failure where reperfusion is required to improve the disease conditions; such event is also referred to as compensatory angiogenesis (Carmeliet 2003).

It is evident from the aforementioned roles in normal physiology that angiogenesis requires regulation in order to switch the process on and off when required. Certain naturally occurring inhibitors of this process prevent the formation of neovasculature under pathologic conditions (Folkman and Shing 1992). It can therefore be concluded that a dynamic relationship must exist between growth factors and inhibitors, and if disrupted could lead to disease.

1.3 Angiogenesis in Disease

Several chronic inflammatory diseases are, in fact, causative agents of excessive angiogenesis as part of the pathology (Carmeliet 2003). In such conditions, the balance between stimulant and inhibitory angiogenic factors tilt towards the stimulant outcome leading to 'angiogenic switch,' resulting in profound angiogenesis (Carmeliet 2005). These diseases include atherosclerosis, rheumatoid arthritis, diabetes, psoriasis, endometriosis and cancer. To treat such disorders, therapeutic inhibition of angiogenesis can be an option of improvement (Carmeliet 2003).

Multiple studies revealed that the immune system is critically impacted by angiogenesis. Studies have shown that during angiogenesis exposure of endothelial cells to angiogenic factors such as FGF-2 and VEGF down-regulates adhesion molecule expression on the endothelial cells. In addition, endothelial adhesion molecules controlled by cytokines such as TNF is severely hampered by angiogenic stimulus which interferes with activation and adhesion of leukocytes to endothelial tissues at the site of neovascularization (Carmeliet 2003). This was an important finding as it suggested that those chronic inflammatory disorders which cause angiogenesis were capable of escaping immune surveillance of the body. Such chronic inflammatory diseases are also capable of causing angiogenesis by introducing angiogenesis stimulatory factors.

In cases of rheumatoid arthritis, the rheumatoid synovial endothelium is enriched with VEGF an FGF-2 which are angiogenesis-promoting cytokines (Carmeliet 2003). clinical trials in attempt to treat rheumatoid arthritis deals with prescribing TNF- α-blocking antibodies which in turn revealed a significant decrease in serum VEGF levels, decreasing the synovial vascularity (Carmeliet 2003). Growth factors like PlGF which plays an angiogenic role in disease without activating quiescent vessels in healthy tissues become an attractive target in the development of a safe anti-angiogenic treatment (Luttun et al. 2004).

In addition, many leukocyte subtypes were found to produce multiple angiogenic factors such as VEGF, PDGF and FGF and various interleukins and proteinases. For example, neutrophils and natural killer cells are implicated in cyclical uterine angiogenesis and in abnormal angiogenesis in endometriosis. This revealed that different leukocytes have the ability to promote many angiogenic processes. Nevertheless leukocytes also generate angiogenesis inhibitors. Hence their overall role in initiating or terminating angiogenesis depends on the temporal and spatial balance of the released modulators. Mast cells are also found to release angiogenic factors when they encounter allergens and pathogens in the skin and mucosa (Mori et al. 2013).

In conditions such as ischemic heart disease or preeclampsia, the angiogenic switch is insufficient which leads to endothelial cell dysfunction, vessel malformation or regression and revascularisation, therefore in such conditions promoting revascularization or therapeutic angiogenesis can be beneficial (Carmeliet 2005). Currently the angiogenic factors such as VEGF and FGF are the genes of interest to be used as a treatment for disorders such as ischaemic tissues.

1.4 Angiogenesis in Malignancy

Cancer is a broad term collectively used to describe a number of diseases in which abnormal cells divide uncontrollably and have the ability to invade other tissues. Cancerous cells are able to spread to other parts of the body via lymph and blood systems. In the normal state, our body's plethora of cell types is constantly undergoing growth, division and eventually replacement when they die or become damaged. The origin of cancer can be attributed to a genetic mutation within cell DNA that affects the normal cell growth and division. Normal cell death does not occur and instead unwarranted new cell formation and growth occurs leading to the development of a mass of tissue known as a tumour. It is important to note that not all tumours are necessarily cancerous. Benign tumours are non-cancerous, do not spread to other parts of the body and can be removed without reoccurrence. Malignant tumours however are much more dangerous and can metastasise to other parts of the body (Nishida et al. 2006). Metastasis is the term used to describe cancer spread and the potential of cancer cells to leave the original site, penetrate into blood vessels and lymphatics and move through blood vessels to other organs (Seaman et al. 2007). Years of cancer research and meticulous studies have clearly shown that the angiogenesis phenomenon plays a pivotal role in the spread of cancer and is a fundamental concept involved in cancer metastasis; specifically the transition of benign tumours to those that become malignant. It was of previous thought prior to the 1960s that blood supply to tumours was actually carried out by vasodilation of pre-existing blood vessels. The new theories however support the angiogenesis phenomenon as the key blood supply for growth of tumours.

Angiogenesis, tumour growth and metastasis are also very closely related. In terms of malignancies, the extra blood vessels sustain the growth of tumours and in fact malignant tissue development, expansion and its spread are all angiogenesis-dependent (Folkman and Shing 1992). For a lethal tumour mass to develop, it is not only the proliferation of the malignant cells that is required but also the formation of new blood vessels (Folkman 2006). In a malignancy, endothelial cells of the neovasculature grow in a disorderly manner presenting with many structural shortcomings. Their role as a functional barrier is lost, and they cannot withstand the increased interstitial pressure, which then alters the blood flow and flux of leukocytes reaching the tumour site (Folkman 1996). Thus, angiogenesis not only sustains the tumour cells by meeting their increasing metabolic needs. They also, resulting from their incomplete structural build (i.e. a defective basement membrane and lack of normal perivascular connective tissue), provide these cells with many advantages; most importantly an easy penetration of metastatic tumour cells into circulation. Tumour cells can simply fall into the lumen and spread to distant sites and organs through the blood flow (Papetti and Herman 2002). Certain genetic changes promote the growth of malignant cells. So-called oncogenes – genetic material that carry the potential to induce cancer – give rise to uncontrolled proliferation and hindered apoptosis of the tumour cells (Papetti and Herman 2002). For every one endothelial cell there are approximately a hundred tumour cells, surrounding the vessels in a cylindrical fashion (Folkman 1995, 1996).

Cancer cells are able to release molecules that in fact act as signals to activate angiogenesis. These include many different many proteins and many other molecules. Among these, the two most angiogenic proteins are Vascular Endothelial Growth Factor (VEGF) and Basic Fibroblast Growth Factor (bFGF). Oncogenes induce angiogenesis primarily through an increased expression of the vascular endothelial growth factors (VEGF) in those cells (Folkman 1996, 2006). The tumour cells themselves are able to synthesize both VEGF and bFGF before they are secreted into surrounding tissue. From here, they bind to receptors (specific proteins) that are located on the outer surface of endothelial cells. Subsequent binding of protein to appropriate receptors initiates a signal transmission from a series of relay proteins into the endothelial cell nucleus (Munoz-Chapuli et al. 2004). This is a stimulus for a gene group to then make products required for endothelial cell growth.

Activated endothelial cells secrete a matrix called metalloproteinases (MMPs) into the surrounding tissue. MMPs are a distinct class of degradative enzymes and are responsible for breakdown of extracellular matrix that is primarily consisted of proteins and polysaccharides. This degradation is required for the proliferation and migration of endothelial cells. Over time, the endothelial cells begin to divide as they proliferate and organize into hollow tubes that progressively advance into a mature network of blood vessels (Lamalice et al. 2007).

Given the fact that angiogenesis is a key step in the development and spread of cancer, blocking angiogenesis seems a plausible solution to stop or slow down tumour growth. Numerous studies have and are currently being undertaken to find synthetic and natural angiogenesis inhibitors. Angiogenesis inhibitors obstruct upon the binding multistep binding process of the signalling molecules such as VEGF and bFGF.

One of the most well-known of the angiogenesis inhibitors is a drug called *Bevacizumab* or *Avastin* (Samant and Shevde 2011; Tanne 2011). *Bevacizumab* is a monoclonal antibody – a laboratory synthesised molecule that is made in such a way that they mimic the antibodies that the human body naturally produces as part of the immune system. The antibody binds to VEGF and blocks signalling of the molecule, leading to supressed formation of new blood vessel growth (angiogenesis). Reduced nutrient supply to the tumour can slow or stop its growth. The U.S. Food and Drug Administration (FDA) approves of Bevacizumab to be used alone for glioblastoma (tumours that originate from connective tissue in brain) and when in combination with other drugs for treatment of non-small cell lung cancers, metastatic colorectal cancer as well as metastatic renal cell cancer (Samant and Shevde 2011; Tanne 2011). Other FDA approved antiangiogenic drugs are *Sorafenib* (for hepatocellular carcinoma and kidney cancer), *Sunitinib* (kidney cancer and neuroendocrine tumours), *Pazopanib* (kidney cancer) and *Evorilimus* (kidney cancer and neuroendocrine tumours) (Samant and Shevde 2011; Tanne 2011). Alternative theories also suggest that *Bevacizumab* induces more stability within the tumour blood vessels (usually leaky) thus allowing for more effective penetration of chemotherapy into cancer cells. Other drugs such as *Endostatins* inhibits angiogenesis directly rather than through the signalling pathway. This drug cause apoptosis of endothelial cells or destruction of proliferating endothelial cells (Samant and Shevde 2011). As outlined previously, breakdown of the extracellular matrix is essential for endothelial cell proliferation and migration. Therefore drugs that can target the MMP's within the matrix can also act at angiogenesis inhibitors.

It is important to distinguish between angiogenesis inhibitors and conventional anticancer drugs. As the name suggests, angiogenesis inhibitors inhibit the growth of blood vessels but does not necessarily kill tumours. Instead, they have a high chance of preventing tumours from growing so it is suggested that that the therapy must be spanned over a long period to have any major effect.

In some cases, the inhibitors will be more effective in combination with other remedies such as chemotherapies. Like other medications there are reports of common side effects for these inhibitors such as high blood pressure, fatigue, rash, dry and itchy skin, hand-foot syndrome (tender, thickened areas of skin), diarrhoea, wound healing problems/cuts reopening and low blood counts (O'Reilly et al. 1997).

In conclusion, angiogenesis plays a key role in the development of cancer. Upon this observation, a suggested strategy against tumour growth is to not only target the tumour cells alone but also the endothelial cells of the newly formed blood vessels – a technique that has so far shown to be effective in the battle against cancer (Folkman 2006). This book attempts to discuss and summarise the role of important regulators of angiogenesis in regards to normal physiology, disease and malignancy.

References

Adair TH, Montani J-P (2010) Overview of angiogenesis. In: Rafael S (ed) Angiogenesis. Morgan & Claypool Life Sciences, San Rafael: California, USA

Albini A, Tosetti F, Li VW, Noonan DM, Li WW (2012) Cancer prevention by targeting angiogenesis. Nat Rev Clin Oncol 9(9):498–509. doi:10.1038/nrclinonc.2012.120

Burri PH, Djonov V (2002) Intussusceptive angiogenesis – the alternative to capillary sprouting. Mol Aspects Med 23(6):1–27. doi:10.1016/s0098-2997(02)00096-1

Carmeliet P (2003) Angiogenesis in health and disease. Nat Med 9(6):653–660. doi:10.1038/nm0603-653 nm0603-653 [pii]

Carmeliet P (2005) Angiogenesis in life, disease and medicine. Nature 438(7070):932–936. doi:10.1038/nature04478

DeWitt N (2005) Angiogenesis. Nature 438(7070):931–931

Felmeden DC, Blann AD, Lip GY (2003) Angiogenesis: basic pathophysiology and implications for disease. Eur Heart J 24(7):586–603

Folkman J (1995) Angiogenesis in cancer, vascular, rheumatoid and other disease. Nat Med 1(1):27–31

Folkman J (1996) New perspectives in clinical oncology from angiogenesis research. Eur J Cancer 32a(14):2534–2539

Folkman J (2006) Angiogenesis. Annu Rev Med 57:1–18. doi:10.1146/annurev.med.57.121304.131306

Folkman J, Shing Y (1992) Angiogenesis. J Biol Chem 267(16):10931–10934

Hillen F, Griffioen AW (2007) Tumour vascularization: sprouting angiogenesis and beyond. Cancer Metastasis Rev 26(3–4):489–502. doi:10.1007/s10555-007-9094-7

Hong H, Chen F, Zhang Y, Cai W (2014) New radiotracers for imaging of vascular targets in angiogenesis-related diseases. Adv Drug Deliv Rev 76:2–20. doi:10.1016/j.addr.2014.07.011

Lamalice L, Le Boeuf F, Huot J (2007) Endothelial cell migration during angiogenesis. Circ Res 100(6):782–794. doi:10.1161/01.res.0000259593.07661.1e

Liekens S, De Clercq E, Neyts J (2001) Angiogenesis: regulators and clinical applications. Biochem Pharmacol 61(3):253–270

Luttun A, Autiero M, Tjwa M, Carmeliet P (2004) Genetic dissection of tumor angiogenesis: are PlGF and VEGFR-1 novel anti-cancer targets? Biochim Biophys Acta 1654(1):79–94. doi:10.1016/j.bbcan.2003.09.002

Mori S, Tran V, Nishikawa K, Kaneda T, Hamada Y, Kawaguchi N, Fujita M, Takada YK, Matsuura N, Zhao M, Takada Y (2013) A dominant-negative FGF1 mutant (the R50E mutant) suppresses tumorigenesis and angiogenesis. PLoS One 8(2):e57927. doi:10.1371/journal.pone.0057927

Munoz-Chapuli R, Quesada AR, Angel Medina M (2004) Angiogenesis and signal transduction in endothelial cells. CMLS Cell Mol Life Sci 61(17):2224–2243. doi:10.1007/s00018-004-4070-7

Nishida N, Yano H, Nishida T, Kamura T, Kojiro M (2006) Angiogenesis in cancer. Vasc Health Risk Manag 2(3):213–219. doi:10.2147/vhrm.2006.2.3.213

O'Reilly MS, Boehm T, Shing Y, Fukai N, Vasios G, Lane WS, Flynn E, Birkhead JR, Olsen BR, Folkman J (1997) Endostatin: an endogenous inhibitor of angiogenesis and tumor growth. Cell 88(2):277–285. doi:10.1016/s0092-8674(00)81848-6

Papetti M, Herman IM (2002) Mechanisms of normal and tumor-derived angiogenesis. Am J Physiol Cell Physiol 282(5):C947–C970. doi:10.1152/ajpcell.00389.2001

Samant RS, Shevde LA (2011) Recent advances in anti-angiogenic therapy of cancer. Oncotarget 2(3):122–134

Seaman S, Stevens J, Yang MY, Logsdon D, Graff-Cherry C, St Croix B (2007) Genes that distinguish physiological and pathological angiogenesis. Cancer Cell 11(6):539–554. doi:10.1016/j.ccr.2007.04.017

Senger DR, Davis GE (2011) Angiogenesis. Cold Spring Harb Perspect Biol 3(8):a005090. doi:10.1101/cshperspect.a005090

Smith SK (2001) Angiogenesis and reproduction. BJOG 108(8):777–783

Tanne JH (2011) FDA cancels approval for bevacizumab in advanced breast cancer. BMJ 343:d7684. doi:10.1136/bmj.d7684

Tonnesen MG, Feng X, Clark RA (2000) Angiogenesis in wound healing. J Invest Dermatol Symp Proc Soc Invest Dermatol Inc Eur Soc Dermatol Res 5(1):40–46. doi:10.1046/j.1087-0024.2000.00014.x

Chapter 2
Adenosine Triphosphate-Binding Cassette (ABC) Lipid Transporters

Abstract Adenosine triphosphate-binding cassette (ABC) transmembrane proteins are a family of transporters moving variety of molecules across extracellular and intracellular membranes. Categorized into seven major subfamilies of transporters ABCA to ABCG, mostly involved in lipid transport and homeostasis. ABCA1 and ABCG1 are the two members of this family known to be involved and interactive in the process of angiogenesis and mainly the stability of the circulatory system. ABCA1 acts as the transporter in the process of cholesterol and phospholipid efflux pump forming high density lipoproteins (HDL) and as HDL levels have been correlated with improved cardiovascular health, ABCA1 has an essential role in maintaining the circulatory system. ABCG1 located at intracellular vesicles and endosomes of the cell membrane will promotes cholesterol efflux for removal by mature HDL and macrophages. ABCG1 is accountable for intracellular cholesterol homeostasis with a protective role in the cardiovascular system. There were limited evidence in the relationship between ABCA1 and ABCG1 and cancer. The majority of research of ABC transport in cancer was mainly derived from its role in cholesterol efflux and reverse cholesterol transport.

Keywords Adenosine triphosphate-binding cassette • ABC lipid transporter • Angiogenesis • Normal physiology • Disease • Malignancy

2.1 ABC Lipid Transporter

The family of adenosine triphosphate-binding cassette (ABC) transmembrane protein is responsible for transporting a variety of molecules across extracellular and intracellular membranes at the expense of energy which is in the form of adenosine triphosphate (ATP) (Santamarina-Fojo et al. 2000). There are seven major subfamilies in the ABC transporter: ABCA to ABCG. Most of them are involved in cellular transport of lipid and homeostasis which heavily rely on angiogenesis for function and passage (Wagner et al. 2014). From this family of lipid transporters, ABCA1 and ABCG1 have been reported in more details with their role in maintaining the angio-vascular system and therefore their roles are reviewed in this chapter.

A. Salajegheh, *Angiogenesis in Health, Disease and Malignancy*,
DOI 10.1007/978-3-319-28140-7_2

2.2 ABC Lipid Transporter in Normal Physiology

2.2.1 ABCA1

According to Wagner et al. (2014), ABCA1 human gene is 149-kb in size and located on the human chromosome 9q22-q31. It is expressed throughout the human body with the highest expression in liver, placenta, testes, and adrenal glands then followed by moderate expression in small intestine, lung, and adipose (Wagner et al. 2014; Yvan-Charvet et al. 2010; Lou et al. 2014). The gene encodes for ABCA1 transmembrane protein which is a full-size transporter that functions as cholesterol and phospholipid efflux pump forming high density lipoproteins (HDL). HDL is responsible for the cholesterol and fats removal in cells as well as within arterial wall and transport to the liver for excretion or metabolic use (Yvan-Charvet et al. 2010). Higher innate HDL levels have been correlated with improved cardiovascular health (Yokoyama 2006) and henceforth ABCA1 has an essential role in maintaining the circulatory system. Yokohama (2006) of Nagoya City University reported that cholesterol, cytokines fatty acids and cyclic adenosine monophosphate (cAMP) are the factors that affect the expression of ABCA1 protein.

2.2.2 ABCG1

The human ABCG1 gene is about 98 kb in size and can be found on chromosome 21q22.3 (Wagner et al. 2014). Wagner et al. (2014) observed that it is expressed all over the body with greatest expression in adrenal glands, spleen, lung, placenta, heart and liver. The gene encodes the intracellular half-size transmembrane protein of ABCG1 which homodimerise or heterodimerise with other half-size transporter to be fully functional (Tarling and Edwards 2011). ABCG1 protein is located at the intracellular vesicles and endosomes of the cell membrane surface and promotes cholesterol and phospholipid efflux from cells for removal by mature HDL and macrophages (Wagner et al. 2014). It is accountable for intracellular cholesterol homeostasis with a protective role in the cardiovascular system (Tarling 2013; Wagner et al. 2014). Impairment of ABCG1 intracellular sterol transporter has various consequences depending on cell and tissue affected. The abnormal level of ABCG1 gene expression is correlated to atherosclerosis, pulmonary inflammation, diabetes and aggressive suicidal behaviours.

2.3 ABC Lipid Transporters in Disease and Malignancy

Disruptions in regulation of ABCA1 gene can lead to a number of pathologies. Overexpression of ABCA1 gene induces multi-drug resistance. Bachmeier et al. (2009) has shown that genetically modified melanoma cell culture with overexpressed ABCA1 is resistant to anti-inflammatory antioxidant cucurmin. When

ABCA1 is down regulated, its primary function in cellular removal of cholesterol is also down regulated. This promotes thickening of the blood vessels (atherogenesis) (Yokoyama 2006) and can lead to atherosclerosis, clot formation and aged-related macular degeneration (AMD). AMD is a medical condition that can lead to a loss of vision in the macula because of damage to the retina. It is a major cause of visual impairment in the elderly community. In the more severe form of AMD, blood vessels grow from the choroid behind the retina and detaches the light-sensitive layer (de Jong 2006). Treatments and medication for AMD are aimed to hinder angiogenesis in between the layers of the eye. Apte (2013) has observed that elevated ABCA1 has a protective effect in regulating angiogenesis in eye disease.

Mutation in ABCA1 gene has been linked to Tangier disease, a rare autosomal dominant disease characterised by the near absence of HDL in blood and accumulation of excess cholesterol in cells (Ordovas 2000; Santamarina-Fojo et al. 2000). People with this disease have orange tonsils, high triglyceride levels in the plasma, and are more susceptible to atherosclerosis (Wagner et al. 2014). They have malfunctioning in the efflux mechanism of cholesterol in cells thus leads to accumulation of excess cellular cholesterol and the inability to form HDL, otherwise commonly known as 'good cholesterol' (Ordovas 2000).

The ABCG1 knockout mouse study by Kennedy et al. (2005) revealed the role of ABCG1 in preventing cholesterol accumulation in macrophages within multiple tissues, especially in the lung. In contrast, they found overexpression of the gene seems to protect the mice from dietary-induced fat build-up. Miroshnikova et al. (2014) research observed lower level of ABCG1 in patients with arthrosclerosis as well as in severe artery stenosis. Out et al. (2007) and Yvan-Charvet et al. (2007) and reported that combined deletion of ABCG1 and ABCA1 genes in knockout mice resulted in significant lipid accumulation within cell and accelerates the progression of atherosclerosis; thus demonstrating the transporters' pivotal role in the prevention of hyperlipidaemia and impeding atherosclerosis.

As part of its role in intracellular lipid regulation, ABCG1 transporter indirectly regulates T-cell lymphocyte proliferation (Bensinger et al. 2008; Out et al. 2007). Yvan-Charvet et al. (2010) suggested that ABCG1 influence anti-inflammatory and immunosuppressive response in the body as its involvement with macrophage second-handedly influence the responses of inflammatory cytokines and chemokines as well as lymphocytic infiltration (Yvan-Charvet et al. 2010). Baldan et al. (2008) reported that ABCG1 is also associated with pulmonary inflammatory response. In their study with ABCG1 knockout mice, there was a distinctive inflammation response of macrophage and lymphocyte proliferation in the lung. It was suggested that sterol accumulation in the lungs would interrupt the distribution and absorption of pulmonary surfactant (lipoprotein) naturally available in the lungs that aids in the maintenance of the alveoli. They concluded that ABCG1 seems to be a protective factor to pulmonary inflammatory diseases (Baldan et al. 2008).

Recent studies have established the correlation between ABCG1 and glucose metabolism. (Sturek et al. 2010) research presented that expression of ABCG1 affects mouse pancreatic beta cells and therefore it is suggested that the gene is also involved in the pathology process of diabetes (Sturek et al. 2010). In their study, ABCG1-deleted mice have high plasma glucose level and were unable to produce insulin response after a glucose challenge. Mauldin et al. (2006) showed decreased ABCG1 expression in Type 2 diabetic rats compared to healthy rats (Mauldin 2009). Lou et al. (2014) compared the expression of ABCG1 in a group of rats with diabetes and small intestine chronic inflammation and the healthy control group. Their results indicated an overall decreased in the expression of ABCG1 in the inflammatory cells of the small intestine in diabetic rats. Hence, ABCG 1 gene might to be a potential therapeutic agent for type2 diabetes (Lou et al. 2014).

Since changes in cholesterol homeostasis has been associated human manners in aggression, violence (Sahebzamani et al. 2013) and suicidal behaviour (Gietl et al. 2007), genes involved in intracellular sterol regulation might be considered as contributors to these traits. Gietl et al. (2007) research indicated an association between ABCG1 and hostility and suicidal behaviour of 571 of attempted suicider, successful suicider and healthy control. However, the underlying mechanism is yet to be determined (Gietl et al. 2007).

There were limited evidence in the relationship between ABC gene and cancer. The majority of research of ABC transport in cancer was mainly derived from its role in cholesterol efflux and reverse cholesterol transport.

Cholesterol "sensor" liver X receptor (LXR) is a key regulator of both ABCA1 and ABCG1 transporters. LXR promotes cellular efflux by inducing the expression of ABCA1 and ABCG1 (Calkin and Tontonoz 2010; Lee and Plutzky 2006). El Roz et al. (2012) reported that treatment with LXR agonists inhibited proliferation and induced apoptosis of breast cancer cell. His *in vitro* study detected elevated expression of ABCG1 and a significant decrease in cellular cholesterol suggests a correlation between ABCG1 expression level and breast cancer cells.

ABCG1 role in immune response is further reinforced in recent research by Duygu Sag (2013) that demonstrated the transporter's role as a novel regulator in cancer immunity. His study showed that the absence of ABCG1 in knockout mice inhibits tumour growth via regulation of immune response within the cancer and provided a link between diet, sterol homeostasis, angiogenesis and cancer (Duygu Sag 2013).

References

Apte RS (2013) Inhibition of choroidal neovascularization. US Patent US 20130317090 A1, 28 Nov 2013

Bachmeier BE, Iancu CM, Killian PH, Kronski E, Mirisola V, Angelini G, Jochum M, Nerlich AG, Pfeffer U (2009) Overexpression of the ATP binding cassette gene ABCA1 determines resistance to Curcumin in M14 melanoma cells. Mol Cancer 8:129. doi:10.1186/1476-4598-8-129

Baldan A, Gomes AV, Ping P, Edwards PA (2008) Loss of ABCG1 results in chronic pulmonary inflammation. J Immunol (Baltimore, MD: 1950) 180(5):3560–3568

Bensinger SJ, Bradley MN, Joseph SB, Zelcer N, Janssen EM, Hausner MA, Shih R, Parks JS, Edwards PA, Jamieson BD, Tontonoz P (2008) LXR signaling couples sterol metabolism to proliferation in the acquired immune response. Cell 134(1):97–111. doi:10.1016/j.cell.2008.04.052

Calkin AC, Tontonoz P (2010) Liver X receptor signaling pathways and atherosclerosis. Arterioscler Thromb Vasc Biol 30(8):1513–1518. doi:10.1161/atvbaha.109.191197 de Jong PT (2006) Age-related macular degeneration. N Engl J Med 355(14):1474–1485. doi:10.1056/NEJMra062326

Duygu Sag CC (2013) ABCG1 as a novel link between cholesterol homeostasis and tumor immunity. J Immunol 190(53):2031

El Roz A, Bard JM, Huvelin JM, Nazih H (2012) LXR agonists and ABCG1-dependent cholesterol efflux in MCF-7 breast cancer cells: relation to proliferation and apoptosis. Anticancer Res 32(7):3007–3013

Gietl A, Giegling I, Hartmann AM, Schneider B, Schnabel A, Maurer K, Moller HJ, Rujescu D (2007) ABCG1 gene variants in suicidal behavior and aggression-related traits. Eur Neuropsychopharmacol 17(6–7):410–416. doi:10.1016/j.euroneuro.2006.11.004

Kennedy MA, Barrera GC, Nakamura K, Baldan A, Tarr P, Fishbein MC, Frank J, Francone OL, Edwards PA (2005) ABCG1 has a critical role in mediating cholesterol efflux to HDL and preventing cellular lipid accumulation. Cell Metab 1(2):121–131. doi:10.1016/j.cmet.2005.01.002

Lee C-H, Plutzky J (2006) Liver X receptor activation and high-density lipoprotein biology: a reversal of fortune? Circulation 113(1):5–8. doi:10.1161/circulationaha.105.590273

Lou J, Zhou H, Li C, Hu L, Lu X, Li J, Yao H, Li W, Zhang X, Xu M (2014) ABCA1 and ABCG1 expression in the small intestine of type 2 diabetic rats. Lab Med 45(1):17–24

Mauldin JP (2009) Macrophage ABCG1 expression and regulation in type 2 diabetes. ASSAY Drug Dev Technol 7(1):101

Mauldin JP, Srinivasan S, Mulya A, Gebre A, Parks JS, Daugherty A, Hedrick CC (2006) Reduction in ABCG1 in Type 2 diabetic mice increases macrophage foam cell formation. J Biol Chem 281(30):21216–21224. doi:10.1074/jbc.M510952200

Miroshnikova VV, Demina EP, Mayorov NV, Davydenko VV, Kurjanov PS, Vavilov VN, Vinogradov AG, Denisenko AD, Schwarzman AL (2014) The expression of ABCG1 transporter gene in peripheral blood mononuclear cells of patients with atherosclerosis. Cell Tissue Biol 8(4):337–343. doi:10.1134/s1990519x14040063

Ordovas JM (2000) ABC1: the gene for Tangier disease and beyond. Nutr Rev 58(3 Pt 1):76–79

Out R, Hoekstra M, Meurs I, de Vos P, Kuiper J, Van Eck M, Van Berkel T (2007) Po2-44 Abcg1 protects against atherosclerotic lesion development. Atheroscler Suppl 8(1):29. doi:10.1016/s1567-5688(07)71054-1

Sahebzamani FM, D'Aoust RF, Friedrich D, Aiyer AN, Reis SE, Kip KE (2013) Relationship among low cholesterol levels, depressive symptoms, aggression, hostility, and cynicism. J Clin Lipidol 7(3):208–216. doi:10.1016/j.jacl.2013.01.004

Santamarina-Fojo S, Peterson K, Qiu Y, Freeman L, Cheng J-F, Osorio J, Remaley A, Yang X-P, Haudenschild C, Prades C, Chimini G, Blackmon E, Francois T, Duverger N, Rubin EM, Rosier M, Denèfle P, Fredrickson DS, Brewer HBJ (2000) Complete genomic sequence of the human ABCA1 gene: analysis of the human and mouse ATP-binding cassette a promoter. Proc Natl Acad Sci U S A 97(14):7987–7992

Sturek JM, Castle JD, Trace AP, Page LC, Castle AM, Evans-Molina C, Parks JS, Mirmira RG, Hedrick CC (2010) An intracellular role for ABCG1-mediated cholesterol transport in the regulated secretory pathway of mouse pancreatic beta cells. J Clin Invest 120(7):2575–2589. doi:10.1172/jci41280

Tarling EJ (2013) Expanding roles of ABCG1 and sterol transport. Curr Opin Lipidol 24(2):138–146. doi:10.1097/MOL.0b013e32835da122

Tarling EJ, Edwards PA (2011) ATP binding cassette transporter G1 (ABCG1) is an intracellular sterol transporter. Proc Natl Acad Sci 108(49):19719–19724. doi:10.1073/pnas.1113021108

Wagner EM, Basso F, Kim CS, Amar MJA (2014) ABC lipid transporters. In: AccessScience. McGraw-Hill Education. http://dx.doi.org/10.1036/1097-8542.801530. Accessed 15 Sept 2015

Yokoyama S (2006) ABCA1 and biogenesis of HDL. J Atheroscler Thromb 13(1):1–15

Yvan-Charvet L, Ranalletta M, Wang N, Han S, Terasaka N, Li R, Welch C, Tall AR (2007) Combined deficiency of ABCA1 and ABCG1 promotes foam cell accumulation and accelerates atherosclerosis in mice. J Clin Invest 117(12):3900–3908. doi:10.1172/jci33372

Yvan-Charvet L, Wang N, Tall AR (2010) Role of HDL, ABCA1, and ABCG1 transporters in cholesterol efflux and immune responses. Arterioscler Thromb Vasc Biol 30(2):139–143. doi:10.1161/atvbaha.108.179283

Chapter 3
Angiogenin

Abstract Angiogenin (ANG) as a member of ribonuclease family has the potential for induction of angiogenesis. Targeting endothelial and smooth muscle cells, ANG stimulates formation and division of new blood vessel. Angiogenin forms a complex with actin which is located in endothelial cell surface and able to activate plasmin catalytic activity hence degrading the basement membrane and extracellular matrix and eventually will lead to initiation of neovascularization. Angiogenin plays a role in vascularization not only of malignancies but also in non-malignant pathologies. Formation of a complex of ANG with fibulin-1 is an important event in the processes like cell proliferation, migration, adhesion and stabilization of new-forming blood vessel walls. Recent studies suggest an association of ANG gene variants with Motor Neuron Disease (MND) and Parkinson's Disease (PD) although the wild-type ANG has a neuroprotective role. Cancer viability and spread is also supported by the expression of ANG in many malignancies. Angiogenin drives tumour angiogenesis, promote cancer growth and metastasis by induction of ribosomal RNA transcription that enables angiogenesis initiation by VEGF and other vascular growth factors.

Keywords Angiogenin • Angiogenesis • Normal physiology • Disease • Malignancy

3.1 Angiogenin

Angiogenin (ANG) is part of the ribonuclease family which is responsible for inducing angiogenesis. It binds to endothelial and smooth muscle cells as a stimulus for new blood vessel creation and division during a variety of physiological states including inflammation, wound healing and antibacterial activity (Tello-Montoliu et al. 2006).

As stated by Tello-Montoliu et al. angiogenin is a 14 124 Da soluble protein, initially isolated and characterised from the conditioned medium of cultured HT-29 human colon adenocarcinoma cells based solely on its angiogenic activity (Tello-Montoliu et al. 2006). mRNA of Angiogenin is expressed in a broad range of cells including tumour cells along with normal epithelial cells, peripheral blood cells, and fibroblasts.

A. Salajegheh, *Angiogenesis in Health, Disease and Malignancy*,
DOI 10.1007/978-3-319-28140-7_3

Angiogenin is a plasma protein with ribonucleolytic angiogenic activities. Unlike the rest of the RNase super family, the ribonucleolytic activity of ANG is unusually low, despite having all the necessary functionality. This appears to be due to partial blockage of the pyrimidine base binding site (Gao and Xu 2008; Tello-Montoliu et al. 2006). The biological activity of ANG is dependent on its ribonucleolytic activity, which is far lower than that of RNase A.

3.2 Angiogenin in Normal Physiology

Even though its ribonucleolytic activity is apparently weaker than other RNases, it is critical for angiogenesis and other functions. The critical structural difference between angiogenin and the other RNases are apparent in the missing of a fourth disulphide bond in the former, resulting in stimulating endothelial cells to create new vessels (Tello-Montoliu et al. 2006). After binding to the specific receptor on the angiogenin-responsive endothelial cell and smooth muscle cell membranes, the receptor-angiogenin complexes are rapidly endocytosed and translocated to the nucleus and accumulate in the nucleolus which is also thought to be necessary for its angiogenic process (Gao and Xu 2008; Tello-Montoliu et al. 2006). Nuclear translocation is essential for cell proliferation since it is considered a third messenger and enhances gene activation and transcription events, and angiogenic activity will be eliminated if the nuclear translocation of angiogenin is hindered.

In addition, Tello-Montoliu et al. have also showed another portion of the angiogenin bound actin located in the endothelial cell surface can stimulate a tissue type plasminogen activator (t-PA) to produce plasmin from plasminogen (Tello-Montoliu et al. 2006). In the presence of the actin–angiogenin complex, the overall catalytic activity of plasmin increases significantly, which thus degrades the basement membrane and extracellular matrix, leading to an initial step of neovascularization. Furthermore, as per results found by Zhang et al. fibulin-1, an extracellular matrix and plasma glycoprotein, binds to ANG, suggesting that the ANG-fibulin-1 complex plays an important role in cell proliferation, migration, adhesion and modulates stabilization of new-forming blood vessel walls (Zhang et al. 2008).

Signal transduction pathways activated by ANG interact with the endothelial cell surface and are involved in the production of extracellular signal-related kinase1/2 (ERK1/2) as well as protein kinase B/Akt. Activation of these proteins enables cell proliferation associated with ongoing angiogenic activities (Gao and Xu 2008).

3.3 Angiogenin in Disease

Interestingly enough angiogenin has been linked to both Motor Neuron Disease (MND) and Parkinson's Disease (PD), however the direct association is still not clearly understood. Initially it was demonstrated in MND that certain angiogenin mutations are correlated with both the sporadic and familial forms of MND (Kieran et al. 2008). This is combined with the recent discovery of gene variants associated with PD (Steidinger et al. 2013). Recent study by Van Es and colleagues have revealed the increased frequency of ANG variants compared to controls at 0.46 % of MND and 0.45 % of PD respectively (van Es et al. 2014).

This is all in stark contrast to the neuroprotective role displayed by wild-type ANG. Kieran and Sebastia et al. investigated the effect of ANG on motor neuron survival in vitro and in vivo (Kieran et al. 2008). It has been suggested that defects in hypoxic signalling, excitotoxicity and/or endoplasmic reticulum stress induced neuronal injury are responsible for the pathogenesis of MND. In their paper, Kieran and colleagues demonstrated in vitro protection of ventral horn motor neurons that had undergone AMPA mediated excitotoxicity and ER stress induced by Tunicamycin. They further demonstrated that wild-type ANG protected NSC34 cells (hybrid embryonic mouse spinal cord and neuroblastoma cell) as opposed to ANG K401 mutant a MND associated variant.

Finally Kieran et al. trialed ANG treatment in SOD1 mice (a knockout that displays MND characteristics) (Kieran et al. 2008). Even after symptom onset treatment provided a significant survival advantage of 135 days compared to 122.8 days. This correlates well to a recent study by Van Es et al. which displayed elevated levels of serum angiogenin in MND which suggests the activation of a compensatory mechanism (van Es et al. 2014). However no such rise was observed in Parkinson's patients. This may reflect a disruption of normal compensatory roles due to alpha-synuclein protein pathology or localized expression confined to the basal ganglia.

ANG has displayed a neuroprotective role in PD. Steidinger et al demonstrated a down-regulation in ANG expression in alpha-synuclein transgenic mice and that angiogenin in vitro provided protection in dopaminergic cells against the neurotoxin 1-methyl-4-phenylpyridinium (MPP +) and rotenone which induces a Parkinsonian syndrome (Steidinger et al. 2011).

These results are consistent with a neuroprotective role of wild-type ANG and disease states with mutations. Further studies will be required to elucidate the various mechanisms involved in both pathology and physiology.

3.4 Angiogenin in Malignancy

Angiogenin was first located in human colon adenocarcinoma cells (Li and Hu 2010). Since then it has been discovered in a wide variety of cancers. These include gastric, pancreatic, hepatocellular carcinomas, breast and neuroendocrine (Tello-Montoliu et al. 2006).

Angiogenin plays an important role in tumour metastasis (Li and Hu 2010). It functions through two pathways in support of ongoing cancer viability and spread. Firstly it drives tumour angiogenesis a necessary step for increasing bulk and metastasis. A secure blood supply is essential for cancer growth. Angiogenin induces ribosomal RNA (rRNA) transcription to enable angiogenesis initiated by VEGF and other vascular growth factors. This has been further confirmed by ANG inhibition ceasing angiogenesis even in the presence of these other factors (Li and Hu 2010). Secondly it appears to work in the cancer cell nucleus as a constitutive gene where it upregulates rRNA transcription. Experiments in PC-3 cells displayed not only a reduction in this, but also an impact on ribosome biogenesis, cell proliferation and tumourigenecity both in vivo and in vitro (Li and Hu 2010).

There has been in some tumors a clear correlation between levels of ANG and prognosis. This has been most clearly seen in prostate cancer with dramatic upregulation in tumour tissue (Li and Hu 2010). It was even able to differentiate between benign prostate hypertrophy and malignant prostate cancer. Admittedly the correlation is not as clear in breast cancer, but further investigations will help us understand the role ANG plays in cancer. Furthermore, the development and use of antiangiogenic drugs for the treatment of cancers has increased due to a considerable amount of clinical research over many decades. Improvements hold the promise of treatment for some malignant diseases through better understanding of the molecular and cellular mechanisms controlling tumor angiogenesis and the response to antiangiogenic therapies (Tello-Montoliu et al. 2006).

References

Gao X, Xu Z (2008) Mechanisms of action of angiogenin. Acta Biochim Biophys Sin (Shanghai) 40(7):619–624

Kieran D, Sebastia J, Greenway MJ, King MA, Connaughton D, Concannon CG, Fenner B, Hardiman O, Prehn JH (2008) Control of motoneuron survival by angiogenin. J Neurosci 28(52):14056–14061. doi:10.1523/jneurosci.3399-08.2008

Li S, Hu GF (2010) Angiogenin-mediated rRNA transcription in cancer and neurodegeneration. Int J Biochem Mol Biol 1(1):26–35

Steidinger TU, Standaert DG, Yacoubian TA (2011) A neuroprotective role for angiogenin in models of Parkinson's disease. J Neurochem 116(3):334–341. doi:10.1111/j.1471-4159.2010.07112.x

Steidinger TU, Slone SR, Ding H, Standaert DG, Yacoubian TA (2013) Angiogenin in Parkinson disease models: role of Akt phosphorylation and evaluation of AAV-mediated angiogenin expression in MPTP treated mice. PLoS One 8(2):e56092. doi:10.1371/journal.pone.0056092

Tello-Montoliu A, Patel JV, Lip GY (2006) Angiogenin: a review of the pathophysiology and potential clinical applications. J Thromb Haemost 4(9):1864–1874. doi:10.1111/j.1538-7836.2006.01995.x

van Es MA, Veldink JH, Schelhaas HJ, Bloem BR, Sodaar P, van Nuenen BF, Verbeek M, van de Warrenburg BP, van den Berg LH (2014) Serum angiogenin levels are elevated in ALS, but not Parkinson's disease. J Neurol Neurosurg Psychiatry 85(12):1439–1440. doi:10.1136/jnnp-2013-307168

Zhang H, Gao X, Weng C, Xu Z (2008) Interaction between angiogenin and fibulin 1: evidence and implication. Acta Biochim Biophys Sin (Shanghai) 40(5):375–380

Chapter 4
Angiopoietins

Abstract Angiopoietins are a family of growth factors genes that stimulate the process of angiogenesis. From the family of four, angiopoietin -1 (Ang-1) and angiopoietin -4 (Ang-4) are identified of being responsible for maturation, stabilisation and quiescence of vessels. Angiopoietins-2 (Ang-2) and angiopoietin-3 (Ang-3) are mainly involved in induction of vascular regression, cell death and inflammation. Ang-1 and Ang-2 are commonly determinants of endothelial permeability and peri-endothelial cell contact while Ang-3 and Ang-4 are not yet fully described for their role in angiogenesis. Ang-1 is found within perivascular cells and the vasculature within smooth muscle cells and pericytes and plays important roles in developmental angiogenesis, stability in established vasculature and pathological angiogenesis. Ang-2 is produced mainly by granular masses within endothelial cells of blood vessels, in parenchymal cells of the extravascular tissue and some malignant cells. In malignancies, Ang-1 has been shown as a promoter of vessel stabilization in the later stages of developmental angiogenesis and tumours' stabilisation and perfusion. Ang-2 is identified as one of the crucial factors in tumour angiogenesis like VEGF and when VEGF inhibition is not a sufficient approach additional Ang-2 inhibition has shown some positive outcome.

Keywords Angiopoietins • Ang-1 • Ang-2 • Angiogenesis • Normal physiology • Disease • Malignancy

4.1 Angiopoietins

Angiopoietins are vascular growth factors that promote the process of angiogenesis. Ang-1 and Ang-4 are responsible for vessel maturation, stabilisation and quiescence whilst angiopoietins-2 (Ang-2) and Ang-3 induce vascular regression, cell death and inflammation (Yuan et al. 2009). Ang-3 and Ang-4 are not yet fully understood and are considered to be lesser factors in angiogenesis than Ang-1 and Ang-2. Ang-1 and Ang-2 are reciprocally determinants of endothelial permeability and peri-endothelial cell contact.

Angiopoietins are glycoproteins that consist of four main regions. They have an N-terminal domain which is where numerous angiopoietins can form multimers

A. Salajegheh, *Angiogenesis in Health, Disease and Malignancy*,
DOI 10.1007/978-3-319-28140-7_4

(Thomas et al. 2013). They also have a central coiled domain and a linker region. Most importantly, they have a C-terminal domain responsible for binding to their angiopoietin receptor.

4.2 Angiopoietins in Normal Physiology

Ang-1 is secreted by fibroblasts, smooth muscle cells and pericytes for paracrine signalling. However, Ang-2 is secreted predominantly by granular masses called Weibel-Palade bodies within endothelial cells of blood vessels for autocrine signalling. It is also secreted for paracrine signalling by parenchymal cells of the extravascular tissue and some malignant cells.

Ang-1 is a polypeptide that consists of 498 amino acids weighing 57 kDa. Ang-2 is slightly shorter with 496 amino acids. These two polypeptides can aggregate at the super clustering region and form any amount of multimer. An aggregation of at least 4 polypeptides (tetramer) is required to activate angiopoietin receptor 2, Tyrosine kinase with immunoglobulin-like and EGF-like domains, or TIE-2 receptor. All four angiopoietins are functional ligands of the TIE- 1 and TIE- 2 receptor which is found exclusively on the cell surface of endothelial cells. TIE- 1 and TIE- 2 receptors induce the tyrosine kinase signalling pathway. This pathway entails the phosphorylation of the TIE-2 receptor which then activates phosphatidylinositol 3-kinase and more downstream intracellular enzymes (Cascone and Heymach 2012). Binding of angiopoietin to TIE-2 facilitates heterodimerisation of TIE-1 and TIE-2 to enhance signal transduction. The function of the TIE-1 receptor is not yet fully understood.

Angiopoietin-1 (Ang-1) is found within perivascular cells, more specifically the vasculature within smooth muscle cells and pericytes (Singh et al. 2011). Ang-1 binds to Tie2 tyrosine kinase receptor found within endothelial cells and Ang-1-Tie2 complex is suggested to play important roles in developmental angiogenesis, stability in established vasculature and pathological angiogenesis.

Ang-1 binding with the tyrosine kinase receptor acts as a signalling mechanism essential for regulating blood vessel development and stability of mature cells. In order for Ang-2 to have a similar effect on cells, instead of its traditional role of promoting inflammation and regression, it must be in the presence of VEGF. Ang-2 interacts with VEGF pathway to induce neovascularisation (Barton et al. 2005).

Ang-1 is an agonist for the tyrosine kinase receptor whereas Ang-2 is an antagonist. However, in some cases, Ang-2 can induce phosphorylation of the receptor depending on the cell type, stimulation and cell confluence. Ang-1 and Ang-2 are almost always present in blood serum and are usually found at equilibrium in a healthy person (Eroglu et al. 2013). The relative level of Ang-1 to Ang-2 is quite low except in incidences of inflammation and regression. They are always needed as there is simultaneous blood vessel proliferation and regression occurring in the body. Cells with a greater need for nutrients and oxygen will express Ang-1whereas redundant cells will express Ang-2. Although Ang-2 has a seemingly negative effect on blood vessels, it is necessary for vascular remodelling. Ang-2 destabilises blood vessel by loosening endothelial cell connections and breaking down the basal lamina and extracellular matrix. This is a necessary process in both vascular regression and sprouting. For an existing blood vessel to create another branch, an area of the endothelium must be transitioned to a more plastic, amended state to allow space for cell sprouting (Yuan et al. 2009). Therefore, Ang-2 has a direct influence on vascular regression but is essential in vascular remodelling. For this reason, Ang-2 is elevated in tumours and some diseases.

4.3 Angiopoietins in Disease

The vasculature is formed during vasculogenesis in the early stages of developmental angiogenesis. During the later stages, Ang-1 controls the diameter, vascular branching (Brindle et al. 2006) and facilitates reorganisation and maturation of the newly formed blood and lymphatic vessels (Brindle et al. 2006; Arita et al. 2014). The significance of Ang-1 in developmental angiogenesis is elucidated through studies of murine embryonic development where transgenic mice with deficient Ang-1 has blood vessels displaying reduced branching, increased dilation, decreased numbers of small vessel and diminished vascular complexity (Arita et al. 2014). Despite the vasculature being formed the mice died between embryonic day (E) 9.5 and E12.5 (Arita et al. 2014; Jeansson et al. 2011). Another murine study examines the role of Ang-1 in vascular development by deleting Ang-1 at each embryonic day and analyses its effects. Embryo at E10.5 shows vascular abnormalities in several organs, notably the liver and kidney, and widespread vessel dilation (affecting mainly the veins) as well as atrial dilation was detected (Fukuhara et al. 2009). In the liver, it has a greater number of vessels whereas the kidney shows more significant defects to the glomerular basement membrane, the microvasculature beds of the glomeruli displays disruption of the endothelial cell attachment. Death in these Ang-1 knockout mice occurred between E17.5 and postnatal day (P0) (Fukuhara et al. 2009). These studies suggest that despite having formed vasculature via vasculogenesis, reorganisation and maturation of the vessels are dependent on Ang-1 as seen by its removal in murine studies causing poor patterning, vascular deficiencies and death.

Once vascular formation, remodelling, and maturation are complete, mechanisms must exist to maintain homeostasis. Evidence suggests cell-cell adhesions and Ang-1-Tie2 signalling cooperatively mediates vascular homeostasis (Koh 2013). This includes control of endothelial permeability, suppression of vascular inflammation and cellular survival. Junctions within the endothelial cells (EC) barrier allow passage of plasma and cellular material between the circulation and the surrounding cells. Ang-1 helps maintain this barrier by regulating the accumulation of the cell-cell adhesion proteins between the junctions, which is mainly comprised of vascular endothelial cadherin (VE-cadherin) (Fagiani and Christofori 2013; Eklund and Saharinen 2013). Ang-1 acts to strengthen the adhesion between

VE-cadherin and platelet endothelial cell adhesion molecule-1 (PECAM-1) reducing permeability (Koh 2013). The Ang-1-Tie2 pathway also counteracts VEGF induced permeability through several mechanisms including sequestration of non-receptor Tyr kinase Src (Src) via mammalian diaphanous (mDia), modulating the interaction between vascular endothelial phosphotyrosine phosphatase (VE-PTP), Tie2 and VE-cadherin, interfering with the phosphatidylinositol 3-kinase (IP3) pathway and suppression of the EC thrombin-induced intracellular calcium transients (Eklund and Saharinen 2013).

The EC barrier mechanisms are also responsible for regulating the EC permeability caused by inflammation (Fagiani and Christofori 2013), however other Ang-1 induced anti-inflammatory controls exist. Ang-1 inhibits the expression of multiple adhesion molecules on the surface of the EC (Fagiani and Christofori 2013), specifically intercellular adhesion molecule-1 (ICAM-1), vascular cell adhesion molecule-1 (VCAM-1) and E-selectin, which are necessary for leukocyte attachment and infiltration into the vasculature (Arita et al. 2014). Additionally, lipopolysaccharide treated mice displays decreased pulmonary expression of ICAM-1, VCAM-1, and E-selectin suggesting that inflammatory response is limited via Ang-1 mediation (Arita et al. 2014). In addition to inhibiting endothelial-leukocyte interaction, Ang-1 also inhibits the activity and expression of tissue factor and may partially suppress inflammation through ABIN-2 (intracellular protein) inhibited NF-κB (Fagiani and Christofori 2013). ABIN-2 is recruited through Tie2 activation by Ang-1 and also aids cellular survival by preventing EC apoptosis (Brudno et al. 2013).

Endothelial cell survival is enhanced through the phosphorylation of AKT by the Ang-1 activation of PI3 pathway (Koh 2013). This mechanism inhibits the pro-apoptotic forkhead transcription factor (FKHR) in EC and promotes quiescence in mature vessels (Arita et al. 2014). Ang-1 induced Krüppel-like factor 2 (KLF2), a type of zinc-finger transcription factor, also exhibits protective functions, vascular stabilisation and anti-inflammatory mechanisms and is activated by the AKT-MEF2 signalling pathway (Arita et al. 2014; Eklund and Saharinen 2013). Ang-1 activation of the Tie2 receptors and the subsequent signalling of PI3-kinase/AKT pathway help maintain endothelial integrity and cellular survival. Blood vessel survival in response to Ang-1 is demonstrated *in vivo* where mice were exposed to radiation as apoptosis to the intestinal and lung endothelial cells was prevented following administration of Ang-1 (Arita et al. 2014). Despite previous research for Ang-1 maintaining homeostasis and quiescence in mature blood vessels, it was unexpectedly determined that the deletion of Ang-1 after E13.5 had no impact on vascular homeostasis and was not required under normal conditions in mice (Fukuhara et al. 2009). This challenged the current theories of Ang-1 and its effects on vascular homeostasis and quiescence and suggests potential Ang-1 independent functions of the Tie2 receptors (Fagiani and Christofori 2013).

Vascular homeostasis is preserved through a fine balance between stabilization and angiogenesis. During injury or the disease process, this balance may shift. Hypoxia is one of the main reasons that can cause a shift towards pro-angiogenesis. Previous research has focused on mediating pro-angiogenic factors including VEGF-A, hepatocyte growth factor (HGF), and basic fibroblast growth factor (bFGF) (Eklund and Saharinen 2013). However, its benefits in neovascularization have been trumped by unfavourable side effects such as instability of new vasculature, accelerated inflammation, and excessive fibrosis thus creating limitations for therapeutic use. Recent research with Ang-1 has shown promise while minimizing such side effects. Supplemental use of Ang-1 has been illustrated to induce vascular enlargement of small vessels and even angiogenesis (with overexpression) to relieve ischemia in the brain, articular joints, kidneys, the heart and the limbs (Eklund and Saharinen 2013). Further studies have indicated that sequential administration of various growth factors would improve initial neovascularization and its subsequent maturation to create a more robust vasculature. As both VEGF and Ang-2 are pro-angiogenic factors, initial administration facilitates an early angiogenic response and spawns new vessel formation via sprouting and pericyte separation from the endothelium, while successive introduction of Ang-1 and platelet derived growth factor (PDGF) promotes vessel maturation (Huang et al. 2009). Administration of Ang-1 and PDGF must be timed accordingly as premature delivery may inhibit VEGF and Ang-2 induced angiogenesis. Together this cascade of factors along with sequential timing can help promote new vessel formation and its appropriate maturation in ischemic states.

4.4 Angiopoietins in Malignancy

Establishing adequate vasculature is also a necessity for tumour growth. Moderate success in vascular targeting therapy using the bevacizumab, an anti-VEGF antibody, has shown its ability to prolong the survival of advanced cancer patients (Saharinen et al. 2011). Despite its ability to block VEGF, the increased expression of Ang-1 has been correlated with the eventual progression of disease (Saharinen et al. 2011). Much like ability of Ang-1 to promote vessel stabilization in the later stages of developmental angiogenesis, it plays a similar role in tumours by stabilisation and perfusion. Alone, Ang-1seems to have minimal effects in tumour vasculature, however during VEGF blockage its up-regulation reduces tumour hypoxia, prevents vessel ablation, promotes dilation, and protects the vasculature from overall regression (Thomas et al. 2013; Saharinen et al. 2011). Other reports of Ang-1 and its role in tumour-associated angiogenesis have yielded conflicting reports. One study concluded a 70 % reduction in tumour growth in squamous cell carcinoma, three times reduction in breast cancer, and a decrease in weight and volume in liver tumours with Ang-1 overexpression (Brudno et al. 2013). In contrast, other studies revealed increased tumour growth in cervical, glioma (Saharinen et al. 2011) and plasma cell tumours (Brudno et al. 2013). The discrepancy in research data suggests Ang-1 may either promote or inhibit growth based on the experimental context/conditions. Different experiments may exhibit promotion of EC, encourage perfusion, and promote Ang-1 up-regulation whereas others may inhibit such actions which may be explained by the differences in reaction based on cancer types and their neoplastic vascular beds. Despite the research on Ang-1 and its function in tumour-associated angiogenesis, its role remains unclear and a direct relationship with malignancy has yet to be proven.

Anti-angiogenic therapy for treating cancer has been widely investigated. The established agents target the signalling pathway of VEGF as it is a major factor in angiogenesis. Bevacizumab is a monoclonal antibody that binds to VEGF to inhibit its interaction with its receptor, VEGF-R (Mazzieri et al. 2011). Other drugs such as Sorafenib and Sunitinib act to inhibit VEGFR itself by occupying the receptor domain, another tyrosine kinase receptor, without eliciting a response (Mazzieri et al. 2011). This technique is initially effective but is not a curative approach. These techniques were tested on several mouse tumours and were shown to induce tumour shrinkage. However, the tumour managed to revascularise by upregulating other proangiogenic factors, increasing metastatic activity and recruiting proangiogenic bone marrow-derived cells. For these reasons, VEGF inhibition is not a sufficient approach to long-term tumour suppression. Ang-2 inhibition is an alternative that is showing some positive results (Holash et al. 1999).

Ang-2 and VEGF are crucial factors in tumour angiogenesis. VEGF is released from tumour cells which combine with Ang-2 to prompt angiogenesis (Holash et al. 1999). The angiogenic process typically follows three general steps: (1) the small tumour integrates with an existing blood vessel for nourishment, (2) Ang-2 induces regression of the tumour's internal vasculature until it becomes avascular and apoptosis begins and (3) VEGF is up regulated to promote extensive angiogenesis at the external margin of the tumour. This process is necessary for the initial growth of certain tumours and outlines the significance of Ang-2 in tumour growth. Therefore, an increase in Ang-2 and VEGF correspond positively with metastasis and tumour development (Holash et al. 1999; Fiedler and Augustin 2006).

Ang-2 inhibition can be a perfect candidate for anti-cancer therapy. However, it must exceed the limitations exhibited by VEGF inhibition. Additionally, inhibition of Ang-2 must not have an adverse effect on normal physiological vascular remodelling. To investigate the efficacy of Ang-2 inhibition, two human antibodies, LC06 and LC08, were generated to inhibit the binding between Ang-2 and the TIE-2 receptor (Yuan et al. 2009). LC06 had a greater affinity for Ang-2 over Ang-1. The antibodies also recognise both rodent and human Ang-2 to allow for animal testing and relevant results. The Ang-2 inhibition presented clear necrosis and tumour growth inhibition in orthotopic and subcutaneous models. These effects were accompanied by decreased intratumoral microvascular density with less branches and increased pericyte coverage. However, LC08 lead to regression of blood vessels in the mouse trachea but LC06 had no obvious effects on tumour vasculature. The results show the potential for Ang-2 inhibitors for cancer treatment but also potential toxicity.

There are other monoclonal antibodies and fc-fusion proteins being trialled as of 2013 (Yuan et al. 2009). The polypeptide AMG386 targets both Ang-1 and Ang-2 and is being evaluated in phase III clinical trials. A highly selective trap molecule for Ang-2 CVX-060, and MEDI3617 a fully human antibody for Ang-2, have entered phase I clinical testing. However, these agents have already presented the common side-effects of back pain, upper abdominal pain, proteinuria, fatigue, peripheral oedema, insomnia. While Ang-2 provides another front against anti-angiogenic therapy, they are unlikely to be developed until the functional consequences of inhibiting Ang-1 and Ang-2 are better understood.

References

Arita Y, Nakaoka Y, Matsunaga T, Kidoya H, Yamamizu K, Arima Y, Kataoka-Hashimoto T, Ikeoka K, Yasui T, Masaki T, Yamamoto K, Higuchi K, Park JS, Shirai M, Nishiyama K, Yamagishi H, Otsu K, Kurihara H, Minami T, Yamauchi-Takihara K, Koh GY, Mochizuki N, Takakura N, Sakata Y, Yamashita JK, Komuro I (2014) Myocardium-derived angiopoietin-1 is essential for coronary vein formation in the developing heart. Nat Commun 5:4552. doi:10.1038/ncomms5552

Barton WA, Tzvetkova D, Nikolov DB (2005) Structure of the angiopoietin-2 receptor binding domain and identification of surfaces involved in Tie2 recognition. Structure (London, England: 1993) 13(5):825–832. doi:10.1016/j.str.2005.03.009

Brindle NP, Saharinen P, Alitalo K (2006) Signaling and functions of angiopoietin-1 in vascular protection. Circ Res 98(8):1014–1023. doi:10.1161/01.res.0000218275.54089.12

Brudno Y, Ennett-Shepard AB, Chen RR, Aizenberg M, Mooney DJ (2013) Enhancing microvascular formation and vessel maturation through temporal control over multiple pro-angiogenic and pro-maturation factors. Biomaterials 34(36):9201–9209. doi:10.1016/j.biomaterials.2013.08.007

Cascone T, Heymach JV (2012) Targeting the angiopoietin/Tie2 pathway: cutting tumor vessels with a double-edged sword? J Clin Oncol Off J Am Soc Clin Oncol 30(4):441–444. doi:10.1200/jco.2011.38.7621

Eklund L, Saharinen P (2013) Angiopoietin signaling in the vasculature. Exp Cell Res 319(9):1271–1280. doi:10.1016/j.yexcr.2013.03.011

Eroglu Z, Stein CA, Pal SK (2013) Targeting angiopoietin-2 signaling in cancer therapy. Expert Opin Investig Drugs 22(7):813–825. doi:10.1517/13543784.2013.793306

Fagiani E, Christofori G (2013) Angiopoietins in angiogenesis. Cancer Lett 328(1):18–26. doi:10.1016/j.canlet.2012.08.018

Fiedler U, Augustin HG (2006) Angiopoietins: a link between angiogenesis and inflammation. Trends Immunol 27(12):552–558. doi:10.1016/j.it.2006.10.004

Fukuhara S, Sako K, Noda K, Nagao K, Miura K, Mochizuki N (2009) Tie2 is tied at the cell-cell contacts and to extracellular matrix by angiopoietin-1. Exp Mol Med 41(3):133–139. doi:10.3858/emm.2009.41.3.016

Holash J, Wiegand SJ, Yancopoulos GD (1999) New model of tumor angiogenesis: dynamic balance between vessel regression and growth mediated by angiopoietins and VEGF. Oncogene 18(38):5356–5362. doi:10.1038/sj.onc.1203035

Huang J, Bae JO, Tsai JP, Kadenhe-Chiweshe A, Papa J, Lee A, Zeng S, Kornfeld ZN, Ullner P, Zaghloul N, Ioffe E, Nandor S, Burova E, Holash J, Thurston G, Rudge J, Yancopoulos GD, Yamashiro DJ, Kandel JJ (2009) Angiopoietin-1/Tie-2 activation contributes to vascular survival and tumor growth during VEGF blockade. Int J Oncol 34(1):79–87

Jeansson M, Gawlik A, Anderson G, Li C, Kerjaschki D, Henkelman M, Quaggin SE (2011) Angiopoietin-1 is essential in mouse vasculature during development and in response to injury. J Clin Invest 121(6):2278–2289. doi:10.1172/jci46322

Koh GY (2013) Orchestral actions of angiopoietin-1 in vascular regeneration. Trends Mol Med 19(1):31–39. doi:10.1016/j.molmed.2012.10.010

Mazzieri R, Pucci F, Moi D, Zonari E, Ranghetti A, Berti A, Politi LS, Gentner B, Brown JL, Naldini L, De Palma M (2011) Targeting the ANG2/TIE2 axis inhibits tumor growth and metastasis by impairing angiogenesis and disabling rebounds of proangiogenic myeloid cells. Cancer Cell 19(4):512–526. doi:10.1016/j.ccr.2011.02.005

Saharinen P, Eklund L, Pulkki K, Bono P, Alitalo K (2011) VEGF and angiopoietin signaling in tumor angiogenesis and metastasis. Trends Mol Med 17(7):347–362. doi:10.1016/j.molmed.2011.01.015

Singh H, Tahir TA, Alawo DO, Issa E, Brindle NP (2011) Molecular control of angiopoietin signalling. Biochem Soc Trans 39(6):1592–1596. doi:10.1042/bst20110699

Thomas M, Kienast Y, Scheuer W, Bahner M, Kaluza K, Gassner C, Herting F, Brinkmann U, Seeber S, Kavlie A, Welschof M, Ries S, Weidner KM, Regula JT, Klein C (2013) A novel angiopoietin-2 selective fully human antibody with potent anti-tumoral and anti-angiogenic efficacy and superior side effect profile compared to Pan-Angiopoietin-1/-2 inhibitors. PLoS One 8(2):e54923. doi:10.1371/journal.pone.0054923

Yuan HT, Khankin EV, Karumanchi SA, Parikh SM (2009) Angiopoietin 2 is a partial agonist/antagonist of Tie2 signaling in the endothelium. Mol Cell Biol 29(8):2011–2022. doi:10.1128/mcb.01472-08

Chapter 5
Angiotropin

Abstract Angiotropin is a macrophage-derived product that exhibits unusual chemical and biological properties. This product is found to be chemically made up of metallopolyribonucleopolypeptide complex that consists of 43 ribonucleotides as the least-common unit of all components and one copper ion per 38 amino acids. From selective enzymatic splitting, it was determined that both the polyribonucleotide and the polypeptides components are essential in angiogenic activity. This monocyte-derived angiogenic substance is one of the molecules out of various different angiogenic and angiostatic factors that allows for the stringent control of the intricate and multi-step angiogenesis process. By the application of the vascularising and nonmitogenic endothelial cell-selective angiotropin, which caused the formation of three-dimensional blood capillaries in the skeletal muscles of rabbits; it was found that angiotropin's capacity for remodelling bioactive blood vessels in muscle tissues can be potentially used for preventing tissue necrosis as well as increasing and supporting tissue functions.

Keywords Angiotropin • Angiogenesis • Normal physiology • Disease • Malignancy

5.1 Angiotropin

Angiotropin is a macrophage-derived product that exhibits unusual chemical and biological properties (Hockel et al. 1988). This product is found to be chemically made up of metallopolyribonucleopolypeptide complex that consists of 43 ribonucleotides as the least-common unit of all components and one copper ion per 38 amino acids (Hockel and Burke 1989). From selective enzymatic splitting, it was determined that both the polyribonucleotide and the polypeptides components are essential in angiogenic activity (Hockel and Burke 1989). Studies have established that angiotropin is a potent angiogenesis factor (Hockel and Burke 1989). and is distinctly different from other known angiogenic polypeptides and fibroblast growth factors (Hockel et al. 1988). These highly purified complexes can be isolated from large-scale cultures of porcine peripheral monocytes, that are activated by the means of concanavalin A, in conditioned serum-free media (Hockel and Burke 1989).

A. Salajegheh, *Angiogenesis in Health, Disease and Malignancy*,
DOI 10.1007/978-3-319-28140-7_5

5.2 Angiotropin in Normal Physiology

This monocyte-derived angiogenic substance is one of the molecules out of various different angiogenic and angiostatic factors that allows for the stringent control of the intricate and multi-step angiogenesis process (Distler et al. 2003). Although a stimulatory molecule of angiogenesis, angiotropin is not mitogenic and hence cannot stimulate cell division and the proliferative activity of capillary endothelial cells (Hockel and Burke 1989). However it can stimulate phenotypic differentiation. endothelial cell migration as well as facilitate and promote tube formation in vitro (Distler et al. 2003); and therefore much research have been focused on utilising its unique features and characteristics for the treatment of diseases as well as reparative medicine, transplant integration, implants and tissue engineering (Wissler 2002).

5.3 Angiotropin in Disease and Malignancy

In studies by Wissler et al. in 1981 and 1986, it was reported strong angiogenic effects in chicken chorioallantoic membrane and rabbit cornea were noticed when the cells underwent stimulation by a substance that had been isolated from peripheral porcine monocytes (Wissler and Renner 1981; Wissler et al. 1986). This substance was later on termed angiotropin because of the responses it could create in cells.

Further studies by Hockel et al. (1988) have determined that angiotropin stimulates the differentiation of capillary endothelial cells in vitro by changing their phenotypic characteristics. The research discovered that endothelial cells changed from a monolayer of epitheloid polygonal cells into migrating "sprout cells" that eventually organised spatially into tube structures. Additionally, it was found that in non-confluent cultures that single capillary endothelial cells did not proliferative but did however undergo migration when stimulated by angiotropin (Hockel et al. 1988).

To further determine and characterise the in vivo effects of angiotropin, based on previous research results and findings, Hockel et el. used the "single bolus-rabbit ear model" (Hockel et al. 1988). Intradermal injections of a single bolus of angiotropin were performed on the dorsal aspect of the skin of the ears of adult, long-eared New Zealand white rabbits (Hockel et al. 1988). After injections, it was found that vascular dilation occurred to all blood vessels except the arteries. This vasodilation caused by angiotropin differed vastly to those of other known vasodilators such as histamine and prostaglandin; such that the more traditional vasodilators had an earlier onset with much shorter effects (Hockel et al. 1988). Similar to the findings of previous studies, in vitro angiotropin demonstrated differentiation of capillary endothelial cells from epitheloid phenotypes into large fibroblastoid cells that migrated and formed tube-like structures (Hockel et al. 1988). In addition it was found that endothelial activation associated with endothelial hyperplasia, epidermal hyperplasia and neovascularization when the cells were induced by angiotropin in a dose-dependent manner occurred (Hockel et al. 1988). The study showed that tissue reactions occurred in near absence of preceding tissue destruction and did not result in any scar formation or tissue necrosis (Hockel et al. 1988). These reactions were found to have striking similarities with the inflammatory-proliferative phase in wound healing (Hockel et al. 1988). From the results, Hockel et al. (1988) found angiotropin to be a promising therapeutic agent that could be utilised in a clinical setting in cases where healing could be a problem.

A more recent study conducted by Wissler (2002) discovered the value of angiotropin in relation to regulation of hemodynamics and muscular tissue function. Wissler investigated the different modes of application of angiotropin into skeletal muscles (Wissler 2002). Angiotropin was isolated from the wound fluids of ischemic or infarcted heart muscles of either dogs and pigs (Wissler 2002). The isolated angiotropin were then focally applied intramuscularly into resting skeletal left hind-leg muscles of non-trained New Zealand rabbits (Wissler 2002). Bio-functions of newly formed capillary patterns of the left hind-leg were compared to the right hind-leg muscles of the rabbits (the control), in terms of tissue homeostasis control mechanisms that could have been affected by hemodynamics and turnover (Wissler 2002). It was found that in the area in which the angiotropin was applied, formation and remodelling of vessels with functionally efficient and active hemodynamics was induced in a matter of minutes (Wissler 2002). The results gathered from the study showed that angiotropin induced capillaries in the skeletal muscles lead to an increase of hemodynamics of about 15 days (Wissler 2002). However, normal hemodynamics were re-established around after 20 days (Wissler 2002). It was also noted that no adverse effects or uncontrolled growth were caused by the application of the angiotropin, and bio-functions in the muscles of the rabbits remained the same (Wissler 2002).

By the application of the vascularising and nonmitogenic endothelial cell-selective angiotropin, which caused the formation of three-dimensional blood capillaries in the skeletal muscles of rabbits; it was found that angiotropin's capacity for remodelling bioactive blood vessels in muscle tissues can be potentially used for preventing tissue necrosis as well as increasing and supporting tissue functions (Wissler 2002). Furthermore, it also indicates the possibility of applying angiotropin in therapeutic aspects where changes in vascular patterns could be problem, such as bypass operations or restenosis after percutaneous transluminal coronary angioplasty (Wissler 2002).

References

Distler JH, Hirth A, Kurowska-Stolarska M, Gay RE, Gay S, Distler O (2003) Angiogenic and angiostatic factors in the molecular control of angiogenesis. Q J Nucl Med 47(3):149–161

Hockel M, Burke JF (1989) Angiotropin treatment prevents flap necrosis and enhances dermal regeneration in rabbits. Arch Surg 124(6):693–698

Hockel M, Jung W, Vaupel P, Rabes H, Khaledpour C, Wissler JH (1988) Purified monocyte-derived angiogenic substance (angiotropin) induces controlled angiogenesis associated with regulated tissue proliferation in rabbit skin. J Clin Invest 82(3):1075–1090. doi:10.1172/jci113664

Wissler J (2002) Engineering of capillary patterns in muscle by a nonmitogenic copper-ribonucleoprotein angiomorphogen [angiotropin CuRNP ribokine]. Ann N Y Acad Sci 961:292–297

Wissler J, Renner H (1981) Inflammation, chemotropisms and morphogenesis-novel leucocyte-derived mediators for directional growth of blood-vessels and regulation of tissue neovascularization. In: Hoppe-Seylers Zeitschrift Fur Physiologische Chemie, vol 3. Walter De Gruyter & Co Genthiner Strasse, Berlin, pp 244–244

Wissler J, Logemann E, Meyer H, Krützfeldt B, Höckel M, Heilmeyer L (1986) Structure and function of a monocytic blood vessel morphogen (angiotropin) for angiogenesis in vivo and in vitro: a copper-containing metallo-polyribonucleo-polypeptide as a novel and unique type of monokine. Protides Biol Fluids 34:517–536

Chapter 6
C-KIT: Tyrosine Kinase Receptors with Potential to Initiate Angiogenesis

Abstract The c-kit gene, also known as CD117 and tyrosine-protein kinase Kit is a proto-oncogene which codes for a protein receptor responding to growth factors. This receptor resides on progenitor cells in homeostatic conditions and is carried on from the stem cell into the specialised daughter cells. The c-kit receptor is commonly found in progenitor cells destined to become endothelial and mast cells. Recent studies investigated the effect of SCF/c-kit on capillary tube formation and its similarity to VEGF and reported the formation, migration and consequent survival of cells in response to SCF/c-kit stimulation. Mutations of the c-kit gene may lead to the complete loss of function or retardation in the expression of the gene and its angiogenic capabilities. Many attempts at inhibiting neoplastic proliferation utilize drugs to target the c-kit gene and inhibitors on the KIT gene which regulates its enzymatic conformation. These drugs are still in early stages, yet offer promise towards therapeutic advantages concerning neoplasia caused by the c-kit gene defect.

Keywords C-KIT • Tyrosine kinase receptors with potential to initiate angiogenesis • Angiogenesis • Normal physiology • Disease • Malignancy

6.1 C-KIT: Tyrosine Kinase Receptors with Potential to Initiate Angiogenesis

The c-kit gene, also known as CD117 and tyrosine-protein kinase Kit (Ferrara and Gerber 2001), is a proto-oncogene which codes for a protein receptor responding to growth factors (Andre et al. 1997). This receptor resides on progenitor cells in homeostatic conditions and is carried on from the stem cell into the specialised daughter cells. The c-kit receptor is commonly found in progenitor cells destined to become endothelial and mast cells (Matsui et al. 2004).

A. Salajegheh, *Angiogenesis in Health, Disease and Malignancy*,
DOI 10.1007/978-3-319-28140-7_6

6.2 C-KIT in Normal Physiology

While our understanding of the expression of this gene is still a work in progress, it is known to acquire a hematopoietic role in both the recruitment and proliferation of endothelial cells from the bone marrow and into the circulation. Bone marrow secretes the ligand Stem Cell Factor (SCF) which activates the c-kit signalling mechanism. SCF is hypothesised to reinforce proliferation and vascularization of endothelial cells and mast cells when bound to the tyrosine kinase receptor, thus promoting the vascular process of angiogenesis under homeostatic conditions. Matsui et al. (2004) conducted an experiment investigating the effect of SCF/c-kit on capillary tube formation in HUVEC and its similarity to VEGF. The report went on to describe the formation, migration and consequent survival of cells in response to SCF/c-kit stimulation (Matsui et al. 2004). Experimental results displayed that the degree of vascularization of umbilical capillary proliferation depended on the application of SCF/c-kit signalling and its concentration. This provides positive evidence of c-kit signalling is utilised as a positive angiogenic factor (Matsui et al. 2004). With the process of blood vessel formation comes the need for relocation and survival of the endothelial cells which form the vessels. Studies by Matsui et al. (2004) show that although c-kit and its signalling factor SCF do indeed support the survival of proliferated endothelial cells in HUVEC, but does not necessarily support the proliferation of these cells. However this could be caused by several factors, including an insufficient concentration of SCF, or possibly an incompatibility of the simple c-kit pathway in comparison to the complex VEGF mechanism to incur proliferation of HUVECs in angiogenesis (Matsui et al. 2004). Hence, the exact effect that SCF has on c-kit is still unknown due to substantial gaps in the experimental design – it is known that SCF supports the survival or existing proliferated cells, but much has yet to be discovered about their potential to initiate angiogenesis. On the other hand, one thing is known for certain: c-kit plays a role in the proliferation of endothelial cells and contributes towards angiogenesis.

6.3 C-KIT in Disease

To delve into how the c-kit gene affects angiogenesis in a diseased environment, the mechanism of its expression must be magnified, again requiring close inspection of miRNAs. Under homeostatic conditions certain miRNAs such as let-7f and miR-27b are better expressed in endothelial cells, including HUVEC, and utilizes pro-angiogenic effects (Urbich et al. 2008). This was proven by Urbich et al. (2008) in the study of blockage of in vitro angiogenesis with 2′-*O*-methyl oligonucleotide inhibitors. Minimal knowledge regarding the exact mechanism of impact of pathological diseases such as cardiovascular disease or diabetes mellitus on angiogenesis has been uncovered, however investigations show promise of certain genes, such as c-kit and their respective miRNAs being affected, leading to an effect on angiogenesis (Urbich et al. 2008). Endothelial cells such as those found in the human umbilical vein has been known to house miR-221 and miR-222, two highly expressed miRNAs which exert anti-angiogenic effects. These effects include inhibition of cell migration, wound healing and vessel formation of endothelial cells, thus impacting the process of angiogenesis when deemed necessary. The expression of these miRNAs are strictly regulated but are found to reduce c-kit expression in haematopoietic progenitor cells.

Under disease circumstances, endothelial cells can be excessively infected with these anti-angiogenic miRNA factors and subsequently unable to form new vasculature during processes such as wound healing thus leading to detrimental effects. Diseases may also cause a depletion of miR-221 and miR-222. This shortage, as seen in the study of Urbich et al. (2008), also affects the expression of other miRNAs in HUVECs, which provides evidence that miRNAs can control the transcription and translation of other miRNAs. Certain miRNAs affected by miR-221 and miR-222 were found to have targeted the tyrosine kinase receptor and interfere with the expression and activity of other miRNAs linked to the c-kit gene and ultimately the role of c-kit in angiogenesis (Urbich et al. 2008).

6.4 C-KIT in Malignancy

Mutations of the c-kit gene may lead to the complete loss of function or retardation in the expression of the gene and its angiogenic capabilities. Possible outcomes include squamous epithelial carcinoma or oral mucosal melanoma (OMM). C-kit's role in OMM tumorigenesis is based on its function in melanocyte development. Although c-kit is expressed in physiological circumstances, excessive expression is correlated to the invasive action of a tumour. Investigations by Rivera et al. (2008) noted that in 18 cases of human OMM, c-Kit mutations were positive in 16 instances. Thus it is clear that mutation of the c-Kit proto-oncogene plays an appropriate role in tumorigenesis of OMM (Rivera et al. 2008).

As noted by Coussens et al. (1999) mutated expression of the c-kit gene in mice can lead to defects in sterility, coat colour abnormalities and formation of neoplasm. These neoplasms included malignant hyperproliferation of keratinocytes and a thickened epidermis with dermal mast cell infiltration of the basement membrane. As already outlined, tumours rely on angiogenesis as a means of spreading through vasculature and acquiring of nutrition. This can be achieved by mast cells with the c-kit receptor attached to its plasma membrane releasing important angiogenic factors such as histamine and VEGF through complex signal transductions. These factors, alongside other mast cell products and growth factors, promote the growth and survival of neoplastic tissues through the initiation of angiogenesis. Furthermore, mutated mast cells are capable of recruiting and activating fibroblasts, epithelial and endothelial cells, as well as enzymes that ultimately promote angiogenesis proliferation and consequently metastasis of neoplastic tissue. It is important to note, however, that mast cells were encountered in environments that were associated with dysplasia (pre-malignancy) rather than in the already developed tumour mass. Studies by Coussens et al. (1999) display the benefits of mast cell assistance to malignant neoplasms; they accumulate at the edges of cancers and help convert normal tissue stroma into neoplastic stroma, hence explaining why mast cells are found in dysplasias (Coussens et al. 1999).

Many attempts at inhibiting neoplastic mast cell proliferation utilize drugs to target the c-kit gene. Bai et al. (2013) discovered two inhibitors on the KIT gene which regulates its enzymatic conformation. These inhibitors, DP-2976 and DP-4851, were examined and found to have been retarded in the mutated c-kit gene. The inactivation of the tumour suppressor component of this gene leads to neoplastic mast cell activation and proliferation. When unaltered, these inhibitors allow apoptosis of neoplastic mast cells, accompanying the reduced spread of malignant cells. This mechanism, although still in early stages, gives promise towards therapeutic advantages concerning neoplasia caused by the c-kit gene defect (Bai et al. 2013).

References

Andre C, Hampe A, Lachaume P, Martin E, Wang XP, Manus V, Hu WX, Galibert F (1997) Sequence analysis of two genomic regions containing the KIT and the FMS receptor tyrosine kinase genes. Genomics 39(2):216–226. doi:10.1006/geno.1996.4482

Bai Y, Bandara G, Ching Chan E, Maric I, Simakova O, Bandara SN, Lu WP, Wise SC, Flynn DL, Metcalfe DD, Gilfillan AM, Wilson TM (2013) Targeting the KIT activating switch control pocket: a novel mechanism to inhibit neoplastic mast cell proliferation and mast cell activation. Leukemia 27(2):278–285. doi:10.1038/leu.2012.218

Coussens LM, Raymond WW, Bergers G, Laig-Webster M, Behrendtsen O, Werb Z, Caughey GH, Hanahan D (1999) Inflammatory mast cells up-regulate angiogenesis during squamous epithelial carcinogenesis. Genes Dev 13(11):1382–1397

Ferrara N, Gerber HP (2001) The role of vascular endothelial growth factor in angiogenesis. Acta Haematol 106(4):148–156

Matsui J, Wakabayashi T, Asada M, Yoshimatsu K, Okada M (2004) Stem cell factor/c-kit signaling promotes the survival, migration, and capillary tube formation of human umbilical vein endothelial cells. J Biol Chem 279(18):18600–18607. doi:10.1074/jbc.M311643200

Rivera RS, Nagatsuka H, Gunduz M, Cengiz B, Gunduz E, Siar CH, Tsujigiwa H, Tamamura R, Han KN, Nagai N (2008) C-kit protein expression correlated with activating mutations in KIT gene in oral mucosal melanoma. Virchows Archiv Int J Pathol 452(1):27–32. doi:10.1007/s00428-007-0524-2

Urbich C, Kuehbacher A, Dimmeler S (2008) Role of microRNAs in vascular diseases, inflammation, and angiogenesis. Cardiovasc Res 79(4):581–588. doi:10.1093/cvr/cvn156

Chapter 7
Cyclin D1 and E1

Abstract Two proto-oncogenes from the Cyclin protein family that are involved in the regulation of the cell cycle are described in this chapter. Cyclin D1 gene or CCND1 is involved in the regulation of the cell cycle and becomes most abundant in the G1 phase in the lead up to DNA replication. The activity of the cyclin D in normal cell physiology contributes to controlled and healthy progression of the cell cycle. Cyclin E1 or CCNE1 is a positive regulator of CDK2 that regulates cell progression from G1 to S phase of the cell cycle. Cyclin E1 is necessary for positive regulation of the cell cycle. Abnormalities with expression of CCNE1 have been suggested as a prognostic tool in some malignancies. Growth promotion is closely regulated and under abnormal circumstances cell growth and sustained division in the absence of the initial induction signal can arise; this proliferation may be due to an oncogene. CCND1 plays several roles in the progression of cancer via its potential contribution to uncontrolled proliferation through its involvement in cell cycle regulation and the promotion of the process of Angiogenesis via its involvement in the regulation of the Vascular Endothelial Growth Factor (VEGF). Understanding the role of Cyclins and their subsequent role in cell division and e angiogenesis associated with malignancies allows for the development of targeted therapy and advances in future medicinal treatment options.

Keywords Cyclin D1 • Cyclin E1 • Angiogenesis • Normal physiology • Disease • Malignancy

7.1 Cyclin D1 and Cyclin E1

The gene CCND1 is located on chromosome 11 in humans and codes for the protein Cyclin D1 (Inaba et al. 1992). Cyclin D1 is considered a "proto-oncogene", which is a normal gene that can become oncogenic upon mutation or over expression (Fu et al. 2004). This is due to its involvement in the regulation of the cell cycle. Cyclin D1 belongs to the 'cyclin' protein family, whose abundance within proliferating cells, undergoing cell cycle, changes with dramatic periodicity (Fu et al. 2004). The cell cycle describes the process by which a cell undergoes division and duplication into two daughter cells, and is divided into the following phases: G0, where the cell

A. Salajegheh, *Angiogenesis in Health, Disease and Malignancy*,
DOI 10.1007/978-3-319-28140-7_7

is quiescent; G1, where the cell increases in size; S, where the replication of DNA occurs; G2, where the cell continues to grow in size; and M, where cell growth stops and mitotic division occurs (Cooper and Hausman 2000). Cyclin D1 becomes most abundant in the G1 phase in the lead up to DNA replication. The cyclin family acts as regulators of the Cyclin-Dependent Kinase (CDK) enzymes. Specifically, Cyclin D1 functions as a regulatory subunit of the enzymes CDK4 or CDK6, where it forms a protein-enzyme complex (cyclin D-Cdk4/6) that is essentially responsible for the cell cycle transition from the G1 phase to the S phase (Cooper and Hausman 2000).

Cyclin E1 or CCNE1 is a gene located at 19q13 and is necessary for positive regulation of the cell-cycle by coding for Cyclin E (Cheng and Eble 2012). Cyclin E functions as a positive regulator of CDK2 which regulates cell progression from G1 to S phase of the cell cycle, a characteristic of proto-oncogenes (Cheng and Eble 2012). It has been shown to be prognostic in some disease when abnormalities with expression are detected (Akli et al. 2004; Akli and Keyomarsi 2003).

7.2 Cyclin D1 and Cyclin E1 in Normal Physiology

In normal cell physiology the activity of the cyclin D-Cdk4/6 complex contributes to controlled and healthy progression of the cell cycle (Fu et al. 2004). Cyclin D1 is known to interact with "Retinoblastoma Tumour Suppressor protein" (pRb) in a way that generates controlled proliferation (Fu et al. 2004). The mechanism by which this occurs has been investigated thoroughly for the last 20 years, and the general putative understanding has been that pRb is progressively multi-phosphorylated by Cyclin D1 in the early G1 phase of the cell cycle, which in turn inactivates the tumor suppressor protein. However, this model of interaction remains unproven biochemically, and recent studies conducted by Narasimha et al. (2014) has found that in a healthy cell, Cyclin D1 in fact exclusively activates the Rb tumor suppressor protein via the process of mono-phosphorylation (Narasimha et al. 2014). Activated Rb essentially functions as a cell cycle repressor; it forms a complex with a member of the E2F family of transcription factors, which acts to inhibit the expression of the genes normally regulated by the E2F factors, which in turn inhibits cells from passing through the G1 phase, thus arresting cell proliferation (Day et al. 2009). In a healthy cell, this mechanism contributes to the healthy cell's effort of ensuring normal proliferation, by preventing tumorous growth (Cooper and Hausman 2000).

CCNE1 plays a key role in regulation of the cell cycle through regulation of cyclin dependent kinase (CDK). In cells undergoing normal mitosis, the G1-synthesis and s phase degredation of CCNE are closely regulated (Yu and Hung 2009), when this regulation is impaired, as seen in overstimulation of cell reproduction, the formation of neoplasms can arise. Communication between the regulation of cyclin E transcription and Cyclin E protein stability results in fine tuning of Cyclin E levels; this emphases its crucial role in regulation of the cell cycle and can predict the deleterious effects of an abnormally high level of expression of Cyclin E which can be seen in many cancers (Sluyser 2005). In late stage G1 of the cell cycle, cyclin E transcription is activated by a series of complexes which relieve repression of the CCNE gene, this allows a G1 arrest and facilitates additional accumulation of CCNE protein; When this accumulation advances to acceptable levels phosphorylation of pRb relieves repression of the S-phase cyclin, cyclin A, and Cdk1, which permits the cell cycle to progress to mitosis (Yu and Hung 2009). Cell division is pivotal to angiogenesis, as growth of new vessels cannot occur without proliferation of the associated tissues. With the continued division of vasculature cells the extension and growth of blood vessels is able to facilitate necessary physiological processes, as well as potential sustainability of proliferative malignancies.

7.3 Cyclin D1 in Disease

This gene appears to play several different roles in the progression of cancer. The two most prominent are: its potential contribution to uncontrolled proliferation as a consequence of its involvement in cell cycle regulation (Chow 2010), and the promotion of the process of Angiogenesis via its involvement in the regulation of the Vascular Endothelial Growth Factor (VEGF) (Pestell and Li 2006).

As to be expected of any gene involved in the regulation of the cell cycle: mutation, amplification, and/or over expression of Cyclin D1 can alter cell cycle progression. In a study conducted by Ewen and Lamb (2004), it was found that over expression of Cyclin D1 in rodent fibroblasts suggested that they had a rate-limiting effect for the progression through the G1 phase of the cell cycle. Cell culture experiments also linked the action of Cyclin D-Cdk4/6 complexes to the function of pRb (Ewen and Lamb 2004). The initial inactivating phosphorylation of pRb during G1 by Cyclin D-Cdk4/6 was found to promote the release of the E2F family of transcription factors from their inactive form in the pRb-E2F complex, which in turn leads to the transcriptional induction of genes such as Cyclin E1 (discussed later in the paper), which are required for cell cycle progression from phase G1 to S (Ewen and Lamb 2004). Note that this study was conducted in 2004, which was before the study conducted by Narasimha et al. (2014), which conversely indicated that the mechanism of Cyclin D1 was to exclusively activate pRb. However, the results obtained by Ewen and Lamb (2004) decidedly purports the ensuing consequence of the over expression of Cyclin D1, which is that of increased inactivation of pRb leading to the promotion of the progression of the cell cycle from G1 to S, thus promoting increased proliferation and oncogenesis.

Furthermore, CCND1 has been found to be involved in the process of vascularization of tumours via the process of angiogenesis. An investigation conducted by Yasui et al. (2006) purported to examine the effect of inhibiting cyclin D1 on tumour-associated neovascularization (angiogenesis). The study suggested that Cyclin D1 may play a role in maintaining Vascular Endothelial Growth Factor (VEGF) expression, which is a vital angiogenic factor contributing to the vascularisation of tumours (Yasui et al. 2006). The inhibitory agent used in the investigation was an adenovirus system for antisense to cyclin D1 (AS CyD1) (Yasui et al. 2006). It was found that xenografts treated with AS CyD1 showed less blood vessel density, and exhibited smaller tumour size in certain colon cancer cell lines (Yasui et al. 2006). In vitro, it was found that AS CyD1 had decreased the expression of VEGF; a substantial decrease in the activity of VEGF promoters was also found (Yasui et al. 2006). Yasui et al. (2006) suggests that AS CyD1 could have potential to be used for targeting cancerous cells and the vascular supply of tumors. Another possible form of treatment via the suppression of angiogenesis by manipulating Cyclin D1 is the synthetic drug TNP-470, an angiogenesis inhibitor that is currently in clinical development for cancer (Kruger and Figg 2000). As explained by Figg and Folkman (2008) in Angiogenesis: An Integrative Approach from Science to Medicine, it was found that tissue treated by TNP-470 induced formation of Cyclin D1 in the endothelial cells. These 'induced' Cyclin D1 proteins were found to form a complex with CDK4 and another protein known as p21, which belongs to a family of CDK inhibitors that antagonises the activation of the Cyclin D1-Cdk4 complex (Figg and Folkman 2008). This antagonising action prevents the phosphorylation of pRb, which in turn prevents the cell cycle progression from phase G1 to S, thus arresting proliferation (Figg and Folkman 2008).

7.4 Cyclin E1 in Malignancy

The genetic damage can occur in a number of ways including retroviral transduction, transposon integration, insertion mutations, point mutations, chromosome translocation, gene amplification and or protein-protein interactions (Caldon and Musgrove 2010). As uncontrolled prolific cell growth is dangerous, it is necessary to tightly regulate the genes responsible. Studying and understanding the function of genes such as CCNE1 are essential for early diagnostic and treatment regiments to be developed. CCNE1 plays a key role in regulation of the cell cycle both in tumorigenesis and angiogenesis (Wang et al. 2009); it is achieved by the overexpression of CCNE and subsequent tumorigenesis by the promotion of G_1 to S-phase transition, increasing CCNE-associated kinase activity and causing genomic instability (Akli et al. 2004; Akli and Keyomarsi 2003). S phase of the cell cycle revolves around DNA synthesis and chromosome duplication and is a common source of DNA mutation and subsequent potential malignancies. CCNE1 regulates cyclin dependent kinase (CDK) and is a cell cycle gene commonly found altered or mutated in expression in human cancer (Caldon and Musgrove 2010).

Because of the essential role Cyclin E1 has in cell division it is a prime candidate for targeted molecular therapy. As previously stated CCNE1 overexpression or deregulation can be seen in a number of malignancies; amplification of CCNE1 is known to be specific for high-grade serous carcinoma vs low grade serous carcinoma or normal ovarian tissue with a frequency of 32.2–36.1 % (Nakayama et al. 2007). In addition CCNE1 expression has been shown to be heterogenous in multiple myeloma cell lines (hMMCLs), and by incubating these with a selective CDK inhibitor known as Seliciclib, apoptosis and down regulation of MCL1 and p27 can occur (Josefsberg Ben-Yehoshua et al. 2012).

The past 20 years have seen much development in the research on the gene CCND1 and the mechanisms of action of Cyclin D1. The investigations conducted on the involvement of Cyclin D1 in the process of angiogenesis has produced very promising results for the field of cancer therapy, such as the synthetic drug TNP-470 (Kruger and Figg 2000; Figg and Folkman 2008) and the antisense to Cyclin D1 (AS CyD1) (Yasui et al. 2006). These developments and the continued research being conducted in this area paint a hopeful picture for the future of cancer supressing treatment via the inhibition of tumour vascularisation.

As discussed, the presence of CCNE1 overexpression in numerous malignancies is evident. Based on various studies there are potential implications for future targeted therapies, strategies including cell cycle deregulation such as the use of cyclin-dependent kinase inhibition (Etemadmoghadam et al. 2010). Increased expression of CCNE1 may have clinical uses in determining patient sensitivity or rate of response to standard treatments associated with primary tumours, this would enable health professionals to pursue the most effective treatment possible available at the earliest availability (Etemadmoghadam et al. 2010). In addition Shapiro anticipated that 'chemotherapeutic agents may sensitize cells to CDK inhibition' (Shapiro 2006); using this information combination targeted therapy may be deemed more effective in the treatment of some of the conditions seen in CCNE1 overexpression. Future research development should focus on efficacy of targeted therapy without necessarily requiring the coexisting conventional therapy thereby promoting the success of single target CDK inhibitors, as the clinical success to date has been disappointing (Malumbres and Barbacid 2009).

References

Akli S, Keyomarsi K (2003) Cyclin E and its low molecular weight forms in human cancer and as targets for cancer therapy. Cancer Biol Ther 2(4 Suppl 1):S38–S47

Akli S, Zheng PJ, Multani AS, Wingate HF, Pathak S, Zhang N, Tucker SL, Chang S, Keyomarsi K (2004) Tumor-specific low molecular weight forms of cyclin E induce genomic instability and resistance to p21, p27, and antiestrogens in breast cancer. Cancer Res 64(9):3198–3208

Caldon CE, Musgrove EA (2010) Distinct and redundant functions of cyclin E1 and cyclin E2 in development and cancer. Cell Div 5:2. doi:10.1186/1747-1028-5-2

Cheng L, Eble JN (2012) Molecular surgical pathology. Springer Science and Business Media, New York

Chow A (2010) Cell cycle control by oncogenes and tumor suppressors: driving the transformation of normal cells into cancerous cells. Nat Educ 3:7

Cooper GM, Hausman RE (2000) The cell. Sinauer Associates, Sunderland

Day PJ, Cleasby A, Tickle IJ, O'Reilly M, Coyle JE, Holding FP, McMenamin RL, Yon J, Chopra R, Lengauer C, Jhoti H (2009) Crystal structure of human CDK4 in complex with a D-type cyclin. Proc Natl Acad Sci U S A 106(11):4166–4170. doi:10.1073/pnas.0809645106

Etemadmoghadam D, George J, Cowin PA, Cullinane C, Kansara M, Gorringe KL, Smyth GK, Bowtell DD (2010) Amplicon-dependent CCNE1 expression is critical for clonogenic survival after cisplatin treatment and is correlated with 20q11 gain in ovarian cancer. PLoS One 5(11):e15498. doi:10.1371/journal.pone.0015498

Ewen ME, Lamb J (2004) The activities of cyclin D1 that drive tumorigenesis. Trends Mol Med 10(4):158–162. doi:10.1016/j.molmed.2004.02.005

Figg W, Folkman J (2008) Angiogenesis: an integrative approach from science to medicine. Springer, New York

Fu M, Wang C, Li Z, Sakamaki T, Pestell RG (2004) Minireview: Cyclin D1: normal and abnormal functions. Endocrinology 145(12):5439–5447. doi:10.1210/en.2004-0959

Inaba T, Matsushime H, Valentine M, Roussel MF, Sherr CJ, Look AT (1992) Genomic organization, chromosomal localization, and independent expression of human cyclin D genes. Genomics 13(3):565–574

Josefsberg Ben-Yehoshua L, Beider K, Shimoni A, Ostrovsky O, Samookh M, Peled A, Nagler A (2012) Characterization of cyclin E expression in multiple myeloma and its functional role in seliciclib-induced apoptotic cell death. PLoS One 7(4):e33856. doi:10.1371/journal.pone.0033856

Kruger EA, Figg WD (2000) TNP-470: an angiogenesis inhibitor in clinical development for cancer. Expert Opin Investig Drugs 9(6):1383–1396. doi:10.1517/13543784.9.6.1383

Malumbres M, Barbacid M (2009) Cell cycle, CDKs and cancer: a changing paradigm. Nat Rev Cancer 9(3):153–166. doi:10.1038/nrc2602

Nakayama K, Nakayama N, Jinawath N, Salani R, Kurman RJ, Shih Ie M, Wang TL (2007) Amplicon profiles in ovarian serous carcinomas. Int J Cancer J Int du Cancer 120(12):2613–2617. doi:10.1002/ijc.22609

Narasimha AM, Kaulich M, Shapiro GS, Choi YJ, Sicinski P, Dowdy SF (2014) Cyclin D activates the Rb tumor suppressor by mono-phosphorylation. eLife 3:e02872. doi:10.7554/eLife.02872

Pestell RG, Li Z (2006) Antisense to cyclin D1 inhibits VEGF-stimulated growth of vascular endothelial cells: implication of tumor vascularization. Clin Cancer Res Off J Am Assoc Cancer Res 12(15):4459–4462. doi:10.1158/1078-0432.ccr-06-0614

Shapiro GI (2006) Cyclin-dependent kinase pathways as targets for cancer treatment. J Clin Oncol Off J Am Soc Clin Oncol 24(11):1770–1783. doi:10.1200/jco.2005.03.7689

Sluyser M (2005) Application of apoptosis to cancer treatment. Springer Science and Business Media, New York

Wang F, Fu XD, Zhou Y, Zhang Y (2009) Down-regulation of the cyclin E1 oncogene expression by microRNA-16-1 induces cell cycle arrest in human cancer cells. BMB Rep 42(11):725–730

Yasui M, Yamamoto H, Ngan CY, Damdinsuren B, Sugita Y, Fukunaga H, Gu J, Maeda M, Takemasa I, Ikeda M, Fujio Y, Sekimoto M, Matsuura N, Weinstein IB, Monden M (2006) Antisense to cyclin D1 inhibits vascular endothelial growth factor-stimulated growth of vascular endothelial cells: implication of tumor vascularization. Clin Cancer Res Off J Am Assoc Cancer Res 12(15):4720–4729. doi:10.1158/1078-0432.ccr-05-1213

Yu D, Hung M-C (2009) Breast cancer chemosensitivity, vol 608. Springer Science and Business Media, New York

Chapter 8
Cluster of Differentiation 71 (CD71)

Abstract Cluster of Differentiation 71 protein (CD71), also known as the Transferrin receptor protein 1 (TfR1) is a protein encoded by the TFRC gene located on chromosome 3q29. It is found as a cell surface membrane protein in humans and is essential for iron delivery to the cells. TfR1 expression is up-regulated or down-regulated in response to cell's iron need. A regulatory protein that regulates the interaction between Transferrin and Transferrin receptor is known as HFE1 and its mutation is thought to increase the risk of Alzheimer Disease. TfR1 knockdown cells have been also found to enhance VEGF production and angiogenesis in vivo. CD71 is used as a marker in breast cancer and has also been linked to many other cancers such as lungs, pancreas, lymph nodes and colon. The gene is currently being explored for its diagnostic and therapeutic opportunities. Currently, CD71 is a new target for the potential treatment of human lymphoma & leukaemia.

Keywords Cluster of differentiation 71 • CD71 • Angiogenesis • Normal physiology • Disease • Malignancy

8.1 Cluster of Differentiation 71 (Cd71) – An Iron Transferrin Receptor Protein

Cluster of Differentiation 71 protein (CD71), also known as the Transferrin receptor protein 1 (TfR1) is a protein encoded by the TFRC gene located on chromosome 3q29 (Aisen 2004). It is found as a cell surface membrane protein in humans and is essential for iron delivery to the cells (Aisen 2004). Iron plays an important role in many cellular functions such as oxygen transport, DNA synthesis and enzyme function (Speeckaert et al. 2010). Iron homeostasis is maintained by Transferrin receptor-1 (TfR1), ferritin and Transferrin (Tf). Transferrin present in blood acts as an iron transporter and is important for iron distribution. Iron-Tf-TfR forms a complex and enters the cells by membrane endocytosis (Speeckaert et al. 2010).

A. Salajegheh, *Angiogenesis in Health, Disease and Malignancy*,
DOI 10.1007/978-3-319-28140-7_8

8.2 Cluster of Differentiation 71 (Cd71) in Normal Physiology

Transferrin receptor (CD71), ferritin and transferrin have been identified as specialised proteins that maintain iron homeostasis as they control the transport, storage and uptake of free iron in tissues (Yefimova et al. 2002). Imbalance in iron homeostasis can lead to the disruption of normal iron distribution and the generation of reactive oxygen species causing cell damage. Retinal neuropathies have also been shown to accompany iron homeostatic disorders in rats (Yefimova et al. 2002). During aging, iron imbalance leads to its accumulation in selective brain regions causing neurodegenerative diseases (Benarroch 2009). The presence of Cd71 antibodies in the serum is used as a measure indicating proliferation in leukaemia and lymphomas (Oudemans et al. 1986).

TfR1 expression is up-regulated or down-regulated in response to cell's iron need (Testa et al. 1993). Iron deficiency within the cells lead to the binding of iron regulatory protein (IRP) to the iron responsive element (IRE) in the downstream untranslated region (3′UTR) of TfR1 mRNA0 (Testa et al. 1991). TfR1 expression is upregulated and the uptake of iron is increased. Similarly, to decrease iron storage, IRP bind to the upstream untranslated region (5′UTR) of ferritin mRNA and to its expression is downregulated. CD71 is expressed on most dividing cells as well as on alveolar macrophage and brain endothelium (Testa et al. 1991). This indicates the significance of the Cd71 protein in cell growth, proliferation, and nutrition supply and the potential consequences of deregulated Cd71 expression on cell surfaces in terms of pathology and cancer development.

8.3 Cluster of Differentiation 71 (Cd71) in Disease and Malignancy

TfR1 knockdown cells have been found to enhance VEGF production and angiogenesis in vivo (Eckard et al. 2010). In addition, a study conducted in 2008, showed iron depletion by desferoxaminemesylate (DFO), an iron chelator, increased the levels of hypoxia inducible factor-1 alpha (HIF-1a) (Dongiovanni et al. 2008). This is an important transcription factor in VEGF regulation, necessary for angiogenesis induction, tumour angiogenesis and metastasis (Dongiovanni et al. 2008). This study did not address the direct link between iron deficiency and tumour angiogenesis but was the basis of another research conducted in 2010 to address this link (Eckard et al. 2010). In an experiment performed in MDA-MB-231, a triple negative breast cancer cell-line, shRNA molecules designed against TfR1 were used to knockdown their expression (Eckard et al. 2010). TfR1 act as gatekeeper for iron uptake and therefore knocking down their expression is equivalent to restricting iron entry into the cell and thus mimics iron deficiency. The results of this study show that cellular iron deficiency plays an important role in the regulation of VEGF,

HIF-1a stabilization, and in the promotion of angiogenesis. The study concludes that cancer patients with systemic iron deficiency and anaemia have an increased risk of tumour recurrence (Eckard et al. 2010).

In cancers, over-expression of *TFRC* is associated with malignancy as it facilitates higher iron uptake required for cell division (Jiang et al. 2010). CD71 is also used as a marker in breast cancer and the protein expression of CD71 is well-characterized using microarrays in patients with invasive breast carcinoma (Habashy et al. 2010). The expression of TfR1 in breast cancers has proven to be up by 5 times in malignancy tissue in comparison to normal tissue. This over expression of TFRC has been linked with the increased proliferative capacity of tumours (Habashy et al. 2010). CD71 is a candidate marker for a subcategory of ER+/luminal-like breast cancer. This cancer is resistant to tamoxifen, an oestrogen antagonist and is characterized by its poor outcome (Habashy et al. 2010). The importance of CD71 as a prognostic biomarker for the treatment of this breast cancer phenotype remains largely unexplored and unknown (Habashy et al. 2010). Over expression of endogenous TfR1 has also been linked to many other cancers such as lungs, pancreas, lymph nodes and colon (Peer et al. 2007). This reflects the need for iron as a cofactor in DNA synthesis enzymes which ultimately results in cell proliferation and rapid cell division (Peer et al. 2007).

Mutation in HFE1, a regulatory protein that regulates the interaction between Transferrin and Transferrin receptor is thought to increase the risk of Alzheimer Disease (AD) (Benarroch 2009). In animal models of Parkinson Disease (PD), over-expression of iron-sequestering ferritin protein and iron chelation has shown to be effective (Benarroch 2009). No disease causing mutations have been identified in the gene but there are missense-coding variants that might have some functional consequences (Borie et al. 2002). Trisomy of chromosome 3, gaining 3q27-qter or a whole 3q arm have been noticed in various malignancies such as haematopoietic ones and solid tumours (Borie et al. 2002). The gene is not involved in cancer-associated translocations. Also the TRFC variant S142G can modify its association with HFE C282Y mutation linked to cancer susceptibility for colorectal cancer, hepatocellular carcinoma, breast cancer, leukaemia and multiple melanomas (Borie et al. 2002).

The gene is currently being explored for its diagnostic and therapeutic opportunities (Liu et al. 2014). CD71 is a new target for the potential treatment of human lymphoma & leukaemia. Future research on the survival of solid organ graft and experiments to improve their survival by the use of monoclonal anti-CD71 antibody to suppress antigen and mitogen responses are currently being conducted (Liu et al. 2014). A particular study investigated the importance of CD71 as a flow cytometric marker in the diagnosis of acute leukaemia (Blanchard et al. 2000). They found both cells CD34 and CD71 gradually increased during the process from myeloid dysplasia to apparent leukemic (Liu et al. 2014). Subsequently, the presence of leukemic cell subsets with variable levels of CD71 and CD34 may be useful for further understanding the dynamic process associated with clonal development seen in leukaemia (Liu et al. 2014).

References

Aisen P (2004) Transferrin receptor 1. Int J Biochem Cell Biol 36(11):2137–2143. doi:10.1016/j.biocel.2004.02.007

Benarroch EE (2009) Brain iron homeostasis and neurodegenerative disease. Neurology 72(16):1436–1440. doi:10.1212/WNL.0b013e3181a26b30

Blanchard J, Li X, Braun M, Milewski S, Cicalese L, Benedetti E, Rastellini C (2000) Administration of anti-Cd71 mab prolongs survival of heterotopic heart and skin graft.: Abstract# 321 Poster Board #-Session: P168-I. Transplantation 69(8):S197

Borie C, Gasparini F, Verpillat P, Bonnet AM, Agid Y, Hetet G, . . . Grandchamp B (2002) Association study between iron-related genes polymorphisms and Parkinson's disease. J Neurol 249(7):801–804. doi:10.1007/s00415-002-0704-6

Dongiovanni P, Valenti L, Ludovica Fracanzani A, Gatti S, Cairo G, Fargion S (2008) Iron depletion by deferoxamine up-regulates glucose uptake and insulin signaling in hepatoma cells and in rat liver. Am J Pathol 172(3):738–747. doi:10.2353/ajpath.2008.070097

Eckard J, Dai J, Wu J, Jian J, Yang Q, Chen H, ... Huang X (2010) Effects of cellular iron deficiency on the formation of vascular endothelial growth factor and angiogenesis. Iron deficiency and angiogenesis. Cancer Cell Int 10:28. doi:10.1186/1475-2867-10-28

Habashy HO, Powe DG, Staka CM, Rakha EA, Ball G, Green AR, ... Gee JM (2010) Transferrin receptor (CD71) is a marker of poor prognosis in breast cancer and can predict response to tamoxifen. Breast Cancer Res Treat 119(2):283–293. doi:10.1007/s10549-009-0345-x

Jiang XP, Elliott RL, Head JF (2010) Manipulation of iron transporter genes results in the suppression of human and mouse mammary adenocarcinomas. Anticancer Res 30(3):759–765

Liu Q, Wang M, Hu Y, Xing H, Chen X, Zhang Y, Zhu P (2014) Significance of CD71 expression by flow cytometry in diagnosis of acute leukemia. Leuk Lymphoma 55(4):892–898. doi:10.3109/10428194.2013.819100

Oudemans PB, Brutel de la Riviere G, Hart GA, van Heerde P, Scholte G, Vroom TM (1986) Determination of transferrin receptors on frozen sections of malignant B-cell lymphomas by immunofluorescence with a monoclonal antibody. Cancer 58(6):1252–1259

Peer D, Karp JM, Hong S, Farokhzad OC, Margalit R, Langer R (2007) Nanocarriers as an emerging platform for cancer therapy. Nat Nanotechnol 2(12):751–760. doi:10.1038/nnano.2007.387

Speeckaert MM, Speeckaert R, Delanghe JR (2010) Biological and clinical aspects of soluble transferrin receptor. Crit Rev Clin Lab Sci 47(5–6):213–228. doi:10.3109/10408363.2010.550461

Testa U, Kuhn L, Petrini M, Quaranta MT, Pelosi E, Peschle C (1991) Differential regulation of iron regulatory element-binding protein(s) in cell extracts of activated lymphocytes versus monocytes-macrophages. J Biol Chem 266(21):13925–13930

Testa U, Pelosi E, Peschle C (1993) The transferrin receptor. Crit Rev Oncog 4(3):241–276

Yefimova MG, Jeanny JC, Keller N, Sergeant C, Guillonneau X, Beaumont C, Courtois Y (2002) Impaired retinal iron homeostasis associated with defective phagocytosis in Royal College of Surgeons rats. Invest Ophthalmol Vis Sci 43(2):537–545

Chapter 9
Connective Tissue Growth Factor (*CTGF*)

Abstract Connective tissue growth factor (CTGF) is a member of the C-terminal cystein-rich protein (CCN) intercellular signaling protein family and functions in the extracellular matrix (ECM). Ultimately, CTGF allows other pericellular moieties to bind to its structural modules which can trigger various responses via signaling molecules. CTGF is activated in endothelial cells via basic fibroblast growth factor (bFGF) or vascular endothelial growth factor (VEGF). Mediated through cell surface integrins, CTGF plays a significant role in promoting endothelial cell growth, migration and adhesion and survival in vitro. The maintenance of CTGF levels in the body is an extremely important mechanism in homeostasis and an imbalance can result in disease or cancer. Many studies have demonstrated a positive correlation between the level of CTGF expression and the degree of malignancy in cancers. As a fibrogenic master switch for EMT, an imbalance of CTGF levels has been also associated with the development of variety of diseases.

Keywords Connective tissue growth factor • CTGF • Angiogenesis • Normal physiology • Disease • Malignancy

9.1 CTGF (Connective Tissue Growth Factor)

Connective tissue growth factor (CTGF) is a member of the C-terminal cystein-rich protein (CCN) intercellular signaling protein family and functions in the extracellular matrix (ECM). Ultimately, CTGF allows other pericellular moieties to bind to its structural modules which can trigger various responses via signaling molecules (Brigstock 2002). CTGF is activated in endothelial cells via basic fibroblast growth factor (bFGF) or vascular endothelial growth factor (VEGF). Mediated through cell surface integrins, CTGF plays a significant role in promoting endothelial cell growth, migration and adhesion and survival *in vitro* (Kuiper et al. 2008). Furthermore, CTGF is secreted by vascular endothelial cells and is induced by agents such as angiotensin II, endothelin-1 and glucocorticoids (Kuiper et al. 2008). Apart from having significant roles in injury repair, wound healing, skeletal development and angiogenesis, CTGF also provides directional growth of endothelial cells within the vasculature via inducing chemotactic movement (inducing

A. Salajegheh, *Angiogenesis in Health, Disease and Malignancy*,
DOI 10.1007/978-3-319-28140-7_9

directional cell migration) and chemokinetic movement (random cell movement) (Hishikawa et al. 1999). Thus, considering the basic functions of CTGF aforementioned, it must be stressed that CTGF plays an extremely important role in the progression of disease and cancers and should therefore be investigated in depth (Braig et al. 2011).

9.2 Connective Tissue Growth Factor (CTGF) in Normal Physiology

The structure of CTGF is significantly related to its function and can be attributed to many aspects relating to angiogenesis. One of the four modular domains encoded in CTGF is the von Willebrand factor type- C repeat (VWC) module which bone morphogenetic protein (BMP) and TGF-β can bind onto (Maeda et al. 2009). CTGF acts as a mediator by binding and inhibiting BMP or enhancing TGF-β to regulate cell proliferation and differentiation. Another significant modular domain is the TSP-1 module which can be bonded by low density lipoprotein receptor related protein (LRP); enabling TSP-1 to bind to extracellular matrix proteins such as matrix metalloproteinases and integrins (Maeda et al. 2009).

9.3 Connective Tissue Growth Factor (CTGF) in Disease

An imbalance of CTGF levels has been associated with the development of variety of diseases. It is concluded from several studies that CTGF is a renal risk marker for Type I diabetes and vascular disease. Over-expression of CTGF results in ECM accumulation and induces Epithelial-to-mesenchymal transition (EMT), which serves three main functions. Two of the main functions of EMT are contribution to wound healing and cancer development, which leads to cells becoming invasively migratory (Braig et al. 2011). Additionally, it is also involved in the promotion of fibrosis, this process initiated by triggering the transition from tubuloepithelial cells to myfibroblast cells. In many studies, CTGF has been proposed as "*a fibrogenic master switch*" for EMT (Gressner and Gressner 2008). The over-expression of CTGF, a pro-fibrogenic factor, can lead to excessive extracellular matrix deposition, type 1 collagen fibers and hence, fibrosis. Ultimately, Fibrosis can lead to many diseases, especially those of the kidney, such as proliferative diabetic retinopathy, pulmonary fibrosis, keloids, systemic sclerosis, atherosclerosis, biliary atresia, cataracts and myocardial fibrosis (Abraham 2008).

9.4 Connective Tissue Growth Factor (CTGF) in Malignancy

Understanding the dynamic relationship between CTGF and VEGF is a vital aspect in CTGF's role in angiogenesis and the CTGF/VEGF relationship strongly corresponds with activity such as metastases. VEGF, binding to CTGF modules, can be released as the bound active form via proteolysis by the matrix metalloproteinases (MMP's). MMP secretion and activation occurs during wound repair and tissue remodeling (Hashimoto et al. 2002; Inoki et al. 2002). As a result, if VEGF were to remain bound to CTGF, angiogenic activity would significantly decrease as VEGF would be unable to exert its pro-angiogenic effects onto its receptor. Therefore, CTGF can be considered as a modulator of VEGF bioavailability (Gressner and Gressner 2008).

Another factor contributing to modulating VEGF expression is its up-regulation via activated hypoxia inducible factors (HIF-1), aiding metastasis via increased angiogenic activity. Additionally, an increase in VEGF expression also stimulates CTGF to drive tumour growth and metastasis (Abraham 2008).

Many studies have demonstrated a positive correlation between the level of CTGF expression and the degree of malignancy in cancers, notably breast and colorectal cancer (Hishikawa et al. 1999; Ladwa et al. 2011). Additionally, a significant number of studies have concluded that CTGF, produced by tumour cells, promote angiogenesis via signaling in the VEGF pathway, leading to tumour growth. However, what is also interesting to note is that the over-expression of the CTGF gene in cells can also induce apoptosis, which has been evident in Human Breast Cancer Cell Line MCF-7 (Hishikawa et al. 1999).

The contribution of CTGF in the promotion of bone metastasis in breast cancer (osteotropic metastasis) has been extensively researched. Osteotropic metastasis occurs due to growth factors secreted by tumours which interact with CTGF (Ivkovic et al. 2003). Additionally, CTGF has been linked to oesophageal tumours and gastric cancer as a high expression of CTGF ultimately leads to poor prognosis and metastases (Xie et al. 2010). Wnt signalling pathway proteins can also induce CTGF expression which promotes metastasis. This can be demonstrated in an investigation of Melanoma cells. The migratory and invasive ability is significantly reduced where CTGF expression is inhibited. Contrastingly, the migratory and invasive actions were promoted when normal melanocytes exhibited an increase in CTGF level (Braig et al. 2011).

The maintenance of CTGF levels in the body is an extremely important mechanism in homeostasis and an imbalance can result in disease or cancer. In a study of Hepatocellular cancer, it was found that TGF-β1 up-regulates CTGF (Brigstock 2002). An inhibition of this up-regulation can diminish tumour growth, vascular invasion and the progression of metastatic disease. However, it was also concluded that patients with high CTGF expressers generally had poor prognosis. Furthermore, CTGF has been found to be accompanied by MMP- 13 expression for its role in promoting cell migration. CTGF over-expression (generally occurring in B cells) can lead to the activation of β- catenin (an element in Wnt protein canonical signaling), an increased expression of MMPs and the induction of apoptosis (Kuiper et al. 2008).

References

Abraham D (2008) Connective tissue growth factor: growth factor, matricellular organizer, fibrotic biomarker or molecular target for anti-fibrotic therapy in SSc? Rheumatology (Oxford) 47(Suppl 5):v8–v9. doi:10.1093/rheumatology/ken278

Braig S, Wallner S, Junglas B, Fuchshofer R, Bosserhoff AK (2011) CTGF is overexpressed in malignant melanoma and promotes cell invasion and migration. Br J Cancer 105(2):231–238. doi:10.1038/bjc.2011.226

Brigstock DR (2002) Regulation of angiogenesis and endothelial cell function by connective tissue growth factor (CTGF) and cysteine-rich 61 (CYR61). Angiogenesis 5(3):153–165

Gressner OA, Gressner AM (2008) Connective tissue growth factor: a fibrogenic master switch in fibrotic liver diseases. Liver Int 28(8):1065–1079. doi:10.1111/j.1478-3231.2008.01826.x

Hashimoto G, Inoki I, Fujii Y, Aoki T, Ikeda E, Okada Y (2002) Matrix metalloproteinases cleave connective tissue growth factor and reactivate angiogenic activity of vascular endothelial growth factor 165. J Biol Chem 277(39):36288–36295. doi:10.1074/jbc.M201674200

Hishikawa K, Oemar BS, Tanner FC, Nakaki T, Luscher TF, Fujii T (1999) Connective tissue growth factor induces apoptosis in human breast cancer cell line MCF-7. J Biol Chem 274(52):37461–37466

Inoki I, Shiomi T, Hashimoto G, Enomoto H, Nakamura H, Makino K, Ikeda E, Takata S, Kobayashi K, Okada Y (2002) Connective tissue growth factor binds vascular endothelial growth factor (VEGF) and inhibits VEGF-induced angiogenesis. FASEB J 16(2):219–221. doi:10.1096/fj.01-0332fje

Ivkovic S, Yoon BS, Popoff SN, Safadi FF, Libuda DE, Stephenson RC, Daluiski A, Lyons KM (2003) Connective tissue growth factor coordinates chondrogenesis and angiogenesis during skeletal development. Development 130(12):2779–2791

Kuiper EJ, Van Nieuwenhoven FA, de Smet MD, van Meurs JC, Tanck MW, Oliver N, Klaassen I, Van Noorden CJ, Goldschmeding R, Schlingemann RO (2008) The angio-fibrotic switch of VEGF and CTGF in proliferative diabetic retinopathy. PLoS One 3(7):e2675. doi:10.1371/journal.pone.0002675

Ladwa R, Pringle H, Kumar R, West K (2011) Expression of CTGF and Cyr61 in colorectal cancer. J Clin Pathol 64(1):58–64. doi:10.1136/jcp.2010.082768

Maeda A, Nishida T, Aoyama E, Kubota S, Lyons KM, Kuboki T, Takigawa M (2009) CCN family 2/connective tissue growth factor modulates BMP signalling as a signal conductor, which action regulates the proliferation and differentiation of chondrocytes. J Biochem 145(2):207–216. doi:10.1093/jb/mvn159

Xie J-J, Xu L-Y, Wu J-Y, Shen Z-Y, Zhao Q, Du Z-P, Lv Z, Gu W, Pan F, Xu X-E, Xie D, Li E-M (2010) Involvement of CYR61 and CTGF in the fascin-mediated proliferation and invasiveness of esophageal squamous cell carcinomas cells. Am J Pathol 176(2):939–951. doi:10.2353/ajpath.2010.090118

Chapter 10
COX 10

Abstract COX10 gene, also known as 'heme A: farnesyltransferase gene', is the gene that encodes for heme A: farnesyltransferase enzyme. The gene was initially discovered in yeast, and the enzyme encoded by it was found to be essential for the expression of cytochrome c oxidase (COX) because its role is to faresylate protoheme (heme B) to heme O during the process of synthesizing heme A. Heme A is an important prosthetic group of COX. Since COX (a.k.a. Complex IX) is the enzyme that transports electrons from cytochrome c molecules to oxygen in the mitochondrial electron transport chain, it is only natural that the alteration of COX10 influences the aerobic respiration of cells. The deficiency I COX10 gene causes the cytochrome C oxidase deficiency, which is one of the critical causes of respiratory chain defect. This would lead to various diseases in humans. COX10 has an imperative role in the mitochondrial electron transport chain in aerobic respiration, and hence it can be concluded that hypoxia-regulated miR-210 triggers the metabolic switch from aerobic to anaerobic respiration in cancer cells via down-regulation of COX10. This, in turn, promotes the survival of tumour cells under hypoxic microenvironment.

Keywords COX10 • Angiogenesis • Normal physiology • Disease • Malignancy

10.1 COX 10

COX10 gene, also known as 'heme A:farnesyltransferase gene', is the gene that encodes for heme A:farnesyltransferase enzyme. The gene was initially discovered in yeast, and the enzyme encoded by it was found to be essential for the expression of cytochrome c oxidase (COX) because its role is to faresylate protoheme (heme B) to heme O during the process of synthesizing heme A (Moraes et al. 2004). Heme A is an important prosthetic group of COX.

A. Salajegheh, *Angiogenesis in Health, Disease and Malignancy*,
DOI 10.1007/978-3-319-28140-7_10

10.2 COX 10 in Normal Physiology

Since COX (a.k.a. Complex IX) is the enzyme that transports electrons from cytochrome c molecules to oxygen in the mitochondrial electron transport chain, it is only natural that the alteration of COX10 influence the aerobic respiration of cells (Diaz et al. 2006). To avoid confusion, it is worth mentioning that the 'heme A:farnesyltransferase', due to its involvement with several other types of heme prosthetic groups, has aliases such as 'Protoheme IX Fanesyltransferase' or 'Heme O Synthase'. There are also many genes other than COX10 which are encoded for cytochrome c oxidase, from both cell nucleus and the mitochondrial DNA. COX10 is a nuclear gene, meaning that it is located in the nucleus of cells (Valnot 2000).

10.3 COX 10 in Disease

It has been reported that an abnormality in COX10 gene results in cytochrome c oxidase deficiency, which is one of the well known causes of respiratory chain defect (Diaz et al. 2006). By utilization of homozygosity mapping in a large African family suffering from COX deficiency and subsequent gene screening, the research team in Columbia University identified a homozygous missense mutation of the COX10 gene as the contributing factor of the disease (Valnot 2000).

COX10 is officially the third gene that has been found to have a link with COX deficiency, where two other genes are SURF1 and SCO2 (Casarin et al. 2012). Although a defect in these three genes can cause isolated COX deficiency accompanied by inadequate expression of its subunits, there were noticeable differences in their molecular influence on COX. For instance, the SURF1 gene defect leads to the reduction of a range of COX subunit levels except subunits Va and Vb, whereas SCO2 defect displays reduction in subunits I and II. COX10 mutation, in particular, affected subunit II to a large extent, but other subunit levels remained relatively stable, suggesting that COX subunits other subunit II do not require heme synthesis to be stabilized (Valnot 2000). There were variations in terms of clinical phenotypes as well, where SURF1 and SCO2 mutations were linked to Leigh's disease and encephalocardiomyopathy respectively, while those who possess COX10 defect presented tubulopathy and leukodystrophy (Valnot 2000).

After the establishment of a link between COX10 defect and tubulopathy and lekodystrophy, further study has been conducted to find out its relevance in other clinical conditions. The study from McGrill University and other institutes have shown that COX10 mutations can lead to a whole range of conditions associated with early onset isolated COX deficiency (Antonicka 2003). They investigated two individuals with isolated COX deficiency, where one of them had anaemia and classical Leigh syndrome, and the other presented with anaemia, sensorinueral hearing loss and fatal infantile hypertrophic cardiomyopathy. These conditions presented were identified as clinical manifestation of COX10 mutations, in addition to tubulopathy and lekodystrophy. Anaemia, in particular, was seen in both patients with COX10 mutation, providing a possible link between the deficiency in heme O synthesis and physiological erythropoiesis (Antonicka 2003).

In order to find out the genetic defect present in the patients, the research team used a technique called ‘functional complementation’ using retrovirus as the vector (Antonicka 2003). The complementary DNA coding for COX10 has been inserted into the cultured fibroblasts derived from the two patients, and the subsequent restoration of their COX activity was observed. To confirm the actions of COX assembly genes other than COX10 (SURF1, SCO1, SCO2, COX17 and more), similar experiments were done on these genes as well, but they all failed to reduce COX deficiency (Antonicka 2003). Furthermore, they attempted transduction of fibroblasts from other patients suffering from SURF1 and SCO2 defects with COX10 gene, but there was no meaningful alleviation of residual COX activity in those cells. These observations strongly suggest that the COX deficiency of the subjects is ascribed to COX10 gene mutation, and that there is very little or no functional redundancy among COX assembly genes investigated (Antonicka 2003).

The mechanism by which different COX gene defects cause various clinical phenotypes in the specific tissues are largely unknown. It has been suspected that the mitochondrial content of different tissues is the major factor associated with the expression of those gene mutations, in addition to the severity and pattern of the COX deficiency. As to the gene itself, there might be various alleles in the same COX10 gene that can lead to different clinical phenotypes (Antonicka 2003).

10.4 COX 10 in Malignancy

Research team from the University of Texas investigated the relationship between the hypoxia-regulated microRNA-210 and the compromised mitochondrial activity in cancer cells, which led them to an astonishing discovery on COX10 gene and ISCU. When tumour cells develop, their rapid growth and mitotic divisions eventually exceed the level that can be sustained by the local blood supply, resulting in a portion of the tumour receiving inadequate oxygen supply compared to other healthy tissues in body. This phenomenon is called ‘tumour hypoxia’. It has been known that, under a hypoxic microenvironment, a range of signaling pathways mediated by HIF-1α occur to trigger biological events such as increased blood vessel formation and mitochondrial dysfunction (Chen et al. 2010).

Under the influence of HIF-1α, a micro-RNA called miR-210 was found to have important functions in the protection of breast cancer cells from hypoxia-induced apoptosis and in cell cycle regulations. Micro RNAs are not coding RNAs but they can silence the expression of specific genes by affecting their transcription process. The researchers attempted to find the target genes of miR-210 using a technology called ‘in silico prediction algorithm’ and obtained a number of likely candidate genes including COX10 and ISCU (Chen et al. 2010). They discovered that mRNA levels of these two genes are consistently reduced in the presence of miR-210, suggesting the possibility that COX10 and ISCU are the target genes of miR-210 (Chen et al. 2010). As discussed earlier, COX10 has an imperative role in the mitochondrial electron transport chain in aerobic respiration, and hence it can be concluded that hypoxia-regulated miR-210 triggers the metabolic switch from aerobic to anaerobic respiration in cancer cells via downregulation of COX10. This, in turn, promotes the survival of tumour cells under hypoxic microenvironment (Chen et al. 2010).

References

Antonicka H (2003) Mutations in COX10 result in a defect in mitochondrial heme A biosynthesis and account for multiple, early-onset clinical phenotypes associated with isolated COX deficiency. Hum Mol Genet 12(20):2693–2702. doi:10.1093/hmg/ddg284

Casarin A, Giorgi G, Pertegato V, Siviero R, Cerqua C, Doimo M, Basso G, Sacconi S, Cassina M, Rizzuto R, Brosel S, Davidson M, DiMauro S, Schon EA, Clementi M, Trevisson E, Salviati L (2012) Copper and bezafibrate cooperate to rescue cytochrome c oxidase deficiency in cells of patients with sco2 mutations. Orphanet J Rare Dis 7:21–21. doi:10.1186/1750-1172-7-21

Chen Z, Li Y, Zhang H, Huang P, Luthra R (2010) Hypoxia-regulated microRNA-210 modulates mitochondrial function and decreases ISCU and COX10 expression. Oncogene 29(30):4362–4368. doi:10.1038/onc.2010.193

Diaz F, Fukui H, Garcia S, Moraes CT (2006) Cytochrome c oxidase is required for the assembly/stability of respiratory complex I in mouse fibroblasts. Mol Cell Biol 26(13):4872–4881. doi:10.1128/MCB.01767-05

Moraes CT, Diaz F, Barrientos A (2004) Defects in the biosynthesis of mitochondrial heme c and heme a in yeast and mammals. Biochimica et Biophysica Acta (BBA) Bioenergetics 1659(2–3):153–159. doi:10.1016/j.bbabio.2004.09.002

Valnot I (2000) A mutation in the human heme A:farnesyltransferase gene (COX10) causes cytochrome c oxidase deficiency. Hum Mol Genet 9(8):1245–1249. doi:10.1093/hmg/9.8.1245

Chapter 11
Cysteine-Rich 61 (CYR61)

Abstract Cysteine-rich 61 (CYR61) is a member of the CCN family of genes, (CCN1) is an angiogenic factor. The CCN family (CYR61, nov, ctgf) are important matricellular regulatory factors, which are involved in internal and external signalling of the cell. CYR61 is a secreted, heparin-binding protein that is encoded by the CYR61 gene. The CYR61 protein contains a highly preserved cysteine-rich primary sequence that is organised into discrete domains similar to insulin-like growth factor binding protein. CYR61 expression is regulated by various agents including growth factors, cytokines, steroid hormones and some drugs. CYR61 has the ability to induce apoptosis and cellular senesce. CYR61 acts as an extracellular matrix-associated signalling molecule, inducing fibroblast apoptosis through adhesion receptors, interaction with cell surface integrin and heparin sulphate proteoglycans, and is able to exert its functions and promote adhesion of endothelial cells. In rheumatoid arthritis CYR61 was found to play a critical role in proliferation in fibroblast-like synoviocytes (FLS) and also likely to contribute to the hyperplasia of synovial lining cells and eventuating in joint destruction. CYR61 is also an inducer of angiogenesis, which is vital for the supply of nutrients and oxygen to nourish the growing and developing tumour.

Keywords Cysteine-rich 61 • CYR61 • Angiogenesis • Normal physiology • Disease • Malignancy

11.1 Cysteine-Rich 61

Cysteine-rich 61 (CYR61) is a member of the CCN family of genes, (CCN1) is an angiogenic factor. The CCN family (CYR61, nov, ctgf) are important matricellular regulatory factors, which are involved in internal and external signalling of the cell (Perbal 2004). CYR61 is a secreted, heparin-binding protein that is encoded by the CYR61 gene (Absenger et al. 2004).

A. Salajegheh, *Angiogenesis in Health, Disease and Malignancy*,
DOI 10.1007/978-3-319-28140-7_11

The CYR61 protein contains a highly preserved cysteine-rich primary sequence that is organised into discrete domains similar to insulin-like growth factor binding protein (Perbal 2004). The primary identification of CYR61 was as a protein in fibroblasts of mice as a serum-inducible gene (Jay et al. 1997). CYR61 encodes a protein of 381 amino acids (Absenger et al. 2004). CYR61 is composed of four structural domains; the IGF binding protein, the Von Willebrand Factor C, the Thrombospondin Type-I and a C-terminal Cysteine Knot Like Domain. The location of CYR61 has been identified at chromosome 1p22-p31 (Mo and Lau 2006).

CYR61 expression is regulated by various agents including growth factors, cytokines, steroid hormones and some drugs. Regulation of CYR61 transcription is achieved by these inducers through different signalling transduction pathways. Interaction with the individual cell or communication with the surrounding cells through autocrine-paracrine mechanism is achieved by the CYR61 gene (Chen and Du 2007). CYR61 has the ability to induce apoptosis and cellular senesce (Todorovic et al. 2005). CYR61 acts as an extracellular matrix- associated signalling molecule, inducing fibroblast apoptosis through adhesion receptors, interaction with cell surface integrin and heparin sulphate proteoglycans, and is able to exert its functions and promote adhesion of endothelial cells (Grzeszkiewicz et al. 2002).

11.2 Cysteine-Rich 61 in Normal Physiology

In normal physiology the role of CYR61 is to link cell surface and extracellular matrix. CYR61 also influences several cellular activities during normal development including proliferation, migration, differentiation, apoptosis and angiogenesis (Jun and Lau 2011).

In normal embryonic development CYR61 is essential for vascular integrity, cardialseptal morphogenesis, and placenta blood vessel transformation. CYR61 is highly expressed in trophoblastic giant cells and also in trophoblasts of the ectroplacental cone. This helps to promote the growth towards the embryo of the uterine vessel (Mo et al. 2002). A study by the Department of Surgery, the Ohio State University, USA, disrupted the CYR61 gene in mice. The findings showed that mice that were exempt from the CYR61 gene suffered embryonic death, due to "placental vascular insufficiency and compromised vessel integrity". This study establishes CYR61 as essential in the regulation of vascular development (Mo et al. 2002).

In adulthood CYR61 is also associated with chronic inflammatory diseases such as rheumatoid arthritis, atherosclerosis, diabetes-related nephropathy, retinopathy and various cancers. CYR61 expression is related to many tissues undergoing inflammation, wound healing, fibrogenesis and their regulation in adults (Lau 2011).

Wound healing in general relates to the following three phases, inflammatory response, Extra cellular matrix deposition, the formation of granulation tissue and tissue remodelling. CCN1 is highly expressed in myofibroblasts of granulation tissues during cutaneous wound healing and CYR61 may promote the proliferation of blood vessels – neovascularisation – during wound repair (Lau 2011).

Myofibroblasts proliferate and synthesise extracellular matrix to retain tissue integrity and increase wound closure. Studies show that during proliferation of myofibroblasts in the granulation tissue, CCN1 influences fibroblasts into senescence and controls the expression of antifibrotic genes thus restricting fibrosis during tissue repair. Conversely, if extracellular matrix in myofibroblasts is not upheld, excessive matrix deposition can lead to fibrosis, loss of tissue function and scarring (Jun and Lau 2010b).

At sites of wound healing and inflammation CYR61 is expressed and holds a significant role however, CCN1 proteins are often deregulated when these processes become chronic, and progress to pathological conditions (Jun and Lau 2011). Consequently, numerous studies have found that CYR61 is expressed and associated with chronic diseases involving tissue repair, inflammation and wound healing (Jun and Lau 2010a).

11.3 Cysteine-Rich 61 in Disease and Malignancy

The development of a number of pathological conditions is a result of abnormal angiogenesis. The expression of the CYR61 gene is highly associated with vascular diseases such as restenosis, atherosclerosis and later-stage human breast cancer (Mo et al. 2002). Other conditions include ovarian cancer, gastric cancer and glioma (Jeong et al. 2014). Thus, expression of CYR61 is suggested to promote tumour growth and vascularisation (Mo et al. 2002). However, CYR61 has presented as a tumour suppressor in prostate cancer, endometrial cancer and non-small lung cancer (Jeong et al. 2014). Thus, many studies have been undertaken to investigate the role CYR61 in angiogenesis and cancer.

CYR61 is an inducer of angiogenesis, which is vital for the supply of nutrients and oxygen to nourish the growing and developing tumour (Jeong et al. 2014). CYR61 is a powerful angiogenic inducer *in vivo* and various studies propose CYR61 promotes cancer cell proliferation, adhesion, invasion, survival, metastasis and angiogenesis, which is achieved by the activation of integrin in endothelial cells (Sun et al. 2008). Clinically CYR61 expression correlates with the tumour stage, the size of the tumour, positive lymph node and poor prognosis in various cancers such as squamous cell carcinoma, prostate cancer, breast cancer and gastric adenocarcinoma (O'Kelly et al. 2008). Furthermore, CYR61 expression in tumour cells promotes vascularisation and growth (Babic et al. 1998).

CCN1 protein has found to be at much higher levels in breast cancer samples when compared to non-tumour tissues (O'Kelly et al. 2008). CYR61 is upregulated in human breast cancer cells that are invasive and metastatic (Tsai et al. 2002). This research delivers evidence that the CYR61 gene is able to 'induce E2-independence and antiestrogen-resistance,' with invasiveness *in vitro* and tumourigenesis *in vivo*. Consequently both of which are characteristic of a phenotype of aggressive breast cancer. Thus, indicating CYR61 is a tumour promoting factor and a significant regulator for the progression of breast cancer (Tsai et al. 2002).

In rheumatoid arthritis CYR61 was found to play a critical role in proliferation in fibroblast-like synoviocytes (FLS) and also likely to contribute to the hyperplasia of synovial lining cells and eventuating in joint destruction. FLS are key features of the hyperplastic synovial pannus, an abnormal layer of fibrovascular tissue or granulation tissue – and FLS invade bone and cartilage during rheumatoid arthritis (Zhang et al. 2009).

CCN1 expression in the human kidney was confined to podocytes in normal adults and embryonic glomeruli which filters blood in the capillary loop stage. The results of this study show that CCN1 can act on glomerular cells to modulate glomerular remodelling, and in diseased kidneys is down regulated. This study suggests expression of CYR61 is decreased in podocytes particularly in kidney disease that involves severe meningeal expansion (Sawai et al. 2007). By inducing unilateral ureteral obstruction in a study on mice, the CYR61 expression was investigated in a progressive renal fibrosis model. Indicating CYR61 contributes to chronic kidney injury's inflammatory process. Furthermore, CYR61 blockade was able to reduce kidney fibrosis in early phases, however, could not be sustained (Lai et al. 2013).

CYR61 is also an important pancreatic cancer marker as CYR61 expression increases with advancing pancreatic disease (Haque et al. 2011). Similarly, promotion of growth of glioma cells was also specified by CYR61 (Xie et al. 2004). In primary glioma cell tumors CYR61 was found overexpressed when compared with normal tissue. In this study the levels of expression of the gene directly correlated with the grade of tumour and consequently with patient survival (Sabile et al. 2012).

A recent study found CYR61 protein levels are up regulated and can be detected in urine samples of patients that have cancers of epithelial origin. Cancers including breast cancer, ovarian cancer, adenocarcinoma, gastrointestinal cancer, lung cancer, skin cancer and other known cancers that effect epithelial cells throughout the body. This study found that the amount of CYR61 protein detected in a urine sample were higher in patients that have malignant forms of cancers, and thus directly correlates with disease severity. Therefore this study suggests that CYR61 urine tests could be used to predict the presence of and the metastatic potential of cancer and thus, involved in cancer diagnosis and prognosis (Moses and Zhang 2010).

However, in 2001 a study involving CYR61 was found to show CYR61as antiproliferative and an antitumorogenic agent in non-small cell lung cancer studies. The expression of CYR61 decreased in human lung tumour samples (Tong et al. 2001). Dependent on the cell types and cancer type it is thought that CYR61 may exert different or opposing functions in carcinogenesis either as an oncogene or tumour suppressor.

The expression of CYR61 in most tumour cells along with its distinct ability to promote vascularisation and tumour growth suggest the CYR61 gene may be a tumour progression marker. Dependent on the cancer type, it is evident that the functions of CYR61 are numerous and varied. The CYR61 gene acts as either an oncogene or a tumour suppressor gene in carcinogenesis. CYR61 demonstrates many different cellular activities, some of which include cellular survival, its ability to trigger cellular apoptosis, enhancement of cellular proliferation, and inducing cellular arrest. It has also been linked to both the promotion of tumour growth and suppression of tumour development, depending on the cell type.

References

Absenger Y, Hess-Stumpp H, Kreft B, Kratzschmar J, Haendler B, Schutze N, Regidor PA, Winterhager E (2004) Cyr61, a deregulated gene in endometriosis. Mol Hum Reprod 10(6):399–407. doi:10.1093/molehr/gah053

Babic AM, Kireeva ML, Kolesnikova TV, Lau LF (1998) CYR61, a product of a growth factor-inducible immediate early gene, promotes angiogenesis and tumor growth. Proc Natl Acad Sci U S A 95(11):6355–6360

Chen Y, Du XY (2007) Functional properties and intracellular signaling of CCN1/Cyr61. J Cell Biochem 100(6):1337–1345. doi:10.1002/jcb.21194

Grzeszkiewicz TM, Lindner V, Chen N, Lam SC, Lau LF (2002) The angiogenic factor cysteine-rich 61 (CYR61, CCN1) supports vascular smooth muscle cell adhesion and stimulates chemotaxis through integrin alpha(6)beta(1) and cell surface heparan sulfate proteoglycans. Endocrinology 143(4):1441–1450. doi:10.1210/endo.143.4.8731

Haque I, Mehta S, Majumder M, Dhar K, De A, McGregor D, Van Veldhuizen PJ, Banerjee SK, Banerjee S (2011) Cyr61/CCN1 signaling is critical for epithelial-mesenchymal transition and stemness and promotes pancreatic carcinogenesis. Mol Cancer 10:8. doi:10.1186/1476-4598-10-8

Jay P, Berge-Lefranc JL, Marsollier C, Mejean C, Taviaux S, Berta P (1997) The human growth factor-inducible immediate early gene, CYR61, maps to chromosome 1p. Oncogene 14(14):1753–1757. doi:10.1038/sj.onc.1200986

Jeong D, Heo S, Sung Ahn T, Lee S, Park S, Kim H, Park D, Byung Bae S, Lee SS, Soo Lee M, Kim CJ, Jun Baek M (2014) Cyr61 expression is associated with prognosis in patients with colorectal cancer. BMC Cancer 14:164. doi:10.1186/1471-2407-14-164

Jun JI, Lau LF (2010a) Cellular senescence controls fibrosis in wound healing. Aging 2(9):627–631

Jun JI, Lau LF (2010b) The matricellular protein CCN1 induces fibroblast senescence and restricts fibrosis in cutaneous wound healing. Nat Cell Biol 12(7):676–685. doi:10.1038/ncb2070

Jun JI, Lau LF (2011) Taking aim at the extracellular matrix: CCN proteins as emerging therapeutic targets. Nat Rev Drug Discov 10(12):945–963. doi:10.1038/nrd3599

Lai CF, Chen YM, Chiang WC, Lin SL, Kuo ML, Tsai TJ (2013) Cysteine-rich protein 61 plays a proinflammatory role in obstructive kidney fibrosis. PLoS One 8(2):e56481. doi:10.1371/journal.pone.0056481

Lau LF (2011) CCN1/CYR61: the very model of a modern matricellular protein. Cell Mol Life Sci 68(19):3149–3163. doi:10.1007/s00018-011-0778-3

Mo FE, Lau LF (2006) The matricellular protein CCN1 is essential for cardiac development. Circ Res 99(9):961–969. doi:10.1161/01.res.0000248426.35019.89

Mo FE, Muntean AG, Chen CC, Stolz DB, Watkins SC, Lau LF (2002) CYR61 (CCN1) is essential for placental development and vascular integrity. Mol Cell Biol 22(24):8709–8720

Moses MA, Zhang B (2010) Cyr61 as a biomarker for diagnosis and prognosis of cancers of epithelial origin. Google Patents

O'Kelly J, Chung A, Lemp N, Chumakova K, Yin D, Wang HJ, Said J, Gui D, Miller CW, Karlan BY, Koeffler HP (2008) Functional domains of CCN1 (Cyr61) regulate breast cancer progression. Int J Oncol 33(1):59–67

Perbal B (2004) CCN proteins: multifunctional signalling regulators. Lancet 363(9402):62–64. doi:10.1016/s0140-6736(03)15172-0

Sabile AA, Arlt MJ, Muff R, Bode B, Langsam B, Bertz J, Jentzsch T, Puskas GJ, Born W, Fuchs B (2012) Cyr61 expression in osteosarcoma indicates poor prognosis and promotes intratibial growth and lung metastasis in mice. J Bone Miner Res 27(1):58–67. doi:10.1002/jbmr.535

Sawai K, Mukoyama M, Mori K, Kasahara M, Koshikawa M, Yokoi H, Yoshioka T, Ogawa Y, Sugawara A, Nishiyama H, Yamada S, Kuwahara T, Saleem MA, Shiota K, Ogawa O, Miyazato M, Kangawa K, Nakao K (2007) Expression of CCN1 (CYR61) in developing, normal, and diseased human kidney. Am J Physiol Renal Physiol 293(4):F1363–F1372. doi:10.1152/ajprenal.00205.2007

Sun ZJ, Wang Y, Cai Z, Chen PP, Tong XJ, Xie D (2008) Involvement of Cyr61 in growth, migration, and metastasis of prostate cancer cells. Br J Cancer 99(10):1656–1667. doi:10.1038/sj.bjc.6604712

Todorovic V, Chen CC, Hay N, Lau LF (2005) The matrix protein CCN1 (CYR61) induces apoptosis in fibroblasts. J Cell Biol 171(3):559–568. doi:10.1083/jcb.200504015

Tong X, Xie D, O'Kelly J, Miller CW, Muller-Tidow C, Koeffler HP (2001) Cyr61, a member of CCN family, is a tumor suppressor in non-small cell lung cancer. J Biol Chem 276(50):47709–47714. doi:10.1074/jbc.M107878200

Tsai MS, Bogart DF, Castaneda JM, Li P, Lupu R (2002) Cyr61 promotes breast tumorigenesis and cancer progression. Oncogene 21(53):8178–8185. doi:10.1038/sj.onc.1205682

Xie D, Yin D, Wang HJ, Liu GT, Elashoff R, Black K, Koeffler HP (2004) Levels of expression of CYR61 and CTGF are prognostic for tumor progression and survival of individuals with gliomas. Clin Cancer Res 10(6):2072–2081

Zhang Q, Wu J, Cao Q, Xiao L, Wang L, He D, Ouyang G, Lin J, Shen B, Shi Y, Zhang Y, Li D, Li N (2009) A critical role of Cyr61 in interleukin-17-dependent proliferation of fibroblast-like synoviocytes in rheumatoid arthritis. Arthritis Rheum 60(12):3602–3612. doi:10.1002/art.24999

Chapter 12
Epidermal Growth Factor

Abstract Epidermal Growth factor, also known as EGF is a naturally occurring stimulatory growth factor in the human body which plays a major role in cell proliferation, differentiation, and growth. The EGF belongs to the receptor tyrosine kinase (RTK) family, and as such, is activated by binding to particular ligands. Once the EGF is activated they quickly become internalized to become early endosomes and eventually either sorted via endosomes–lysosomes for degradation or recycled via the plasma membrane, these mechanisms are believed to terminate the EGF actions. The EGF can be found in various resins throughout the body, including human platelets, plasma, urine, saliva, milk and macrophages. EGF has been also involved in promoting wound healing by stimulating epidermal and dermal repair, confirming that it is a powerful dose-dependent mitogen on the formation of granulation tissue. The development of new capillaries in tumours directly relate to the release of angiogenic factors related to EGF and its multitude of receptors. As such, the intensity of neovascularisation directly correlates with the development of metastasis and thus the number of EGF, EGFR and VEGF. Therefore, various current and developing drugs target these receptors in hoping to eventually find a cure for the EGF associated tumours such as lung and rectal cancers.

Keywords Epidermal growth factor • EGF • Angiogenesis • Normal physiology • Disease • Malignancy

12.1 Epidermal Growth Factor

Epidermal Growth factor, also known as EGF is a naturally occurring stimulatory growth factor in the human body which plays a major role in cell proliferation, differentiation, and growth (Taylor et al. 1972; Harris 2003). The EGF belongs to the receptor tyrosine kinase (RTK) family, and as such, is activated by binding to particular ligands. Once the EGF is activated they quickly become internalized to become early endosomes and eventually either sorted via endosomes–lysosomes for degradation or recycled via the plasma membrane, these mechanisms are believed to terminate the EGF actions.

A. Salajegheh, *Angiogenesis in Health, Disease and Malignancy*,
DOI 10.1007/978-3-319-28140-7_12

Epidermal Growth Factors (EGF) was first isolated from the submaxillary glands of mice (Dagogo-Jack et al. 1986). It was discovered that EGF affects precocious opening of eye lids and the eruption of central incisors in newborn mice (Cohen 2008). EGF is composed of a single polypeptide chain containing 53 amino acids and has a low molecular weight of 6045 Da (Taylor et al. 1972). Various tissues in the body produce EGF such as kidneys, small intestines, pancreas, pituitary gland, mammary glands, and submaxillary glands (Mroczkowski et al. 1988). The EGF can be found in various resins throughout the body, including human platelets, plasma, urine, saliva, milk and macrophages (Cartlidge and Elder 1989).

There are six types of EGF- like growth factors: Epidermal Growth Factor (EGF), amphiregulin (AMP), Tumour growth factor α (TGF-α), pox virus growth factors, neuregulin (NRG), betacelluline (BTC), and epiregulin (EPR) (Dreux et al. 2006). These EGF-like ligands have high affinity for both EGF and ErbB receptors. Thus, upon binding, they will elicit its biological effects. Numerous different types of human cells, such as vascular endothelial cells, lens cells, granulosa cells, fibroblasts, corneal cells, epidermal cells and various human cancer cells possess specific cell membrane receptors for EGF. This denotes the ubiquitous qualities of this EGF and its importance in the human body (Dreux et al. 2006; Normanno et al. 2006).

12.2 Epidermal Growth Factor in Normal Physiology

Epidermal growth factor (EGF) is a 53 amino acid polypeptide that functions as a potent mitogenic factor, and is involved in the regulation of cell growth, proliferation and differentiation across various physiological systems (Harris 2003). EGF is mapped to chromosome 4q and alternate splicing creates multiple transcript variants. In addition, testing through immunohistochemical localization in rodents, EGF expression was dominantly located in the exocrine glands of the gastrointestinal tract, in the serous acini of the nasal cavity, in the submaxillary glands, and in the apical membrane of the thick ascending limb of the kidney (Harris 2003). Hypomagnesaemia type 4 can develop due to faults in EGF, while growth and advancement of specific cancers are directly attributed to dysregulation of EGF (Kloth et al. 2010). More importantly, EGF acts through binding to epidermal growth factor receptor, a high affinity cell surface receptor, which ultimately leads to angiogenesis EGF (Dunn et al. 2000; Harris 2003; Ellis 2004). Categorized a human epidermal receptor, more specifically HER1, epidermal growth factor receptor (EGFR) is known to modulate normal cell growth and differentiation (Ellis 2004). EGFR is also a tyrosine kinase that is expressed by most normal cells, particularly ones with an endothelial origin and found in malignant tissues (Harris 2003; Ellis 2004). The activation of tyrosine kinase leads to a cascade of signal transduction events, resulting in various biochemical changes occurring within the cell: an increase in intracellular calcium levels, increase protein synthesis and glycolysis, and an increase in gene expression such as the gene coding for EGF receptors. Ultimately, all these events cause DNA synthesis and cell proliferation (Wahl et al. 1989).

EGF is widely expressed in mesenchymal cells, epithelial cells, neuronal cells, and growing embryos. It is thought to be a potent mitogen, a substance that promotes mitosis, and has been determined to be an important regulator of growth and development (Mroczkowski et al. 1988). In addition, EGF is found to aid in wound healing and homeostasis in normal adult tissue. It has been reported that membrane bound EGF plays a role in cell to cell attachment as well (Nishida et al. 1992). Also some researchers suggested that the presence of EGF in milk implicate that EGF may contribute towards neonatal development (Donnelly et al. 1992; Dvorak et al. 2003). EGF is a known ligand for EGFR and is most commonly co-expressed with EGFR in several types of cancer. Together, they both play a role in autocrine and paracrine functions that lead to dysregulated EGFR activation and unregulated tumor growth (Dunn et al. 2000; Ellis 2004).

12.3 Epidermal Growth Factor in Disease and Malignancy

Many studies show the importance of this relation including a study showing that EGF complemented by mouse fibroblasts overexpressing EGFR causes cellular modification (Ellis 2004). In relation, cellular transformations are shown to be reliant on both the levels of EGFR and the ligand, which is attributed to fact that in cells expressing moderate levels of EGFR, highly expressed growth factors are still weak modifying agents (Ellis 2004). EGFR activation obeys three essential steps, which are (1) ligand binding, (2) receptor dimerization, through EGFR homodimerization or heterodimerization with other HER family members, and (3) activation of the receptor tyrosine kinase by means of intramolecular phosphorylation (Dunn et al. 2000; Schlessinger 2002; Ellis 2004). EGFR activation is followed by rapid endocytosis and degradation or recycling of the receptor and the ligand, EGF in this case (Marti and Wells 2000; Ellis 2004). EGF binding to EGFR is a main pathway for inducing angiogenesis in cancerous cells, and therefore EGF is recognized to be pro-angiogenic (Harris 2003; Ellis 2004). Further research has shown that EGF prevents apoptosis in cells that express EGFR, and blocking EGFR would cause apoptosis promotion in a variety of different cancer cell lines (Ellis 2004). This is accomplished by the influence EGF has on the expression of cyclin D1, an important mediator of cell cycle progression in human prostate carcinoma cells (Ellis 2004).

EGF is a pro-angiogenic growth factor. Many studies have identified different relationships between EGF and other growth factors and receptors. Dunn et al. looked at the role EGF played in angiogenesis by examining gliomas (Dunn et al. 2000). Gliomas arise from glial cells commonly found in the brain and conform to the fundamentals of angiogenesis, which include (1) growing rapidly beyond 1–2 mm, (2) nutrient delivery, via producing a system of blood vessels, and (3) developing methods of waste removal (Dunn et al. 2000). The study illustrated that tumor-secreted growth factors, including EGF, can display direct and indirect effects on glioma angiogenesis. Furthermore, they are shown to stimulate endothelial cell proliferation, mediate the expression of key proteases on endothelial cells necessary for angiogenesis, and regulate the expression of vascular endothelial growth factor VEGF (Dunn et al. 2000).

Tong et al. examined the, EGF-mediated, neutrophil gelatinase-associated lipocalin (NGAL) regulation of cells responsible for pancreatic ductal adenocarcinoma (PDAC) (Tong et al. 2010). In a previous study, they established that NGAL overexpression considerably blocked invasion and angiogenesis of PDAC cells (Tong et al. 2010). NGAL expression is related with increased tumour progression and intrusiveness in esophageal squamous cell carcinoma and breast cancer when it is up-regulated (Tong et al. 2010). However, in ovarian, colon, and pancreatic cancer, NGAL acts as an anti-tumor and anti-metastatic factor (Tong et al. 2010). Furthermore, NGAL is an important biomarker for identifying early-stage cancer. Their study demonstrated that NGAL expression was clearly related with the differentiation of tumour cells, and was considerably down-regulated after EGF treatment coinciding with an associated decline of E-cadherin expression in PDAC cells. Moreover, the down-regulation of E-cadherin, a transmembrane protein, decreased NGAL expression in PDAC cells (Tong et al. 2010). In contrast, overexpression of E-cadherin caused an increase in NGAL expression, and partially released inhibition of NGAL expression by EGF (Tong et al. 2010). EGF also partially reduced NGAL promoter activity by blocking NF-κB, a transcription factor, activation via E-cadherin (Tong et al. 2010). As a result, the down-regulation of NGAL through EGF would result in angiogenic activities to continue in proliferation uninhibited (Tong et al. 2010).

In relation, another study by Soufla et al. examined endometrial cancer, which is known to be one of the most common gynaecological malignancies (Soufla et al. 2008). Angiogenesis is a significant factor in the development and progression of most human carcinomas, and endometrial cancer is no exception. More specifically, EGF acts as a mitogen in promoting endothelial cell proliferation and differentiation (Soufla et al. 2008). In this study, they examined the mRNA expression patterns of EGF, IGF-I, and FGF-2 to their specific roles in angiogenesis. The results showed no observed correlation between EGF, IGF-I, and FGF transcript levels in normal endometrium, while FGF-2 mRNA was significantly correlated with EGF and IGF-I transcript levels in the malignant endometrium (Soufla et al. 2008). Furthermore, cross-talk between EGF-mediated and IGF-mediated signalling pathways triggered by disruptions in mRNA co-expression pattern supported their hypothesis of it being involved in promoting endometrial cancer through endothelial cell proliferation and differentiation (Soufla et al. 2008).

As previously mentioned, it is noted that EGF affects VEGF and a study by Petit et al. showed that the oncogenicity of EGF and its receptor may play a role in being mediated through promotion of angiogenesis by up-regulating VEGF, which is a key stimulator of angiogenesis, inducing proliferation, differentiation, and migration of endothelial cells (Petit et al. 1997). Moreover, the vascular permeability is increased by VEGF further stimulating the production of many types of proteases taking part in the modification of the extracellular matrix (Petit et al. 1997). Studies have also shown that EGF stimulation of glioma cells repeatedly increased the production of VEGF by these cells (Petit et al. 1997; Dunn et al. 2000; van Cruijsen et al. 2005; Soufla et al. 2008; Tong et al. 2010). The conditioned medium of the stimulated glioma cells showed an induced activation of human umbilical vein endothelial cells (HUVECs) (van Cruijsen et al. 2005). Furthermore, studies have shown that EGF augmented mRNA of VEGFR in many gastric cancer cell lines. In addition, EGF increased the production of VEGF and basic fibroblast growth factor (bFGF) in bladder carcinoma cell lines (van Cruijsen et al. 2005). In pancreatic and gastric cell lines, EGF stimulation initiated an increased production of neuropilin-1, a co-receptor of VEGFR-2, causing an increase in the affinity of specific isoforms of VEGF to VEGFR-2 (van Cruijsen et al. 2005). This suggests that EGF and its receptor contribute indirectly to signaling in angiogenesis (van Cruijsen et al. 2005).

In contrast, anti-angiogenic drugs combined with anti-EGFR drugs allow for more effective inhibition of angiogenesis (van Cruijsen et al. 2005). A dual inhibition drug of EGFR and VEGFR, named AEE788, inhibiting receptor tyrosine kinase ErbB-1 and ErbB-2, VEGFR-1 and VEGFR-2, showed anti-proliferative activity against a variety of EGFR overexpressing tumor cell lines and EGF stimulated HUVECs (van Cruijsen et al. 2005). Moreover, pre-clinical models of this agent had noteworthy antitumor activity, stimulated apoptosis of tumor-related endothelial cells, and decreased tumor's microvessel density (MVD) (van Cruijsen et al. 2005).

The process of angiogenesis is vital in defeating cancer and EGF inhibition is one way to reduce angiogenic response. Scicli et al. conducted a study focused on the Cytochrome P4504A enzyme (CYP4A), which is known to metabolize arachidonic acid to form 20-HETE, a compound that plays a key role in the regulation of myogenic tone and blood flow auto-regulation (Scicli et al. 2004). With limited previous studies examining the function of CYP4A in the regulation of angiogenesis, this study looked into finding whether inhibition of CYP4504A combined with the highly selective inhibitor N-hydroxy-N′-(4-butyl-2-methylphenol) formamidine (HET0016) had any effect on angiogenesis (Scicli et al. 2004). The study was conducted on the stroma of rat corneas, adjacent to the temporal limbus. The results showed that EGF alone had a vessel length and a vascularised area of 258 ± 19 pixels and $89{,}074 \pm 1209$ pixels respectively (Scicli et al. 2004). EGF, combined with formamidine, had a marked reduction in vessel length and area of vascularisation, 73.8 ± 7.6 pixels and $17{,}944 \pm 4277$ pixels respectively (Scicli et al. 2004). In other words, the study was successful in showing that formamidine, a selective inhibitor of CYP4A, can be an inhibitor of angiogenesis. The results therefore propose that a factor formed by CYP4A, predictably 20-HETE, is a significant downstream mediator of angiogenesis through its inducing of tyrosine kinase-activated receptors including EGF (Scicli et al. 2004).

The hyperactivity of EGF has been observed in certain cancers, in all of which there have been abnormal functions and correlated mutations in the receptor (Vallbohmer and Lenz 2005). Rectal and lung cancers in particular are believed to have a strong correlation with EGF hyperactivity (Vallbohmer and Lenz 2005; Uhm et al. 2009). Epidermal growth factor receptors (EGFR) has been one of the main targets for cancer therapy (Balagula et al. 2011). Main examples are therapeutics directed against EGFR including cetuximab for colon cancer, and erlotinib and gefitinib for lung cancer (Roe et al. 2006; Akerley et al. 2009). Additional approaches to inhibiting EGFR include the use of small molecules as ligands to inhibit the cytoplasmic active site of EGFR, which essentially inhibits the activation of the EGFR, subsequently inhibiting the production of essential downstream proteins for cancer cells i.e. halting the signal cascade required for tumour growth, proliferation and migration (Chang et al. 2004; Thomas et al. 2007; Lurje and Lenz 2009; Uhm et al. 2009).

Many studies have shown the complexity of EGF and its relationship to angiogenesis. Further research has also differentiated between EGF and additional EGF-like growth factors to further enhance the understanding and isolation of causes of angiogenesis in cancer cells (Harris 2003). As a result, its interaction with EGFR is the most prominent in inducing angiogenesis in tumours, but also seen as an area where further research to inhibit angiogenesis must be undertaken and ultimately reduce the effects of cancer. Currently, many researchers aim at targeting the inhibition of EGF receptor as a cancer therapy for metastatic colorectal cancer. Through the blocking of EGF receptors, their stimulation will be decreased. Therefore, angiogenesis can be prevented and growth of the tumour will be limited due to the insufficient supply of nutrients and oxygen. Animal trials have verified inhibitory effect of EGF receptor antagonists such as panitumumab and cetuximab. Studies have shown that EGF receptor antagonists have high efficacy in advanced colorectal cancer, especially in conjunction with radiation and chemotherapy (Khong et al. 2013).

References

Akerley W, Boucher KM, Bentz JS, Arbogast K, Walters T (2009) A phase II study of erlotinib as initial treatment for patients with stage IIIB-IV non-small cell lung cancer. J Thorac Oncol 4(2):214–219. doi:10.1097/JTO.0b013e3181943bb9

Balagula Y, Garbe C, Myskowski PL, Hauschild A, Rapoport BL, Boers-Doets CB, Lacouture ME (2011) Clinical presentation and management of dermatological toxicities of epidermal growth factor receptor inhibitors. Int J Dermatol 50(2):129–146. doi:10.1111/j.1365-4632.2010.04791.x

Cartlidge SA, Elder JB (1989) Transforming growth factor alpha and epidermal growth factor levels in normal human gastrointestinal mucosa. Br J Cancer 60(5):657–660

Chang GC, Yang TY, Chen KC, Yin MC, Wang RC, Lin YC (2004) Complications of therapy in cancer patients: Case 1. Paronychia and skin hyperpigmentation induced by gefitinib in advanced non-small-cell lung cancer. J Clin Oncol 22(22):4646–4648. doi:10.1200/jco.2004.02.168

Cohen S (2008) Origins of growth factors: NGF and EGF. J Biol Chem 283(49):33793–33797. doi:10.1074/jbc.X800008200

Dagogo-Jack S, Atkinson S, Kendall-Taylor P (1986) Demonstration of significant epidermal growth factor activity in murine thyroid tissue. J Endocrinol 110(1):67–72

Donnelly MM, Hoath SB, Pickens WF (1992) Early metabolic consequences of epidermal growth factor administration to neonatal rats. Am J Physiol 263(5 Pt 1):E920–E927

Dreux AC, Lamb DJ, Modjtahedi H, Ferns GA (2006) The epidermal growth factor receptors and their family of ligands: their putative role in atherogenesis. Atherosclerosis 186(1):38–53. doi:10.1016/j.atherosclerosis.2005.06.038

Dunn IF, Heese O, Black PM (2000) Growth factors in glioma angiogenesis: FGFs, PDGF, EGF, and TGFs. J Neurooncol 50(1/2):121–137. doi:10.1023/a:1006436624862

Dvorak B, Fituch CC, Williams CS, Hurst NM, Schanler RJ (2003) Increased epidermal growth factor levels in human milk of mothers with extremely premature infants. Pediatr Res 54(1):15–19. doi:10.1203/01.PDR.0000065729.74325.71

Ellis LM (2004) Epidermal growth factor receptor in tumor angiogenesis. Hematol Oncol Clin North Am 18(5):1007–1021. doi:10.1016/j.hoc.2004.06.002

Harris R (2003) EGF receptor ligands. Exp Cell Res 284(1):2–13. doi:10.1016/s0014-4827(02)00105-2

Khong TL, Thairu N, Larsen H, Dawson PM, Kiriakidis S, Paleolog EM (2013) Identification of the angiogenic gene signature induced by EGF and hypoxia in colorectal cancer. BMC Cancer 13:518. doi:10.1186/1471-2407-13-518

Kloth DD, Iacovelli L, Arbuckle R, McIntosh AC (2010) The escalating role of epidermal growth factor receptor inhibitors in cancer management: clinical considerations for the health system pharmacist. Pharm Ther 35(4):219–229

Lurje G, Lenz H-J (2009) EGFR signaling and drug discovery. Oncology 77(6):400–410

Marti U, Wells A (2000) The nuclear accumulation of a variant epidermal growth factor receptor (EGFR) lacking the transmembrane domain requires coexpression of a full-length EGFR. Mol Cell Biol Res Commun MCBRC 3(1):8–14. doi:10.1006/mcbr.2000.0177

Mroczkowski B, Reich M, Whittaker J, Bell GI, Cohen S (1988) Expression of human epidermal growth factor precursor cDNA in transfected mouse NIH 3T3 cells. Proc Natl Acad Sci U S A 85(1):126–130

Nishida T, Nakamura M, Murakami J, Mishima H, Otori T (1992) Epidermal growth factor stimulates corneal epithelial cell attachment to fibronectin through a fibronectin receptor system. Invest Ophthalmol Vis Sci 33(8):2464–2469

Normanno N, De Luca A, Bianco C, Strizzi L, Mancino M, Maiello MR, Carotenuto A, De Feo G, Caponigro F, Salomon DS (2006) Epidermal growth factor receptor (EGFR) signaling in cancer. Gene 366(1):2–16. doi:10.1016/j.gene.2005.10.018

Petit AM, Rak J, Hung MC, Rockwell P, Goldstein N, Fendly B, Kerbel RS (1997) Neutralizing antibodies against epidermal growth factor and ErbB-2/neu receptor tyrosine kinases down-regulate vascular endothelial growth factor production by tumor cells in vitro and in vivo: angiogenic implications for signal transduction therapy of solid tumors. Am J Pathol 151(6):1523–1530

Roe E, Garcia Muret MP, Marcuello E, Capdevila J, Pallares C, Alomar A (2006) Description and management of cutaneous side effects during cetuximab or erlotinib treatments: a prospective study of 30 patients. J Am Acad Dermatol 55(3):429–437. doi:10.1016/j.jaad.2006.04.062

Schlessinger J (2002) Ligand-induced, receptor-mediated dimerization and activation of EGF receptor. Cell 110(6):669–672

Scicli AG, Chen P, Edwards P, Roman R (2004) Blockade of CYP4A inhibits angiogenic responses to VEGF, BFGF and EGF. Cardiovasc Pathol 13(3):57. doi:10.1016/j.carpath.2004.03.167

Soufla G, Sifakis S, Spandidos DA (2008) FGF2 transcript levels are positively correlated with EGF and IGF-1 in the malignant endometrium. Cancer Lett 259(2):146–155. doi:10.1016/j.canlet.2007.10.002

Taylor JM, Mitchell WM, Cohen S (1972) Epidermal growth factor. Physical and chemical properties. J Biol Chem 247(18):5928–5934

Thomas MB, Chadha R, Glover K, Wang X, Morris J, Brown T, Rashid A, Dancey J, Abbruzzese JL (2007) Phase 2 study of erlotinib in patients with unresectable hepatocellular carcinoma. Cancer 110(5):1059–1067. doi:10.1002/cncr.22886

Tong Z, Chakraborty S, Sung B, Koolwal P, Kaur S, Aggarwal BB, Mani SA, Bresalier RS, Batra SK, Guha S (2010) Epidermal growth factor down-regulates the expression of neutrophil gelatinase-associated lipocalin (NGAL) through E-cadherin in pancreatic cancer cells. Cancer 117(11):2408–2418. doi:10.1002/cncr.25803

Uhm JE, Park BB, Ahn MJ, Lee J, Ahn JS, Kim SW, Kim HT, Lee JS, Kang JH, Cho JY, Song HS, Park SH, Sohn CH, Shin SW, Choi JH, Park K (2009) Erlotinib monotherapy for stage IIIB/IV non-small cell lung cancer: a multicenter trial by the Korean Cancer Study Group. J Thorac Oncol 4(9):1136–1143. doi:10.1097/JTO.0b013e3181b270a7

Vallbohmer D, Lenz HJ (2005) Epidermal growth factor receptor as a target for chemotherapy. Clin Colorectal Cancer 5(Suppl 1):S19–S27

van Cruijsen H, Giaccone G, Hoekman K (2005) Epidermal growth factor receptor and angiogenesis: opportunities for combined anticancer strategies. Int J Cancer 117(6):883–888. doi:10.1002/ijc.21479

Wahl MI, Nishibe S, Suh PG, Rhee SG, Carpenter G (1989) Epidermal growth factor stimulates tyrosine phosphorylation of phospholipase C-II independently of receptor internalization and extracellular calcium. Proc Natl Acad Sci U S A 86(5):1568–1572

Chapter 13
Epidermal Growth Factor Domain-Like 7 (EGFL7)

Abstract Epidermal growth factor domain-like 7 (EGFL7) is a gene that codes for an endothelial cell protein. In humans the EGFL7 gene is located on chromosome 9; it contains two Epidermal Growth Factor-Like Domains. The protein encoded from this gene play an important role in regulating vasculogenesis and angiogenesis in humans. EGFL7 gene is also a secreted angiogenic signalling factor that is highly conserved in vertebrates. Current studies show that this gene may be involved in the proliferation of tumour cells. What makes the EGFL7 gene unique is the fact that it is expressed by endothelial cells, thus of mesodermal origin, and acts exclusively on endothelial cells. While other angiogeneic signalling factors are not expressed from endothelial cell types. EGFL7 is a gene that is exclusively expressed by endothelial cells; it is expressed at low levels in normal quiescent blood vessels, but is up- regulated in tumorigenesis. EGFL7 regulates angiogenesis by controlling cell adhesion, the spatial arrangement of ECM and modulation of the Notch signalling pathway. It impacts vascular development in both hypoxic and hyperoxic environments, and is determined to be an agent of tumour angiogenesis. There are many aspects of the EGFL7 gene that can be explored as future therapeutics for cancer and vascular damage.

Keywords Epidermal growth factor domain-like 7 (EGFL7) • Angiogenesis • Normal physiology • Disease • Malignancy

13.1 Epidermal Growth Factor Domain-Like 7 (EGFL7)

Epidermal growth factor domain-like 7 (EGFL7) is a gene that codes for an endothelial cell protein. In humans the EGFL7 gene is located on chromosome 9; it contains two Epidermal Growth Factor-Like Domains. The protein encoded from this gene play an important role in regulating vasculogenesis and angiogenesis in humans. EGFL7 gene is also a secreted angiogenic signalling factor that is highly conserved in vertebrates. Current studies show that this gene may be involved in the proliferation of tumour cells. What makes the EGFL7 gene unique is the fact that it is expressed by endothelial cells, thus of mesodermal origin, and acts exclusively on

A. Salajegheh, *Angiogenesis in Health, Disease and Malignancy*,
DOI 10.1007/978-3-319-28140-7_13

endothelial cells. While other angiogeneic signalling factors are not expressed from endothelial cell types (Nichol and Stuhlmann 2012).

13.2 Epidermal Growth Factor Domain-Like 7 in Normal Physiology

EGFL7 is an extracellular matrix protein that is involved in blood vessel formation, through the regulation and support of endothelial cell adhesion and growth (Schmidt et al. 2007). Endothelial cells are the principal cells which forms the lining of blood vessels. During embryogenesis, and physiologic and pathologic angiogenesis high levels of EGFL7 gene is found in proliferating endothelial cells. Especially during the endothelial regeneration phase after blood vessel injury. This gene is expressed at low levels in healthy dormant adult blood vessels in comparison to when it is inside the embryo (Nichol and Stuhlmann 2012).

EGFL7 is a chemo attractant for embryonic endothelial cells and regulates endothelial cell adhesion. However this adhesion property is weaker than other cell-cell adhesion factors, thus there is a hypothesis that EGFL7 favours endothelial motility (Nichol and Stuhlmann 2012). Hence the cells can easily adjust themselves until they are positioned properly. If the EGFL7 gene is absent, the cells tend to clump together and form oversized clots, which could result in impaired migration of endothelial cells and delayed vascularization.

The remodelling of extracellular matrix (ECM) and modulation of its rigidity by EGFL7 gene is another way to promote endothelial cell migration and invasion. EGFL7 represses the formation of elastin by inhibiting the enzyme lysyl oxidase, which is responsible for the conversion of tropoelastin to elastin, which gives the ECM flexibility to adjust itself (Nichol and Stuhlmann 2012). Thus, this gene is crucial in promoting endothelial cell functions during angiogenesis, as it influences their behaviour by signalling the ECM to create a favourable environment for angiogenesis to take place.

The EGFL7 gene is also responsible for the spatial arrangement of cells during vascular tube assembly. During primary vessel network formation, angioblasts arrange themselves into vascular cords which are later transformed into vascular tubes. The transition from cord to tube is controlled by cell polarisation, redistribution of junctional proteins, thus leading to changes in cell shape. EGFL7 is secreted by the endothelial cells into ECM, ECM molecules is detected by adjacent endothelial cells, and within enlarged angiogenic sprouts. Therefore this suggests that the endothelial cells containing the EGFL7 gene plays a role in detecting sprouting boundaries (Nikolic et al. 2010).

One of the important ways the EGFL7 regulates angiogenesis is through the modulation of the Notch signalling pathways. These pathways coordinate cellular behaviour during angiogenesis. Endothelial cells with activated Notch signalling, will remain within the parent blood vessel, while the endothelial cells without Notch signalling will eventually leave the parent blood vessel and not contribute to the formation of further vasculature (Siekmann and Lawson 2007). EGFL7 experts its effects by signalling notch receptors on endothelial cells. The signalling could be in both autocrine (EGFL7 is secreted from the same cells that it is signalling to) and paracrine (signalling to notch receptor in neighbouring cells) manner, to enhance vessel maturation. EGFL7 binds to an area in the Notch receptor involved in ligand-mediated receptor activation, thus it acts as an antagonist for Notch signalling (Schmidt et al. 2009).

13.3 Epidermal Growth Factor Domain-Like 7 in Disease and Malignancy

EGFL7 appears to play a part in repair after vascular damage, and ischemia. In an experiment done on mice, EGFL7 expression is induced after arterial vascular injury. An increase in EGFL7 levels under hypoxic conditions appears to stimulate and angiogenic response. An elevated EGFL7 expression in neonatal rat brain as a result of hypoxia, showed an increase in vascular density, resulting in a protective effect against hypoxic –ischemic conditions. By contrast, inhibition of EGFL7 occurs after hyperoxic exposure in both neonatal lungs and endothelial cells. Reduced expression of this gene is associated with endothelial cell death. Up-regulation of EGFL7 in endothelial cells acts as a protection against hyperoxia-induced apoptosis, by inhibiting genes that are closely associated with mitochondria-induced apoptosis (Nichol and Stuhlmann 2012).

EGFL7 can have a negative effect when it comes to tumorigenesis. Overexpression of EGFL7 gene is observed in various tumours, including kidney, breast, hepatocellular carcinomas, and malignant gliomas, leading to future cancer development in these areas (Nichol and Stuhlmann 2012). The adhesion properties of endothelial cells due to the presence of EGFL7 are weaker than others in order to favour endothelial motility. However a consequence of this is when tumour starts to develop, the adhesion of lymphocytes to the endothelium is also reduced. This increases the rate of tumour escape from the body's immunity, resulting in a more rapid tumour proliferation. It should be noted that EGFL7 gene has no effect on immune cells themselves (Pinte and Soncin 2012). EGFL7's main effect is to repress the tumour endothelium activation, so the immune cells remain isolated in the blood circulation, thus they are unable to infiltrate into the tumour mass (Nikolic et al. 2010). Because EGFL7 overexpression results in abnormal vessel formation and modelling, it is possible that it stimulates tumour angiogenesis. It does so by contributing to irregularity shaped and leaky vessels typical of tumour vasculature characteristics.

Through experimentation on mice it was also observed that expression of EGFL7 gene alters blood vessel permeability, and decreases expression of Tie-2, which is an angiopoietin receptor, they are required for angiogenesis (Martin et al. 2008).

This reinforces the fact that the endothelium structure is altered by the presence of EGFL7. Lack of solid vessel structure is commonly found in tumours, and could possibly favour further proliferation through metastasis (Nikolic et al. 2010).

There are many aspects of the EGFL7 gene that can be explored as future therapeutics for cancer and vascular damage. The escape of tumour cells from the body's immune system is definitely a vital process to consider when it comes to targeting cancer progression and metastasis. However because EGFL7 is down regulated in normal dormant blood vessels, interfering with the properties of this gene may create adverse side effects in normal organs, and interfere with angiogenesis. Hence any new therapies should be extensively tested before starting human trials.

EGFL7 could also be target for any diseases associated with hypoxic-ischemic conditions, such as atherosclerosis, stroke, and coronary artery disease. The protective properties of this gene in hyperoxia environments can be used to prevent endothelial cell damage in patients who require long term oxygen treatment (Nichol and Stuhlmann 2012).

A multidimensional clinical experiment was carried out to identify the optimal dose of anti-EGFL7, an antiangiogenesis agent. Anti-EGFL7 inhibits two important processes that EGFL7 is responsible for; endothelial cell adhesion, and hyperoxia induced apoptosis. As well as increased anti- Vascular Endothelial Growth Factor (VEGF) "medicated vascular damage in various murine tumour models". The experiment involved using genetically engineered mice with advanced lung cancer. It was found that the anti-EGFL7 enhanced the overall survival rate in a progression-free mode of the mouse from anti-VEGF therapy in a "dose-dependent" manner. Furthermore a circulating progenitor cell type that was regulated by EGFL7 is also identified. The evaluation of these cells to the anti-EGFL7 therapy in both tumour bearing mice and cancer patients during a phase I clinical trial, along with repeated testing, provides enough information to allow rational selection of the dose required for phase II clinical trials (Johnson et al. 2013). The aim is to be able to conclude on an optimal effective dose range, and further develop this therapy in hopes of it becoming a viable cancer treatment option.

Conversely, EGFL7 is a gene that is exclusively expressed by endothelial cells; it is expressed at low levels in normal quiescent blood vessels, but is up- regulated in tumorigenesis. EGFL7 regulates angiogenesis by controlling cell adhesion, the spatial arrangement of ECM and modulation of the Notch signalling pathway. It impacts vascular development in both hypoxic and hyperoxic environments, and is determined to be an agent of tumour angiogenesis. Currently there are many experiments and clinical trials being conducted with regards to EGFL7 and its role as a potential cancer treatment in vascular diseases, due to its unique role in tumour growth and regulation of angiogenesis.

References

Johnson L, Huseni M, Smyczek T, Lima A, Yeung S, Cheng JH, Molina R, Kan D, De Maziere A, Klumperman J, Kasman I, Zhang Y, Dennis MS, Eastham-Anderson J, Jubb AM, Hwang O, Desai R, Schmidt M, Nannini MA, Barck KH, Carano RA, Forrest WF, Song Q, Chen DS, Naumovski L, Singh M, Ye W, Hegde PS (2013) Anti-EGFL7 antibodies enhance stress-induced endothelial cell death and anti-VEGF efficacy. J Clin Invest 123(9):3997–4009. doi:10.1172/jci67892

Martin V, Liu D, Fueyo J, Gomez-Manzano C (2008) Tie2: a journey from normal angiogenesis to cancer and beyond. Histol Histopathol 23(6):773–780

Nichol D, Stuhlmann H (2012) EGFL7: a unique angiogenic signaling factor in vascular development and disease. Blood 119(6):1345–1352. doi:10.1182/blood-2011-10-322446

Nikolic I, Plate KH, Schmidt MH (2010) EGFL7 meets miRNA-126: an angiogenesis alliance. J Angiogenes Res 2(1):9. doi:10.1186/2040-2384-2-9

Pinte S, Soncin F (2012) Egfl7 promotes tumor escape from immunity. Oncoimmunology 1(3):375–376. doi:10.4161/onci.18964

Schmidt M, Paes K, De Maziere A, Smyczek T, Yang S, Gray A, French D, Kasman I, Klumperman J, Rice DS, Ye W (2007) EGFL7 regulates the collective migration of endothelial cells by restricting their spatial distribution. Development 134(16):2913–2923. doi:10.1242/dev.002576

Schmidt MH, Bicker F, Nikolic I, Meister J, Babuke T, Picuric S, Muller-Esterl W, Plate KH, Dikic I (2009) Epidermal growth factor-like domain 7 (EGFL7) modulates Notch signalling and affects neural stem cell renewal. Nat Cell Biol 11(7):873–880. doi:10.1038/ncb1896

Siekmann AF, Lawson ND (2007) Notch signalling and the regulation of angiogenesis. Cell Adh Migr 1(2):104–106

Chapter 14
Erythropoietin-Producing Hepatocellular Receptors A: Ephrin A1, Ephrin A2 and Ephrin A3

Abstract There are many cellular mechanisms in the human body vital for its function. One of these cellular mechanisms responsible for regulating cell movement, rearrangement during neuronal, vascular and epithelial development, signalling between cells and many more are the Receptor Tyrosine Kinases (RTK); specifically, Erythropoietin-producing hepatocellular receptors, or Eph receptors, which contain tyrosine kinase domains. They have a high binding affinity to their ephrin ligands and are very important members of these cellular mechanisms. The ligands and receptors of the Eph/ephrin RTK family are located on plasma membranes; hence a very important factor in Eph-ephrin binding is close contact between cells for binding to occur due to their cellular location. The Eph receptors and their ligands are subclassed into two categories – Class A and Class B, by their structural domains. Due to ephrin-A1's diverse involvement in many organ systems around the body, it is studied as a potential therapeutic target to find the treatment for many abnormal conditions. The role of EphA2 in angiogenesis of tumours has a direct link to aggressive tumour vasculature and invasion. Lastly, the genetic alterations of the EphA3 in research suggest the gene contributes to the initiation and metastasis of various tumours.

Keywords Erythropoietin-producing hepatocellular receptors A • Ephrin A1 • Ephrin A2 • Ephrin A3 • Angiogenesis • Normal physiology • Disease • Malignancy

14.1 Erythropoietin-Producing Hepatocellular Receptors, or Eph Receptors

There are many cellular mechanisms in the human body vital for its function. One of these cellular mechanisms responsible for regulating cell movement, rearrangement during neuronal, vascular and epithelial development, signalling between cells and many more are the Receptor Tyrosine Kinases (RTK); specifically, Erythropoietin-producing hepatocellular receptors, or Eph receptors, which contain tyrosine kinase domains (Pasquale 1997). They have a high binding affinity to their ephrin ligands and are very important members of these cellular mechanisms. The ligands and receptors of the Eph/ephrin RTK family are located on plasma

A. Salajegheh, *Angiogenesis in Health, Disease and Malignancy*,
DOI 10.1007/978-3-319-28140-7_14

membranes; hence a very important factor in Eph-ephrin binding is close contact between cells for binding to occur due to their cellular location. The Eph receptors and their ligands are subclassed into two categories – Class A and Class B, by their structural domains. There are six Ephrin-A class proteins which contain an extracellular domain with high affinity for EphA receptors. These extracellular domains are attached to the outer plasma membrane by a glycosylphosphatidylinositol (GPI) anchor. Likewise, there are three ephrin-B class proteins that also contain extracellular domains with high affinity for EphB receptors, but instead of a GPI linkage, these possess a lipid-soluble transmembrane domain with an extension that goes into the cytoplasm (Pasquale 2008).

Several exceptions among the ephrin ligands exist which bind relatively well with their non-respective receptors, and an ephrin ligand can also bind to multiple receptors. Examples include ephrin-A5 ligand binding to EphA2 receptor or ephrin-A4 ligand binding to EphB receptors (Beauchamp and Debinski 2012). The ephrin ligands activate the tyrosine kinase domain of the Eph receptors by binding to them, and hence induce a signal in the receptors. Interestingly, there exists a unique feature in those ligands and receptors where the receptors also induce a signal in the ligands as well when bound. The first is referred to as 'forward' signalling, and the second is referred to as 'reverse' signalling, and this process is referred to as 'bidirectional' signalling (Pasquale 2008). This feature allows the Eph-ephrin RTK family to achieve many of their functions.

14.2 Ephrin-A1

Among this broad RTK family, the ephrin ligand ephrin-A1, and its primary receptor EphA2, are particularly well studied (Wykosky and Debinski 2008). According to studies, ephrin-A1 ligand appears to have the highest affinity for the EphA2 receptor. Their binding mechanism follows the 'lock and key' model, and once bound EphA2 receptors go through tyrosine autophosphorylation. Recent studies divulged a glycosylation on the Asn-26 of ephrin-A1. This has been shown to be incredibly important regarding the interaction between ephrin-A1 and EphA2 as further research showed that this glycosylation site on Asn-26 of ephrin-A1 interacted with carbohydrates and Asp-78 and Lys-136 on EphA2. Upon testing enzymatically deglycosylated ephrin-A1 ligand, it was found that it could not sufficiently activate the EphA2 receptor and could not induce bidirectional signalling between the complex when bound (Ferluga et al. 2013). Furthermore, in conditions where ephrin-A1 is deficient is also adequate to down-regulate EphA2 phosphorylation (Wykosky and Debinski 2008).

Ephrin-A1 was discovered in 1990 in human umbilical vein endothelial cells (HUVECs) as a tumour necrosis factor (TNF) inducing protein. By 1994, it was found to be a ligand binding on the EphA2 receptor which was, at that time, the only known receptor of the RTK receptors since their discovery in 1987 (Beauchamp and Debinski 2012). Ephrin-A1 is coded by the EFNA1 gene located on Chromosome 1

(Pasquale 1997). The gene itself is important for angiogenesis in activating certain enzymes and arranging vascular endothelial cells for angiogenesis to occur properly and for tumour neovascularisation (Ogawa et al. 2000). Ephrin-A1 ligands are involved in many physiological functions, including expressions on immune cells, skin cells, and cardiovascular and angiogenic processes. However, there is increasing research and evidence to support their involvement in various tumours and cancers. Current research is not yet adequate to explain all the mechanisms associated with this pair. There are still many questions to be answered regarding their roles, but due to their wide involvement around the human body, they are known to be good therapeutic targets in future studies.

14.2.1 Ephrin-A1 in Normal Physiology

Ephrin-A1 ligands are expressed diversely around the body. With relation to immune cells, ephrin-A1 ligands bind to CD8+ T cells and stimulates cell movement as well as inducing T cell receptor (TCR) signalling (Hjorthaug and Aasheim 2007). They are also found on CD4+ T cells and are thought to produce effects through reverse signalling by binding with antibodies to suppress TCR responses (Pasquale 2008). Ephrin-A1 is also involved in the cardiovascular system in cardiac valve formation. Numerous studies conducted on embryonic hearts have revealed its involvement in cardiac valve formation. One line of evidence shows that, in ephrin-A1 engineered embryos possessing a lack of neural crest cell action, the state or degree of cells present in the area where the blood flows out of the heart, which also forms heart valves (also called endocardial cushions), is greatly improved. Another line of evidence is that when the same effect as the first line of evidence has been produced but without sufficient amount of ephrin-A1, it was seen to have been assisted by undifferentiated cells, rather than other specialised proteins (Frieden et al. 2010). Research with mice has also shown expression of ephrin-A1 ligands and EphA2 receptors on skin epithelia. Eph/ephrin signalling has been known to have been involved with regulating skin epithelial homeostasis. It has been studied in mice skin that EphA2 and ephrin-A1 are present complementarily in mice skin. EphA2 was seen to be most present in the spinous and granular layers and less present in the basal layers of mice skin whereas for ephrin-A1, it was more prominent in the basal layers. Studying their patterns together implies that most of the interactions between EphA2 and ephrin-A1 take place at the basal layer to aid in cell proliferation (Miao and Wang 2009). The ephrin-A1 and EphA2 pair has also attracted attention for its involvement in managing vascular morphogenesis and angiogenesis.

Angiogenesis is the process by which new vasculature is formed from existing blood vessels from a series of complicated processes (Wykosky and Debinski 2008). According to several studies, ephrin-A1 produces angiogenic effects by activating EphA2 tyrosine receptors via phosphorylation on cells lining the blood vessel endothelia. Mutations in the binding region, or areas close to the membrane of EphA2 suppress vascular organisation induced by ephrin-A1. Furthermore, ephrinA1 on tumour cells can generate vascular endothelial growth factor (VEGF) signals which promotes growth of new vasculature, leading to angiogenesis and spread of tumour (Beauchamp and Debinski 2012).

14.2.2 Ephrin-A1 in Disease and Malignancy

Although their exact roles in tumours and cancers have not been fully established yet, it has been made clear that Eph receptors and their ephrin ligands are significantly involved in tumour or cancer in forms of both growth and suppression. It has been previously mentioned that a glycosylation site was found on ephrin-A1. As well as being crucial for the binding affinity between ephrin-A1 and EphA2, it has also been shown that it is important for stimulating EphA2 so that the receptor acts as a tumour suppressor (Ferluga et al. 2013). On the other hand, there are also other studies that show ephrin-A1 and EphA2 acting as tumour-promoters. For instance, up-regulation of ephrin-A1 and EphA2 during melanoma progression as well as a overexpression of both has been shown to be related to poor patient survival in ovarian cancer, and similarly for bladder, gastric and cervical cancers. Interestingly, this is not the case for all tumours or cancers. In some other cancers, it has been shown that ephrin-A1 and EphA2 are present in varying levels. An example would be that EphA2 was seen to be concentrated in many glioblastoma multiforme (GBM) tumours and breast cancers, but ephrinA1 was shown to be less expressed in both. However, it was also found that when genes coding for cells containing high levels of human ephrin-A1 and EphA2 was transfected to cancer cells, it resulted in suppression of malignant properties through phosphorylation and down-regulation of EphA2 (Beauchamp and Debinski 2012). The results of these studies only emphasises that the mechanisms of ephrin-A1/EphA2 in cancer cells is of a highly complex nature.

Due to ephrin-A1's diverse involvement in many organ systems around the body, it is studied as a potential therapeutic target to find the treatment for many abnormal conditions. Although still in its early stages, some recent studies show that immobilised ephrin-A1 was used to form a tubular organisation on surface-modified hydrogels by making use of ephrin-A1's detachment properties for the purpose of preventing cell-cell attachments or interactions (Saik et al. 2011). This approach is a possible commencement point for future cancer therapies. Furthermore, as EphA2 receptors and their ephrin ligands are crucial for vascular development, they also present chances of therapeutically targeting cancer-associated angiogenesis. The approach is not yet certain, but the idea is to use VEGF to induce ephrin-A1 which will then stimulate EphA2, which in turn will use the EphA2 to obstruct VEGF (Kandouz 2012). Currently, other speculations include inhibiting tumour-promoting effects or improving tumour-suppressing effects through interfering with RNA, immunotherapy, vector gene therapy, small molecule inhibitors and nanoparticles (Giaginis et al. 2014). Another strategy is by using EphA2 as a means of transporting substances in the form of drugs or immune cells to the tumour cells (Wykosky and Debinski 2008). However the delicacy of the matter is that if ephrin-A1 or EphA2 or both are used to reach the tumour cells, they may still interact with tumour cells in such a way that induces angiogenesis and promotes spread of the tumour. Hence, instead of using ephrin-A1 or EphA2 on their own, a biotechnologically engineered form of ephrin-A1-Fc or EphA2-Fc are used where ephrin-A1 obtained from mice is attached to the Fc chain of human antibody IgG by proteins (Jing et al. 2012).

14.3 Ephrin-A2

EphA2 is a member of the Eph group, the largest group of receptors in the receptor tyrosine kinases (RTK) family. The Eph group is further classified into two subclasses; EphA (EphA1-10) and EphB (EphB1-6). These subclasses share a degree of specificity by ways of containing an amino-acid loop on their extracellular surface (Tandon et al. 2011; Cheng et al. 2002). EphA and EphB receptors contain a single transmembrane domain, while the extracellular region consists of a N-terminal ligand binding domain, a cysteine rich motif and two fibronectin III repeats (Brantley-Sieders et al. 2004). The Eph receptors have a high affinity to binding to their respective class, cell surface bound, ephrin ligands. That is, the EphA receptors bind to GPI anchored Class-A ephrin ligands and EphB receptors bind to the trans-membrane bound Class-B ligands respectively (Brantley-Sieders et al. 2004). The interaction of Ephrin ligands with their respective Eph receptors is associated with a multitude of cancers as well as its involvements in many physiological functions including the nervous system, neuronal development, brain development, angiogenesis of tumour formation and angiogenesis during embryonic development (Tandon et al. 2011; Ferluga et al. 2013). The EphA2 receptor has been studied extensively as this receptor in particular has been found to be overexpressed in human cancers; and specifically its involvement with the angiogenesis of tumours (Biao-xue et al. 2011).

14.3.1 Ephrin-A2 in Normal Physiology

EphA2 was first identified in 1990 in the exploration of highly conserved regions of protein tyrosine kinases in human epithelium. The human EphA2 receptor is located on chromosome 1 where it encodes 976 amino acids to produce a receptor tyrosine kinase (Ferluga et al. 2013). The mRNA expression of EphA2 is observed in skin, bone marrow, thymus, uterus, testis, prostate, urinary bladder, kidney, small intestine, colon, spleen, liver and brain (Tandon et al. 2011).

EphA2 is a transmembrane protein that has an important role in signal transduction in normal physiology and is primarily found in adult human epithelial cells (Tawadros et al. 2012; Walker-Daniels et al. 2003). However the specific cellular functions in normal epithelia is not well understood, but its involvement in tumour models suggests its involvement in angiogenesis, migration, survival and regulation of cell growth (Walker-Daniels et al. 2003). Furthermore, although the reasoning is unknown, it is important to note that ligand binding is not essential for EphA2 activity (Walker-Daniels et al. 2003; Landen et al. 2006). In malignant cells EphA2 is overexpressed and accumulates on the cell surface to remain unphosphorylated (Brantley-Sieders et al. 2004). EphA2 binds to the ephrin ligands A1, A2, A3, A4 and A5 (Landen et al. 2006). As previously mentioned, the glycosylation on ephrin A1 plays a critical role in the activation of the EphA2 receptor. The ligand binding of EphA2 undergoes tyrosine autophosphorylation followed by internalisation and degradation, which several point mutants have shown a high affinity to (Tandon et al. 2011).

The function of the Eph receptor to ephrin ligands in angiogenesis is still yet to be fully understood, however the activation of EphA2 by ephrinA1 has been established to suggest a role in tumour angiogenesis (Ferluga et al. 2013). This pairing is consistently expressed in the endothelium of tumours that are angiogenic-dependant. Furthermore, the Ephrin A1 ligand is expressed predominantly in the tumour tissue while the EphA2 receptor is mainly expressed in the vasculature of the tumour. Research has shown the activation of EphA2 receptor increases microvascular endothelial cells, but inhibits vascular smooth muscle to retract the perivascular supporting cells, to allow endothelial cells to proliferate and migrate towards angiogenesis (Cheng et al. 2002). In vivo studies have found EphA2 can mediate endothelial cell network formation and VEGF mediated angiogenesis. Clinical studies have shown that EphA2 overexpression may directly regulate VEGF levels in tumours (Brantley-Sieders et al. 2004). In 2014 Lennon et al. confirmed the interaction between EphA2 and CD44 signalling pathways having a critical role, specifically in low molecular weight hyaluronan mediated angiogenesis (Lennon et al. 2014).

14.3.2 EphA2 in Disease and Malignancy

RTK's, especially EphA2 are overexpressed and display increased activity to play an essential role in cell signalling pathways in carcinogenesis (Ferluga et al. 2013). Interaction of the EphA2 receptor activated by ephrin-A1 ligand causes deregulated sequential events that are associated with carcinogenesis. However, the activation of EphA2 by ephrinA1 both suppresses and activates cell signalling (Brantley-Sieders et al. 2004). The overexpression of EphA2 in malignant cells is not phosphorylated and thus functions differently to tyrosine-phosphorylated EphA2 epithelial cells (Walker-Daniels et al. 2003). The failure to efficiently bind to a ligand inhibits the phosphorylation of EphA2 resulting in pathological changes in the cell. As the accumulation of non-phosphorylated receptors increases in the cell there is a loss of cell to cell contact, which causes oncogenic signalling. As a result, angiogenesis and proliferation causes metastasis in the tissues (Brantley-Sieders et al. 2004).

Furthermore, molecular interactions between signalling proteins and EPHA2 are associated with metastasis through the modulation of cytoskeleton dynamics, cell adhesion, proliferation and differentiation (Tandon et al. 2011). EphA2 is overexpressed in breast, prostate, urinary bladder, skin, lung, ovary, oesophagus and brain cancers. The exact reason for the overexpression of EphA2 is unknown, however there is a direct link to the survival and maintenance of tumours as a result of high levels of EphA2 found in advanced forms of cancer. Conclusively, there is increasing evidence to support that EphA2 contributes towards angiogenesis, oncogenesis and tumour metastasis.

The role of EphA2 in angiogenesis of tumours has a direct link to aggressive tumour vasculature and invasion (Tandon et al. 2011). As a result, low survival rates place significance on therapeutic manipulation of the protein to predict clinical outcomes. EphA2 overexpression and the change to oncogenic function both provide an opportunity for intervention as an antiangiogenic target (Brantley-Sieders et al. 2004). Targeting the overexpression of the receptor in vivo and in vitro carcinoma by EphA2 antibodies, RNAi, adenoviral vectors, immunotherapy, small molecule inhibitors and by other approaches have demonstrated promising results. There is also potential to decrease the malignant tumour survival by modulating the microenvironment to disrupt the regulation of angiogenic activity associated with the overexpression of EphA2. There is over a decade of research using multiple approaches to successfully target the receptor in cancer models with promising results (Tandon et al. 2011; Biao-xue et al. 2011). Despite the emerging research and efficacy of targeted therapies there is still more understanding of the complex nature of this specific target. There is not one single therapeutic treatment that has shown complete efficacy in decreasing tumour vasculature or invasion. This warrants further research into not just EphA2 but to also consider contributing pathological stimuli and synergistic risk factors such as lifestyle and environmental factors (Tandon et al. 2011).

14.4 Ephrin-A3

Eph receptors are family of RTK family that may be future prognostic targets in cancer treatment (Keane et al. 2012; Merlos-Suárez and Batlle 2008; Lackmann et al. 2014; Guan et al. 2011). Their complex bi-signalling processes controls intracellular mediated process that result in cellular growth, apoptosis, movement dictating cellular adhesion or repulsion (Keane et al. 2012; Pasquale 2010; Guan et al. 2011). It is widely expressed in embryological neural and vasculature development and decreases in expression in adult tissues (Keane et al. 2012; Day et al. 2013; Lackmann et al. 2014). Abnormal reappearance in postembryonic tissue indicates malignancy, reflecting consistent correlations between its expansive metastatic behaviour and its overexpression (Keane et al. 2012). The reality of clinical data of EphA-3 antibody directed therapies such as KB004 remain yet to be further established through multiple trials and further testing, but remain promising and highly anticipated (Ashton et al. 2012).

14.4.1 *Ephrin-A3 in Normal Physiology*

Categorically, Ephrin-A3 receptor is a subfamily of Ephrin, the largest class of growth factor regulated Receptor Tyrosine Kinases (RTK), which plays an important role in a wide range of functional processes (Keane et al. 2012; Merlos-Suárez and Batlle 2008; Héroult et al. 2006). Many of these processes regulate and maintain morphological developmental processes, tumorogenesis, and neovascular cell signalling pathways (Keane et al. 2012; Merlos-Suárez and Batlle 2008; Héroult et al. 2006). Ephrin-A3 possesses organisational influences to modulate cellular adhesion, migration and morphology through ligand binding complexes and oligomerization (Lackmann et al. 2014).

Ephrin ligands and their respectful receptors are divided up into two subgroups A and B, where Eph class A receptors preferably bind to ephrinA ligands, and likewise for the B subclass, in accordance to its ligand specificity with a few exceptions (Merlos-Suárez and Batlle 2008; Guan et al. 2011). EphA3 receptors bind with high affinity ligands to ephrin-A2 and ephrin-A5 and with lower affinities to ephrin-A3 and ephrin-A4 (Merlos-Suárez and Batlle 2008). Unlike other RTK subfamilies, Ephrin A3 protein ligand is bound to the membrane through a glycosyl-phosphatidylinositol (GPI) anchor (Merlos-Suárez and Batlle 2008). When activated via ligand binding, intracellular bidirectional signalling pathways are initiated in both ligand and receptors, resulting in either repulsion or attraction of interacting cells (Xi et al. 2012b). Cell repulsion occurs through activation of GTPase hydrolase enzymes, forming protein tyrosine phosphatases (PTP) inducing signals that are in favour of cell rounding and cytoskeletal contraction. On the other hand, cell attraction is also possible and occurs when conditions favour cell adhesion and migration (Pasquale 2008).

Mechanisms involving cell migration and boundary formations are mediated via Ephrin involvement, and consequently are found at significantly higher levels during embryogenesis in the axial muscles, lungs, kidney, and heart (Keane et al. 2012). It plays an important role in segmentation and patterning of mesenchymal and endothelial developing tissues, maintaining its boundaries through repulsive actions during Ephrin signalling by modulating Jak/Stat pathway (Xi et al. 2012b). This is evident through clinical studies achieved on mice, where expressions of EphA3 in the retina were found to be responsible for retinotectal mapping through its regulation of cell axon movement (Ortalli et al. 2012). Patterning of tissue boundaries during embryological development via control axon guidance occurs through kinase-dependant epithelial-mesenchymal transition (EMT) (Keane et al. 2012; Stephen et al. 2007).

Evidence suggests EPH receptors are found to mediate the process of tumour neovascularisation in the vasculature morphogenesis of embryonic vasculature in the process of vasculogenesis (Héroult et al. 2006). Eph-A3 receptors is critically involved in hematologic development by influencing the migration of hematopoietic cells as it inhibits mitogen-activated protein kinase (MAPK) pathway signalling important in platelets, epidermis and vasculature (Miao et al. 2001). This is confirmed through clinical studies which look at expression patterns and defects of EphA3 deficient and normal mice (Stephen et al. 2007). Results display that EphA3 knockout mice developed irregular developmental defects in atrioventricular valves and septa, resulting in 75 % lethality due to pulmonary oedema (Stephen et al. 2007). Impaired valve formation resulted from irregular EMT and failure of axon guidance of mesenchymal cells into endocardial cushions (Stephen et al. 2007).

Another study evaluated the involvement of EphA3 in tumour angiogenesis in gastric carcinoma. A strong positive correlation between VEGF and EphA3 expression was discovered, suggesting EphA3 may play a role in VEGF signalling and promoting angiogenesis in gastric cancer (Xi et al. 2012a). In addition, a longer survival rate was associated with a lower expression of EphA3 in gastric cancer (Xi et al. 2012a).

14.4.2 EphA3 in Disease and Malignancy

Implications of genetic alterations of the EphA3 in research suggest the gene contributes to the initiation and metastasis of various tumours (Xi et al. 2012b; Day et al. 2013). However, there exists great variability in its expression in different tumour cell lines amongst various tissues (Keane et al. 2012). This reflects on the degree of complexity of the ligand-independent nature of Ephrin-A3, as it operates in dual contradictory roles in malignancy (Keane et al. 2012).

When activated by soluble-Fc ligands, Ephrin receptors can suppress tumour cell proliferation by forcing Eph forward signalling (Pasquale 2010). Eph forward signalling then acts to inhibit oncogene signalling pathways (Pasquale 2010). Cancer cells minimise the effects of this suppression in growth by altering the relative expressions of receptors to be higher than ligands, which results in asymmetrical bi-directional Eph forward and reverse signalling (Pasquale 2010).

Conversely, it is possible for Ephrin-dependent Eph receptors to have lost its tumour suppressor ability due to mutations and upon activation, they adopt an oncogenic ability. What results from this is inadequate RTK regulation, which frequently leads to the progression of malignant tumours (Pasquale 2010). The precise nature of the effects of mutations beyond the molecular level on Eph/Ephrin receptors in tumourogenesis remains inconclusive and is unclear due to its complex cell context dependency for many signalling pathways (Pasquale 2010). No single model explaining the function and role of Eph/Ephrin in cancer can encompass and explain its variability of expression (Keane et al. 2012). Overexpression of EphA3 has been found in breast, lung, leukaemia, lymphoma, melanoma, sarcoma, and gastric carcinoma and is down-regulated in hematopoietic and prostate carcinoma cell lines (Xi et al. 2012b).

Based on several findings, it was proposed that the expression of EphA3 could be predictive of self-sustained undifferentiated cancer behaviour, through neural cell differentiation which is a result of positively regulating the MAPK pathway (Day et al. 2013). However, an investigation on the function and expression of EphA3 in brain cancer, Glioblastoma (GMB), suggests EphA3 kinase achieves its tumorogenic potential by blocking and decrease MAPK signalling (Day et al. 2013). In this experiment, they discovered a strong positive inverse correlation of survival of GMB subtypes with EphA3 expression, where a twofold increase of EphA3 expression resulted in decreased survival (Day et al. 2013).

The same clinical data revealed higher expressions of EphA3 receptors were present in undifferentiated, highly proliferative tumorogenic cells; which were then down-regulated to induce differentiation and suppress proliferative potential (Day et al. 2013). Based on their results, they explored the potential of EphA3-targeted therapy on the tumour initiating cell compartment as a therapeutic treatment. In their findings, successful elimination of less differentiated, tumorogenic cells was achieved (Guan et al. 2011). It was targeted on the mice using high doses of anti-EphA3 monoclonal antibody therapy of Lutetium-177 (Day et al. 2013). The subsequent results proved that tumour cells were actively maintained in their undifferentiated, tumorigenic state through high expression of EphA3 (Day et al. 2013).

Interestingly, other studies reveal that soluble forms of Ephrin-A receptors (EphA2-Fc and EphA3-Fc) function to block receptor activation in response to Ephrin stimulation, preventing blood vessel recruitment (Brantley et al. 2002). Depriving tumours of essential oxygen, nutrient and growth factors by inhibiting endothelial cell migration, it prevents angiogenesis and tumour progression *in vivo* (Brantley et al. 2002). However, due to the lack of specificity of Eph-A receptors, it was not explicitly specified in this research if the expression of EphA3 was directly involved in the blood vessels of both normal and tumour tissues (Brantley et al. 2002). In fact, no prior clinical investigations were able to verify the presence of EphA3 within endothelial cells of tumour vasculature until 2009 (Vail et al. 2014).

Elevated EphA3 expression in B- and all T-cell lymphoid tumour cell lines and its absence in corresponding normal cells enable it to act as a potential therapeutic target for cancer vaccines (Keane et al. 2012; Guan et al. 2011). Theories have been presented regarding the essential importance of EphA3 in inflammation as well as modulating chemotaxis responses in T cells, and is consequently involved in the progression of lymphoid malignancy (Guan et al. 2011). A clinical case study analysing LK63 cells discovered that cellular repulsion occurred by preventing EphA3 phosphorylation inhibiting signalling mechanisms which result in maintaining Eph/Ephrin bonds. Further studies by Guan et al. (2011) revealed that EphA3 was responsible for copy number variation (CNV) in the genome of leukaemia patients, which is associated with many haematological malignancies. This was confirmed by Ashton et al. (2012) who investigated EphA3 knockdown in leukemic stem cells of the bone marrow. They identified EphA3 as a cooperative response gene (CRG), which regulates leukaemia stem cell survival and growth.

Growing recognition of the presence of EphA3 on newly forming solid tumours and its involvement in various forms of leukaemia, Kalabios commenced therapeutic intervention associated with KB004, a monoclonal antibody of EphA3, as a potential target for cancer treatment (Keane et al. 2012; Lackmann et al. 2014). Previously known as EphA3-activating mAb IIIA4, KB004 has recently been ongoing phase I/II clinical trials in leukaemia and haematological cancers (Vail et al. 2014; Keane et al. 2012; Lackmann et al. 2014). KB004 binds and activates EphA3, initiating events in a similar manner to ephrin-A5, suppressing tumour survival by stimulating apoptosis as well as disrupting the cellular microenvironment (Keane et al. 2012).

References

Ashton JM, Balys M, Neering SJ, Hassane DC, Cowley G, Root DE, Miller PG, Ebert BL, McMurray HR, Land H (2012) Oncogene cooperativity analysis reveals a novel set of genes that regulate the in vivo growth and survival of leukemia stem cells. Cell Stem Cell 11(3):359

Beauchamp A, Debinski W (2012) Ephs and ephrins in cancer: ephrin-A1 signalling. Semin Cell Dev Biol 23(1):109–115. doi:10.1016/j.semcdb.2011.10.019

Biao-xue R, Xi-guang C, Shuan-ying Y, Wei L, Zong-juan M (2011) EphA2-dependent molecular targeting therapy for malignant tumors. Curr Cancer Drug Targets 11(9):1082–1097

Brantley DM, Cheng N, Thompson EJ, Lin Q, Brekken RA, Thorpe PE, Muraoka RS, Cerretti DP, Pozzi A, Jackson D, Lin C, Chen J (2002) Soluble Eph A receptors inhibit tumor angiogenesis and progression in vivo. Oncogene 21(46):7011–7026. doi:10.1038/sj.onc.1205679

Brantley-Sieders D, Schmidt S, Parker M, Chen J (2004) Eph receptor tyrosine kinases in tumor and tumor microenvironment. Curr Pharm Des 10(27):3431–3442

Cheng N, Brantley DM, Chen J (2002) The ephrins and Eph receptors in angiogenesis. Cytokine Growth Factor Rev 13(1):75–85

Day BW, Stringer BW, Al-Ejeh F, Ting MJ, Wilson J, Ensbey KS, Jamieson PR, Bruce ZC, Lim YC, Offenhauser C, Charmsaz S, Cooper LT, Ellacott JK, Harding A, Leveque L, Inglis P, Allan S, Walker DG, Lackmann M, Osborne G, Khanna KK, Reynolds BA, Lickliter JD, Boyd AW (2013) EphA3 maintains tumorigenicity and is a therapeutic target in glioblastoma multiforme. Cancer Cell 23(2):238–248. doi:10.1016/j.ccr.2013.01.007

Ferluga S, Hantgan R, Goldgur Y, Himanen JP, Nikolov DB, Debinski W (2013) Biological and structural characterization of glycosylation on ephrin-A1, a preferred ligand for EphA2 receptor tyrosine kinase. J Biol Chem 288(25):18448–18457. doi:10.1074/jbc.M113.464008

Frieden LA, Townsend TA, Vaught DB, Delaughter DM, Hwang Y, Barnett JV, Chen J (2010) Regulation of heart valve morphogenesis by Eph receptor ligand, ephrin-A1. Dev Dyn 239(12):3226–3234. doi:10.1002/dvdy.22458

Giaginis C, Tsoukalas N, Bournakis E, Alexandrou P, Kavantzas N, Patsouris E, Theocharis S (2014) Ephrin (Eph) receptor A1, A4, A5 and A7 expression in human non-small cell lung carcinoma: associations with clinicopathological parameters, tumor proliferative capacity and patients' survival. BMC Clin Pathol 14(1):8. doi:10.1186/1472-6890-14-8

Guan M, Liu L, Zhao X, Wu Q, Yu B, Shao Y, Yang H, Fu X, Wan J, Zhang W (2011) Copy number variations of EphA3 are associated with multiple types of hematologic malignancies. Clin Lymphoma Myeloma Leuk 11(1):50–53. doi:10.3816/CLML.2011.n.006

Héroult M, Schaffner F, Augustin HG (2006) Eph receptor and ephrin ligand-mediated interactions during angiogenesis and tumor progression. Exp Cell Res 312(5):642–650. doi:http://dx.doi.org/10.1016/j.yexcr.2005.10.028

Hjorthaug HS, Aasheim HC (2007) Ephrin-A1 stimulates migration of CD8+CCR7+ T lymphocytes. Eur J Immunol 37(8):2326–2336. doi:10.1002/eji.200737111

Jing X, Miwa H, Sawada T, Nakanishi I, Kondo T, Miyajima M, Sakaguchi K (2012) Ephrin-A1-mediated dopaminergic neurogenesis and angiogenesis in a rat model of Parkinson's disease. PLoS One 7(2):e32019. doi:10.1371/journal.pone.0032019

Kandouz M (2012) The Eph/Ephrin family in cancer metastasis: communication at the service of invasion. Cancer Metastasis Rev 31(1–2):353–373. doi:10.1007/s10555-012-9352-1

Keane N, Freeman C, Swords R, Giles FJ (2012) EPHA3 as a novel therapeutic target in the hematological malignancies. Expert Rev Hematol 5(3):325–340. doi:10.1586/ehm.12.19

Lackmann M, Scott AM, Bebbington CR, Yarranton GT, Murone C, To C (2014) EphA3 antibodies for the treatment of solid tumors. Google Patents

Landen CN Jr, Lu C, Han LY, Coffman KT, Bruckheimer E, Halder J, Mangala LS, Merritt WM, Lin YG, Gao C, Schmandt R, Kamat AA, Li Y, Thaker P, Gershenson DM, Parikh NU, Gallick GE, Kinch MS, Sood AK (2006) Efficacy and antivascular effects of EphA2 reduction with an agonistic antibody in ovarian cancer. J Natl Cancer Inst 98(21):1558–1570. doi:10.1093/jnci/djj414

Lennon FE, Mirzapoiazova T, Mambetsariev N, Mambetsariev B, Salgia R, Singleton PA (2014) Transactivation of the receptor-tyrosine kinase ephrin receptor A2 is required for the low molecular weight hyaluronan-mediated angiogenesis that is implicated in tumor progression. J Biol Chem 289(35):24043–24058. doi:10.1074/jbc.M114.554766

Merlos-Suárez A, Batlle E (2008) Eph–ephrin signalling in adult tissues and cancer. Curr Opin Cell Biol 20(2):194–200. doi:http://dx.doi.org/10.1016/j.ceb.2008.01.011

Miao H, Wang B (2009) Eph/ephrin signaling in epithelial development and homeostasis. Int J Biochem Cell Biol 41(4):762–770. doi:10.1016/j.biocel.2008.07.019

Miao H, Wei BR, Peehl DM, Li Q, Alexandrou T, Schelling JR, Rhim JS, Sedor JR, Burnett E, Wang B (2001) Activation of EphA receptor tyrosine kinase inhibits the Ras/MAPK pathway. Nat Cell Biol 3(5):527–530. doi:10.1038/35074604

Ogawa K, Pasqualini R, Lindberg RA, Kain R, Freeman AL, Pasquale EB (2000) The ephrin-A1 ligand and its receptor, EphA2, are expressed during tumor neovascularization. Oncogene 19(52):6043–6052. doi:10.1038/sj.onc.1204004

Ortalli AL, Fiore L, Di Napoli J, Rapacioli M, Salierno M, Etchenique R, Flores V, Sanchez V, Carri NG, Scicolone G (2012) EphA3 expressed in the chicken tectum stimulates nasal retinal ganglion cell axon growth and is required for retinotectal topographic map formation. PLoS One 7(6):e38566. doi:10.1371/journal.pone.0038566

Pasquale EB (1997) The Eph family of receptors. Curr Opin Cell Biol 9(5):608–615

Pasquale EB (2008) Eph-ephrin bidirectional signaling in physiology and disease. Cell 133(1):38–52. doi:10.1016/j.cell.2008.03.011

Pasquale EB (2010) Eph receptors and ephrins in cancer: bidirectional signalling and beyond. Nat Rev Cancer 10(3):165–180

Saik JE, Gould DJ, Keswani AH, Dickinson ME, West JL (2011) Biomimetic hydrogels with immobilized ephrinA1 for therapeutic angiogenesis. Biomacromolecules 12(7):2715–2722. doi:10.1021/bm200492h

Stephen LJ, Fawkes AL, Verhoeve A, Lemke G, Brown A (2007) A critical role for the EphA3 receptor tyrosine kinase in heart development. Dev Biol 302(1):66–79. doi:10.1016/j.ydbio.2006.08.058

Tandon M, Vemula SV, Mittal SK (2011) Emerging strategies for EphA2 receptor targeting for cancer therapeutics. Expert Opin Ther Targets 15(1):31–51. doi:10.1517/14728222.2011.538682

Tawadros T, Brown MD, Hart CA, Clarke NW (2012) Ligand-independent activation of EphA2 by arachidonic acid induces metastasis-like behaviour in prostate cancer cells. Br J Cancer 107(10):1737–1744. doi:10.1038/bjc.2012.457

Vail ME, Murone C, Tan A, Hii L, Abebe D, Janes PW, Lee FT, Baer M, Palath V, Bebbington C, Yarranton G, Llerena C, Garic S, Abramson D, Cartwright G, Scott AM, Lackmann M (2014)

Targeting EphA3 inhibits cancer growth by disrupting the tumor stromal microenvironment. Cancer Res 74(16):4470–4481. doi:10.1158/0008-5472.can-14-0218

Walker-Daniels J, Hess AR, Hendrix MJ, Kinch MS (2003) Differential regulation of EphA2 in normal and malignant cells. Am J Pathol 162(4):1037–1042. doi:10.1016/s0002-9440(10)63899-0

Wykosky J, Debinski W (2008) The EphA2 receptor and ephrinA1 ligand in solid tumors: function and therapeutic targeting. Mol Cancer Res 6(12):1795–1806. doi:10.1158/1541-7786.mcr-08-0244

Xi HQ, Wu XS, Wei B, Chen L (2012a) Aberrant expression of EphA3 in gastric carcinoma: correlation with tumor angiogenesis and survival. J Gastroenterol 47(7):785–794. doi:10.1007/s00535-012-0549-4

Xi HQ, Wu XS, Wei B, Chen L (2012b) Eph receptors and ephrins as targets for cancer therapy. J Cell Mol Med 16(12):2894–2909. doi:10.1111/j.1582-4934.2012.01612.x

Chapter 15
Erythropoietin-Producing Hepatocellular Receptors B: Ephrin B2, Ephrin B4

Abstract The ephrin receptor family and their ligands, the ephrins, are vital for numerous developmental events and the maintenance of organism function. Specifically, the ephrin family mediates embryonic growth processes such as axon guidance, boundary formation and cell migration. While during adulthood, the key targets comprise nervous system function including potentiation, central and peripheral stem cell differentiation and angiogenesis. Ephrin-B2 is a transmembrane, receptor tyrosine kinase that is transcribed by the EFNB2 gene in humans. Binding of Ephrin-B2 is important for cellular adhesion, migration and repulsion for adequate development of the vasculature, epithelium and neural system. Eph-B4 is a protein encoded by the EphB4 gene in humans and is implicated in blood vessel development, morphogenesis, remodelling and permeability. Ephrin-B2 and EphB4 binding are one of the more heavily studied interactions with regards to angiogenesis.

Keywords Erythropoietin-producing hepatocellular receptors B • Ephrin B2 • Ephrin B4 • Angiogenesis • Normal physiology • Disease • Malignancy

15.1 Erythropoietin-Producing Hepatocellular Receptors B

The ephrin receptor family and their ligands, the ephrins, are vital for numerous developmental events and the maintenance of organism function (Himanen et al. 2007). Specifically, the ephrin family mediates embryonic growth processes such as axon guidance, boundary formation and cell migration. While during adulthood, the key targets comprise nervous system function including potentiation, central and peripheral stem cell differentiation and angiogenesis (Genander and Frisen 2010).

The ephrin receptors are the largest known subclass of the tyrosine kinase receptor family. Based primarily on their sequence homology and affinity for specific ligands, the ephrin receptors are subdivided into Eph-A and Eph-B class receptors. The Eph-A class are affixed to the membrane (via glycosylphosphatidylinositol), while the Eph-B class, including EphB2 and B4 are transmembrane proteins (Himanen et al. 2007).

A. Salajegheh, *Angiogenesis in Health, Disease and Malignancy*,
DOI 10.1007/978-3-319-28140-7_15

15.2 Ephrin B2

Ephrin-B2 is a transmembrane, receptor tyrosine kinase that is transcribed by the EFNB2 gene in humans. Ephrin-B2 binds non-selectively to different types of Eph receptors on neighbouring cells (Pasquale 2005). Ephrin ligands were identified in the late 1980s and have been found to be important in different physiological and pathophysiological processes such as angiogenesis, embryonic development and tumorigenesis. More specifically binding of Ephrin-B2 is important for cellular adhesion, migration and repulsion for adequate development of the vasculature, epithelium and neural system (Abengozar et al. 2012).

15.2.1 Ephrin-B2 in Normal Physiology

While it is known that ephrin-B2 plays an important role in early angiogenic remodelling and lymph vessel formation, exactly how ephrin-B2 exerts its effects during the developmental stages remains unclear (Abengozar et al. 2012; Germain and Eichmann 2010).

Similarly to Ephrin A1, when Ephrin B2 binds to an Eph Receptor signalling will occur in two directions; forward and reverse signalling (Carmeliet and Jain 2011). A number of studies are looking into the promiscuous associations between Ephrin-B2 and EPHB4, EPHA4 and EPHA3 receptors respectively (Abengozar et al. 2012; Qin et al. 2010; Ibrahim and Hess 2012).

Ephrin-B2 and EphB4 binding are one of the more heavily studied interactions with regards to angiogenesis. One mechanism thought to occur during vasculogenesis is ephrin-B2 and EphB4 of the vascular plexus which prevents cells from intermingling. A study using zebrafish found that Ephrin-B2 and EphB4 are responsible for directing the migration of progenitor cells from precursor vessels. This is also important in segregating arterial and venous segregation (Herbert et al. 2009). A study in 2010 found that ephrin-B2 promotes VEGFR-2 and VEGFR-3 endocytosis. In the study ephrin-B2 was surgically removed from mouse cells which resulted in angiogenic and lymphangiogenic defects. The authors found that the defects were because downstream signalling pathways involving RAC1, AKT and ERK ½ were compromised (Wang et al. 2010; Gaengel and Betsholtz 2013). Another study found that Ephrin-B2 and EPHB4 binding is crucial to heart morphogenesis and angiogenesis. It was found that the binding interaction is a key regulator for cellular adhesion and migration in these developmental pathways. In 2012 Nakayama et al. confirmed the established findings that Ephrin-B2 regulates VEGFR-mediated signalling. Further to this the team performed an extensive analysis on VEGFR endocytosis regulation by isolating interactions of ephrin-B2. One protein that was identified was a clathrin-associated sorting protein, Disabled 2 (DAB2), which is important for cargo selection during endocytosis. PAR3, the polarity protein facilitated an indirect association between DAB2 and ephrin-B2. Both VEGFR2 and VEGFR3 are known to interact with DAB2 and PAR3 so it is possible that VEGFR endocytosis is regulated through the interactions of ephrin-B2–PAR3–DAB2 (Nakayama et al. 2013).

15.2.2 Ephrin-B2 in Disease

MicroRNA (miRNA), a class of noncoding 21- to 25- nucleotide RNA, can negatively regulate gene expression post-transcriptionally. If these miRNA's are upregulated during pregnancy then the placental Ephrin-B2 is significantly down-regulated. A study conducted by Wang et al. in 2012 found that this extensive down-regulation can contribute to the pathogenesis of preeclampsia (Wang et al. 2012).

A 2005 study by Bonaparte et al. discovered that Ephrin-B2 ligand can serve as a functional receptor for the Hendra virus and the Nipah virus. These two viruses cause fatal disease in both animals and humans and are known to infect cells through membrane fusion that is mediated by attachment to glycoproteins. One human cell line that was used in the study and also encoded human ephrin-B2 (EFNB2) was seen render the cells permissive for cell fusion as well as infection by live virus (Bonaparte et al. 2005).

While diabetes can induce stem/progenitor cell dysfunction, a study in 2012 has found that treating peripheral blood mononuclear cells (PB-MNCs) with Ephrin-B2 can reverse the process. When the diabetic mice (with hindlimb ischaemia) were injected with Ephrin-B2, the Ephrin bound to the PB-MNCs and increased cell adhesion and transmigration of PB-MNCs. This in turn raised the number of circulating vascular progenitor cells in the diabetic mice and increased the ability of endogenous bone marrow MNCs to differentiate into cells with an endothelial phenotype and enhance their proangiogenic potential (Broqueres-You et al. 2012). This finding opens the possibility for clinical development of an innovative and accessible strategy in diabetic patients with critical ischaemic diseases.

15.2.3 Ephrin-B2 in Malignancy

The EphB4/ephrin-B2 interaction has been reported to have an important role in tumor angiogenesis and growth. Kimura et al. (2009) study found that soluble ephrin-B2-Fc (a fusion protein that is made up of the extracellular domain of ephrin-B2 and the Fc portion of human IgG1) molecules were able to induce maturation of vessels in the tumour and this in turn suppressed the growth of head and neck squamous cell carcinoma xenografts (Kimura et al. 2009).

Another key player in tumour angiogenesis is the reverse signalling that occurs through ephrins. When EphB4 receptors (lacking the kinase domain and tagged with EGFP) are expressed in breast cancer cells of a mouse xenograft model increased tumour angiogenesis is reported (Noren et al. 2004). It is possible to suggest this is because the EphB4 binds to the ephrin-B2 ligand which is expressed in the vasculature and stimulates angiogenesis through ephrin-B2 reverse signalling.

15.3 Ephrin-B4

The action of ephrin receptor family and their ligands, the ephrins, are particularly significant in angiogenesis, with ephrin receptor B4 (Eph-B4) known to be essential. Eph-B4 is a protein encoded by the EphB4 gene in humans and is implicated in blood vessel development, morphogenesis, remodelling and permeability (Bai et al. 2014).

The ephrin receptors are the largest known subclass of the tyrosine kinase receptor family. Based primarily on their sequence homology and affinity for specific ligands, the ephrin receptors are subdivided into Eph-A and Eph-B class receptors. The Eph-A class are affixed to the membrane (via glycosylphosphatidylinositol), while the Eph-B class, including Eph-B4 are transmembrane proteins (Himanen et al. 2007).

15.3.1 Ephrin-B4 in Normal Physiology

Eph-B4 with respect to the ephrin family displays specific, but overlapping patterns of expression, suggesting both distinctive and redundant actions. Its expression is typically higher during embryonic growth processes compared to adulthood. In human embryonic tissues, Eph-B4 is found in high abundance, particularly in the heart, lung, kidney, thymus, gut and trapezius muscle (Zhou 1998). While in human adult tissues, Eph-B4 is again found in relatively high concentration, with the chief organs comprising the liver, lung, kidney, intestine, placenta, muscle and heart (Zhou 1998). Notably, its expression in the nervous system (especially the brain), during both embryonic growth processes and adulthood, is relatively low, contrasting most ephrin receptors.

The signalling pathway replicates the typical tyrosine kinase receptor model. Since both Eph-B4 and its ligands are membrane-associated molecules, the binding and activation processes to initiate signalling occur only through direct cell to cell contact (Chen et al. 2008). This interaction results in a high affinity linkage to the extracellular, globular domain, which then leads to activation of the intracellular, cytoplasmic domain. Here, phosphorylation of the tyrosine residues in the key juxtamembrane region eventuates (Lisabeth et al. 2013). This permits the up-regulation of tyrosine kinase catalytic activity and either propagation or repression of downstream signal transduction cascades.

Furthermore, Eph-B4 and the other ephrin receptors display uniqueness in their bi-directional signalling ability, unlike most of tyrosine kinase receptor family.

Following cell to cell contact, both the receptor-expressing cell and the opposing ephrin-expressing cell (forward and reverse signalling respectively), have the capability to instigate signalling cascades (Marquardt et al. 2005). Ephrin-B2 is a known ligand of Eph-B4, where Ephrin-B2 is linked to reverse signalling and Eph-B4 forward signalling. The functional significance of this mechanism is yet to be completely understood although its importance in angiogenesis has been substantiated (Martiny-Baron et al. 2004).

The formation of early vasculature occurs via vasculogenesis and angiogenesis. Vasculogenesis describes the initial in situ transformation of endothelial cells from mesodermal precursors, the angioblasts. This result in the formation of key, primitive blood vessels principally related to the heart. Angiogenesis then follows with the formation of new blood vessels from this pre-existing vasculature established through vasculogenesis (Helbling et al. 2000). Comprehensive gene knockout experiments have confirmed the significance of Eph-B4 in this early vascular development (Fuller et al. 2003). Eph-B4 functions as a marker for correct arterio-venous formation. When Eph-B4 is deficient, the physiological response is ultimately expressed as premature embryonic lethality with disrupted arterio-venous remodelling and improper maturation (Fuller et al. 2003). Equally, targeted gene knockout of Ephrin-B2 displays a similar phenotype to Eph-B4 knockout (Krasnoperov et al. 2010). Although, it is thought that Eph-B4 forward signalling is more relevant than Ephrin-B2 reverse signalling in relation to early angiogenesis (Davies et al. 2010).

Eph-B4 and Ephrin-B2 are not only essential in proper blood vessel development and artery-vein demarcation, but controlling morphogenesis, remodelling, permeability and establishing boundary formation between arterial and venous domains (Genander and Frisen 2010). In terms of their expression, Eph-B4 is limited to venous endothelial cells and Ephrin-B2 to arterial endothelial cells. This pattern of expression is important as it is a primary determinant in the correct balance between the principal afferent-efferent vessels and consequent maturation of their specific capillary plexus (Fuller et al. 2003; Hamada et al. 2003).

15.3.2 Eph-B4 in Disease and Malignancy

The ephrin receptors and their ligands are typically up-regulated in tumours and are associated with their growth, invasiveness and metastasis. Examples of such tumours include melanoma, neuroblastoma, malignant glioma, carcinoma of the pancreas, breast, colon, prostate, lung, gastroinstential tract, ovaries, oesophagus, liver and thyroid (Xi et al. 2012). However, decreased levels of expression in some tumours have also been reported (Xi et al. 2012).

This growth, invasiveness and metastasis of the cancer are reliant on its ability to recruit new blood vessels and transform the simple, existing vasculature into a complex, rapidly evolving network. This functions as a source of oxygen, nutrients and various factors that are essential for its survival (Cheng et al. 2002). Using breast cancer as a specific example, Eph-B4 has been proven to offer a survival advantage.

It decreases the effect of intrinsic cell death mechanisms and enhances the expression of anti-apoptotic factors (Xi et al. 2012). Furthermore, gene knockout experiments (that disrupt Eph-B4) have shown that breast cancer viability, including migration and invasiveness in relation to angiogenesis is reduced (Kumar et al. 2006).

It is thought that Ephrin-B2 specifically supports angiogenesis in Eph-B4-positive cancers. Their expression on tumour vascular endothelial cells is considered essential for sprouting, proliferation and capillary plexus formation (Krasnoperov et al. 2010). The exact mechanism is unknown, however it is thought these molecules may act as contact-dependent organising factors to direct incoming blood vessels and to ultimately promote cancerous growth (Xi et al. 2012). Furthermore, Eph-B4 atypically found on tumour cells themselves can communicate with Ephrin-B2 endothelial cells, further supporting tumour angiogenesis (Mosch et al. 2010).

Tumour resistance to VEGF inhibitors can be attributed, at least partially, to the activation of the EphB4/Ephrin-B2 pathway (Welti et al. 2013; Mosch et al. 2010). Upon binding, Ephrin-B2 will generate a reverse signal to decrease vessel number and increase lumen diameter, and this will assist tumour progression. Ephrin-B2 is a downstream target of DLL4-Notch signalling and VEGF can effectively induce Ephrin-B2 but inhibits EphB4 (Scehnet et al. 2009). However, in DLL4 tumours VEGF levels are reduced because of the reduction in hypoxic conditions (Williams et al. 2006). So the findings of Li et al. in 2011 showed that when tumour cells are engineered to overexpress DLL4 will develop large mature vessels, resistant to VEGF blockade. In addition to this, it was found that if Notch signalling could be inhibited, then sensitivity to antiangiogenic drugs (in a xenograft model) could be restored (Li et al. 2011).

Ultimately, Eph-B4 and Ephrin-B2 do not function in complete isolation but rather interact with other angiogenic factors to potentiate effects. For example, other ephrin receptors may interact with their ligands on adjacent endothelial cells to enhance angiogenesis through distinct mechanisms, possibly contributing to synergistic action (Cheng et al. 2002). These molecules have the potential to become potent targets for therapeutic treatment and function as markers for clinical evaluation of tumour prognosis. However, the complexity of their pathology complicates strategies that target Eph-B4 and other members of the ephrin family but ongoing research continues.

References

Abengozar MA, de Frutos S, Ferreiro S, Soriano J, Perez-Martinez M, Olmeda D, Marenchino M, Canamero M, Ortega S, Megias D, Rodriguez A, Martinez-Torrecuadrada JL (2012) Blocking ephrinB2 with highly specific antibodies inhibits angiogenesis, lymphangiogenesis, and tumor growth. Blood 119(19):4565–4576. doi:10.1182/blood-2011-09-380006

Bai J, Wang YJ, Liu L, Zhao YL (2014) Ephrin B2 and EphB4 selectively mark arterial and venous vessels in cerebral arteriovenous malformation. J Int Med Res 42(2):405–415. doi:10.1177/0300060513478091

Bonaparte MI, Dimitrov AS, Bossart KN, Crameri G, Mungall BA, Bishop KA, Choudhry V, Dimitrov DS, Wang LF, Eaton BT, Broder CC (2005) Ephrin-B2 ligand is a functional receptor for Hendra virus and Nipah virus. Proc Natl Acad Sci U S A 102(30):10652–10657. doi:10.1073/pnas.0504887102

Broqueres-You D, Lere-Dean C, Merkulova-Rainon T, Mantsounga CS, Allanic D, Hainaud P, Contreres JO, Wang Y, Vilar J, Virally M, Mourad JJ, Guillausseau PJ, Silvestre JS, Levy BI (2012) Ephrin-B2-activated peripheral blood mononuclear cells from diabetic patients restore diabetes-induced impairment of postischemic neovascularization. Diabetes 61(10):2621–2632. doi:10.2337/db11-1768

Carmeliet P, Jain RK (2011) Molecular mechanisms and clinical applications of angiogenesis. Nature 473(7347):298–307. doi:10.1038/nature10144

Chen Y, Fu AK, Ip NY (2008) Bidirectional signaling of ErbB and Eph receptors at synapses. Neuron Glia Biol 4(3):211–221. doi:10.1017/s1740925x09990287

Cheng N, Brantley DM, Chen J (2002) The ephrins and Eph receptors in angiogenesis. Cytokine Growth Factor Rev 13(1):75–85

Davies MH, Stempel AJ, Hubert KE, Powers MR (2010) Altered vascular expression of EphrinB2 and EphB4 in a model of oxygen-induced retinopathy. Dev Dyn 239(6):1695–1707. doi:10.1002/dvdy.22306

Fuller T, Korff T, Kilian A, Dandekar G, Augustin HG (2003) Forward EphB4 signaling in endothelial cells controls cellular repulsion and segregation from ephrinB2 positive cells. J Cell Sci 116(Pt 12):2461–2470. doi:10.1242/jcs.00426

Gaengel K, Betsholtz C (2013) Endocytosis regulates VEGF signalling during angiogenesis. Nat Cell Biol 15(3):233–235. doi:10.1038/ncb2705

Genander M, Frisen J (2010) Ephrins and Eph receptors in stem cells and cancer. Curr Opin Cell Biol 22(5):611–616. doi:10.1016/j.ceb.2010.08.005

Germain S, Eichmann A (2010) VEGF and ephrin-B2: a bloody duo. Nat Med 16(7):752–754. doi:10.1038/nm0710-752

Hamada K, Oike Y, Ito Y, Maekawa H, Miyata K, Shimomura T, Suda T (2003) Distinct roles of ephrin-B2 forward and EphB4 reverse signaling in endothelial cells. Arterioscler Thromb Vasc Biol 23(2):190–197

Helbling PM, Saulnier DM, Brandli AW (2000) The receptor tyrosine kinase EphB4 and ephrin-B ligands restrict angiogenic growth of embryonic veins in Xenopus laevis. Development 127(2):269–278

Herbert SP, Huisken J, Kim TN, Feldman ME, Houseman BT, Wang RA, Shokat KM, Stainier DY (2009) Arterial-venous segregation by selective cell sprouting: an alternative mode of blood vessel formation. Science 326(5950):294–298. doi:10.1126/science.1178577

Himanen JP, Saha N, Nikolov DB (2007) Cell-cell signaling via Eph receptors and ephrins. Curr Opin Cell Biol 19(5):534–542. doi:10.1016/j.ceb.2007.08.004

Ibrahim G, Hess AR (2012) Differential expression of EphB4 and ephrin-B2 in highly aggressive versus poorly aggressive human melanoma tumor cells. BIOS 83(4):127–135. doi:10.1893/0005-3155-83.4.127

Kimura M, Kato Y, Sano D, Fujita K, Sakakibara A, Kondo N, Mikami Y, Tsukuda M (2009) Soluble form of ephrinB2 inhibits xenograft growth of squamous cell carcinoma of the head and neck. Int J Oncol 34(2):321–327

Krasnoperov V, Kumar SR, Ley E, Li X, Scehnet J, Liu R, Zozulya S, Gill PS (2010) Novel EphB4 monoclonal antibodies modulate angiogenesis and inhibit tumor growth. Am J Pathol 176(4):2029–2038. doi:10.2353/ajpath.2010.090755

Kumar SR, Singh J, Xia G, Krasnoperov V, Hassanieh L, Ley EJ, Scehnet J, Kumar NG, Hawes D, Press MF, Weaver FA, Gill PS (2006) Receptor tyrosine kinase EphB4 is a survival factor in breast cancer. Am J Pathol 169(1):279–293. doi:10.2353/ajpath.2006.050889

Li JL, Sainson RC, Oon CE, Turley H, Leek R, Sheldon H, Bridges E, Shi W, Snell C, Bowden ET, Wu H, Chowdhury PS, Russell AJ, Montgomery CP, Poulsom R, Harris AL (2011) DLL4-Notch signaling mediates tumor resistance to anti-VEGF therapy in vivo. Cancer Res 71(18):6073–6083. doi:10.1158/0008-5472.can-11-1704

Lisabeth EM, Falivelli G, Pasquale EB (2013) Eph receptor signaling and ephrins. Cold Spring Harb Perspect Biol 5(9):a009159. doi:10.1101/cshperspect.a009159

Marquardt T, Shirasaki R, Ghosh S, Andrews SE, Carter N, Hunter T, Pfaff SL (2005) Coexpressed EphA receptors and ephrin-A ligands mediate opposing actions on growth cone navigation from distinct membrane domains. Cell 121(1):127–139. doi:10.1016/j.cell.2005.01.020

Martiny-Baron G, Korff T, Schaffner F, Esser N, Eggstein S, Marme D, Augustin HG (2004) Inhibition of tumor growth and angiogenesis by soluble EphB4. Neoplasia 6(3):248–257. doi:10.1593/neo.3457

Mosch B, Reissenweber B, Neuber C, Pietzsch J (2010) Eph receptors and ephrin ligands: important players in angiogenesis and tumor angiogenesis. J Oncol 2010:135285. doi:10.1155/2010/135285

Nakayama M, Nakayama A, van Lessen M, Yamamoto H, Hoffmann S, Drexler HC, Itoh N, Hirose T, Breier G, Vestweber D, Cooper JA, Ohno S, Kaibuchi K, Adams RH (2013) Spatial regulation of VEGF receptor endocytosis in angiogenesis. Nat Cell Biol 15(3):249–260. doi:10.1038/ncb2679

Noren NK, Lu M, Freeman AL, Koolpe M, Pasquale EB (2004) Interplay between EphB4 on tumor cells and vascular ephrin-B2 regulates tumor growth. Proc Natl Acad Sci U S A 101(15):5583–5588. doi:10.1073/pnas.0401381101

Pasquale EB (2005) Eph receptor signalling casts a wide net on cell behaviour. Nat Rev Mol Cell Biol 6(6):462–475. doi:10.1038/nrm1662

Qin H, Noberini R, Huan X, Shi J, Pasquale EB, Song J (2010) Structural characterization of the EphA4-Ephrin-B2 complex reveals new features enabling Eph-ephrin binding promiscuity. J Biol Chem 285(1):644–654. doi:10.1074/jbc.M109.064824

Scehnet JS, Ley EJ, Krasnoperov V, Liu R, Manchanda PK, Sjoberg E, Kostecke AP, Gupta S, Kumar SR, Gill PS (2009) The role of Ephs, Ephrins, and growth factors in Kaposi sarcoma and implications of EphrinB2 blockade. Blood 113(1):254–263. doi:10.1182/blood-2008-02-140020

Wang Y, Nakayama M, Pitulescu ME, Schmidt TS, Bochenek ML, Sakakibara A, Adams S, Davy A, Deutsch U, Luthi U, Barberis A, Benjamin LE, Makinen T, Nobes CD, Adams RH (2010) Ephrin-B2 controls VEGF-induced angiogenesis and lymphangiogenesis. Nature 465(7297):483–486. doi:10.1038/nature09002

Wang W, Feng L, Zhang H, Hachy S, Satohisa S, Laurent LC, Parast M, Zheng J, Chen DB (2012) Preeclampsia up-regulates angiogenesis-associated microRNA (i.e., miR-17, -20a, and -20b) that target ephrin-B2 and EPHB4 in human placenta. J Clin Endocrinol Metab 97(6):E1051–E1059. doi:10.1210/jc.2011-3131

Welti J, Loges S, Dimmeler S, Carmeliet P (2013) Recent molecular discoveries in angiogenesis and antiangiogenic therapies in cancer. J Clin Invest 123(8):3190–3200. doi:10.1172/jci70212

Williams CK, Li JL, Murga M, Harris AL, Tosato G (2006) Up-regulation of the Notch ligand Delta-like 4 inhibits VEGF-induced endothelial cell function. Blood 107(3):931–939. doi:10.1182/blood-2005-03-1000

Xi HQ, Wu XS, Wei B, Chen L (2012) Eph receptors and ephrins as targets for cancer therapy. J Cell Mol Med 16(12):2894–2909. doi:10.1111/j.1582-4934.2012.01612.x

Zhou R (1998) The Eph family receptors and ligands. Pharmacol Ther 77(3):151–181

Chapter 16
Ets-1

Abstract Initially discovered in, and named after, an avian erthyroblastos virus (E26), the Ets-1 is one of most studied genes in the ETS family and its involvement in a range of physiological and pathological processes has been studied. Ets-1 is involved in a wide range of biological functions. Transcription factors are responsible for the regulation of specific genes and thus have vital roles in controlling cell function. The ETS family of transcription factors in particular, are key regulators of the endothelial expression of genes and angiogenesis and have been found to be involved in other processes such as development, proliferation and apoptosis. Ets-1 factors function both upstream and downstream of various genes which promote angiogenesis. Genes promoting angiogenesis such as the ones coding for hepatocyte growth factor (HGF) and vascular endothelial growth factor (VEGF) are regulated by Ets-1 factors and the overexpression of Ets-1 has been linked to the increased expression of these genes as well. In addition to this, Ets-1 is involved in regulating the invasiveness, metastatic potential and progression of tumours by regulating genes which influence cancer cell behaviour. Potential cancer treatment strategies using gene silencing therapies targeting the Ets-1 gene found that the treatment also resulted in downregulation of matrix metalloproteinase-1 and urokinase plasminogen activators indicating Ets-1 gene silencing as a potential strategy in some cancer treatment.

Keywords Ets-1 • Angiogenesis • Normal physiology • Disease • Malignancy

16.1 Ets-1

Ets-1 codes for c-Ets1, which is one of the best characterised members of the ETS family of transcription factors and is involved in a wide range of biological functions (Randi et al. 2009). Transcription factors are responsible for the regulation of specific genes and thus have vital roles in controlling cell function (Russell and Garrett-Sinha 2010). The ETS family of transcription factors in particular, are key regulators of the endothelial expression of genes and angiogenesis (Randi et al. 2009). In addition to this, ETS transcription factors have been found to be involved in other processes such as development, proliferation and apoptosis (Shaikhibrahim

A. Salajegheh, *Angiogenesis in Health, Disease and Malignancy*,
DOI 10.1007/978-3-319-28140-7_16

and Wernert 2012). Initially discovered in, and named after, an avian erthyroblastos virus (V-ets avian erythroblastosis virus E26 oncogene homolog 1), the Ets-1 is one of most studied genes in the ETS family and its involvement in a range of physiological and pathological processes has been studied (Russell and Garrett-Sinha 2010; Shaikhibrahim and Wernert 2012). All ETS transcription factors are known to possess a conserved DNA binding domain shared known as the Ets domain, which allows for redundancy in the ETS factors that are able to regulate certain target genes (Wei et al. 2009; Russell and Garrett-Sinha 2010). Target genes for Ets-1 transcription factors have vital roles in regulating angiogenesis in adult (Randi et al. 2009). Here, the actions of Ets-1 in physiology and cancer will be examined and potential cancer treatment strategies targeting Ets-1 will be reviewed.

16.2 Ets-1 in Normal Physiology

Ets-1 is expressed in a variety of tissues and is involved in many physiological processes (Dittmer 2003; Shaikhibrahim and Wernert 2012). Among its many roles, Ets-1 is involved in angiogenesis and its expression influences the physiology of the vascular system. The involvement of Ets-1 in angiogenesis has been demonstrated in various studies through the regulatory effect Ets-1 transcription factors have on key genes involved in mediating angiogenesis in endothelial cells (Dejana et al. 2007; Randi et al. 2009; Oettgen 2010). Importantly, Ets-1 is transiently expressed in endothelial cells and vascular smooth muscle cells after activation by angiogenic stimuli (Dittmer 2003). Hypoxia induced factors and various growth factors have also been known to stimulate Ets-1 expression (Dejana et al. 2007). In the absence of these stimuli however, Ets-1 is normally expressed at low levels (Oettgen 2010). Ets-1 factors function both upstream and downstream of various genes which promote angiogenesis. Genes promoting angiogenesis such as the ones coding for hepatocyte growth factor (HGF) and vascular endothelial growth factor (VEGF) are regulated by Ets-1 factors and the overexpression of Ets-1 has been linked to the increased expression of these genes as well. Other genes involved in angiogenesis that Ets-1 transcription factors are known to regulate include VEGF receptor, fibroblast growth factor, chemokine CCl2, urokinase and various matrix metalloproteinase (MMP) genes (Oettgen 2010). In addition to its role in normal angiogenesis, the expression of Ets-1 has also been associated in the formation of new blood vessels during chronic inflammation and the angiogenesis of tumours (Randi et al. 2009).

Other biological processes Ets-1 is involved in include embryonic development and haematopoietic development. Ets-1 is known to be expressed at sites of active blood vessel development early during embryonic development through studies with model mouse embryos systems (Dejana et al. 2007; Oettgen 2010). A study by Wei et al. (2009) in particular has found that Ets-1 is important for cell survival during angiogenesis in the embryonic development of mice. Ets-1 also has a role in the immune system as it is also expressed in lymphoid tissues and regulates the functions of immune cells (Dittmer 2003; Russell and Garrett-Sinha 2010). For instance, expression of Ets-1 has been found to occur in all stages of the development of T- and B-cells, in natural killer cells and erythroid cells during their differentiation (Dittmer 2003).

16.3 Ets-1 in Disease

In addition to its involvement in cancers, variants of the Ets-1 gene have been associated with diseases involving the immune system (Russell and Garrett-Sinha 2010). Genome wide association studies (GWAS) have shown variants of the Ets-1 gene to be associated with the autoimmune disorders systemic lupus erythematosus and celiac disease (Dubois et al. 2010; Yang et al. 2010). Ets-1 is vital to the differentiation of lymphocytes and is important in regulating immune function and consequently preventing autoimmune disease (Russell and Garrett-Sinha 2010). Variants of the Ets-1 genes may lead to decreased Ets-1 transcription factor production which may promote autoimmune disease similar to how it does in mouse models (Russell and Garrett-Sinha 2010).

16.4 Ets-1 in Malignancy

A classic example of an oncogene, inappropriate expression of Ets-1 is involved in the early progression of a variety of cancers (Holterman et al. 2010). Ets-1 is involved in angiogenesis, which is vital in tumour development and metastasis (Lefter et al. 2008). In addition to this, Ets-1 is involved in regulating the invasiveness, metastatic potential and progression of tumours by regulating genes which influence cancer cell behaviour (Wan et al. 2013). Genes with these functions include the genes for matrix metalloproteinases, Urokinase plasminogen activators and integrin receptors (Desgrosellier and Cheresh 2010; Gialeli et al. 2011; Tang and Han 2013). Matrix metalloproteinases are involved in the degredation of the extracellular matrix (ECM) which is a key even in invasion (Gialeli et al. 2011). Urokinase plasminogen activators play a role in metastasis as they are involved in the degredation and regeneration of the basement membrane and ECM (Tang and Han 2013). Integrin receptors are used by cells to regulate functions involved in the metastasis, progression and initiation of tumours (Desgrosellier and Cheresh 2010). Ets-1 expression regulates a variety of genes influencing cancer cell behaviour and has been linked to tumour angiogenesis in a wide variety of human cancers (Oettgen 2010).

Extensive research has shown Ets-1 expression to be present in a wide variety of human cancers. Frequently, poorer cancer prognoses are frequently associated with increased expression of Ets-1 in the tumour cells of cancer patients (Oettgen 2010). Ets-1 expression has been found to occur frequently in many human cancers such as: Ovarian and breast cancer, oesophageal squamous cell carcinoma, endometrial and cervical carcinoma, gastric cancer and colorectal cancer (Tsutsumi et al. 2005; Nagarajan et al. 2009; Pallai et al. 2012; Verschoor et al. 2013; Peng et al. 2014). Due to the involvement of Ets-1 in cancer, it has become a target for potential clinical cancer treatments.

Potential cancer treatment strategies which target the Ets-1 gene or its products include the use of gene silencing therapies (Lefter et al. 2008; Oettgen 2010). For example, a study by Lefter et al. (2009) attempted the use of a gene therapy approach to downregulate Ets-1 in vitro with pancreatic tumour cell lines and in vivo with nude mouse models. This was done to investigate whether the suppression of Ets-1 transcription was efficient enough to treating pancreatic adenocarcinoma. They used an adenoviral vector encoding for the DNA binding domain of Ets-1 which was suggested to compete with ETS-1 for DNA binding and consequently decrease the binding of Ets-1. Although it did not influence the proliferation rate of pancreatic tumour cell line in vitro, treatment with the adenoviral vector was found to inhibit the growth of human pancreatic adenocarcinomas grown in nude mice (Lefter et al. 2008). Additionally, Lefter et al. (2009) found that the treatment resulted in the downregulation of the expression of matrix metalloproteinase-1 and urokinase plasminogen activators. Combined, the results found by Lefter et al. (2009) indicate a potential Ets-1 gene silencing strategy for the treatment of pancreatic cancer.

References

Dejana E, Taddei A, Randi AM (2007) Foxs and Ets in the transcriptional regulation of endothelial cell differentiation and angiogenesis. Biochim Biophys Acta (BBA) Rev Cancer 1775(2):298–312. doi:http://dx.doi.org/10.1016/j.bbcan.2007.05.003

Dema B, Fehrmann RSN, Fernandez-Arquero M, Fiatal S, Grandone E, Green PM, Groen HJM, Gwilliam R, Houwen RHJ, Hunt SE, Kaukinen K, Kelleher D, Korponay-Szabo I, Kurppa K, MacMathuna P, Maki M, Mazzilli MC, McCann OT, Mearin ML, Mein CA, Mirza MM, Mistry V, Mora B, Morley KI, Mulder CJ, Murray JA, Nunez C, Oosterom E, Ophoff RA, Polanco I, Peltonen L, Platteel M, Rybak A, Salomaa V, Schweizer JJ, Sperandeo MP, Tack GJ, Turner G, Veldink JH, Verbeek WHM, Weersma RK, Wolters VM, Urcelay E, Cukrowska B, Greco L, Neuhausen SL, McManus R, Barisani D, Deloukas P, Barrett JC, Saavalainen P, Wijmenga C, van Heel DA (2010) Multiple common variants for celiac disease influencing immune gene expression. Nat Genet 42(4):295–302. doi:http://www.nature.com/ng/journal/v42/n4/suppinfo/ng.543_S1.html

Desgrosellier JS, Cheresh DA (2010) Integrins in cancer: biological implications and therapeutic opportunities. Nat Rev Cancer 10(1):9–22. doi:10.1038/nrc2748

Dittmer J (2003) The biology of the Ets1 proto-oncogene. Mol Cancer 2:29. doi:10.1186/1476-4598-2-29

Gialeli C, Theocharis AD, Karamanos NK (2011) Roles of matrix metalloproteinases in cancer progression and their pharmacological targeting. FEBS J 278(1):16–27. doi:10.1111/j.1742-4658.2010.07919.x

Holterman CE, Franovic A, Payette J, Lee S (2010) ETS-1 oncogenic activity mediated by transforming growth factor α. Cancer Res 70(2):730–740. doi:10.1158/0008-5472.can-09-2090

Lefter LP, Dima S, Sunamura M, Furukawa T, Sato Y, Abe M, Chivu M, Popescu I, Horii A (2008) Transcriptional silencing of ETS-1 efficiently suppresses angiogenesis of pancreatic cancer. Cancer Gene Ther 16(2):137–148

Lefter LP, Dima S, Sunamura M, Furukawa T, Sato Y, Abe M, Chivu M, Popescu I, Horii A (2009) Transcriptional silencing of ETS-1 efficiently suppresses angiogenesis of pancreatic cancer. Cancer Gene Ther 16(2):137–148. doi:10.1038/cgt.2008.65

Nagarajan P, Parikh N, Garrett-Sinha LA, Sinha S (2009) Ets1 induces dysplastic changes when expressed in terminally-differentiating squamous epidermal cells. PLoS One 4(1):e4179. doi:10.1371/journal.pone.0004179

Oettgen P (2010) The role of ets factors in tumor angiogenesis. J Oncol 2010:767384. doi:10.1155/2010/767384

Pallai R, Bhaskar A, Sodi V, Rice LM (2012) Ets1 and Elk1 transcription factors regulate cancerous inhibitor of protein phosphatase 2A expression in cervical and endometrial carcinoma cells. Transcription 3(6):323–335. doi:10.4161/trns.22518

Peng C, Gao H, Niu Z, Wang B, Tan Z, Niu W, Liu E, Wang J, Sun J, Shahbaz M, Agrez M, Niu J (2014) Integrin alphavbeta6 and transcriptional factor Ets-1 act as prognostic indicators in colorectal cancer. Cell Biosci 4(1):53

Randi AM, Sperone A, Dryden NH, Birdsey GM (2009) Regulation of angiogenesis by ETS transcription factors. Biochem Soc Trans 37:1248–1253. doi:10.1042/bst0371248

Russell L, Garrett-Sinha LA (2010) Transcription factor Ets-1 in cytokine and chemokine gene regulation. Cytokine 51(3):217–226. doi:http://dx.doi.org/10.1016/j.cyto.2010.03.006

Shaikhibrahim Z, Wernert N (2012) ETS transcription factors and prostate cancer: the role of the family prototype ETS-1. Int J Oncol 40(6):1748–1754. doi:10.3892/ijo.2012.1380

Tang L, Han X (2013) The urokinase plasminogen activator system in breast cancer invasion and metastasis. Biomed Pharmacother 67(2):179–182. doi:http://dx.doi.org/10.1016/j.biopha.2012.10.003

Tsutsumi S, Kuwano H, Nagashima N, Shimura T, Mochiki E, Asao T (2005) Ets-1 expression in gastric cancer. Hepato-Gastroenterology 52(62):654–656

Verschoor ML, Verschoor CP, Singh G (2013) Ets-1 global gene expression profile reveals associations with metabolism and oxidative stress in ovarian and breast cancers. Cancer Metab 1(1):17. doi:10.1186/2049-3002-1-17

Wan SM, Peng P, Guan T (2013) Ets-1 regulates its target genes mainly by DNA methylation in human ovarian cancer. J Obstet Gynaecol 33(8):877–881. doi:10.3109/01443615.2013.820268

Wei G, Srinivasan R, Cantemir-Stone CZ, Sharma SM, Santhanam R, Weinstein M, Muthusamy N, Man AK, Oshima RG, Leone G, Ostrowski MC (2009) Ets1 and Ets2 are required for endothelial cell survival during embryonic angiogenesis. Blood 114(5):1123–1130. doi:10.1182/blood-2009-03-211391

Yang W, Shen N, Ye D-Q, Liu Q, Zhang Y, Qian X-X, Hirankarn N, Ying D, Pan H-F, Mok CC, Chan TM, Wong RWS, Lee KW, Mok MY, Wong SN, Leung AMH, Li X-P, Avihingsanon Y, Wong C-M, Lee TL, Ho MHK, Lee PPW, Chang YK, Li PH, Li R-J, Zhang L, Wong WHS, Ng IOL, Lau CS, Sham PC, Lau YL, Asian Lupus Genetics C (2010) Genome-wide association study in Asian populations identifies variants in ETS1 and WDFY4 associated with systemic lupus erythematosus. PLoS Genet 6(2):e1000841. doi:10.1371/journal.pgen.1000841

Chapter 17
Fibrin

Abstract Fibrin is a fibrous, non-globular protein which is formed by the polymerisation of its precursor fibrinogen by thrombin. Fibrin deposition has also been associated in angiogenesis during tissue repair, with fibrin matrices acting as a seal and scaffold for directing tissue repair. After injury, platelets possessing thrombin receptors move to the site and bind with the available thrombin molecules in the serum. The activated thrombin then converts soluble fibrinogen into fibrin. This helps in blood clotting to seal open wounds. Excessive generation of fibrin can lead to thrombosis. Excessive activation of coagulation and extravascular proteolysis are linked to a number of malignancies, These actions are not only adverse effects of cancer but also important mechanisms of tumour growth, vascularization and metastasis. Although knowledge into the role of blood clotting proteins in tumour progression and metastasis is not extensive, it has been established that fibrin, fibrinogen and products of their degradation are associated with the pathophysiology of many cancers. The recent developments in this area of research render it a vital area of focus to develop novel target cascade pathways and target factors for the regulation of tumourigenesis. Future investigations could lead to the developments of pioneering therapeutic advancements in the field of clinical oncology.

Keywords Fibrin • Angiogenesis • Normal physiology • Disease • Malignancy

17.1 Fibrin

Fibrin is an elastic fibrous, non-globular protein. It is a primary extracellular constituent of blood clots and it is produced from a water soluble plasma glycoprotein called fibrinogen. Fibrinogen is activated from its inactive form by a protease enzyme called thrombin. The reactions that lead to Fibrin clot formation are commonly described as a cascade, in which the product of each step is an enzyme or cofactor needed for following reactions to proceed efficiently. Fibrin works as a binding agent and plays an important role as a provisional matrix during wound healing and tissue remodeling. The entire clotting cascade can be divided into three portions of (a). Extrinsic pathway, (b). Intrinsic pathway and (c). Common pathway (Mackman et al. 2007). When bleeding occurs, platelet clump forms at the site of

A. Salajegheh, *Angiogenesis in Health, Disease and Malignancy*,
DOI 10.1007/978-3-319-28140-7_17

injury, forming a weak plug; neighbouring blood vessels constrict, reducing blood flow in the area, and protein fibrin assembles into a tough network that clots forming an insoluble blockage called a "sturdy scab" (Goodsell 2006).

Prothrombin activator from PF3 (a blood clotting factor formed from platelets and exposure to tissue factor in damaged tissue) along with calcium and other clotting factors in blood plasma leads to activation of thrombin, then fibrinogen, then Fibrin. Inactive fibrinogen is activated to form Fibrin. Fibrin networks should be assembled only in localized areas of the wound, ensuring continuing blood flow to other parts of the body; a cascade of specialized proteins controls this (Goodsell 2006). Once activated, fibrinogen (soluble) is converted to Fibrin (insoluble) that is produced in response to bleeding. Fibrin monomers join to form long Fibrin threads that entangle platelets, building up a spongy mass that gradually hardens and contracts to form the blood clot. Fibrin forms a temporary patch on the wound site. The hardening process is stabilized by Fibrin-stabilizing factor XIII (Sidelmann et al. 2000; Weisel 2005; Goodsell 2006; Laurens et al. 2006).

Fibrin plays a major role in wound healing. Soluble plasma fibrin move to the injury site and form a blood clot when in contact with air. This stops the bleeding and protects the underlying tissue. As the wound continues to repair itself, fibroblasts secrete proteases that dissolve the plasma protein and replace it with cellular fibronectin and assemble it into an insoluble matrix. Platelet Tissue Factor is a protein that is responsible for the formation of thrombin which stimulates activation of fibrinogen into fibrin. However it has been observed that in cancer patients these proteins were overregulated due to medication and chemotherapy leading to excess clotting and possible risk of thrombosis which might lead to death (Undas and Ariens 2011; Wolberg 2012).

17.2 Fibrin in Normal Physiology

Fibrin, derived from the proteolytic cleavage of fibrinogen, is an insoluble protein imperative in the process of vascular injury and repair (Wolberg 2012). In a case of endothelial injury, fibrin molecules rapidly combine to stabilize the platelet plug and form a 3-dimensional network of platelets and fibrin fibres (Undas and Ariens 2011). This provisional matrix not only prevents blood loss, providing critical biophysical and biochemical support to the blood clot but also plays a fundamental role in tissue repair, leukocyte cell adhesion and endothelial migration during angiogenesis (Weisel 2007; Undas and Ariens 2011).

In repair-associated angiogenesis, growth factors such as vascular endothelial growth factor (VEGF) and basic fibroblast growth factor (bFGF) first stimulate endothelial cells to proliferate and migrate into the extracellular matrix (Feng et al. 2013). Proteolytic enzymes of the plasminogen activator/plasmin system and the matrix metallo-proteinases then begin the invasion of these endothelial cells into the extracellular matrix (Laurens et al. 2006; Feng et al. 2013). Thereafter, the cells proliferate and elongate and vessel stabilization is finally attained by interaction with pericytes and the reconstitution of the basement membrane (Laurens et al. 2006). Fibrin has a direct impact on the angiogenic process as it constitutes the extracellular matrix (Hinsbergh et al. 2001). The architecture of the fibrin fibers facilitates cell-matrix interactions by providing binding sites or cell receptors, especially integrin to stimulate the successful invasion of endothelial cells into the matrix (Hinsbergh et al. 2001). Moreover the structure of the matrix also determines the rate and extent of the proteolytic degradation of this scaffold by invading cells (Hinsbergh et al. 2001). A study by Folkman (2003) investigated the angiogenic response by fixing fibrin-containing chambers subcutaneously in guineas pigs. Within a 4-day period, vessel formation was evident in the chambers through the surface pores (Folkman 2003). These findings confirm the effectiveness of a fibrin scaffold as a substrate for the invasion of EC and development of new capillary-like structures.

17.3 Fibrin in Disease

Producing too much Fibrin can lead to a condition called thrombosis. Thrombosis is the blockage of RBCs caused by agglutination of excess RBCs, platelets, polymerized or calcified Fibrin and other components. It is formation of a blood clot inside a blood vessel, which in turn obstructs the flow of blood through the circulatory system. Insufficient production of Fibrin leads to hemorrhaging, which means blood loss from the circulatory system (Sidelmann et al. 2000). The most familiar hereditary abnormality related to Fibrin is the genetic disorder that causes hemophilia. For patients who are presenting a deficiency of fibrinogen and are prone to hemorrhage, bleeding can be corrected by an infusion of fresh frozen plasma, or concentrations of fibrinogen (Al-Mondhiry and Ehmann 1994; Sidelmann et al. 2000). Bleeding can occur internally from blood vessels, externally from a break in the skin, and through a natural opening such as the mouth, nose, ear, urethra etc. (Standeven et al. 2005; Undas and Ariens 2011).

Liver disease or dysfunction can interrupt production of inactive fibrinogen or form abnormal fibrinogen (dysfibrinogenaemia) which can affect Fibrin formation. Hereditary disorders involving Fibrin: deficiency of fibrinogen, abnormal fibrinogen, lack of factor XIII. This disorders result in difficulties in stopping bleeding. Concerns of reduced, absent, or dysfunctional Fibrin is likely to render patients as haemophiliacs (Biswas et al. 2014; Chai-Adisaksopha et al. 2014). Examples of hereditary abnormalities of fibrinogen (the gene carried in chromosome 4) are of both measurable and qualitative in nature and are called, Afibrinogenaemia, Hypofibrinogenaemia, Dysfibrinogenaemia and Hypodysfibrinogenaemia (Al-Mondhiry and Ehmann 1994).

New emerging evidence suggests that thrombotic vascular diseases may be caused due to unique structure of Fibrin clot, which can be affected by genetic variants of fibrinogen and factor XIII. Fibrin clot formation is a dynamic event in haemostasis and thrombosis, which accompany severe inflammation, and cancer invasion or metastasis. The covalent cross-linking of Fibrin through activation of Factor XIII (FXIIIa) stops leakage of circulatory system by forming a rigid and elastic structure. Cross-linked Fibrin is less susceptible to proteolytic or mechanical disruption (Idell et al. 1995; Im 2004; Standeven et al. 2005). Fibrin presents a complex milieu of signals to embedded cells, many of which are not well understood. A study has been done to reestablish Fibrin as a protein that has a well-characterized structure and biochemistry, along with its ability to support angiogenesis specifically. Although Fibrin's structure-function relationships have been studied for decades, opportunities to engineer new and improved synthetic hydrogels can be achieved by further manipulating Fibrin's design. Most cancer patients exhibit hypercoagulability which causes serious problems such as venous thromboembolism. It promotes angiogenesis which helps to cause the pathogenesis of tumour growth and metastasis. Clotting-dependent induction of tumour angiogenesis is mainly facilitated by TF-induced formation of thrombin following deposition of cross-linked Fibrin. A cross-linked Fibrin network builds an interim matrix that enables blood vessel infiltration. Therefore, TF may be useful to become an innovative target for cancer therapy (Fernandez et al. 2004; Ceccarelli and Putnam 2014).

17.4 Fibrin in Malignancy

Over decades, information has been collected that proposes that Fibrin assists in the formation and growth of tumours. Recent methods have illustrated that Fibrin is part of the connective tissue stroma in tumour but in only a few types of tumour. Therapeutic intervention researches using drugs that hinder thrombin activity or boost fibrinolysis generated positive clinical effects in at least one such tumour type. These positive findings confirm that a cause-and-effect relationship exists between thrombin generation with Fibrin formation and tumour growth. This suggests that a cogent foundation proves that future drug intervention trials is needed to target certain reactions related to specific tumour types, which means producing specific drugs for specific tumours (Costantini and Zacharski 1992).

Since erosive tumours are apt to cause micro-haemorrhages, even early benign tumours detected may be violently malignant (Kołodziejczyk and Ponczek 2013). Tumours that are erosive are also more damaging and thus result in Fibrin clot formation. When cancer clusters wear down adjacent normal or tumour vessels, micro-haemorrhage may be resulted, and Fibrin clots are instantly formed to stop the bleeding. These Fibrin clots are then replaced by collagenous stroma, similar to normal wound healing (Dvorak 1986). Normally, more aggressive cancers possess more abundant cancer stroma, due to frequent haemorrhages at many places within or adjacent to the tumour tissue. In this case, Fibrin clot formation continues asymptomatically as long as cancer cells survive and expand from a tiny tumour to the advanced stage. The deposition of Fibrin in non-malignant disease is invariably accompanied by numerous symptoms related to the particular pathology, such as infarction or inflammation (Yasunaga et al. 2011; Matsumura 2012).

Excessive activation of coagulation and extravascular proteolysis are linked to a number of malignancies (Kołodziejczyk and Ponczek 2013; Morin and Tranquillo 2013). These actions are not only adverse effects of cancer but also important mechanisms of tumour growth, vascularization and metastasis (Morin and Tranquillo 2013). Although knowledge into the role of blood clotting proteins in tumour progression and metastasis is not extensive, it has been established that fibrin, fibrinogen and products of their degradation are associated with the pathophysiology of many cancers (Kołodziejczyk and Ponczek 2013). Their involvement in pathological angiogenesis is of great significance. Vascularization for the increase supply of oxygen and nutrients is an essential feature for the proliferation and progression of tumours (Papetti and Herman 2001). Under pathological conditions, the initial stage of blood vessel construction is the formation of a fibrin gel. Development of the fibrin gel is a result of local activation of coagulation and increased movement by tissue factor and cancer pro-coagulant (secretions from tumour cells) of fibrinogen and other haemostatic proteins from plasma into the extravascular region (Kołodziejczyk and Ponczek 2013). This process is an advantage as the fibrin gel structure allows a foundation around the tumour tissue for migrating endothelial cells, fibroblasts and macrophages to harbour and promote growth (Kołodziejczyk and Ponczek 2013). Available data has also reported the effects of fibrin derivatives, specifically the action of fragment E. Bootle-Wilbraham et al. (2001) found a significant increase in the pro-angiogenic effects of VEGF and bFGF in the presence of fragment E, which again encourages angiogenesis and stimulates proliferation, migration and differentiation of endothelial cells (Bootle-Wilbraham et al. 2001). This would be highly unfavourable in the presence of a cancer-induced fibrin clot, as it would strengthen the resistance and increase the persistence of the tumour (Kołodziejczyk and Ponczek 2013).

Cytotoxic immunoconjugate therapy is a novel concept recently introduced as a cancer strategy. Focusing on the breakdown of cancer-induced fibrin clots, an anti-fibrin chimeric antibody was created as a transport vehicle for an anti-cancer agent (ADA). Due to its selective nature in only interacting with fibrin, the immunoconjugate was able to accumulate in the fibrin scaffold of tumour cells and provide sustained release of ADA. Success of the therapy was observed in a mouse model, where the ADA diffused effectively into the malignant tissue and vessels and caused fibrin matrix destruction (Yasunaga et al. 2011). Cancer immunotherapy is another recent treatment aimed at removing the fibrin matrix barrier protecting tumour cells. Investigation into prostate cancer cells revealed that by blocking disulfide exchange between fibrinogen and albumin (a necessary reaction in the formation of a fibrin scaffold) through administration of four-valent sodium selenite, formation of a fibrin matrix complex was inhibited. Tumour cells were therefore exposed to and recognized by the immune system and eliminated (Lipinski 2010). Although progress in interventions are evident, the role of haemostatic factors, particularly fibrin, fibrinogen and their degradations is not yet sufficient. Improved knowledge and understanding will reduce pathogenic angiogenesis and contribute to facilitate better management and treatment of various cancer tumours.

References

Al-Mondhiry H, Ehmann WC (1994) Congenital afibrinogenemia. Am J Hematol 46(4):343–347. doi:10.1002/ajh.2830460416

Biswas A, Ivaskevicius V, Thomas A, Oldenburg J (2014) Coagulation factor XIII deficiency. Diagnosis, prevalence and management of inherited and acquired forms. Hamostaseologie 34(2):160–166. doi:10.5482/hamo-13-08-0046

Bootle-Wilbraham CA, Tazzyman S, Thompson WD, Stirk CM, Lewis CE (2001) Fibrin fragment E stimulates the proliferation, migration and differentiation of human microvascular endothelial cells in vitro. Angiogenesis 4(4):269–275. doi:10.1023/a:1016076121918

Ceccarelli J, Putnam AJ (2014) Sculpting the blank slate: how fibrin's support of vascularization can inspire biomaterial design. Acta Biomater 10(4):1515–1523. doi:10.1016/j.actbio.2013.07.043

Chai-Adisaksopha C, Rattarittamrong E, Norasetthada L, Tantiworawit A, Nawarawong W (2014) Younger age at presentation of acquired haemophilia A in Asian countries: a single-centre study and systematic review. Haemophilia 20(3):e205–e210

Costantini V, Zacharski LR (1992) The role of fibrin in tumor metastasis. Cancer Metastasis Rev 11(3–4):283–290. doi:10.1007/bf01307183

Dvorak HF (1986) Tumors: wounds that do not heal. Similarities between tumor stroma generation and wound healing. N Engl J Med 315(26):1650–1659. doi:10.1056/nejm198612253152606

Feng X, Tonnesen MG, Mousa SA, Clark RAF (2013) Fibrin and collagen differentially but synergistically regulate sprout angiogenesis of human dermal microvascular endothelial cells in 3-dimensional matrix. Int J Cell Biol 2013:1–11. doi:10.1155/2013/231279

Fernandez P, Patierno S, Rickles F (2004) Tissue factor and fibrin in tumor angiogenesis. Semin Thromb Hemost 30(1):31–44. doi:10.1055/s-2004-822969

Folkman J (2003) Fundamental concepts of the angiogenic process. Curr Mol Med 3(7):643–651. doi:10.2174/1566524033479465

Goodsell DS (2006) Fibrin. RCSB Protein Data Bank. doi:10.2210/rcsb_pdb/mom_2006_11

Hinsbergh VWM, Collen A, Koolwijk P (2001) Role of fibrin matrix in angiogenesis. Ann N Y Acad Sci 936(1):426–437. doi:10.1111/j.1749-6632.2001.tb03526.x

Idell S, Pueblitz S, Emri S, Gungen Y, Gray L, Kumar A, Holiday D, Koenig KB, Johnson AR (1995) Regulation of fibrin deposition by malignant mesothelioma. Am J Pathol 147(5):1318–1329

Im JH (2004) Coagulation facilitates tumor cell spreading in the pulmonary vasculature during early metastatic colony formation. Cancer Res 64(23):8613–8619. doi:10.1158/0008-5472.can-04-2078

Kołodziejczyk J, Ponczek MB (2013) The role of fibrinogen, fibrin and fibrin(ogen) degradation products (FDPs) in tumor progression. Contemp Oncol (Pozn) 17(2):113–119. doi:10.5114/wo.2013.34611

Laurens N, Koolwijk P, De Maat MPM (2006) Fibrin structure and wound healing. J Thromb Haemost 4(5):932–939. doi:10.1111/j.1538-7836.2006.01861.x

Lipinski B (2010) Prostate cancer vaccines, fibrin and selenium: a conceptual review ~!2010-05-31~!2010-06-16~!2010-08-07~! TOPCANJ 3(1):69–73. doi:10.2174/1876822901003010069

Mackman N, Tilley RE, Key NS (2007) Role of the extrinsic pathway of blood coagulation in hemostasis and thrombosis. Arterioscler Thromb Vasc Biol 27(8):1687–1693. doi:10.1161/atvbaha.107.141911

Matsumura Y (2012) Cancer stromal targeting (CAST) therapy. Adv Drug Deliv Rev 64(8):710–719. doi:10.1016/j.addr.2011.12.010

Morin KT, Tranquillo RT (2013) In vitro models of angiogenesis and vasculogenesis in fibrin gel. Exp Cell Res 319(16):2409–2417. doi:10.1016/j.yexcr.2013.06.006

Papetti M, Herman IM (2001) Mechanisms of normal and tumor-derived angiogenesis. AJP Cell Physiol 282(5):C947–C970. doi:10.1152/ajpcell.00389.2001

Sidelmann JJ, Gram J, Jespersen J, Kluft C (2000) Fibrin clot formation and lysis: basic mechanisms. Semin Thromb Hemost 26(06):605–618. doi:10.1055/s-2000-13216

Standeven KF, Ariëns RAS, Grant PJ (2005) The molecular physiology and pathology of fibrin structure/function. Blood Rev 19(5):275–288. doi:10.1016/j.blre.2005.01.003

Undas A, Ariens RAS (2011) Fibrin clot structure and function: a role in the pathophysiology of arterial and venous thromboembolic diseases. Arterioscler Thromb Vasc Biol 31(12):e88–e99. doi:10.1161/atvbaha.111.230631

Weisel JW (2005) Fibrinogen and fibrin. In: Fibrous proteins: coiled-coils, collagen and elastomers. Elsevier, CA, USA. doi:10.1016/s0065-3233(05)70008-5

Weisel JW (2007) Structure of fibrin: impact on clot stability. J Thromb Haemost 5:116–124. doi:10.1111/j.1538-7836.2007.02504.x

Wolberg AS (2012) Determinants of fibrin formation, structure, and function. Curr Opin Hematol 19(5):349–356. doi:10.1097/moh.0b013e32835673c2

Yasunaga M, Manabe S, Matsumura Y (2011) New concept of cytotoxic immunoconjugate therapy targeting cancer-induced fibrin clots. Cancer Sci 102(7):1396–1402. doi:10.1111/j.1349-7006.2011.01954.x

Chapter 18
Fibroblast Growth Factors (Acidic: FGF-1; Basic: FGF-2) and Its Receptors (FGFR)

Abstract Fibroblast growth factors (FGF) are structurally related family of 22 molecules. FGFs bind with high affinity to heparin sulphate proteoglycans (HSPGs) which are situated in the extracellular matrix on the surfaces of most cells. This pool of FGFs acts as a reservoir for regulatory growth factors which are released in a regulated manner by heparanases. Members of FGFs from FGF-1 to FGF-10 bind to high-affinity tyrosine kinase FGF receptors (FGFR) denoted FGFR-1 through to FGFR-4. FGFR-1 is the most commonly expressed receptor on endothelial cells and FGFR-2 shown to be in certain circumstances and FGFR-3 and 4 are yet to be seen in the endothelium. FGF signalling is responsible for regulating multiple events within the human body which includes: mesenchymal-epithelial signalling and the development of multiple organ systems in embryonic development, cell proliferation, differentiation and survival, and angiogenesis and wound healing in disorders and mainly tumours.

Keywords Fibroblast growth factors • Acidic FGF • FGF-1 • Basic FGF • FGF-2 • Fibroblast growth factor receptor • FGFR • Angiogenesis • Normal physiology • Disease • Malignancy

18.1 Fibroblast Growth Factors (FGF)

Fibroblast growth factors (FGF) are structurally related family of 22 molecules. FGFs bind with high affinity to heparin sulphate proteoglycans (HSPGs) which are situated in the extracellular matrix on the surfaces of most cells. This pool of FGFs acts as a reservoir for regulatory growth factors which are released in a regulated manner by heparanases (Beenken and Mohammadi 2009). Members of FGFs from FGF-1 to FGF-10 bind to high-affinity tyrosine kinase FGF receptors (FGFR) denoted FGFR-1 through to FGFR-4. FGFR-1 is the most commonly expressed receptor on endothelial cells and FGFR-2 shown to be in certain circumstances and FGFR-3 and 4 are yet to be seen in the endothelium (Mori et al. 2013; Liu et al. 2014). Alternative splicing of the FGFR mRNA allows for the variant FGFR receptors which allow a range of receptor–ligand interactions. It was also discovered that the disruption of the genes encoding FGFR-1 or FGFR-2 leads to embryonic death

A. Salajegheh, *Angiogenesis in Health, Disease and Malignancy*,
DOI 10.1007/978-3-319-28140-7_18

before gastrulation (Beenken and Mohammadi 2009). FGF signalling is responsible for regulating multiple events within the human body which includes: mesenchymal-epithelial signalling and the development of multiple organ systems in embryonic development including nervous system, proliferation, differentiation and survival, and angiogenesis and wound healing in disorders and mainly tumours (Mori et al. 2013; Ornitz and Itoh 2001; Blaber et al. 1996).

18.1.1 Fibroblast Growth Factors in Normal Physiology

FGF-1 is also known as acidic fibroblast growth factor which is located at 5q31.3 and is widely expressed in developing and adult tissues (Beenken and Mohammadi 2009). Additionally FGF-1 is known to play the more important role in endothelial cell proliferation along with FGF-2. FGF-1 is also involved in controlling biological processes such as survival, migration and differentiation of variety of cell types (Beenken and Mohammadi 2009). Unlike most growth factors, FGF-1 lacks cytoplasmic sequences for extracellular export, meaning that they are not secreted from its producer cells (Griffioen and Molema 2000). In general terms, damaged cells release FGF-1 by exocytosis which is independent from the 'normal' exocytosis pathway of any molecule involving packaging of the Golgi apparatus (Mori et al. 2013). The released FGF-1 then can bind to one of the four FGFRs to elicit endothelial cell proliferating effect and allows motility of endothelial cells through paracrine signalling (Cross and Claesson-Welsh 2001). Immunohistochemical staining has shown that colorectal and gastric tissues in both normal and tumour express FGF-1 and its immunoreactivity is mainly cytoplasmic. High levels of FGF-1 are detected in hepatocellular carcinoma whereas it cannot be in normal hepatic tissue (Griffioen and Molema 2000). FGF-1 functions as a mitogen for numerous cell types in vitro and is implicated in a range of physiological processes such as development, morphogenesis, wound healing and angiogenesis. However FGF-1/FGF-2 double knockout mice didn't exhibit any of phenotypic abnormalities which described that FGF-1 does not have any role in the phenotypes of an individual (Griffioen and Molema 2000). FGF-1 binds to integrin $\alpha\nu\beta3$ and the integrin-binding site of FGF-1 overlaps with the heparin-binding site. When they bind FGF-1 induces integrin-FGF1-FGFR1 ternary complex which is considered to be the mechanism of promoting angiogenesis by FGF-1 (Griffioen and Molema 2000). Along with angiogenic activities, FGF-2 inhibits osteoblast mineralization. Interaction of FGFs with four distinct FGF receptors on cells of endodermal, mesodermal, and ectodermal origin, deriving changes in migration, function and proliferation (Liu et al. 2014).

18.1.2 Fibroblast Growth Factors in Disease and Malignancy

Many previous studies were conducted to explain whether FGF Receptor (FGFR) signaling is necessary for normal vascular homeostasis or pathological angiogenesis (Carmeliet and Jain 2011). ECs express FGFRs on their cell surfaces and are activated via FGFR signaling - FGF-mediated intrinsic tyrosine kinase signal transduction (Reuss and von Bohlen und Halbach 2003). According to Oladipupo et al. (2014), the deletion of ECs by using Tie2-Cre transgenic mice, which is a derivative of endothelial cell-specific receptor tyrosine kinase promotor/enhancer for Cre recombinase expression, impaired FGFR1/2-mediated wound healing process after skin or ocular disorders due to ischaemia. This suggests the role of EC-FGFR signaling is specific for tissue repair and injury-induced angiogenesis and not for vascular homeostasis.

Vascular remodeling is one of the neuroprotective mechanisms to increase and restore levels of cerebral blood flow (CBF) in ischaemic stroke (Issa et al. 2005; Marti et al. 2000). Thus, initiation of angiogenesis (vasculogenesis and arteriogeneis) is pivotal for the long-term recovery of stroke patients (Navaratna et al. 2009). Among the family of FGFs, ten have been identified in the brain. Neurons are found to synthesise FGF-2, FGF-5 and FGF-9 while astrocytes primarily synthesise FGF-2 (Reuss and von Bohlen und Halbach 2003). In 2005, Issa and the colleagues (2005) investigated the expression levels of FGF-2 mRNA and protein in brain tissues of 10 acute ischaemic stroke patients who survived for between 24 h and 43 days. The results of the study showed increased mRNA expression and protein levels of FGF-2 in both infarcted core and penumbral region in comparison to normal contralateral hemisphere of patients. Between FGF-2 expression levels in the infarcted core region and the penumbra, higher levels of FGF-2 were identified in the penumbra. This is because the brain cells in the penumbra are salvageable while most neurons in the infarcted core undergo necrosis that leads to irreversible cell death (Fann et al. 2013). Hence, the study has revealed that the up-regulation of FGF-2 is important for stimulating angiogenesis and neuroprotection after stroke.

More recently, FGFR-2 isoforms such as IIIb and IIIc have been identified. The expression levels of these isoforms have been linked to VEGF, which stimulates over-growth of blood vessels during tumor progression (Matsuda et al. 2014). This finding is significant as VEGF is found to enhance ECs against death signals and thus promote resistance against angiogenesis inhibitors (Ferrara and Kerbel 2005). The current study by Matsuda et al. (2014) suggests inhibiting these isoforms during the pancreatic ductal adenocarcinoma decreases VEGF levels and hence delay the progression of pancreatic cancer. Consequently, this provides a rationale that FGFs may have multiple abilities to either decrease disease progression via promoting healing mechanisms or communicate with other angiogenic factors to augment an injury. Therefore, specific targeting of the EC FGFR activation during injuries is one of the key directions for the development of future therapeutics.

FGFs play pivotal roles in the pathogenesis of various epithelial cancers including lung, breast, bladder, melanoma, pancreas and liver (Corn et al. 2013). The possible mechanisms are mainly related to the abnormal FGFR, but these mechanisms are not related to angiogenesis. Meanwhile, for the most angiogenesis cases are mainly found in the prostate cancer, thus FGFs are considered as a good target for anti-angiogenesis in cancer treatment. In 2013, Cron and his colleagues conducted clinical trials based on the inhibition of FGF/FGFR signaling pathway by using tyrosine kinase inhibitors (TKI), monoclonal antibodies and FGF ligand traps. The results showed inhibition of FGF/FGFR signaling pathway but it is still under investigation whether this promotes anti-angiogenic responses.

FGF-1 is implicated in various malignancies such as gastrointestinal tumour and breast cancer. In human colorectal and gastric cancers, cancer associated fibroblasts expressing fibroblast activation protein (FAP) promotes the fibroblast cells to produce FGF-1 and activates FGFR-3 of colon cancer cells in vitro causing increased cell migration and invasion (Griffioen and Molema 2000).

In cases of breast cancer, cancer adjacent cells express active FGF-1 protein in combination with the high expression of mRNA of FGF-1 within the benign neoplastic and hyperplastic tissue which leads to angiogenesis and progression of the tumour from benign to malignant. In addition, studies with animal models revealed that aberrant FGF signalling promotes tumour development by increased cell proliferation and survival additional to promoting angiogenesis at the site of the tumour (Mori et al. 2013). FGF-1 mRNA copy number was found to be significantly correlated with the expression levels of proteins contributing to such proliferations. Both the FGF-1 mRNA and protein levels were found to be causing increased tumour angiogenesis and autocrine stimulation of cancer cells there by giving compelling evidence that FGF-1 plays a very significant role in tumourigenesis and tumour angiogenesis (Mori et al. 2013). Similarly, in cases of lymphoproliferative diseases, such as acute myeloid leukemia, acute lymphoblastoid leukemia and lymphoma, which are dependent on angiogenesis as well, revealed elevated expression of FGF as well as VEGF (Griffioen and Molema 2000). This again demonstrated the important role of FGF-1 in promoting angiogenesis, which is generally present in disorders such as tumours and chronic inflammatory diseases. Recently, additional studies addressing the role of VEGF and FGF in leukocyte vessel wall interactions showed that in general a strong induction of angiogenesis inhibits leukocyte adhesion to the affected site (Al Sabti 2007). This phenomenon demonstrated the inability for the immune system to counteract against the tumour cells which is caused by increased angiogenetic events caused by factors such as FGF-1.

FGF-1 however is favoured in tissue healing processes as well due to its angiogenic functions. In cases of cardiac ischemia, FGF-1 is released from the myocardial tissue into pericardial in cases of myocardial ischemia. By this FGF-1 contributes to functional preservation for myocardium damage. Therefore treatment of cardiac ischemia with FGF-1 by extravascular delivery system has shown to increase coronary flow and therefore proved to be a treatment option in ischemic cases (Beenken and Mohammadi 2009). Due to this the level FGF-1 in pericardial fluid is associated with myocardial ischemia which can be used as a diagnostic tool.

The promotion of angiogenesis by FGF-1 has led to the ongoing research for a treatment option in cancer focusing on the inhibition or altering of FGF-1. These therapeutic interventions for cancer can include targeting the FGFR tyrosin kinases in order to inhibit FGF in its action of tumourigenesis and tumour angiogenesis (Al Sabti 2007). In addition recent finding suggest that dominant negative mutant of FGF-1, such as R50E can suppress tumourigenesis indirectly through suppressing angiogenesis in vivo by acting as an antagonist to FGFR. R50E doesn't bind to $\alpha\nu\beta3$ but still binds to FGFR-1 and heparin. In addition R50E doesn't induce integrin-FGF-1-FGFR1 ternary complex as FGF-1 does and this therefore causes suppression of angiogenesis and tumourigenesis (Lieu et al. 2011). This suggests a new potential anti-angiogenesis and anti-cancer therapeutic agent by suppressing FGF-1.

In recent research, fusion of extracellular region of FGFR1 with Fc region of IgG1- FGF-Trap has been identified and reported to have the potential to inhibit the tumour growth and angiogenesis in renal cell carcinoma (Caki-1 cell line) *in vivo* (Li et al. 2014). Besides, Ronca and his colleagues reported that FGF antagonist, Pentraxin-3 (PTX3) resulted in anti-angiogenic and anti-neoplastic activity in prostate cancer. According to their results, the recombinant PTX3 protein inhibited cell proliferation and angiogenic activity *in vitro* and also inhibited the tumour growth *in vivo* (Ronca et al. 2013a). Moreover, the same group has found that PTX3 in melanoma inhibiting the FGF-2/FGFR pathway plays a critical role in melanoma progression and the blockage of autocrine/paracrine induction of tumour cells proliferation and migration via angiogenesis (Ronca et al. 2013b).

FGF-2 has been found within the dentine-pulp complex (Roberts-Clark and Smith 2000). During tooth development, FGF-2 was responsible for inducing endothelial cells and odontoblasts differentiation, angiogenic dentinal tubule formation and dentine matrix secretion (Tweden et al. 1989; Russo et al. 1998; Tsuboi et al. 2003). In the study by Madan and Kramer, FGF-2 was not only involved in crown development of teeth, it was also believed to be involved in signaling for epithelio-mesenchymal interactions associated with root development in all stages of tooth formation (Madan and Kramer 2005).

Levels of FGF-2 were observed in all areas of periodontal ligament (PDL). As a proangiogenic stimulator, FGF-2 promotes wound healing through their interaction with FGF receptors (FGFR) in the PDL by proliferation (Sako and Hosomichi 2010). In the human clinical trial conducted by Kitamura et al. (2008), FGF-2 was used for the regeneration of periodontal tissue. In this randomised double-blinded controlled study, although their results of clinical attachment level and alveolar bone level gain did not show statistically significant differences between the group treated with conventional periodontal treatment and the group with use of FGF-2 in periodontal treatment, it did show some efficiency with periodontal tissue regeneration when FGF-2 was used. Although this study had no clinical problems related to the safety of administering FGF-2, the study was conducted with small sample size. So more trials are needed before this therapeutic intervention to be given to the public. In recent study, Sakoda et al. (2012) investigated enamel matrix derivative (EMD) and its effect for periodontal regeneration and periodontal soft tissue wound healing. Their results suggested that EMD induced periodontal wound healing angiogenesis via TGF-β1 and FGF-2 in human gingival fibroblasts to stimulate the production of VEGF. Consequently, further research on the mechanism of FGF-2 expression in dental healing will benefit as a future therapeutic target to enhance outcome and prognosis of dental procedures.

18.2 Fibroblast Growth Factors Receptors

Fibroblast growth factors (FGF's) can cause multiple regulatory effects on a single gene by communicating with heparin-sulfate proteoglycans and tyrosine kinase FGF receptors (FGFRs) on the cell surface (Presta et al. 2005). There are different FGFR's each with slightly differing structure and function, FGFR1 is composed of three extracellular immunoglobulin-like domains with an acid box (D1-3), a cytoplasmic tyrosine kinase domain and a transmembrane domain. The isoforms IIIb and IIIc are formed when there is alternate splicing in the juxamembrane D3 domain (Hu and Bouloux 2010). HSPG core proteins interact with pro and negative regulators of angiogenesis, regulating the biological activity of FGFR1 (Presta et al. 2005). This correlation with angiogenesis leads to postulation regarding FGFR1's role in cancer and disease (Presta et al. 2005).

18.2.1 Fibroblast Growth Factors Receptors in Normal Physiology

FGFR1 is released into the cytosol by the Integrative nuclear FGF receptor-1 signaling (INFS) gene regulatory mechanism (Hu and Bouloux 2010). Through FRAP analysis, three nuclear FGFR1 populations were exposed; a fast mobile group, chromatin bound FGFR1, and lastly a group within the nuclear matrix. Intracellular messengers, growth factor receptors, and cell surface neurotransmitters are activated stimulating the release, which eventually leads to the ability to control cell development (Hu and Bouloux 2010). FGFR1 and its ligand FGF2 participate in gene regulation at different chromosomal loci once they are cotransported in the nucleus (Hu and Bouloux 2010). CREB-binding proteins (CBP) receive the signals from FGFR1 via the INFS signaling mechanism, and act as a gene activation-gating factor essential for the transcription process (Hu and Bouloux 2010). The "feed-forward-and-gate" signaling mechanism allows coordinated gene activation through the coupled activation of CBP by INFS (Hu and Bouloux 2010). Overall, this analysis found that the FGFR1 gene plays both a direct and global role in gene regulation, ruled by the movement of proteins and their collisions with nuclear structures and other proteins (Hu and Bouloux 2010).

The multi-steps of angiogenesis allow the formation of new blood vessels to occur in both pathological and normal physiological conditions detrimental to health (Presta et al. 2005). Pro-angiogenic activity of the FGFs is regulated by the availability of high affinity free extracellular matrix molecules and receptors that lie on the endothelial cell surface (Presta et al. 2005). Phosphorylation of FGFR1 activates mitogen-activated protein kinase (MAPK) and other parallel signaling pathways, leading to proliferation of the endothelial cell. The urokinase-type plasminogen activator [uPA] -inducing activity of FGF2 correlates with this auto phosphorylation of specific tyrosine residues (Presta et al. 2005). The number of cell receptors increases, in turn affecting one of the first phases of angiogenesis; extracellular matrix degradation. Matrix metalloproteinases (MMPs) are activated when plasminogen is converted to plasmin by uPA aiding in the degradation of matrix proteins like fibrin (Presta et al. 2005). The next phase, endothelial cell migration is stimulated when the cytoplasmic tail of FGFR1 enables the gene to stimulate movement into the cells via the Wortmanin-sensitive pathway activation (Presta et al. 2005). Unlike FGF1, the FGF8b isoform and FGF10s stimulation of endothelial cell migration, FGF2 activates chemotaxis via the MAPK signaling pathway (Presta et al. 2005). The destination of migrating cells and subsequent cellular interactions are monitored and restricted by communication between ECM molecules (Presta et al. 2005). The maturation of new vessels is primarily regulated by FGFs ability to stimulate interactions and the production of ECM components (Presta et al. 2005).

18.2.2 Fibroblast Growth Factors Receptors in Malignancy

Tumourigenesis, the creation of cancer, is assisted by the biological activity of the gene FGFR1. In the prostate, it has been hypothesized that it disrupts the communication between the stromal and epithelial cells (Miles et al. 2008). Prostate cancer progression caused by FGFR1 activation can be monitored by use of the Juxtaposition of CID and kinase-1 (JOCK-1) model (Miles et al. 2008). The relationship between FGFR1 and the prostate epithelium and the resulting effects have been investigated using a murine model and the JOCK-1 system (Miles et al. 2008). It was found that new vessel growth is initiated by an activated FGFR1 gene, but the sustainability of these vessels was not found to be dependent on it (Miles et al. 2008). While prolonged activation of FGFR1 was linked with tumor progression, down regulation of the gene was not found to halt or reverse the process (Miles et al. 2008). Further, it appears that there is a threshold after which the tumor progression becomes independent of FGFR1 expression, therefore consideration of the spatial and temporal molecular mechanisms and their roles in the contribution of the disease need to be addressed (Miles et al. 2008).

The effects of FGFR1 in lung cancer became a popular topic of research when it was found to be hypothetically linked to prognosis (Yang et al. 2014). Several studies determined that FGFR1 was up regulated in squamous cell lung cancer biopsy tissue, although further evaluation of its link with patient prognosis is required (Yang et al. 2014). FGF1 overexpression has also been investigated in non-small cell lung carcinoma, however no statistically significant differences were evident between the two cancer subtypes (Yang et al. 2014). The hazardous ratio average for the studies showed the gene copy number being insignificant on the survival of the patients in non-small-cell lung cancer and squamous cell lung cancer (Yang et al. 2014). Even though the abundance of FGFR1 in SQCC was higher than in NSCLC the influence had no statistical significance (Yang et al. 2014). Although its links to cancer prognosis remain unclear, FGF1 remains a key target for antitumor therapies due to its key regulatory function in angiogenesis (Yang et al. 2014).

The direct effects of FGFR1 and its role in non-neoplastic disease can been seen through Kallmann Syndrome, the human genetic disorder characterized by disrupted FGFR1 signaling (Yang et al. 2014). Mutations within FGFR1 are connected with altered olfactory and reproductive phenotypes and midline defects significantly dental agenesis (Yang et al. 2014). The loss of function comes from disrupted tyrosine kinase activity or receptor folding through the heterozygous FGFR1 mutation (Yang et al. 2014). Without accurate quantitative regulation of the FGFR1 gene signaling, the normal development of a person is compromised (Yang et al. 2014).

Vascular endothelial growth factors, inflammatory cytokines and FGF's communication helps control the modulation of blood vessel growth in these different pathological conditions, but the role of other genes and their contribution to tumour vascularization also needs to be considered (Presta et al. 2005).

References

Al Sabti H (2007) Therapeutic angiogenesis in cardiovascular disease. J Cardiothorac Surg 2:49. doi:10.1186/1749-8090-2-49

Beenken A, Mohammadi M (2009) The FGF family: biology, pathophysiology and therapy. Nat Rev Drug Discov 8(3):235–253

Blaber M, DiSalvo J, Thomas KA (1996) X-ray crystal structure of human acidic fibroblast growth factor. Biochemistry 35(7):2086–2094. doi:10.1021/bi9521755

Carmeliet P, Jain RK (2011) Molecular mechanisms and clinical applications of angiogenesis. Nature 473(7347):298–307. doi:10.1038/nature10144

Corn PG, Wang F, McKeehan WL, Navone N (2013) Targeting fibroblast growth factor pathways in prostate cancer. Clin Cancer Res 19(21):5856–5866. doi:10.1158/1078-0432.CCR-13-1550

Cross MJ, Claesson-Welsh L (2001) FGF and VEGF function in angiogenesis: signalling pathways, biological responses and therapeutic inhibition. Trends Pharmacol Sci 22(4):201–207

Fann DY, Lee SY, Manzanero S, Tang SC, Gelderblom M, Chunduri P, Bernreuther C, Glatzel M, Cheng YL, Thundyil J, Widiapradja A, Lok KZ, Foo SL, Wang YC, Li YI, Drummond GR, Basta M, Magnus T, Jo DG, Mattson MP, Sobey CG, Arumugam TV (2013) Intravenous immunoglobulin suppresses NLRP1 and NLRP3 inflammasome-mediated neuronal death in ischemic stroke. Cell Death Dis 4:e790. doi:10.1038/cddis.2013.326

Ferrara N, Kerbel RS (2005) Angiogenesis as a therapeutic target. Nature 438(7070):967–974. doi:10.1038/nature04483

Griffioen AW, Molema G (2000) Angiogenesis: potentials for pharmacologic intervention in the treatment of cancer, cardiovascular diseases, and chronic inflammation. Pharmacol Rev 52(2):237–268

Hu Y, Bouloux PM (2010) Novel insights in FGFR1 regulation: lessons from Kallmann syndrome. Trends Endocrinol Metab 21(6):385–393. doi:10.1016/j.tem.2010.01.004

Issa R, AlQteishat A, Mitsios N, Saka M, Krupinski J, Tarkowski E, Gaffney J, Slevin M, Kumar S, Kumar P (2005) Expression of basic fibroblast growth factor mRNA and protein in the human brain following ischaemic stroke. Angiogenesis 8(1):53–62. doi:10.1007/s10456-005-5613-8

Kitamura M, Nakashima K, Kowashi Y, Fujii T, Shimauchi H, Sasano T, Furuuchi T, Fukuda M, Noguchi T, Shibutani T, Iwayama Y, Takashiba S, Kurihara H, Ninomiya M, Kido J, Nagata T, Hamachi T, Maeda K, Hara Y, Izumi Y, Hirofuji T, Imai E, Omae M, Watanuki M, Murakami S (2008) Periodontal tissue regeneration using fibroblast growth factor-2: randomized controlled phase II clinical trial. PLoS One 3(7):e2611. doi:10.1371/journal.pone.0002611

Li D, Wei X, Xie K, Chen K, Li J, Fang J (2014) A novel decoy receptor fusion protein for FGF-2 potently inhibits tumour growth. Br J Cancer 111(1):68–77. doi:10.1038/bjc.2014.282

Lieu C, Heymach J, Overman M, Tran H, Kopetz S (2011) Beyond VEGF: inhibition of the fibroblast growth factor pathway and antiangiogenesis. Clin Cancer Res 17(19):6130–6139. doi:10.1158/1078-0432.ccr-11-0659

Liu C, Cui Y, Luan J, Zhou X, Liu Z, Han J (2014) Fibroblast growth factor-2 inhibits mineralization of osteoblast-like Saos-2 cells by inhibiting the functioning of matrix vesicles. Drug Discov Ther 8(1):42–47

Madan AK, Kramer B (2005) Immunolocalization of fibroblast growth factor-2 (FGF-2) in the developing root and supporting structures of the murine tooth. J Mol Histol 36(3):171–178. doi:10.1007/s10735-005-2684-1

Marti HJ, Bernaudin M, Bellail A, Schoch H, Euler M, Petit E, Risau W (2000) Hypoxia-induced vascular endothelial growth factor expression precedes neovascularization after cerebral ischemia. Am J Pathol 156(3):965–976. doi:10.1016/s0002-9440(10)64964-4

Matsuda Y, Yoshimura H, Suzuki T, Uchida E, Naito Z, Ishiwata T (2014) Inhibition of fibroblast growth factor receptor 2 attenuates proliferation and invasion of pancreatic cancer. Cancer Sci 105(9):1212–1219. doi:10.1111/cas.12470

Miles RJ, Price DK, Figg WD (2008) Temporal-mediated FGFR1 independence: implications for targeting candidate molecules in prostate cancer. Cancer Biol Ther 7(8):1180–1181

Mori S, Tran V, Nishikawa K, Kaneda T, Hamada Y, Kawaguchi N, Fujita M, Saegusa J, Takada YK, Matsuura N, Zhao M, Takada Y (2013) A dominant-negative FGF1 mutant (the R50E mutant) suppresses tumorigenesis and angiogenesis. PLoS One 8(2):e57927. doi:10.1371/journal.pone.0057927

Navaratna D, Guo S, Arai K, Lo EH (2009) Mechanisms and targets for angiogenic therapy after stroke. Cell Adh Migr 3(2):216–223

Oladipupo SS, Smith C, Santeford A, Park C, Sene A, Wiley LA, Osei-Owusu P, Hsu J, Zapata N, Liu F, Nakamura R, Lavine KJ, Blumer KJ, Choi K, Apte RS, Ornitz DM (2014) Endothelial cell FGF signaling is required for injury response but not for vascular homeostasis. Proc Natl Acad Sci U S A 111(37):13379–13384. doi:10.1073/pnas.1324235111

Ornitz DM, Itoh N (2001) Fibroblast growth factors. Genome Biol 2(3):Reviews3005

Presta M, Dell'Era P, Mitola S, Moroni E, Ronca R, Rusnati M (2005) Fibroblast growth factor/fibroblast growth factor receptor system in angiogenesis. Cytokine Growth Factor Rev 16(2):159–178. doi:10.1016/j.cytogfr.2005.01.004

Reuss B, von Bohlen und Halbach O (2003) Fibroblast growth factors and their receptors in the central nervous system. Cell Tissue Res 313(2):139–157. doi:10.1007/s00441-003-0756-7

Roberts-Clark DJ, Smith AJ (2000) Angiogenic growth factors in human dentine matrix. Arch Oral Biol 45(11):1013–1016

Ronca R, Alessi P, Coltrini D, Di Salle E, Giacomini A, Leali D, Corsini M, Belleri M, Tobia C, Garlanda C, Bonomi E, Tardanico R, Vermi W, Presta M (2013a) Long pentraxin-3 as an epithelial-stromal fibroblast growth factor-targeting inhibitor in prostate cancer. J Pathol 230(2):228–238. doi:10.1002/path.4181

Ronca R, Di Salle E, Giacomini A, Leali D, Alessi P, Coltrini D, Ravelli C, Matarazzo S, Ribatti D, Vermi W, Presta M (2013b) Long pentraxin-3 inhibits epithelial-mesenchymal transition in melanoma cells. Mol Cancer Ther 12(12):2760–2771. doi:10.1158/1535-7163.MCT-13-0487

Russo LG, Maharajan P, Maharajan V (1998) Basic fibroblast growth factor (FGF-2) in mouse tooth morphogenesis. Growth Factors 15(2):125–133

Sako E, Hosomichi J (2010) Alteration of bFGF expression with growth and age in rat molar periodontal ligament. Angle Orthod 80(5):904–911. doi:10.2319/011910-38.1

Sakoda K, Nakajima Y, Noguchi K (2012) Enamel matrix derivative induces production of vascular endothelial cell growth factor in human gingival fibroblasts. Eur J Oral Sci 120(6):513–519. doi:10.1111/j.1600-0722.2012.00999.x

Tsuboi T, Mizutani S, Nakano M, Hirukawa K, Togari A (2003) Fgf-2 regulates enamel and dentine formation in mouse tooth germ. Calcif Tissue Int 73(5):496–501. doi:10.1007/s00223-002-4070-2

Tweden KS, Spadone DP, Terranova VP (1989) Neovascularization of surface demineralized dentin. J Periodontol 60(8):460–466. doi:10.1902/jop.1989.60.8.460

Yang W, Yao Y-W, Zeng J-L, Liang W-J, Wang L, Bai C-Q, Liu C-H, Song Y (2014) Prognostic value of FGFR1 gene copy number in patients with non-small cell lung cancer: a meta-analysis. J Thorac Dis 6(6):803–809. doi:10.3978/j.issn.2072-1439.2014.05.02

Chapter 19
Fibronectin

Abstract Fibronectin is an extracellular matrix glycoprotein that binds to transmembrane proteins and plays a significant role in formation of blood vessel. Fibronectin is also important as a regulator in the process of remodelling in extracellular matrix. In normal situation fibronectin is involved in the process of tissue repair and helps the formation of clot. Elevation of cellular fibronectin and persistent production of extracellular matrix have been identified as fibroproliferative response. This will end in the destruction of the integrity in chronically injured or diseased tissue. With regards to angiogenic role of fibronectin in malignancies, expressions of fibronectin and consequently integrin in cancer have been shown to be associated with cancer cell proliferation, poor patient survival rate, lymph node metastases, facilitating tumour cell invasion.

Keywords Fibronectin • Angiogenesis • Normal physiology • Disease • Malignancy

19.1 Fibronectin

Fibronectin is an extracellular matrix glycoprotein that binds to transmembrane proteins called integrins (von Au et al. 2013). It contributes significantly to blood vessel formation and is an important regulator in extracellular matrix remodelling (von Au et al. 2013; Shi and Sottile 2011). It is composed of two protein heterodimers linked by a pair of disulfide bonds and exists in two major forms; cellular fibronectin (an insoluble form of fibrils) and plasma fibronectin (a soluble form produced by hepatocytes in the liver) (Astrof and Hynes 2009). Plasma fibronectin is important in tissue injury as it plays a key role in platelet function to mediate haemostasis, while cellular fibronectin aids in the reconstitution of damaged tissues (von Au et al. 2013; Astrof and Hynes 2009). Different isotypes of cellular fibronectin exists in human tissue due to the splicing of fibronectin mRNA from a single gene to repeats of Type I, II and III (Aziz-Seible and Casey 2011).

A. Salajegheh, *Angiogenesis in Health, Disease and Malignancy*,
DOI 10.1007/978-3-319-28140-7_19

19.2 Fibronectin in Normal Physiology

The multifunctional roles of fibronectin in angiogenesis, cell adhesion, growth, migration, survival and differentiation have been well established (Astrof and Hynes 2009; Aziz-Seible and Casey 2011; von Au et al. 2013). While extensive studies have associated its fundamental role in biological processes in the body, its physiological properties have been under researched. However, there is a growing body of evidence in the literature suggesting that fibronectin may be involved in regulating cellular behaviour, particularly under pathological circumstances (To and Midwood 2011). In addition, the importance of fibronectin has also been implicated in wound healing and the development of various carcinomas (Fernandez-Garcia et al. 2014; Schaffner et al. 2013).

Under normal physiological conditions, both cellular and plasma fibronectin exists in balanced amounts (Aziz-Seible and Casey 2011). During the early process of tissue repair, circulating plasma fibronectin in the blood migrate to the injury and aids in the formation of a fibrin clot (Shi and Sottile 2011). The process is mediated by a series of mechanisms that allow the incorporation of the fibronectin into the fibrin matrix (Shi and Sottile 2011; Aziz-Seible and Casey 2011). The fibronectin can non-covalently interact with the fibrin or can covalently bind to the fibrin by the activation of the blood coagulation cascade (Gui et al. 2006). Additionally, it can also be bound to the matrix on the platelet surface. These mechanisms stabilize the clot and enhance platelet adhesion and aggregation (Aziz-Seible and Casey 2011; Gui et al. 2006).

Cellular fibronectin plays a key role in the late process of tissue repair as it is an important component in granulation tissue (Aziz-Seible and Casey 2011). Once the fibrin clot is formed, endothelial cells and fibroblasts that have migrated to the site of injury begin to deposit cellular fibronectin into the area. A dense three-dimensional fibrillar network is formed around the fibroblasts and on the cell surface which is essential in maintaining the integrity and structure of the healing tissue. This structural network facilitates the regulation of both the extracellular matrix composition and the deposition of other important extracellular matrix molecules such as collagen, fibrinogen, fibulin and laminin (Aziz-Seible and Casey 2011; Rozario et al. 2009). Due to this three-dimensional structure; cellular adhesion (Friedland et al. 2009), proliferation, spreading (Gui et al. 2006), migration (Rozario et al. 2009), and apoptotic activity (Natal) are all enhanced. It has also been shown that the fibronectin matrix can inhibit growth factors in order to regulate cell signalling events (To and Midwood 2011).

19.3 Fibronectin in Disease

Located at the carboxyl-terminal end of the fibronectin molecule are Type III repeats, the major domain for the binding of heparin molecules in wound repair (Mitsi et al. 2006). Type I repeats are located near the site of the disulfide bonds and is also the second major site for the binding of fibrin. The central region of the fibronectin molecule is made up of Type III homologous repeats that include the spliced EIIIA and EIIIB sequences. Increased levels of EIIIA and EIIIB have been found to be associated with disease and injury (Mitsi et al. 2006).

During injury or disease, there is an increased level of circulating and tissue cellular fibronectin in the body compared to plasma fibronectin (Aziz-Seible and Casey 2011). However, this accumulation due to prolonged tissue damage may lead to further injury in certain conditions (Gui et al. 2006). When the levels of cellular fibronectin remain elevated over time, fibronectin-mediated cellular activity may become amplified. Furthermore, the down-regulation of other signalling mechanisms may eventually become activated. This causes a change in normal physiological cellular activity, shifting it towards a more pathological process (Aziz-Seible and Casey 2011; Gui et al. 2006; To and Midwood 2011).

Elevated levels of cellular fibronectin enhance the persistent production of extracellular matrix (To and Midwood 2011). Due to the accumulation of scar tissue from a chronic injury or disease, a fibroproliferative response is initiated and eventually destroys the integrity of the tissue (To and Midwood 2011). Studies have found this fibroproliferative response to cause disease and affect particularly the pulmonary, renal and hepatic systems (Astrof and Hynes 2009; To and Midwood 2011). Although reversible, some studies have associated increased fibrotic activity to cause organ failures in various circumstances (To and Midwood 2011).

Under prolonged conditions of injury such as chronic liver disease, there is an increased level of cellular fibronectin in plasma (Aziz-Seible and Casey 2011). Additionally, immunohistochemical and RT-PCR analyses demonstrated similar findings as well as fibronectin mRNA present in high amounts in diseased livers. These findings confirm that elevated levels of cellular fibronectin are correlated with chronic injury and disease (Aziz-Seible and Casey 2011). Though it has been recommended to use levels of cellular fibronectin as a biomarker for chronic liver disease (Aziz-Seible and Casey 2011), its lack of specificity to differentially diagnose the condition limits it diagnostic ability. However, cellular fibronectin should remain as a reliable indicator for chronic tissue damage.

It is important to note that different manifestations of increased cellular fibronectin exist in various diseases due to underlying factors associated with each individual disease. Certain parameters relating to the chronic injury or disease need to be taken into account in order to understand the role that cellular fibronectin has on the diseases' progression.

19.4 Fibronectin in Malignancy

Isoforms of fibronectin, particularly EIIIA and EIIIB have been shown to be found in high amounts in tumorigenesis (White et al. 2008). Morphological changes displayed in tumors have been associated with fibronectin degradation and increased fibronectin expression as well as the expression of fibronectin receptor-binding integrins such as α5β1 (To and Midwood 2011; White et al. 2008). Presence of fibronectin leads to the activation of α5β1 integrin (White et al. 2008). Overexpression of this integrin has been shown to be associated with cancer cell proliferation and poor patient survival rate. Furthermore, α5β1 integrin expression has also been correlated with lymph node metastases and provides a differential diagnosis between an adenocarcinoma and a squamous cell carcinoma. The signalling system between fibronectin and α5β1 integrin has been shown to enhance the progression of lung cancer (Ritzenthaler et al. 2008). This is due to the PI3K/AKT/mTOR pathway, a specific mechanism that mediates the fibronectin-integrin effects on proliferation. It is found that fibronectin plays a role in protecting the apoptotic activity of lung cancer cells through the activation of β1 integrin (Ritzenthaler et al. 2008). Metastases in the lung have been correlated with plasma fibronectin facilitating tumour cell invasion of the blood clot during healing (Ritzenthaler et al. 2008; To and Midwood 2011). Also, plasma fibronectin has been reported to protect tumour cells from cytotoxic activity and have also been implicated in the stimulation of androgenic receptors which control the expression of cyclin D, an important component in cell cycle control (To and Midwood 2011).

Recent findings reported the relationship between extracellular matrix fibronectin and its role in the activation of focal adhesion kinases (FAK) (To and Midwood 2011). These are important proteins which determine cell adhesion and have been shown to be involved in cancer cells. It was shown that the inactivation of FAK caused a slower metastatic rate, while the activation was highly dependent on fibronectin expression (Ritzenthaler et al. 2008; To and Midwood 2011). Similarly in studies investigating breast cancer, authors have concluded metastatic activity from breast carcinomas showed fibronectin expression in tumour cells (Fernandez-Garcia et al. 2014).

Collectively, the data showed there to be an increase in tumour progression particularly associated with fibronectin and its binding receptor α5β1 integrin. Future studies should investigate the potential role of fibronectin in diagnostic measures and early intervention of disease. Furthermore, it is important to study if fibronectin may prevent the spread of malignant cells by blocking angiogenesis into the primary tumour.

References

Astrof S, Hynes RO (2009) Fibronectins in vascular morphogenesis. Angiogenesis 12(2):165–175

Aziz-Seible RS, Casey CA (2011) Fibronectin: functional character and role in alcoholic liver disease. World J Gastroenterol 17(20):2482–2499. doi:10.3748/wjg.v17.i20.2482

Fernandez-Garcia B, Eiro N, Marin L, Gonzalez-Reyes S, Gonzalez LO, Lamelas ML, Vizoso FJ (2014) Expression and prognostic significance of fibronectin and matrix metalloproteases in breast cancer metastasis. Histopathology 64(4):512–522. doi:10.1111/his.12300

Friedland JC, Lee MH, Boettiger D (2009) Mechanically activated integrin switch controls alpha-5beta1 function. Science 323(5914):642–644

Gui L, Wojciechowski K, Gildner CD, Nedelkovska H, Hocking DC (2006) Identification of the heparin-binding determinants within fibronectin repeat III1: role in cell spreading and growth. J Biol Chem 281(46):34816–34825

Mitsi M, Hong Z, Costello CE, Nugent MA (2006) Heparin-mediated conformational changes in fibronectin expose vascular endothelial growth factor binding sites. Biochemistry 45(34):10319–10328

Ritzenthaler JD, Han S, Roman J (2008) Stimulation of lung carcinoma cell growth by fibronectin-integrin signalling. Mol Biosyst 4(12):1160–1169. doi:10.1039/b800533h

Rozario T, Dzamba B, Weber GF, Davidson LA, DeSimone DW (2009) The physical state of fibronectin matrix differentially regulates morphogenetic movements in vivo. Dev Biol 327(2):386–398

Schaffner F, Ray AM, Dontenwill M (2013) Integrin α5β1, the fibronectin receptor, as a pertinent therapeutic target in solid tumors. Cancers (Basel) 5(1):27–47

Shi F, Sottile J (2011) MT1-MMP regulates the turnover and endocytosis of extracellular matrix fibronectin. J Cell Sci 124(Pt 23):4039–4050

To WS, Midwood KS (2011) Plasma and cellular fibronectin: distinct and independent functions during tissue repair. Fibrogenesis Tissue Repair 4:21

von Au A, Vasel M, Kraft S, Sens C, Hackl N, Marx A, Stroebel P, Hennenlotter J, Todenhofer T, Stenzl A, Schott S, Sinn HP, Wetterwald A, Bermejo JL, Cecchini MG, Nakchbandi IA (2013) Circulating fibronectin controls tumor growth. Neoplasia 15(8):925–938

White ES, Baralle FE, Muro AF (2008) New insights into form and function of fibronectin splice variants. J Pathol 216(1):1–14

Chapter 20
Granulocyte-Macrophage and Granulocyte Colony Stimulating Factor (GM-CSF and G-CSF)

Abstract There are numerous genes that contribute to the control and regulation of angiogenesis. Colony stimulating factors (CSF) are stimulative cytokine, and Granulocyte–macrophage CSF (GM-CSF) specifically involved in both physiological and pathological circumstances of angiogenesis. GM-CSF is present in most tissues and is associated with the extracellular matrix, the serum and as an integral membrane protein. Granulocyte-colony stimulating factor (G-CSF) is a granulocyte and stem cell stimulant, which mainly triggers the bone marrow to produce and release them into the vascular circulation. G-CSF has also been shown to stimulate tumour growth and angiogenesis in certain scenarios. GM-CSF is more known for its immune interactions, however it has demonstrated significant involvement in angiogenesis. GM-CSF can be beneficial for the host by means of assisting vessels for wound healing however, further study should be undertaken about the suppression of CSF factors when the formation of new blood vessels is considered unfavourable in situations such as aiding tumour growths.

Keywords Granulocyte-macrophage colony stimulating factor • Granulocyte colony stimulating factor • GM-CSF • G-CSF • Angiogenesis • Normal physiology • Disease • Malignancy

20.1 Granulocyte-Macrophage and Granulocyte Colony Stimulating Factor (GM-CSF and G-CSF)

Granulocyte–macrophage colony stimulating factor (GM-CSF) is a stimulative cytokine. GM-CSF is significantly involved in both physiological and pathological circumstances of angiogenesis. GM-CSF is present in most tissues and is also associated with the extracellular matrix, the serum and as an integral membrane protein. GM-CSF is secreted in response to immune activation as well as in response to cytokines that mediate inflammation (Shi et al. 2006). Endothelial cells, T cells, mast cells, macrophages and fibroblasts all produce granulocyte-macrophage colony-stimulating factor (Kimura et al. 2006). GM-CSF was initially discovered within a rodent lung tissue conditioned medium after an injection of lipopolysaccharide. GM-CSF has the ability to stimulate proliferation of rodent bone marrow

A. Salajegheh, *Angiogenesis in Health, Disease and Malignancy*,
DOI 10.1007/978-3-319-28140-7_20

cells whilst in vitro and also generate colonies of macrophages and granulocytes (Burgess et al. 1977). Granulocyte colony stimulating factor (G-CSF) is manufactured by several tissues and cells, which includes various immune cells, endothelium and macrophages (Xu et al. 2000). G-CSF naturally occurs in two forms, 174 and 177 amino acid long proteins with an approximate molecular weight of 19,600 g per mole. 174-chain occurs more frequently, and is commonly used in biotechnology for drug development (Thomas et al. 2002).

20.2 GM-CSF and G-CSF in Normal Physiology

GM-CSF has two main functions. As an immune modulator, GM-CSF influences macrophage activity and the activity and maturation of dendritic immune cells. The other function of GM-CSF includes being an indirect promoter of the formation of new branching blood vessels in angiogenesis (Zhao et al. 2014). GM-CSF can influence angiogenesis in situations from normal physiology including controlling the levels of hematopoietic cells in the blood, spleen and bone marrow, to accelerating wound healing, or even aiding some tumour types in finding oxygen and nutrient supply (Zhao et al. 2014).

GM-CSF targets a large range of cells that are used by the body for wound healing. These involve dendritic cells, fibroblasts, endothelial cells and macrophages. These particular types of cells can also synthesis GM-CSF (Zhao et al. 2014). Granulocyte-macrophage colony stimulating factor can promote dendritic cell differentiation and maturation, increase major histocompatibility complex molecules, co-stimulate molecule expression and enhance the capacity of tumour antigen delivery via dendritic cells (He et al. 2011). It is suggested that in early research GM-CSF can increase vascular injury wounds, by boosting the formation of granular tissue (Zhao et al. 2014). Granulocyte-macrophage colony-stimulating factor accelerates wound healing due to its immune and granulation properties but it also assists the generation of new blood vessels to the previously damaged tissue (Aller et al. 2010). GM-CSF was also shown to mature the newly formed microvessels and decrease their permeability. Speed and quality of wound healing was improved by the treatment of Granulocyte-macrophage colony-stimulating factor (Zhao et al. 2014).

Granulocyte-colony stimulating factor (G-CSF) also known as colony-stimulating factor 3 (CSF 3), is a granulocyte and stem cell stimulant, which triggers the bone marrow to produce and release them into the vascular circulation (Bendall and Bradstock 2014). It is essentially a cytokine and undertakes the role of a hormone. G-CSF is being produced by several tissues, and acts as a type of colony stimulating factor (Xu et al. 2000). G-CSF activates the continuity, propagation, differentiation and function of the precursors and matured neutrophils (Dong and Larner 2000; Xu et al. 2000).

20.3 GM-CSF and G-CSF in Disease and Malignancy

Wound healing is a beneficial affect of the expression of GM-CSF in regards to angiogenesis. On the other hand, a detrimental affect is GM-CSFs' ability to increase angiogenesis in regards to providing oxygen and nutrients for a tumour. Through its' stimulatory properties to generate the expression of vascular endothelial growth factor (VEGF), furthermore leads to the promotion of angiogenesis. Tumour growth and size is restricted by its' ability to obtain nutrient and oxygen supply. GM-CSF stimulates the expression of VEGF, which is a major player in angiogenesis (Horiuchi and Weller 1997; Eubank et al. 2004). Any contribution to the stimulation of angiogenesis will promote oxygen and nutrient delivery to a tumour and therefore support its' growth and increase its' levels of malignancy. As further contribution to tumour growth, GM-CSF, when expressed by tumour cells, increases proliferation and migration of the potentially malignant cells (Wang et al. 2014). Studies both in vivo and in vitro have shown that GM-CSF expression in endothelial cells, tumour cells and keratinocytes increases migration and proliferation both of which contribute to the size and growth of tumours (Gutschalk et al. 2013).

VEGF is not the only other factor on which GM-CSF shows influence. GM-CSF also demonstrates a relationship with endothelial progenitor cells (EPC) (Du et al. 2012). Endothelial progenitor cells are undifferentiated cells that originate from within the bone marrow. Once mobilised, they migrate into the circulatory system and significantly contribute to post natal vessel formation (Du et al. 2012). GM-CSF is responsible for the mobilisation of endothelial progenitor cells into the peripheral circulation, which then promotes angiogenesis in the ischemic tissues (Peplow 2014). This influence that GM-CSF has upon endothelial progenitor cells furthermore indirectly makes a connection to the promotion of angiogenesis through the formation of postnatal blood vessels (Du et al. 2012; Peplow 2014; Zhao et al. 2014).

Another effect of GM-CSF that appears to be unfavourable to the host in regards to illness and disease is its' involvement in Rheumatoid arthritis (Bendtzen et al. 2003; Shi et al. 2006). Research conducted by Cook et al. indicated that when GM-CSF was neutralised and unstimulated in controlled rodents, the destruction of cartilage tissue within the joints by inflammatory means was reduced. As a whole, after Anti-GM-CSF treatment, both the severity and number of affected limbs of arthritis affected mice where reduced (Cook et al. 2012).

GM-CSF has been implemented in numerous clinical circumstances and Investigations. Science and application has focused aspects of applying the use of GM-CSF for an immune adjuvant due to its ability to enhance dendritic cell function, maturation and macrophage activity. GM-CSF has been used to treat neutropenia in patients with cancer whom are undergoing chemotherapy and also in for patients who fit the criteria after having bone marrow transplants. GM-CSF has also been used therapeutically as a contribution to the management of AIDS (Kimura et al. 2006). In a further clinical scenario, A study combining GM-CSF and a DNA vaccine elicited protection against herpes simplex virus infection in the presence of both Th1 and Th2 components (Burgess et al. 1977). During an experiment by He et al. (2011) it was discovered that, to efficiently and specifically lyse tumour cells, GM-CSF was to be primed onto T-cells then modified dendritic cells were to be added to the tumour cells environment. A recent discovery by Daud et al. (2008) found that dendritic cell differentiation, stimulated by GM-CSF, may result in benefits in patients with high- risk melanoma. Many studies discuss that GM-CSF is a regulatory function on mature myeloid cells, in addition to the long known survival-promoting activity that allows it to directly induce differentiation on dendritic cells with distinct functional and phenotypic traits (Conti and Gessani 2008).

GM-CSF is more known for its immune interactions, however it has demonstrated significant involvement in angiogenesis.

G-CSF has been shown to stimulate tumour growth and angiogenesis in certain scenarios (Voloshin et al. 2011). G-CSF might offer some resistance to anti-VEGF treatment. Experiments were done on this facet: clinical mice with B16F1 tumours received recombinant G-CSF, and were then administered control or anti-VEGF antibodies. G-CSF treatment was related to decreased response to anti-VEGF, as tumour mass in G-CSF drugged mice was a lot greater than those in PBS treated ones (Voloshin et al. 2011). Concentrations of G-CSF in tumours is measured, and since it is implicated in differentiation of hematopoietic progenitors and in angiogenesis (as aforementioned), it is ample in tumour-associated fibroblasts. It also recruits CXCR4+ cells to the tumours resulting in increased tumour angiogenesis (Voloshin et al. 2011). G-CSF is also abundant in plasma and tumours in patients with refractory tumours, whilst other cytokines are not detectable or in very low concentrations. Levels of G-CSF in different tumour vary, e.g. greater in LLC tumours than EL4 tumours (Shojaei et al. 2009). Because of this, it can be induced that tumour stroma is a source of G-CSF in EL4 cells. G-CSF is also used for activating granulopoiesis in patients with granulocytopenia (Gora-Tybor et al. 1996). G-CSF was given to patients with acute myeloblastic leukaemia, and this reflected little adverse effects (Motoji et al. 1991). However, since there are still patients with the same condition who respond to G-CSF administration, its usage must cautious. G-CSF has helped patients with acute myeloblastic leukaemia to achieve hematopoietic recovery (Motoji et al. 1991).

Colony stimulating factors and more specifically GM-CSF can be beneficial for the host by means of assisting vessels for wound healing however, further study should be undertaken about the suppression of these factors when the formation of new blood vessels is considered unfavourable in situations such as aiding tumour growths. Tumour size and metastasis is dependant of oxygen and nutrient supply and also the connection to the circulatory system via vessels. Down regulation of GM-CSF in circumstances including tumour growth may well be potentially successful aim in the treatment of some cancers.

References

Aller M-A, Arias J-I, Arias J (2010) Pathological axes of wound repair: gastrulation revisited. Theor Biol Med Model 7:37. doi:10.1186/1742-4682-7-37

Bendall LJ, Bradstock KF (2014) G-CSF: from granulopoietic stimulant to bone marrow stem cell mobilizing agent. Cytokine Growth Factor Rev 25(4):355–367. doi:10.1016/j.cytogfr.2014.07.011

Bendtzen K, Hansen PR, Rieneck K, The Spironolactone/Arthritis Study Group (2003) Spironolactone inhibits production of proinflammatory cytokines, including tumour necrosis factor-α and interferon-γ, and has potential in the treatment of arthritis. Clin Exp Immunol 134(1):151–158. doi:10.1046/j.1365-2249.2003.02249.x

Burgess AW, Camakaris J, Metcalf D (1977) Purification and properties of colony-stimulating factor from mouse lung-conditioned medium. J Biol Chem 252(6):1998–2003

Conti L, Gessani S (2008) GM-CSF in the generation of dendritic cells from human blood monocyte precursors: recent advances. Immunobiology 213(9–10):859–870. doi:10.1016/j.imbio.2008.07.017

Cook AD, Pobjoy J, Steidl S, Durr M, Braine EL, Turner AL, Lacey DC, Hamilton JA (2012) Granulocyte-macrophage colony-stimulating factor is a key mediator in experimental osteoarthritis pain and disease development. Arthritis Res Ther 14(5):R199. doi:10.1186/ar4037

Daud AI, Mirza N, Lenox B, Andrews S, Urbas P, Gao GX, Lee JH, Sondak VK, Riker AI, Deconti RC, Gabrilovich D (2008) Phenotypic and functional analysis of dendritic cells and clinical outcome in patients with high-risk melanoma treated with adjuvant granulocyte macrophage colony-stimulating factor. J Clin Oncol 26(19):3235–3241. doi:10.1200/jco.2007.13.9048

Dong F, Larner AC (2000) Activation of Akt kinase by granulocyte colony-stimulating factor (G-CSF): evidence for the role of a tyrosine kinase activity distinct from the janus kinases. Blood 95(5):1656–1662

Du F, Zhou J, Gong R, Huang X, Pansuria M, Virtue A, Li X, Wang H, Yang X-F (2012) Endothelial progenitor cells in atherosclerosis. Front Biosci 17:2327–2349

Eubank TD, Roberts R, Galloway M, Wang Y, Cohn DE, Marsh CB (2004) GM-CSF induces expression of soluble VEGF receptor-1 from human monocytes and inhibits angiogenesis in mice. Immunity 21(6):831–842. doi:10.1016/j.immuni.2004.10.011

Gora-Tybor J, Krykowski E, Robak T (1996) Treatment of drug-induced agranulocytosis with colony stimulating factors (G-CSF or GM-CSF). Arch Immunol Ther Exp (Warsz) 44(4):255–258

Gutschalk CM, Yanamandra AK, Linde N, Meides A, Depner S, Mueller MM (2013) GM-CSF enhances tumor invasion by elevated MMP-2, -9, and -26 expression. Cancer Med 2(2):117–129. doi:10.1002/cam4.20

He S, Wang L, Sun K, Zhang Y, Li D (2011) Enhanced anti-tumor immunity ex vivo induced by GM-CSF gene transducted dendritic cell vaccine. Chin-Ger J Clin Oncol 10(3):178–182

Horiuchi T, Weller PF (1997) Expression of vascular endothelial growth factor by human eosinophils: upregulation by granulocyte macrophage colony-stimulating factor and interleukin-5. Am J Respir Cell Mol Biol 17(1):70–77. doi:10.1165/ajrcmb.17.1.2796

Kimura K, Shinozaki Y, Jujo S, Shizuma T, Fukuyama N, Nakazawa H (2006) Triiodothyronine acutely increases blood flow in the ventricles and kidneys of anesthesized rabbits. Thyroid 16(4):357–360. doi:10.1089/thy.2006.16.357

Motoji T, Watanabe M, Uzumaki H, Kusaka M, Fukamachi H, Shimosaka A, Oshimi K, Mizoguchi H (1991) Granulocyte colony-stimulating factor (G-CSF) receptors on acute myeloblastic leukaemia cells and their relationship with the proliferative response to G-CSF in clonogenic assay. Br J Haematol 77(1):54–59

Peplow PV (2014) Growth factor- and cytokine-stimulated endothelial progenitor cells in post-ischemic cerebral neovascularization. Neural Regen Res 9(15):1425–1429. doi:10.4103/1673-5374.139457

Shi Y, Liu CH, Roberts AI, Das J, Xu G, Ren G, Zhang Y, Zhang L, Yuan ZR, Tan HS, Das G, Devadas S (2006) Granulocyte-macrophage colony-stimulating factor (GM-CSF) and T-cell responses: what we do and don't know. Cell Res 16(2):126–133. doi:10.1038/sj.cr.7310017

Shojaei F, Wu X, Qu X, Kowanetz M, Yu L, Tan M, Meng YG, Ferrara N (2009) G-CSF-initiated myeloid cell mobilization and angiogenesis mediate tumor refractoriness to anti-VEGF therapy in mouse models. Proc Natl Acad Sci U S A 106(16):6742–6747. doi:10.1073/pnas.0902280106

Thomas J, Liu F, Link DC (2002) Mechanisms of mobilization of hematopoietic progenitors with granulocyte colony-stimulating factor. Curr Opin Hematol 9(3):183–189

Voloshin T, Gingis-Velitski S, Bril R, Benayoun L, Munster M, Milsom C, Man S, Kerbel RS, Shaked Y (2011) G-CSF supplementation with chemotherapy can promote revascularization and subsequent tumor regrowth: prevention by a CXCR4 antagonist. Blood 118(12):3426–3435. doi:10.1182/blood-2010-11-320812

Wang Y, Han G, Wang K, Liu G, Wang R, Xiao H, Li X, Hou C, Shen B, Guo R, Li Y, Chen G (2014) Tumor-derived GM-CSF promotes inflammatory colon carcinogenesis via stimulating epithelial release of VEGF. Cancer Res 74(3):716–726. doi:10.1158/0008-5472.can-13-1459

Xu S, Hoglund M, Hakansson L, Venge P (2000) Granulocyte colony-stimulating factor (G-CSF) induces the production of cytokines in vivo. Br J Haematol 108(4):848–853

Zhao J, Chen L, Shu B, Tang J, Zhang L, Xie J, Qi S, Xu Y (2014) Granulocyte/macrophage colony-stimulating factor influences angiogenesis by regulating the coordinated expression of VEGF and the Ang/Tie system. PLoS One 9(3):e92691. doi:10.1371/journal.pone.0092691

Chapter 21
GAX and HOXA5

Abstract GAX is a homeobox gene, also known as MEOX-2, which inhibits angiogenesis in vascular endothelial cells (EC). The GAX gene is originally derived from embryonic mesoderm and muscle precursors, however, its expression in adults is mostly restricted to the cardiovascular system and kidneys. Increased knowledge of the mechanism of the GAX gene in angiogenesis through recent studies has been a contributing factor to uncovering the enigma of Alzheimer's disease and discovering treatment options for Human Hepatocellular Carcinoma. HOXA5 is another gene where through gaining sufficient knowledge on its function in angiogenesis, may be manipulated for therapeutic treatment options in lung disease, coronary heart disease and breast cancer. The physiology of the HOXA5 have been found to reiterate that of GAX. HOXA5 is a contributing factor in human lung development it may also play specific roles in vasculogenesis and angiogenesis of the developing organs. HOXA5 also plays a significant role in the formation of tumours and vascular integrity. Understanding the mechanisms of these genes in angiogenesis is crucial as angiogenesis is a significant physiological process inherently linked with the progression of human conditions.

Keywords GAX and HOXA5 • Angiogenesis • Normal physiology • Disease • Malignancy

21.1 GAX

GAX is a homeobox gene, also known as MEOX-2, which inhibits angiogenesis in vascular endothelial cells (EC). Vascular endothelial cells are cells that are essential for the proper functioning of angiogenesis (Chen and Gorski 2008). In a healthy individual, there is a balance of proangiogenic and antiangiogenic factors established by various cells in order to maintain homeostasis of blood vessels (Nishida et al. 2006). This precise balance is a key determinant of whether or not the vascular endothelial cells will respond in a proangiogenic manner to a variety of physiologic signals. Under specific circumstances such as menstrual cycles, wound repair, embryogenesis and organogenesis, Proangiogenic factors and vascular endothelial growth factors (VEGF) bind to cell surface receptors. This ultimately leads to the

A. Salajegheh, *Angiogenesis in Health, Disease and Malignancy*,
DOI 10.1007/978-3-319-28140-7_21

attachment of transcription factors, stimulating genes whose regulation is crucial for the proangiogenic phenotype (Hillen and Griffioen 2007).

21.1.1 GAX in Normal Physiology

The GAX gene is originally derived from embryonic mesoderm and muscle precursors, however, its expression in adults is mostly restricted to the cardiovascular system and kidneys. Studies have shown that within the vascular smooth muscle cells, GAX expression is down-regulated by mitogen (proangiogenic or pro-inflammatory) factors and up-regulated by growth-arrested signals (Patel et al. 2005).

GAX exhibits two main mechanisms in vascular endothelial cells. First, GAX inhibits the proliferation of ECs, whereby this mechanism is carried out by the up-regulation of a cyclin-dependent kinase inhibitor, p21 or CDK-interacting protein 1 (p21WAF1/CIP1) (Chen et al. 2007). This is achieved through binding to the promoter and the enhancer of p21WAF1/CIP1. In mitosis, GAX expression arrests the G_1 cell cycle and stops the movement of vascular smooth muscle cells (Charron et al. 2006). GAX expression has been proven to inhibit proliferative stenosis of the arterial lumen proceeding injury (Ferguson and Patterson 2003). The second GAX mechanism is where GAX expression down-regulates a number of nuclear factors such as κB (NF-κB), while inhibiting the angiogenic influence of NF-κB and their action of activation in vascular cells (Chen et al. 2010). Such mechanisms of GAX reiterate the significance of GAX as an angiogenic phenotype inhibitor (Chen et al. 2007; 2010).

The angiogenic phenotype has been revealed to be potentially regulated by a class of short, single stranded noncoding micro RNAs (miRNAs). Here, they inhibit protein expressions in many human diseases including cancer, cardiovascular disease, psoriasis and neurodegenerative disorders (Dong et al. 2013). It is theorised that some microRNAs take part in modulating angiogenesis (Wang and Olson 2009). For example, miR-221 inhibits the proliferation and migration of the vascular endothelial cell, while miR-130a plays an important role in regulating GAX and HOXA5. The angiogenic phenotype is stimulated through the down-regulation of both homeobox genes miR-221 and miR-130a (Chen and Gorski 2008; Chen et al. 2010).

MicroRNA (miR-130a) is a proangiogenic phenotype regulator, thus embodying the ability to regulate GAX activity in vascular endothelial cells (Chen and Gorski 2008). Two miR-130a targeting sites were discovered on the 280-base pair fragments from the *GAX* 3′-untranslated region (3′-UTR), where these targeting sites are essential in the quick down-regulation of GAX expression in the presence of proangiogenic factors. This same 280-base pair sequence forces miR-130a expression whilst simultaneously blocking GAX expression (Chen and Gorski 2008). However, miR-130a functions in a reciprocal conduct in the vascular endothelial cells: it is up-regulated by mitogens and proangiogenic factors. In addition, miR-130a also seems to down-regulate other homeobox genes, including HOXA5, to keep ECs in the resting, quiescent state (Chen and Gorski 2008).

21.1.2 GAX in Disease

Current research has shown a potential to utilise GAX gene as a target for therapeutic methods to treat Alzheimer's disease, a leading causation of dementia in the middle or old aged population (Wu et al. 2005). Alzheimer's patients present symptoms such as confusion, mood swings, language difficulties, short-term memory loss, and compromised sustainability of a range of neurovascular functions (Wu et al. 2005).

As a neurodegenerative disorder, the cause and progression of the disease is poorly understood, thus the currently available treatments are only capable of supressing some of the symptoms. However, by saying that, studies have proven that Alzheimer's disease is characterised by insufficient angiogenesis. Data reveals that GAX, present in the human brain endothelial cells (BECs) in Alzheimer's, has relatively low gene expression of vascular restriction, therefore mediating atypical angiogenesis. GAX also triggers the AFX-1-dependent pro-apoptotic pathway while simultaneously restraining the expression of the LRP clearance receptor for Aβ at blood brain barrier (BBB) (Deane and Zlokovic 2007). This close linkage of GAX and Alzheimer's disease paves a pathway for future treatments of Alzheimer's disease. With the knowledge currently held, it is believed that by promoting angiogenesis and vascular remodelling through restoring GAX expression, AFX-mediated apoptosis will be terminated. In vitro, partial deletion of the GAX gene was performed on mice. As a result, the mice experienced apoptosis, vessel malformation and regression due to their compromised pro-angiogenic ability. This ultimately led to brain capillary density and CBF reduction. Another pathological phenotype is found in the blood brain barrier, where there is very minimal clearance capability due to the drop of LRP levels, resulting in the accumulation of Aβ. Both of these phenomenon are commonly present in patients with Alzheimer's disease (Wu et al. 2005; Deane and Zlokovic 2007).

With the current available knowledge and technology we are not able to cure the disease. However, we believed that with progressing research and clinical studies, the use of GAX may be anticipated as therapeutic target to stop or reverse the progress of Alzheimer's disease.

21.1.3 GAX in Malignancy

GAX expression has been hypothesised to be up-regulated by miR-221, where this is achieved by acting on multiple ZEB2 binding sites, and mediating the down-regulation of GAX (Chen et al. 2010). Laboratory studies have also shown that by down-regulating ZEB2, GAX becomes up-regulated. From this observation, it can be concluded that ZEB2 acts as a direct molecular target for miR-221 in the anti-angiogenic therapy of cancer and other pathophysiology driven by excessive angiogenesis (Chen et al. 2010).

GAX has a postulated role in regulating vascular endothelium cells during angiogenesis. There is also a great potential for GAX to be the molecular target in future antiangiogenic cancer therapy. However, it is practically very difficult to target GAX therapeutically due to the fact that it is antiangiogenic and is also a transcription factor. Nevertheless, recent studies have shown a new possibility to target factors that down-regulate GAX, such as ZEB2 and microRNA-221 (Chen et al. 2010).

As evident in the research regarding GAX in normal physiology, microRNAs harness a close relationship with the development of cancers. One example is the close relationship between tumourigenesis in hepatocellular carcinoma (HCC) and the overexpression of microRNA-301a (miR-301a). miRNAs are non-coding RNA with a small number of nucleotides that control gene expression at the post-transcriptional level (Zhou et al. 2012).

Hepatocellular carcinoma, also known as malignant hematoma, has a third-highest cancer-related death rate globally (Buendia and Neuveut 2015). Patients with HCC may present with symptoms as small as abdominal pain, loss of appetite, yellow-coloured skin and blood clotting abnormalities. Current treatments depend on the size and staging of the tumour, hence surgery may be performed if the tumour shows no evidence of metastasis. In addition, prior to the surgery, chemotherapy is prescribed to reduce the size of the tumour. Radiation therapy may also be useful, however, there is no effective agent for therapeutic treatment for HCC (Buendia and Neuveut 2015).

Studies have revealed that high expression levels of miRNA-301 have been detected in patients with HCC (Zhou et al. 2012). Even though there is very minimal knowledge on the biological behaviour and mechanisms of miR301a, we know that GAX is negatively regulated by miR-201a, and is believed to take part in the complex NF-kB signalling pathway through its expression. This makes microRNA a potential diagnostic marker, which can be used in prospective therapeutic method for HCC. Also, GAX expression can be used clinically to manipulate miRNA-301 behaviours and to elevate NF-kB activity (Zhou et al. 2012).

21.2 HOXA5

21.2.1 HOXA5 in Normal Physiology

Exploring HOXA5, the physiology have been found to reiterate that of GAX. The stimulation or inhibition of angiogenesis is regulated and controlled by specific microRNAs (miRNAs). In angiogenesis, there are two groups of miRNA: those that promote angiogenesis and those that suppress it, hence referred to as pro-angiogenic and anti-angiogenic, respectively. miR-221 is a highly expressed anti-angiogenic miRNA found in endothelial cells, known to decrease endothelial cell proliferation, migration and wound healing. On the other hand, pro-angiogenic miRNAs exhibits the opposite effect as evident in the mechanisms of miR-130a (Mujahid et al. 2013). It has been shown that miR-130a inhibits the homeobox proteins GAX and HOXA5 whilst simultaneously promotes endothelial cell proliferation and migration. Forced expression of miR-130a thus inhibits HOXA5 expression, and it is triggered when there is a need for angiogenesis to occur (Mujahid et al. 2013).

21.2.2 HOXA5 in Disease

HOXA5 is a contributing factor in human lung development and disease, where studies of various organs reveal that HOXA5 may play specific roles in vasculogenesis and angiogenesis of the developing lung (Kachgal et al. 2012; Silfa-Mazara et al. 2014). In the study conducted by Sana Mujahid on the regulation of lung development by miRNAs and HOX genes, it has been found that reduced airway branching was related to a disorganized vascular network due to either increased miR-221 or decreased miR-130a levels. Here, HOXA5 displayed angiostatic or anti-angiogenic effects. Any minor changes in the cellular expression of HOXA5 can lead to significant downstream changes in gene regulation and ultimately leads to angiogenesis (Mujahid et al. 2013). HOXA5 prevents several pro-angiogenic substance expression through inhibiting vessel branching. Thus, modified HOXA5 cellular distribution alters the hierarchy of HOX regulation in lung mesenchyme and results in the arrest of microvascular and airway branch development of lungs (Mujahid et al. 2013).

The inhibition of HOXA5 expression is advantageous and a key factor which influences coronary heart disease once it is rapidly initiated in response to hypoxic or ischemic conditions (Gaur et al. 1995). Coronary heart disease is a condition where the coronary arteries are constricted due to fatty deposits, resulting in decreased blood flow reaching the myocardium. Cases have been observed where new blood vessels have been formed naturally through the inhibition of HOXA5 and selectively triggering other angiogenic cytokines, producing collateral circulation and a flow of supplements through the occluded artery (Gaur et al. 1995, 2001; Jakob et al. 2012). This spontaneous process is being studied in order to apply stimulated angiogenesis to treat blocked vessels (Jakob et al. 2012).

21.2.3 HOXA5 in Malignancy

Angiogenesis is critical for the growth, invasion, and metastasis of human tumours. Studies reveal that HOXA5 play a significant role in the formation of tumours in the human breast (Mielnicki et al. 2001; Boucherat et al. 2009). HOXA5 is absent in the microvessels of invasive ductal breast carcinomas. According to studies performed by Zhu on HOXA5 expression and inhibition of the growth of experimental hemangiomas in the brain, the loss of HOXA5 may lead to vascular instability (Zhu et al. 2009). Quiescent vessels maintain normal HOXA5 levels but activated tumour vessels and proliferating infantile hemangiomas lack the transcription factor (Zhu et al. 2009). Hemangiomas are benign tumours of the vascular endothelium linked with increased activities of angiogenic factors. A disruption in the balance between pro-angiogenic and anti-angiogenic miRNAs may contribute to haemangioma formation. In the same study by Zhu, the growth of cutaneous hemangiomas was blocked by the re-expression of HOXA5 in haemangioma-derived endothelial cells. Permeability of endothelial cells was reduced and adherence to junctions were also stabilized. The study concluded that HOXA5 plays central role in coordinating a stable vascular phenotype (Zhu et al. 2009).

Currently, studies are being undertaken to investigate the role of miR-130a and thus HOXA5 in the regulation of branching morphogenesis, and how they may contribute in the development of many other organs through mediating vascular changes. miR-130a may represent a promising target for anti-angiogenic target of cancer (Chen and Gorski 2008). The therapeutic expression of HOXA5 is being further researched to inhibit cancer through genetic engineering. HOX gene products are highly conserved DNA-binding transcription factors that are capable of both activating and repressing gene expression (Volpe et al. 2008). Subtle changes in HOX cellular expression patterns lead to dramatic tissue development or abnormalities. Studies are being performed in order to discover the mechanism by which HOX genes alter the vasculature and also the effects of oxygen on miRNA levels, hence on the levels of HOX genes (Volpe et al. 2008).

Thus, through the observation of these genes, conclusive evidence can be gathered to advocate the potential of manipulating these genes in the therapeutic industry to either stimulate or inhibit angiogenesis, in our favour, to discover treatment options for various cancers and diseases. Understanding the mechanisms of these genes in angiogenesis is crucial as angiogenesis is a significant physiological process inherently linked with the progression of human conditions.

References

Boucherat O, Guillou F, Aubin J, Jeannotte L (2009) Hoxa5: a master gene with multifaceted roles. Med Sci 25(1):77–82. doi:10.1051/medsci/200925177

Buendia MA, Neuveut C (2015) Hepatocellular Carcinoma. Cold Spring Harbor Perspect Med 5(2):a021444. doi:10.1101/cshperspect.a021444

Charron T, Nili N, Strauss BH (2006) The cell cycle: a critical therapeutic target to prevent vascular proliferative disease. Can J Cardiol 22(Suppl B):41B–55B

Chen Y, Gorski DH (2008) Regulation of angiogenesis through a microRNA (miR-130a) that down-regulates antiangiogenic homeobox genes GAX and HOXA5. Blood 111(3):1217–1226. doi:10.1182/blood-2007-07-104133

Chen Y, Leal AD, Patel S, Gorski DH (2007) The homeobox gene GAX activates p21WAF1/CIP1 expression in vascular endothelial cells through direct interaction with upstream AT-rich sequences. J Biol Chem 282(1):507–517. doi:10.1074/jbc.M606604200

Chen Y, Banda M, Speyer CL, Smith JS, Rabson AB, Gorski DH (2010) Regulation of the expression and activity of the antiangiogenic homeobox gene GAX/MEOX2 by ZEB2 and MicroRNA-221. Mol Cell Biol 30(15):3902–3913. doi:10.1128/mcb.01237-09

Deane R, Zlokovic BV (2007) Role of the blood-brain barrier in the pathogenesis of Alzheimer's disease. Curr Alzheimer Res 4(2):191–197

Dong H, Lei J, Ding L, Wen Y, Ju H, Zhang X (2013) MicroRNA: function, detection, and bioanalysis. Chem Rev 113(8):6207–6233. doi:10.1021/cr300362f

Ferguson JE, Patterson C (2003) Break the cycle: the role of cell-cycle modulation in the prevention of vasculoproliferative diseases. Cell Cycle 2(3):210–218. doi:10.4161/cc.2.3.366

Gaur A, Lemanski LF, Dube DK (1995) Identification and expression of a homologue of the murine HoxA5 gene in the Mexican axolotl (Ambystoma mexicanum). Gene 162(2):249–253

Gaur A, Zajdel RW, Bhatia R, Isitmangil G, Denz CR, Robertson DR, Lemanski LF, Dube DK (2001) Expression of HoxA5 in the heart is upregulated during thyroxin-induced metamorphosis of the Mexican axolotl (Ambystoma mexicanum). Cardiovasc Toxicol 1(3):225–235

Hillen F, Griffioen AW (2007) Tumour vascularization: sprouting angiogenesis and beyond. Cancer Metastasis Rev 26(3–4):489–502. doi:10.1007/s10555-007-9094-7

Jakob P, Doerries C, Briand S, Mocharla P, Krankel N, Besler C, Mueller M, Manes C, Templin C, Baltes C, Rudin M, Adams H, Wolfrum M, Noll G, Ruschitzka F, Luscher TF, Landmesser U (2012) Loss of angiomiR-126 and 130a in angiogenic early outgrowth cells from patients with chronic heart failure: role for impaired in vivo neovascularization and cardiac repair capacity. Circulation 126(25):2962–2975. doi:10.1161/circulationaha.112.093906

Kachgal S, Mace KA, Boudreau NJ (2012) The dual roles of homeobox genes in vascularization and wound healing. Cell Adhes Migr 6(6):457–470. doi:10.4161/cam.22164

Mielnicki LM, Asch HL, Asch BB (2001) Genes, chromatin, and breast cancer: an epigenetic tale. J Mammary Gland Biol Neoplasia 6(2):169–182

Mujahid S, Nielsen HC, Volpe MV (2013) MiR-221 and miR-130a regulate lung airway and vascular development. PLoS One 8(2):e55911. doi:10.1371/journal.pone.0055911

Nishida N, Yano H, Nishida T, Kamura T, Kojiro M (2006) Angiogenesis in cancer. Vasc Health Risk Manag 2(3):213–219

Patel S, Leal AD, Gorski DH (2005) The homeobox gene Gax inhibits angiogenesis through inhibition of nuclear factor-kappaB-dependent endothelial cell gene expression. Cancer Res 65(4):1414–1424. doi:10.1158/0008-5472.can-04-3431

Silfa-Mazara F, Mujahid S, Thomas C, Vong T, Larsson I, Nielsen HC, Volpe MV (2014) Oxygen differentially affects the hox proteins Hoxb5 and Hoxa5 altering airway branching and lung vascular formation. J Cell Commun Signal 8(3):231–244. doi:10.1007/s12079-014-0237-7

Volpe MV, Wang KT, Nielsen HC, Chinoy MR (2008) Unique spatial and cellular expression patterns of Hoxa5, Hoxb4, and Hoxb6 proteins in normal developing murine lung are modified in pulmonary hypoplasia. Birth Defects Res A Clin Mol Teratol 82(8):571–584. doi:10.1002/bdra.20481

Wang S, Olson EN (2009) AngiomiRs – key regulators of angiogenesis. Curr Opin Genet Dev 19(3):205–211

Wu Z, Guo H, Chow N, Sallstrom J, Bell RD, Deane R, Brooks AI, Kanagala S, Rubio A, Sagare A, Liu D, Li F, Armstrong D, Gasiewicz T, Zidovetzki R, Song X, Hofman F, Zlokovic BV (2005) Role of the MEOX2 homeobox gene in neurovascular dysfunction in Alzheimer disease. Nat Med 11(9):959–965. doi:10.1038/nm1287

Zhou P, Jiang W, Wu L, Chang R, Wu K, Wang Z (2012) miR-301a is a candidate oncogene that targets the homeobox gene Gax in human hepatocellular carcinoma. Dig Dis Sci 57(5):1171–1180. doi:10.1007/s10620-012-2099-2

Zhu Y, Cuevas IC, Gabriel RA, Su H, Nishimura S, Gao P, Fields A, Hao Q, Young WL, Yang GY, Boudreau NJ (2009) Restoring transcription factor HoxA5 expression inhibits the growth of experimental hemangiomas in the brain. J Neuropathol Exp Neurol 68(6):626–632. doi:10.1097/NEN.0b013e3181a491ce

Chapter 22
Heparanase

Abstract Heparanase is multifunctional endoglycosidase expressed by platelets, leukocytes and capillary endothelium. Heparanase degrades Heparan Sulfate, one of the proteoglycan constituents of extracellular matrix (ECM). Little is known about physiological function of heparanase, however, heparanase expression or activity was found for hair growth, tissue regeneration, wound healing, embryonic and placenta development. Heparanase plays critical roll in many diseases to bring about abnormal function in physiological condition such as inflammation, kidney disease and thrombosis. Recently heparanase was found a critical role player in the pathophysiology of type I diabetes. Heparanase in malignancy is well characterized for its role in angiogenesis, metastasis and tumour growth. Heparanase degrades HS and subendothelial basement membrane facilitating endothelial cell invasion into ECM facilitating invasion and metastasis. By cleaving HS proteoglycan, by-product such as VEGF, bFGF from ECM and basement membrane is released. This induces new blood vessels to form, enhance tumour growth and metastasis.

Keywords Heparanase • Angiogenesis • Normal physiology • Disease • Malignancy

22.1 Heparanase

Heparanase is multifunctional endoglycosidase expressed by platelets, leukocytes and capillary endothelium that degrade Heparan Sulfate (HS), which is complex saccharide that is one of the proteoglycan constituents of extracellular matrix (ECM), and also found on cell surface (Nasser 2008; Ginath et al. 2014). Little is known about physiological function of heparanase that is yet to be investigated, however, heparanase expression or activity was found for HS turn over, hair growth, tissue regeneration, wound healing, embryonic development and human placenta (Malgouries et al. 2008; Chan et al. 2009; Fux et al. 2009; Coulson-Thomas et al. 2014; Ginath et al. 2014).

A. Salajegheh, *Angiogenesis in Health, Disease and Malignancy*,
DOI 10.1007/978-3-319-28140-7_22

22.2 Heparanase in Normal Physiology

Animal study shown that heparanase expression increased in anagen hair follicles, moreover accelerated with overexpression of Heparanase and decreased in catagen follicles (Malgouries et al. 2008; Coulson-Thomas et al. 2014). Heparanase is also expressed in skin and the wound granulation tissue, result in keratinocyte migration and wound closure observed in vitro. Moreover, topical application of heparanase accelerated wound healing (Zcharia et al. 2005).

22.3 Heparanase in Disease

Heparanase itself does not cause any diseases, however, it plays critical roll in many diseases to bring about abnormal function in physiological condition such as inflammation, kidney disease, type I diabetes and thrombosis (van den Hoven et al. 2007; Fux et al. 2009; Favaro et al. 2010).

Remodeling of HS by heparanase may affect in different aspect of inflammation since HS controls inflammatory responses with multiple roles including sequester chemo/cytokines in ECM or activation of innate immune system by interaction with toll-like receptors (Fux et al. 2009; Goldberg et al. 2013). Anti-inflammatory responses were demonstrated with heparanase inhibiting substances such as heparin (Goldberg et al. 2013). In acute inflammatory response, neutrophils play important effector roles. The heparanase on neutrophils were reported as degradation of endothelial glycocalyx in sepsis-associated lung injury (Fux et al. 2009). In both animal and human studies, consistent with HS in glycocalyx structure facilitate neutrophil adhesion molecules (Goldberg et al. 2013). Also, heparanase knockout mice and mice treated with inhibitor demonstrated endothelial hyperpermeability with facilitation of neutrophil recruitment. On the other hand, transgenic mice that overexpress heparanase demonstrated reduction in neutrophil recruitment and neutrophil adhesion molecule on endothelial surface (Goldberg et al. 2013).

If the acute inflammation not resolved, infiltration of the neutrophil changes to macrophages, dominant player in chronic inflammation. Activated macrophage induces heparanase expression with tumour necrotic factor alpha dependent mechanism and increases post-translational modification of cathepsin L (Goldberg et al. 2013). Cathepsin L is protease that is found in lysosome that activates heparanase.

Heparanase serves role in glomerular diseases, its expression is associated in loss of HS in basement membrane of glomerulus. Transgenic mice with overexpressing heparanase developed proteinuria, which suggest that HS on basement membrane have role of charge-selecting permeability in such way that reduced interaction of HS and the basement membrane or release the fragment from the HS or intracellular signaling through binding to glomeruli (van den Hoven et al. 2007).

Heparanase also plays in thrombosis after vascular injury followed by stent treatment. The anticoagulant property of the HS is from interaction between HS and

Antithrombin-3 (AT-3), inhibit cascade of thrombin and factor Xa. This mechanism is analogue to mechanism of heparin. However, overexpression of heparanase in simple endothelial injury does not increase risk of thrombosis but in stent-induced flow disturbance (Baker et al. 2012).

Recent study have identified new roles in type I diabetes, HS in the basement membrane of islets in pancreas acts as barrier to leukocyte invasion, also increased in expression in HS in the islet of Langham in beta cell producing insulin is critical for cell survival, thus provide protecting the cells in pancreas from the reactive oxygen species (ROS) damage (Parish et al. 2001). Moreover, heparanase induces self reactive T cell and other inflammatory cells to degrade intracellular HS, which make beta cell prone to damage from ROS and lead to cell death. In animal study in mice, the incidence of the induced type I diabetes significantly reduced after administering heparanase inhibitor, PI-88, which preserves HS in beta cell and reduced inflammation in pancreas (Parish et al. 2001; Nasser 2008).

22.4 Heparanase in Malignancy

Although only little is known about physiology in normal cell, pathophysiology of heparanase in angiogenesis, metastasis and tumour growth is well known. Heparanase degrades HS and subendothelial basement membrane facilitating endothelial cell invasion into ECM facilitating invasion and metastasis (Vlodavsky et al. 1999; Vlodavsky et al. 2002). Metastatic potential correlates the expression of heparanase (Parish et al. 2013). In severe metastatic cancer, heparanase could be detected in urine. Heparanase potentially can cross glomeruli basement membrane and degrade HS in the kidney and destroying the function of selective permeability (Vlodavsky et al. 2002; van den Hoven et al. 2007). It also produces indirect angiogenic effects by cleaving HS and release growth factors as by-product such as VEGF, bFGF from ECM and basement membrane (Vlodavsky et al. 2002). Immunohistological stain found that increased expression of the heparanase in proximal area of tumour in colon, pancreas and breast carcinoma (Vlodavsky et al. 1999). This induces new blood vessels to form, enhance tumour growth and metastasis (Vlodavsky et al. 1999, 2002).

Extracellular heparanase can also interact with cellular receptor which lead to pro-angiogenic effects by upregulating VEGF, HGF and stimulating Akt, P13K and p38 MAP kinase signaling pathway result in endothelial cell migration (Parish et al. 2013). In result, at certain stage of angiogenesis would be heparanase dependent therefore, this can be targeted for anti-cancer drug.

Direct role of the Heparanase in the metastatic tumour is through providing conversion of non-metastatic T-cell (non heparanase activity) into metastatic behaviour (Express heparanase activity) (Vlodavsky et al. 2002). In animal studies, mortality of the mice was increased after injecting heparanase overexpressing cell subcutaneously compared to inoculation of lymphoma (Vlodavsky et al. 2002). Moreover, survival year after operation was inversely correlated to lower in survived years in heparanase-positive malignancy compared to higher in survived years in heparanase-negative malignancy (Vlodavsky et al. 2002).

Angiogenesis is the process of generating new capillary blood vessels from pre-existing vessels. Heparanase cleaves HS which is complex saccharide proteoglycan constituents of extracellular matrix. By cleaving HS proteoglycan, by-product such as VEGF, bFGF from ECM and basement membrane is released. This induces new blood vessels to form, enhance tumour growth and metastasis. Also, extracellular heparanase can also interact with cellular receptor which lead to pro-angiogenic effects by upregulating VEGF, HGF. Moreover, cleaving HS in ECM provides space for new blood vessel to form.

References

Baker AB, Gibson WJ, Kolachalama VB, Golomb M, Indolfi L, Spruell C, Zcharia E, Vlodavsky I, Edelman ER (2012) Heparanase regulates thrombosis in vascular injury and stent-induced flow disturbance. J Am Coll Cardiol 59(17):1551–1560. doi:10.1016/j.jacc.2011.11.057

Chan SY, Zhang Y-Y, Hemann C, Mahoney CE, Zweier JL, Loscalzo J (2009) MicroRNA-210 controls mitochondrial metabolism during hypoxia by repressing the iron-sulfur cluster assembly proteins ISCU1/2. Cell Metab 10(4):273–284. doi:10.1016/j.cmet.2009.08.015

Coulson-Thomas VJ, Gesteira TF, Esko J, Kao W (2014) Heparan sulfate regulates hair follicle and sebaceous gland morphogenesis and homeostasis. J Biol Chem 289(36):25211–25226. doi:10.1074/jbc.m114.572511

Favaro E, Ramachandran A, McCormick R, Gee H, Blancher C, Crosby M, Devlin C, Blick C, Buffa F, Li J-L, Vojnovic B, Pires das Neves R, Glazer P, Iborra F, Ivan M, Ragoussis J, Harris AL (2010) MicroRNA-210 regulates mitochondrial free radical response to hypoxia and Krebs cycle in cancer cells by targeting iron sulfur cluster protein ISCU. PLoS One 5(4):e10345. doi:10.1371/journal.pone.0010345

Fux L, Ilan N, Sanderson RD, Vlodavsky I (2009) Heparanase: busy at the cell surface. Trends Biochem Sci 34(10):511–519. doi:10.1016/j.tibs.2009.06.005

Ginath S, Lurie S, Golan A, Amsterdam A, Sandbank J, Sadan O, Kovo M (2014) The expression of heparanase in normal and preeclamptic placentas. J Matern Fetal Neonatal Med 28(13):1589–1593. doi:10.3109/14767058.2014.962506

Goldberg R, Meirovitz A, Hirshoren N, Bulvik R, Binder A, Rubinstein AM, Elkin M (2013) Versatile role of heparanase in inflammation. Matrix Biol 32(5):234–240. doi:10.1016/j.matbio.2013.02.008

Malgouries S, Donovan M, Thibaut S, Bernard BA (2008) Heparanase 1: a key participant of inner root sheath differentiation program and hair follicle homeostasis. Exp Dermatol 17(12):1017–1023. doi:10.1111/j.1600-0625.2008.00739.x

Nasser NJ (2008) Heparanase involvement in physiology and disease. Cell Mol Life Sci 65(11):1706–1715. doi:10.1007/s00018-008-7584-6

Parish CR, Freeman C, Hulett MD (2001) Heparanase: a key enzyme involved in cell invasion. Biochim Biophys (BBA) - Rev Cancer 1471(3):M99–M108. doi:10.1016/s0304-419x(01)00017-8

Parish CR, Freeman C, Ziolkowski AF, He YQ, Sutcliffe EL, Zafar A, Rao S, Simeonovic CJ (2013) Unexpected new roles for heparanase in Type 1 diabetes and immune gene regulation. Matrix Biol 32(5):228–233. doi:10.1016/j.matbio.2013.02.007

van den Hoven MJ, Rops AL, Vlodavsky I, Levidiotis V, Berden JH, van der Vlag J (2007) Heparanase in glomerular diseases. Kidney Int 72(5):543–548. doi:10.1038/sj.ki.5002337

Vlodavsky I, Friedmann Y, Elkin M, Aingorn H, Atzmon R, Ishai-Michaeli R, Bitan M, Pappo O, Peretz T, Michal I, Spector L, Pecker I (1999) Gene cloning, expression and function in tumour progression and metastasis. Nat Med 5(7):793–802. doi:10.1038/10518

Vlodavsky I, Goldshmidt O, Zcharia E, Atzmon R, Rangini-Guatta Z, Elkin M, Peretz T, Friedmann Y (2002) Mammalian heparanase: involvement in cancer metastasis, angiogenesis and normal development. Semin Cancer Biol 12(2):121–129. doi:10.1006/scbi.2001.0420

Zcharia E, Zilka R, Yaar A, Yacoby-Zeevi O, Zetser A, Metzger S, Sarid R, Naggi A, Casu B, Ilan N, Vlodavsky I, Abramovitch R (2005) Heparanase accelerates wound angiogenesis and wound healing in mouse and rat models. FASEB J 19(2):211–221. doi:10.1096/fj.04-1970com

Chapter 23
Hepatocyte Growth Factor

Abstract Hepatocyte Growth Factor (HGF) is a gene primarily involved in the regulation of cell growth, motility and morphogenesis. Specifically, this growth factor acts upon cells of both epithelial and endothelial origin, in addition to haemopoietic stem cells. Typically, its function is achieved through a mechanism in which HGF binds to its corresponding membrane receptor, c-Met or MET, thus triggering the activation of Tyrosine Kinase signalling cascade. This cascade is collectively known as the process of "invasive growth" and gives rise to a sequence of biological responses essential to carry out its function. These responses facilitate cell growth and proliferation, which are essential for embryonic development and wound healing. In addition; the stimulatory effects of HGF on the processes of mitogenesis, cell motility and matrix invasion provide the cytokine with an imperative role in angiogenesis. HGF exerts a marked effect in cell growth, motility and morphogenesis. It demonstrates regulation of angiogenesis through the stimulation of mitogenesis, cell motility and matrix invasion. The presence of angiogenesis in tumour formation, growth and metastasis provide malignant tissue with sufficient nutrient supply to maintain its function.

Keywords Hepatocyte growth factor • HGF • Angiogenesis • Normal physiology • Disease • Malignancy

23.1 Hepatocyte Growth Factor

Hepatocyte Growth Factor (HGF) is a gene primarily involved in the regulation of cell growth, motility and morphogenesis (Blumenschein et al. 2012). Specifically, this growth factor acts upon cells of both epithelial and endothelial origin, in addition to haemopoietic stem cells.

Typically, its function is achieved through a mechanism in which HGF binds to its corresponding membrane receptor, c-Met or MET, thus triggering the activation of Tyrosine Kinase signalling cascade (Nakamura and Mizuno 2010). This cascade is collectively known as the process of "invasive growth" and gives rise to a sequence of biological responses essential to carry out its function. These responses facilitate cell growth and proliferation, which are essential for embryonic development and

A. Salajegheh, *Angiogenesis in Health, Disease and Malignancy*,
DOI 10.1007/978-3-319-28140-7_23

wound healing (Trusolino et al. 2000). In addition; the stimulatory effects of HGF on the processes of mitogenesis, cell motility and matrix invasion provide the cytokine with an imperative role in angiogenesis (Lemmon and Schlessinger 2010).

HGF was initially discovered in rat hepatocytes as a mitogen or 'mitosis-inducing substance' in 1984. Conversely, the c-Met receptor was recognised as a proto-oncogene in the mid 1980s, with the ability to induce malignant formation in the human cell line, if mutated. HGF is formed in embryonic organs from stromal cells and is secreted strictly by mesenchymal cells as a single, inactive polypeptide whilst MET is produced by parenchymal cells (Nakamura and Mizuno 2010). Serine proteases slice the polypeptide into an alpha and beta chain. A disulphide bridge bonding these chains will produce an active heterodimer (Lemmon and Schlessinger 2010).

23.2 The Role of HGF Gene in Normal Physiology

HGF has potent mitotic and angiogenic influence under normal physiological conditions. The action of HGF in the body is most predominantly regulated through any direct activity of vascular endothelial cells. These may include any stimulation due to cell proliferation, migration, invasion or the formation of protease. Cancers of invasive nature in the human body, had revealed over-expression of the HGF gene in comparison to tumours of benign or non-invasive nature (Rosen et al. 1997). The Von Willebrand factor is a blood glycoprotein associated with hemostasis, and typically marks the presence of vascular endothelial cells (Zanetta et al. 2000). In certain studies the level of Von Willebrand factor positively correlated to that of HGF in invasive breast carcinoma and thus, demonstrates the angiogenic influence of HGF, particularly under diseased condition (Rosen et al. 1997).

In foetal development, the neutralisation of the growth factor or destruction of its receptor may have devastating physiological effects including hypoplasia of the developing organs. Thus, proper regulation of HGF and MET is vital for organ growth and development. HGF has the capacity to aid in the regeneration of damaged liver, kidney and lung tissue. Moreover, the gene wields protective function over the heart and brain, both epithelial and non-epithelial organs, through anti-inflammatory and anti-apoptotic signalling (Nakamura and Mizuno 2010). Studies depicting the effects of HGF production have been established using mice as subjects. It had been delineated that the implementation of the anti-HGF antibody had accelerated the destruction of tissue in the test rodents. Furthermore, levels of HGF in blood plasma have been shown to vastly increase during periods of organ disease, and as such, inadequate release or production of HGF will lead to organ failure, whilst its presence is required for the minimisation of disease (Ueda et al. 2000).

23.3 The Role of HGF Gene in Disease and Malignancy

Cancer may be most commonly defined as uncontrolled cell division, leading to tumour formation and the destruction of healthy tissue in the body. The incidence of cancer is becoming increasingly prominent in today's society with estimates of 12.7 million new cases and 7.6 million deaths in just 27 types of tumours in 2008 (Mizuno and Nakamura 2013).

In cancer cells, the abnormal activation of the c-Met receptor by HGF stimulates both tumour growth and angiogenesis. The formation of new blood vessels enables the supply of nutrients to the tumour cell, enabling its growth and the potential to metastasise to other healthy, functioning organs. Under normal conditions, MET can be expressed by stem and progenitor cells (Mazzone and Comoglio 2006). This typically facilitates the generation of new tissues in an embryo or the regeneration of damaged tissue in adulthood, by invasive growth of these stem cells. It has been suggested that the ability of regular stem cells to express the c-Met receptor can be inhibited by the action of cancerous stem cells. Thus, malignancies are enabled to persist and further metastasise. Numerous lines of *in-vitro* studies have widely implicated HGF in the scattering and migration of cancerous cells (Ding et al. 2003).

The production and expression of HGF and MET in malignant tissues occur in stromal and cancer cells respectively. It has been determined that the mutation of c-Met is highly correlated to the incidence of familial renal carcinoma, among other forms of sporadic malignancies. Irrespective to the effect of the MET receptor in cancerous tissue, stromal cell secretion of HGF has also been linked as a prevalent factor in invasive tumour growth. HGF is necessary in the infiltration of neighbouring tissues in malignant cells, including vascular beds across the basement membrane (Mizuno and Nakamura 2013). The events which induce tumour growth and metastasis involving the HGF-MET complex are as follows. For the spread of malignant tissue to occur the tumour cells must first dissociate. Cell to cell adhesion is maintained by substances known as Cadherins and intracellular Catenin molecules (Ding et al. 2003). HGF enables cluster cells to disperse into single cells by endocytosis of E-cadherin. This substance is removed from the cell surface and transported to the cytoplasm, leading to its down-regulation, and in-turn, the loss of cell to cell adhesion. Once the tumor cells have dissociated, they require motility to further complete metastasis. The role of HGF in this component includes the 'nuclear localisation' of β-catenin. This substance is a transcription factor for genes and plays a vital role in cell motility. In order for cancer cells to be able to invade adjacent cells, they must first have the capacity to cross the basement membrane lying between the epithelium and submucosa. To achieve this, the basement membrane must be broken down. HGF plays a role in this phase by two mechanisms. the first involves the phosphorylation of 'Focal Adhesion Kinase' (FAK) whilst the second includes the up-regulation of several forms of MMP (Mizuno and Nakamura 2013). FAK is a gene that induces intracellular signal transduction pathways, which essentially enhance the dissociation of cell contacts with the extracellular matrix, thus promoting cell motility and migration (Mazzone and Comoglio 2006). Collectively, the phosphorylation and up-regulation of FAK and MMP respectively, are necessary components in the metastatic and invasive process of cancer formation (Mizuno and Nakamura 2013).

The HGF-MET complex is required for remote metastasis. HGF triggers the formation of vascular beds and promotes chemokine-induced homing, in addition to protecting cancer cells from apoptosis due to inadequate cell-matrix interactions. As previously stated angiogenesis plays a vital role in the spread of primary tumours to auxiliary organs. Angiogenesis is induced and enhances through the regulation of growth, motility and morphogenesis of endothelial cells, as achieved by HGF. This cytokine enhances the contact between the tumour and endothelial cell via the FAK mechanism previously described (Jiang et al. 1999). Furthermore, HGF is responsible for the down-regulation of 'Occludin', a molecule responsible for cell to cell adhesion, thus enabling cell motility. In turn, cancer vessels have a decreased resistance allowing endothelial cells to be invaded by malignant tissue (Mizuno and Nakamura 2013).

In numerous studies the presence of HGF has shown high prevalence in a vast array of cancers. In Hepatocellular Carcinoma (HCC), HGF had revealed immense over-expression in the microenvironment comparative to levels in the healthy adult liver. The expression of transgenic HGF in mice had enabled the acceleration of chemically induced hepatocarcinogenesis, thus suggestive of the oncogenic influence of HGF. However, the severity of this growth factor in human cases of HCC is yet to be determined (Goyal et al. 2013).

The presence of HGF in head and neck Squamous Cell Carcinoma was also identified with a significant increase of its levels in this tissue in comparison to normal mucosal tissue. This finding was also indicative of a poorly differentiated tumour type as well as a decrease in the chance of survival (Uchida et al. 2001).

Both HGF and c-Met are found to be over-expressed in a high majority of Papillary Thyroid Carcinomas (PTC). Its overexpression is highly correlated to metastasis, as well as recurrence, particularly in children and young adults suffering from the disease. It was found that in response to HGF, tumours in the thyroid showed increased motility and invasiveness, an increase in chemokine production, and finally the induction of dendritic cells and new blood vessels (Koo et al. 2014).

As previously suggested, angiogenesis plays a predominant role in the maintenance and metastasis of tumour cells, supplying them with the nutrients essential for their survival and growth. As angiogenesis can be triggered by the stimulation of the HGF-MET complex, its inhibition may prove as a therapeutic treatment against malignancies. HGF is the only known ligand to the c-Met receptor, thus their binding inhibits further MET activity. As such, HGF inhibitors may be used to obstruct HGF from binding to c-Met (Bottaro et al. 1991). Three such types of HGF inhibitors are in existence and have been trialled, including truncated or shortened HGF, anti-HGF neutralising antibodies, and an uncleaved form of HGF. Certain studies have suggested that at least three neutralising anti-HGF antibodies are required to prevent the MET associated activation of the Tyrosine Kinase mechanism, as deduced from the testing of different HGF epitopes. However, in more recent studies, human monoclonal antibodies have been found with the ability to individually bind and neutralise human HGF, which in effect, has lead to tumour degeneration in mouse subjects. At present, two of the anti-HGF antibodies are available for clinical trialling including AVEO (AV299) and AMGEN (AMG102) (Burgess et al. 2006). Uncleavable HGF carries a single amino-acid substitution, which disables the potential for the HGF molecule to mature. The substance is able to bind to c-Met with a high affinity that blocks the matured HGF from binding, and thus inhibits the biological activity of MET. If uncleaved HGF is expressed either locally or systemically, malignant growth and metastasis may be inhibited (Mazzone et al. 2004).

As a growth factor, HGF contains regenerative properties which enable its role in the development, maintenance, and repair of tissues. Studies have demonstrated that the employment of exogenous HGF has supplied several organs in the body such as the brain, liver, and heart with positive and regenerative influence.

In Japan, a group of researchers had gathered evidence regarding the influence of HGF after transient middle cerebral artery occlusion. HGF had shown to decrease the size of infarct by significant measures, as well as reduce the number of TUNEL-positive cells. In addition, HGF was successful in reducing the production of glial scar tissue and its thickness, as well as heightening synaptogenesis and angiogenesis (Shang et al. 2011).

Another study had explored a method in which bone marrow stromal stem cells (BMSC) could be altered to form cardiomyocyte-like cells, by action of HGF. In this study, HGF was an essential component in the formation of cardiomyocyte-like cells, and these cells were utilised in the treatment of ischemic heart disease (Wen et al. 2010).

Myositis is an uncommon disease in which generally affects the muscles causing weakness in limbs as well as muscle swelling, pain and inflammation (Shang et al. 2011). Research has suggested that HGF is a factor present during muscle regeneration. In cultured myoblasts, it was determined that the addition of HGF had heightened the expression of the regulatory factors required for muscle regeneration, and thus had positive effects when given to patients suffering from myositis (Sugiura et al. 2010).

Conclusively, HGF exerts a marked effect in cell growth, motility and morphogenesis. It demonstrates regulation of angiogenesis through the stimulation of mitogenesis, cell motility and matrix invasion. The presence of angiogenesis in tumour formation, growth and metastasis provide malignant tissue with sufficient nutrient supply to maintain its function. In effect, the presence of HGF-inhibitors may be used to block the HGF-MET mediated pathway for angiogenesis. Studies have demonstrated the high prevalence of HGF in cancerous tissue however the full extent of its influence in the body in both healthy and diseased tissue remains unclear. Its regenerative properties in tissues and organs provide the growth factor with great potential of therapeutic use in today's society. In effect, these genes provide discernible effects on the process of angiogenesis and its role in tumour formation, and hence, manipulation of these genes may 1 day pave a therapeutic resolution to a myriad and malignancies and disease.

References

Blumenschein GR Jr, Mills GB, Gonzalez-Angulo AM (2012) Targeting the hepatocyte growth factor-cMET axis in cancer therapy. J Clin Oncol 30(26):3287–3296. doi:10.1200/jco.2011.40.3774

Bottaro DP, Rubin JS, Faletto DL, Chan AM, Kmiecik TE, Vande Woude GF, Aaronson SA (1991) Identification of the hepatocyte growth factor receptor as the c-met proto-oncogene product. Science (New York, NY) 251(4995):802–804

Burgess T, Coxon A, Meyer S, Sun J, Rex K, Tsuruda T, Chen Q, Ho SY, Li L, Kaufman S, McDorman K, Cattley RC, Sun J, Elliott G, Zhang K, Feng X, Jia XC, Green L, Radinsky R, Kendall R (2006) Fully human monoclonal antibodies to hepatocyte growth factor with therapeutic potential against hepatocyte growth factor/c-Met-dependent human tumors. Cancer Res 66(3):1721–1729. doi:10.1158/0008-5472.can-05-3329

Ding S, Merkulova-Rainon T, Han ZC, Tobelem G (2003) HGF receptor up-regulation contributes to the angiogenic phenotype of human endothelial cells and promotes angiogenesis in vitro. Blood 101(12):4816–4822. doi:10.1182/blood-2002-06-1731

Goyal L, Muzumdar MD, Zhu AX (2013) Targeting the HGF/c-MET pathway in hepatocellular carcinoma. Clin Cancer Res 19(9):2310–2318. doi:10.1158/1078-0432.ccr-12-2791

Jiang WG, Martin TA, Matsumoto K, Nakamura T, Mansel RE (1999) Hepatocyte growth factor/scatter factor decreases the expression of occludin and transendothelial resistance (TER) and increases paracellular permeability in human vascular endothelial cells. J Cell Physiol 181(2):319–329. doi:10.1002/(sici)1097-4652(199911)181:2<319::aid-jcp14>3.0.co;2-s

Koo BS, Kim JM, Seo ST, Yoon YH, Kwon KR, Kim SH, Kwon HW, Bae WJ, Lim YC (2014) Upregulation of HGF and c-MET is associated with subclinical central lymph node metastasis in papillary thyroid microcarcinoma. Ann Surg Oncol 21(7):2310–2317. doi:10.1245/s10434-014-3553-5

Lemmon MA, Schlessinger J (2010) Cell signaling by receptor tyrosine kinases. Cell 141(7):1117–1134. doi:10.1016/j.cell.2010.06.011

Mazzone M, Comoglio PM (2006) The Met pathway: master switch and drug target in cancer progression. FASEB J 20(10):1611–1621. doi:10.1096/fj.06-5947rev

Mazzone M, Basilico C, Cavassa S, Pennacchietti S, Risio M, Naldini L, Comoglio PM, Michieli P (2004) An uncleavable form of pro-scatter factor suppresses tumor growth and dissemination in mice. J Clin Invest 114(10):1418–1432. doi:10.1172/jci22235

Mizuno S, Nakamura T (2013) HGF-MET cascade, a key target for inhibiting cancer metastasis: the impact of NK4 discovery on cancer biology and therapeutics. Int J Mol Sci 14(1):888–919. doi:10.3390/ijms14010888

Nakamura T, Mizuno S (2010) The discovery of hepatocyte growth factor (HGF) and its significance for cell biology, life sciences and clinical medicine. Proc Jpn Acad Ser B Phys Biol Sci 86(6):588–610

Rosen EM, Lamszus K, Laterra J, Polverini PJ, Rubin JS, Goldberg ID (1997) HGF/SF in angiogenesis. Ciba Found Symp 212:215–226; discussion 227–219

Shang J, Deguchi K, Ohta Y, Liu N, Zhang X, Tian F, Yamashita T, Ikeda Y, Matsuura T, Funakoshi H, Nakamura T, Abe K (2011) Strong neurogenesis, angiogenesis, synaptogenesis, and antifibrosis of hepatocyte growth factor in rats brain after transient middle cerebral artery occlusion. J Neurosci Res 89(1):86–95. doi:10.1002/jnr.22524

Sugiura T, Kawaguchi Y, Soejima M, Katsumata Y, Gono T, Baba S, Kawamoto M, Murakawa Y, Yamanaka H, Hara M (2010) Increased HGF and c-Met in muscle tissues of polymyositis and dermatomyositis patients: beneficial roles of HGF in muscle regeneration. Clin Immunol (Orlando, Fla) 136(3):387–399. doi:10.1016/j.clim.2010.04.015

Trusolino L, Cavassa S, Angelini P, Ando M, Bertotti A, Comoglio PM, Boccaccio C (2000) HGF/scatter factor selectively promotes cell invasion by increasing integrin avidity. FASEB J 14(11):1629–1640

Uchida D, Kawamata H, Omotehara F, Nakashiro K, Kimura-Yanagawa T, Hino S, Begum NM, Hoque MO, Yoshida H, Sato M, Fujimori T (2001) Role of HGF/c-met system in invasion and metastasis of oral squamous cell carcinoma cells in vitro and its clinical significance. Int J Cancer J Int du Cancer 93(4):489–496

Ueda T, Takeyama Y, Hori Y, Shinkai M, Takase K, Goshima M, Yamamoto M, Kuroda Y (2000) Hepatocyte growth factor increases in injured organs and functions as an organotrophic factor in rats with experimental acute pancreatitis. Pancreas 20(1):84–93

Wen Q, Zhou CY, Zhou MQ, Luo W, Ma L (2010) Differentiation of 5-azacytidine-induced bone marrow stromal stem cells transduced with hepatocyte growth factor gene. Nan fang yi ke da xue xue bao 30(7):1537–1540

Zanetta L, Marcus SG, Vasile J, Dobryansky M, Cohen H, Eng K, Shamamian P, Mignatti P (2000) Expression of Von Willebrand factor, an endothelial cell marker, is up-regulated by angiogenesis factors: a potential method for objective assessment of tumor angiogenesis. Int J Can J Int du Can 85(2):281–288

Chapter 24
HIF-1α

Abstract Angiogenesis is regulated by the oxygen demand in local tissue, with the hypoxia-inducible factor-1 (HIF-1) being the primary factor mediating this response. HIF-1 is a transcriptional activator which is active in hypoxic environments. It is composed of two subunits, HIF-1α and HIF-1β. Under normal conditions, HIF-1α protein is constantly synthesized and degraded, a protective mechanism to help our body respond rapidly to hypoxic stress. In hypoxia, HIF-1α degradation is inhibited, this leads to the accumulation HIF-1α. The HIF complex then binds to hypoxia response elements resulting rapid increase in oxygen supply. For long-term adaptation, HIF pathway stimulates angiogenesis by regulating several pro-angiogenic genes such as angiopoietin-1, angiopoietin-2, Tie2, vascular endothelial growth factor (VEGF), platelet-derived growth factor (PDGF), basic fibroblast growth factor (bFGF) and monocyte chemoattractant protein-1 (MCP-1). HIF-1α expression in cancer is linked with unfavourable prognosis which is mainly due to therapeutic resistance. There is a hypothesis that increased levels of HIF-1α has a selective advantage for tumor survival. This makes the HIF angiogenic pathway an attractive target for cancer therapies as it regulates several pro-angiogenic pathways.

Keywords HIF-1α • Angiogenesis • Normal physiology • Disease • Malignancy

24.1 HIF-1α

Angiogenesis means the formation of new blood vessels which is a vital process in life, as it re-establishes a dense microvascular network for oxygen and nutrients in the case of tissue injury (Ahluwalia and Tarnawski 2012). Angiogenesis is regulated by the oxygen demand in local tissue, with the hypoxia-inducible factor-1 (HIF-1) being the primary factor mediating this response (Ke and Costa 2006). As the name suggests, HIF-1 is a transcriptional activator which is active in hypoxic environments. It is composed of two subunits, HIF-1α and HIF-1β. HIF-1β is being constantly synthesized and maintained at a steady level regardless of the environment while HIF-1α level is oxygen dependent. Under normal conditions, HIF-1α protein is constantly synthesized and degraded, a protective mechanism to help our body respond rapidly to hypoxic stress (Wang and Semenza 1993). In hypoxia, HIF-1α

A. Salajegheh, *Angiogenesis in Health, Disease and Malignancy*,
DOI 10.1007/978-3-319-28140-7_24

degradation is inhibited, this leads to the accumulation HIF-1α. HIF-1α is then transported from the cytoplasm into the nucleus and here in the nucleus, HIF-1α forms a dimer with HIF-1β which results in the transcriptionally active HIF complex (Hirota and Semenza 2006). The HIF complex then binds to hypoxia response elements (HREs) inside the targeted genes, activating transcription with help from transcriptional co-activators p300 and CBP. The result of this is a rapid increase in oxygen supply through the increased activity of the enzyme inducible nitric oxide synthase. This helps to relax vascular smooth muscle cells which lead to increased blood flow to the hypoxic tissues. In addition, oxygen demand is also decreased to provide short term relief by increasing the utilization of the glycolysis pathway, decreasing cell proliferation and inhibition of mitochondrial respiration (Krock et al. 2011). For long-term adaptation, the HIF pathway stimulates angiogenesis by regulating several pro-angiogenic genes. These include "angiopoietin-1, angiopoietin-2, Tie2, vascular endothelial growth factor (VEGF), platelet-derived growth factor (PDGF), basic fibroblast growth factor (bFGF) and monocyte chemoattractant protein-1 (MCP-1)" (Hirota and Semenza 2006). The net effect is an increase in **permeability of the blood vessels**, proliferation of **endothelial cell**, migration, adhesion, **formation of tubes** (Krock et al. 2011). This achieves in creating denser vascularity and therefore helps diminish hypoxic stress in the area. Hence, with HIF-1's comprehensive involvement in angiogenesis, it's not surprising that HIF-1 has an important role in embryonic development and cancer development.

24.2 HIF-1α in Normal Physiology

Embryonic development is a complex process with a huge demand on oxygen and nutrient supply. The process is complicated not only by the relatively poor levels of oxygen in the uterine environment, the rapidly proliferating tissues also needs a good oxygen delivery system to necessitate the nutrients and energy required for organ development (Krock et al. 2011). In the early phases of embryogenesis, the oxygenation of the embryo can be achieved with simple diffusion, but when the diffusion limit is reached, an oxygen delivery system is vital to meet the increasing demand. HIF-1 is therefore important in embryonic development, where it coordinates vasculogenesis and forms the primitive vascular system (Wang and Semenza 1993). The necessity of this system has been proved in mice trials, where knockout of either HIF-1α or HIF-1β lead to abnormal vascular development, including vascular regression and extensive endothelial cell death, leading to lethality in mice (Semenza 2000). After this primitive vascular system is formed, outgrowths of endothelial cells form from existing vessels in a process called sprouting angiogenesis. VEGF is the main regulator of this process with hypoxia being the principle factor for its expression (Hirota and Semenza 2006).

24.3 HIF-1α in Disease

Another important role of HIF is its involvement in ischemic, hypoxic and inflammatory conditions (Semenza 2000). Increased levels of HIF-1α have been found in areas of organ and tissue damage, including patients with acute coronary artery occlusion and also in animal models of myocardial or cerebral ischemia (Ke and Costa 2006). In this case, we aim to promote HIF-1 activity to increase VEGF expression and formation of new blood vessels in tissue suffering from ischemia rather than inhibiting HIF-1 in cancer therapies (Sumbayev et al. 2010). This is especially important for organs such as the heart and brain, where ischemia can cause irreversible damage. A study on transgenic mice overexpressing HIF-1α has been conducted, showing a marked increase in vascularity without unwanted side effects such as oedema, inflammation or vascular leakage (Ke and Costa 2006). If the regulation of the HIF pathway can be refined, this can confer a major advantage to the treatment of many ischemic vascular diseases, such as myocardial infarction and atherosclerosis where the growth of collaterals can be stimulated to bypass stenotic vessels (Hirota and Semenza 2006). However, caution should be taken before applying this to clinical circumstances, as agents that promote angiogenesis can be beneficial for ischemic vascular diseases but induce undesirable side effects to patients having an occult carcinoma (Ahluwalia and Tarnawski 2012). Hence, much work remains before these therapeutic interventions can be applied to human subjects in clinical settings.

24.4 HIF-1α in Malignancy

Cancer development also relies on angiogenesis, as the growing size of the tumor creates an increasingly hypoxic interior. Without angiogenesis, the tumor will be depleted of oxygen and nutrients and hence inhibit future growth (Semenza 2000). This clues the overexpression of HIF-1α in cancer (Lai et al. 2014), where several genetic alterations block the degradation of HIF-1α (Ke and Costa 2006). Yet, despite active angiogenesis, many hypoxic domains are created as the vessels formed are leaky, irregular and poorly functioned which ultimately leads to HIF-1α stabilization. HIF-1α expression is linked with unfavourable prognosis in cancer which is mainly due to therapeutic resistance. Oxygen is required for the cell toxicity effects of radiation therapy to take place as most of the tumor is a hypoxic environment; the cytotoxic effect of radiation therapy is greatly reduced. Hypoxic environment in the tumor also leads to poor drug delivery which inhibits the effects of chemotherapy. These 2 effects combined together will lead to a poor prognosis in cancer (Krock et al. 2011). A hypothesis suggests increased levels of HIF-1α to have a selective advantage for tumor survival. This is demonstrated in immunohistochemical analysis, where the level of HIF-1α protein increases accordingly with benign tumours, primary malignant tumours and tumor metastases (Ke and Costa

2006). Moreover, several studies also show correlation between increased HIF-1α with patient mortality and treatment resistance, which in conclusion, makes HIF-1 a promising target for new cancer therapeutics (Semenza 2000).

Many cancer treatments aim at depriving the tumor of oxygen and nutrients but lacks in finding a target that can simultaneously block multiple pro-angiogenic pathways. A VEGF inhibitor which has been previously approved by the Food and Drug Administration (FDA) was recently disapproved, as studies suggest that agents which only target single anti-angiogenic pathway can actually advance tumourigenesis (Krock et al. 2011). Hypoxia can encourage the migration and invasive activities of tumor cells, not to mention the cytotoxic effects resulting from anti-angiogenic therapies. Moreover, tumours cells can evade mono-therapy anti-angiogenic treatments by activating other pro-angiogenic pathways (Wang and Semenza 1993). This makes the HIF angiogenic pathway an attractive target for cancer therapies as it regulates several pro-angiogenic pathways.

References

Ahluwalia A, Tarnawski AS (2012) Critical role of hypoxia sensor – HIF-1alpha in VEGF gene activation. Implications for angiogenesis and tissue injury healing. Curr Med Chem 19(1):90–97

Hirota K, Semenza GL (2006) Regulation of angiogenesis by hypoxia-inducible factor 1. Crit Rev Oncol Hematol 59(1):15–26. doi:10.1016/j.critrevonc.2005.12.003

Ke Q, Costa M (2006) Hypoxia-inducible factor-1 (HIF-1). Mol Pharmacol 70(5):1469–1480. doi:10.1124/mol.106.027029

Krock BL, Skuli N, Simon MC (2011) Hypoxia-induced angiogenesis: good and evil. Genes Cancer 2(12):1117–1133. doi:10.1177/1947601911423654

Lai Q, Wang T, Guo Q, Zhang Y, Wang Y, Yuan L, Ling R, He Y, Wang W (2014) Positive correlation between the expression of hEag1 and HIF-1alpha in breast cancers: an observational study. BMJ Open 4(5):e005049. doi:10.1136/bmjopen-2014-005049

Semenza GL (2000) HIF-1: mediator of physiological and pathophysiological responses to hypoxia. J Appl Physiol (Bethesda, MD: 1985) 88(4):1474–1480

Sumbayev VV, Nicholas SA, Gibbs BF (2010) Differential role of hypoxia-inducible factor 1 alpha in toll-like receptor-mediated and allergic inflammatory reactions. World Allergy Organ J 3(10):245–249. doi:10.1097/WOX.0b013e3181f8daa5

Wang GL, Semenza GL (1993) Characterization of hypoxia-inducible factor 1 and regulation of DNA binding activity by hypoxia. J Biol Chem 268(29):21513–21518

Chapter 25
Insulin Like Growth Factor (IGF)

Abstract Insulin growth factors are well known for their mitogenic and angiogenic properties. They have profound influence on the growth and metastasis of tumour by affecting their cell proliferation, angiogenesis, survival and motility, as well as through synergism with other growth factors. Insulin-like growth factor (IGF) molecules form part of an axis, which includes (IGF-I and IGF-II), the two cell surface receptors (IGF-IR and IGF-IIR), a family of IGF binding proteins (IGFBP-1 to IGFBP-6), and proteases. IGF-I has been known to shape the cochlea, whereas IGF-II is required for the development of vital organs such as brain and liver. In malignant lesion, it has been reported that hypoxic regions inside tumours could induce IGF expression and IGF induces the expression of hypoxia-inducible factor alpha which upregulates the expression of VEGF, the key mediator of angiogenesis. IGFs themselves are angiogenic by promoting endothelial migration, differentiation and survival. High levels of circulating IGF has also been linked to increased risk of a variety of cancers including prostate, breast, colorectal, and lung cancers.

Keywords Insulin like growth factor • IGF • Angiogenesis • Normal physiology • Disease • Malignancy

25.1 Insulin Like Growth Factor (IGF)

Insulin like growth factors (IGFs) are well known for their mitogenic and angiogenic properties. They have profound influence on the growth and metastasis of tumour by affecting their cell proliferation, angiogenesis, survival and motility, as well as through synergism with other growth factors. Insulin-like growth factor (IGF) molecules form part of an axis, which includes (IGF-I and IGF-II), the two cell surface receptors (IGF-IR and IGF-IIR), a family of IGF binding proteins (IGFBP-1 to IGFBP-6), and proteases (Stoeltzing et al. 2003; Tuomisto et al. 2004). The focus will primarily be IGF-I and IGF-II. Other components of the axis will also be touched upon in the role they play in physiologic and tumor angiogenesis; and tumours progression including dissemination and metastasis.

IGF-I has been known to shape the cochlea, whereas IGF-II is required for the development of vital organs such as brain and liver (Pan and Kastin 2000; Welch

A. Salajegheh, *Angiogenesis in Health, Disease and Malignancy*,
DOI 10.1007/978-3-319-28140-7_25

and Dawes 2007). Studies reported stunted growth in mice with global knockout of *IGF-I* gene (Bach Leon et al. 2013). It has been documented that hypoxic regions inside tumours could also induce IGF expression and IGF induces the expression of hypoxia-inducible factor alpha which upregulates the expression of VEGF, the key mediator of angiogenesis (Clayton et al. 2011). In addition, IGFs themselves are angiogenic by promoting endothelial migration, differentiation and survival (Shigematsu et al. 1999; Lee et al. 2000; Grulich-Henn et al. 2002). It has also been suggested that IGF-I promotes metastasis of breast carcinoma cells by promoting expression of MMP signals (Mira et al. 1999; Zhang et al. 2004). Further, high levels of circulating IGF has been linked to increased risk of a variety of cancers including prostate, breast, colorectal, and lung cancers (Firth and Baxter 2002; Roddam et al. 2008; Chen et al. 2009; Key et al. 2010; Rinaldi et al. 2010). It is also evident that tumour cells in renal cell carcinoma and colorectal cancer with high level expression of IGF receptors seem to be more resistant to chemotherapy (Yuen et al. 2009).

25.2 Insulin Like Growth Factor (IGF) in Normal Physiology

IGF systems have been found to utilize multiple mechanisms aiding in inducing angiogenesis. The majority of IGF-I activity was reportedly mediated by the IGF-I receptor (IGF-IR), which was found to bind IGF-I with a 3 times higher affinity than IGF-II (Bach Leon et al. 2013). The signalling cascade, which is activated upon by receptor binding to IGF molecule includes the RAS proteins/mitogen-activated protein kinase (MAPK) pathway to initiate mitogenesis, and phosphoinositide 3-kinase (PI-3 K)/protein kinase B (Akt) to inhibit apoptosis (Bach Leon et al. 2013). IGF-I and IGF-II can be associated with the migration and morpho-differentiation of endothelial cells through binding to the sub endothelial extra cellular matrix (ECM) and stabilizing the endothelial cells. This was confirmed by studies using the vascular endothelial IGF receptor knockout (VENIFARKO), which showed significant reduction in retinal angiogenesis, compared to control mice. Studies showed IGF systems are also able to interact with EGFR to induce angiogenesis (Samani et al. 2007).

Insulin-like Growth Factor 1 (IGF-1), formerly called Sulfation factor or Somatomedin C, is a peptide hormone first identified in 1957 by Salmon and Daughaday due to its capacity to stimulate cartilage proteoglycan sulfation in rats (Pescovitz and Eugster 2004). The gene is present on chromosome 12 and is closely related to insulin as it has a similar molecular structure with 70-amino-acid polypeptides. It is produced in all tissue types but primarily in the liver under the influence of Growth Hormone (GH), from where it can then be transported to other tissues (Pescovitz and Eugster 2004). Thus, IGF-1 can be classified as a traditional endocrine hormone or a local growth factor due to its autocrine and paracrine growth stimulation (Jameson and De Groot 2010).

There are low levels of serum IGF-1 during embryonic growth but increases gradually from birth till puberty, then surges during puberty and declines with increasing age (Chong et al. 2007). Thus, the hormone is responsible for determining height and muscle growth during puberty and insufficient IGF-1 production can cause dwarfism. Furthermore, IGF-1 regulates cell proliferation, apoptosis, survival and repair due to its pro-mitogenic nature (Chong et al. 2007) and helps to repair brain, heart, muscle and nerve cells. It promotes growth of new motor neurons, collagen formation, connective tissue production, and cartilage repair. The hormone can also stimulate muscle regeneration to counteract muscle loss by activating human stem cells and, hence, increasing the number and size of cells (Pescovitz and Eugster 2004). Under normal cellular regulation, IGF-1 also aids in body fat reduction as it stops insulin from moving glucose through the body, forcing the use of body fat as the primary energy source.

25.3 Insulin Like Growth Factor (IGF) in Disease

Many studies have been conducted regarding angiogenesis in a variety of organs and conditions. Dobrucki et al. investigated how IGF-1 induces angiogenesis in infarcted rat hearts (Dobrucki et al. 2010). Their study used adeno-associated virus (AAV) as a vector to transfer *IGF-I* gene and control AAV encoding β-galactosidase (AAV-*lacZ*) *in vivo* and was over-expressed in a time dependent manner. AAV-*IGF-I* system provided sustained IGF expression, associated with increased angiogenic activities. To determine the results, immunohistochemical analysis showed that there was a significant increase in angiogenic factors and subsequent increased capillary density in the hearts of rats infected with AAV-*IGF-I* compared to control. *In vitro* analyses were done on various angiogenic factors where as *in vivo* micro SPECT-CT imaging was conducted targeting αv integrin, which increases during angiogenesis. IGF-I considerably increased capillary number compared to control, representing angiogenesis. The study used αv integrin molecules as markers to differentiate between mature and newly formed capillaries. Four weeks later, these markers were significantly increased in comparison to control group (Dobrucki et al. 2010).

Another study showed that a combination of high glucose (hyperglycaemia) and high serum IGF-1 levels stimulate endothelial cell migration and tubular formation (Shigematsu et al. 1999). Thus, IGF-1 plays a role in angiogenesis in people with diabetic complications (Shigematsu et al. 1999).

Lopez et al. studied angiogenic mechanisms in the brain, as a result of traumatic brain injury or otherwise promoted by the actions of IGF-I. The study also showed that retinal neovascularization was influenced by IGF-I (Lopez-Lopez et al. 2004). It was found that vessel growth decreased in mice with low serum levels of IGF. This was further confirmed by IGF-I injections and observing an increase in vasculature.

It was also found that IGF-I levels increased around the brain's traumatic lesion. The increase in vascular density was preceded by an increase in VEGF, both in liver IGF-I deficient (LID) and control group with normal IGF-I levels. This concluded that IGF-I may play part in the induction of VEGF. They utilized mice with liver *IGF-I* gene disrupted, resulting in 60 % decreased serum levels in the body except the brain, where IGF-I levels remained normal. IGF-I was shown to be associated with development of tubular structures on collagen, deemed prerequisites for angiogenesis; whereas these tubular structures remained absent from control group with normal IGF-I (Lopez-Lopez et al. 2004).

Synthetic IGF-1 has been developed for therapeutic use such as in growth disorders however medical supervision is needed due to its potential side effects, such as headache, vomiting, hypoglycaemia and oedema. The heart swelling side effect in particular can lead to higher blood pressure and perhaps even failure. Also, long-term trials of IGF-1 treatments have resulted in a build-up of tolerance to the hormone due to development of an anti-IGF-1 antibody serum concentration (Pescovitz and Eugster 2004).

25.4 Insulin Like Growth Factor (IGF) in Malignancy

Many studies concluded that IGF is responsible for the promotion, survival, proliferation, and invasion of many cancer cells (Sachdev and Yee 2001). Tumor associated angiogenesis is promoted by hypoxic manner in which VEGF synthesis plays a key role; IGF was shown to have a positive influence on VEGF and VEGFR levels (Samani et al. 2007).

Tissue hypoxia plays an important part in the angiogenic process. Studies showed that Hypoxia and IGF systems work reciprocally in inducing angiogenesis (Samani et al. 2007). Hypoxia has been shown to induce *IGF-I* gene expression in the heart and lungs, and that of *IGF-II* gene in human hepatocellular carcinoma HepG2 cell line. IGF-I, however, can recruit the hypoxia-inducible factor 1-alpha (HIF-1α) which can in turn induce other tumor-derived angiogenic factors such as VEGF (Lopez-Lopez et al. 2004). This is done using various mechanisms, one of which IGF-I could utilize MAPK and PI-3 K pathways to directly induce HIF-1α protein synthesis, which would then cause the synthesis of IGF-II, IGFBP-2 and IGFBP-3 (Bach Leon et al. 2013). IGF -II was also found to play a significant role in rhabdomyosarcoma (RMS) angiogenesis by inducing synthesis of VEGF which is the central mediator of angiogenesis (Bach Leon et al. 2013).

IGF axis also includes a family of insulin like growth factors binding proteins (IGFBP). These proteins have been shown to inhibit IGF by binding it as a result of their higher affinity than the IGF receptors (IGF-IR and IGF-IIR). IGFBP-6 was found to work in an independent manner to IGF; it inhibited angiogenesis while still promoting cancer cell migration (Beattie et al. 2006).

Metastasis of tumor cells can be a result of angiogenesis. Factors that limit the rate of metastasis are tumor induced angiogenesis, degradation of extra cellular matrix (ECM), cell movement through tissue barriers, including entry/exit into and from blood vessels; and survival and proliferation within foreign microenvironments. Metastasis of tumor cells to secondary sites has low viability at foreign microenvironment. Tumor cells sometimes manage to survive at a foreign location due to complementation of their properties and favourable growth conditions present at the new microenvironment (Samani et al. 2007).

Studies found that *IGF-I* is expressed in various target organs such as lungs and colon of mice as a major liver metastasis-promoting factor (Moromisato et al. 1996); whereas colorectal carcinoma patients showed increased IGF-II and IGF-IR activities. *IGF-I* gene is highly expressed in the liver, which is the main area of metastasis from colon cancer (Reinmuth et al. 2002).

Protease of matrix metalloproteinase (MMP) family has been implicated to cause tumor invasion in malignant progression and metastasis. Various components of IGF system have been reported in various studies to be able to regulate MMP molecules and thus promote tumor invasion (Samani et al. 2007). IGF-IR has been reported to promote the synthesis of MMP-2, leading to tumor invasion *in vitro*. This was revealed by regulating *IGF-IR* expression which resulted in direct regulation of MMP-2 (Samani et al. 2007). IGF has also been shown to indirectly increase cell migration in breast carcinoma by regulating the production of MMP-9 (Samani et al. 2007).

Numerous studies have been conducted targeting the IGF axis for therapeutics and to fight malignancies. Research on lung cancer has focused on *IGFBP-6* expression, which increases by influence from SEMA3B, a tumor suppressor that competes with VEGF, increasing the anti proliferative effects of IGFBP-6 (Bach Leon et al. 2013). IGF receptors have been targeted by monoclonal antibodies and small tyrosine kinase inhibitors (TKI) which bind IGF receptors and prevent their coupling to IGF molecules. A major barrier to these studies however is the toxicity that has been reported in some cases, especially in patients with hyperglycaemia, as a major side effect for TKI targeting; whereas GH hyper secretion and insulin resistance is a major issue with approaches that used antibodies (Bach Leon et al. 2013).

Due to its potent ability to inhibit apoptosis, over-expression of IGF-1 or its receptor can enhance tumour formation (Jameson and De Groot 2010). High IGF-1 levels are usually associated with increased GH production (Pescovitz and Eugster 2004). This is normal during pregnancy and puberty. However, if the phenomenon occurs at other times, it is indicative of pituitary tumours and can also increase risks of developing cancer due to increased cell proliferation and survival. In older people, there is a much greater chance of cancer-causing genetic mutations occurring in rapidly dividing cells Furthermore, the growth of those mutated cells is stimulated by IGF-1, thus worsening the condition (Pescovitz and Eugster 2004). On the other hand, low IGF-1 concentrations indicate a deficiency of or insensitivity to GH, and such individuals appear to be at less risk of developing cancer (Jameson and De Groot 2010). If the deficiency happens during childhood, it can cause short stature (Laron syndrome) and delayed development. Studies have also shown that deletion of the IGF-1 receptor results in animals growing to a maximum of 40 % of their normal size and presenting with neural, skin and bone defects (Jameson and De Groot 2010).

IGF-1 works by activating its membrane-anchored receptor (IGF-1R), causing a signal transduction cascade (Oh et al. 2002). It is this gene expression profile that is associated with increased risk of cancer progression, especially those related to angiogenesis. Abnormal growth, resulting in various cancers of breast, lungs or prostate, may also be attributed to serum IGF-1 that stays in the body for too long. This is due to tumour cell growth promotion and apoptosis inhibition, leading to cancer invasion and metastasis (Oh et al. 2002).

In the case of breast cancer, IGF-1 promotes malignant transformation of normal breast cells, maintains the malignant phenotype, increases metastatic potential, resists apoptosis and resists cytotoxic drugs (Chong et al. 2007). AKT, also known as Protein Kinase B, is a nuclear messenger that is phosphorylated and activated in the signal transduction pathway of IGF-1 and is responsible for regulating angiogenesis by stimulating the factors HIF-1α and VEGF (Vascular Endothelial Growth Factor), the latter being a potent vascular mitogen (Oh et al. 2002). Some studies also suggest that IGF-1 and oestrogen are cofactors in the same pathway that contribute to breast cancer development, especially since oestrogen sensitizes cells to the mitogenic effect of IGF-1 (Chong et al. 2007).

Other IGF-1-induced genes involved in neovascularisation that promote angiogenesis include cyclooxygenase 2, uPA/uPAR, Fibroblast Growth Factor receptor, transferrin, angiopoietin-1, matrix metalloproteinase-1, metalloproteinase-2, interleukin B1 and ephrin B2 receptor (Oh et al. 2002). Some of these genes stimulate blood vessel maturation and sprouting while others allow the proliferation and migration of new blood vessels. IGF-1 also suppresses genes that inhibit angiogenesis, including plasminogen activator inhibitor-1 and 2 (PAI-1 and PAI-2), tissue inhibitor of matrix metalloproteinase-1 and 2 (TIMP1 and TIMP2) and protease inhibitor-1 (Oh et al. 2002).

Research shows that disruption of IGF-1R or its signal transduction pathway may be the key to cancer intervention to inhibit whole classes of angiogenesis genes, instead of targeting a single gene product (Oh et al. 2002). A study in mice shows that a decrease in serum IGF-1 actually slowed tumourigenesis and inhibition of IGF-1R also suppressed breast cancer metastases (Oh et al. 2002). Consequently, low IGF-1 levels are associated with decreased cancer risk.

Therefore, IGF-1 is an important regulator of growth and development in humans and other animals by mediating cell proliferation, apoptosis and survival. As a result, a deficiency of IGF-1 in the plasma can lead to problems with stature and development. In contrast, its over-expression can potentiate the risk of tumour development and progression by inhibiting apoptosis, stimulating cell proliferation and inducing the formation of angiogenic peptides.

The IGF and its signalling pathways are, therefore, a possible target for cancer therapy as it plays a pivotal role in the proliferation and progression of tumour cells (Zhang et al. 2004). Indeed, several drug candidates are showing high anti-neoplastic activity in vivo both as single agents and in conjunction with other currently approved agents (Goya et al. 2004; Haluska et al. 2006; Ji et al. 2007; Rowinsky et al. 2007). Animal models and preliminary clinical trials had suggested that monoclonal antibodies against IGF-I receptors including AMG479 and figitumumab are showing potential efficacy against certain cancers and sarcomas (Cohen et al. 2005; Wu et al. 2006; Tolcher et al. 2009).

It is notable that IGFs are regulated by at least six members of the insulin-like growth factor binding protein (IGFBP) family and they have been starting to attract interest in the recent years. Several studies have shown that certain members of the IGFBP family including IGFBP-5 and IGFBP-7 are able to modulate tumour angiogenesis through different mechanisms (Rho et al. 2008; Tamura et al. 2009). IGFBP-5 was found to downregulate the expressions of phosphorylated protein kinase B and phosphorylated endothelial NO synthase which are both stimulated by VEGF during angiogenesis (Rho et al. 2008). IGFBP-7 was shown to inhibit angiogenesis by mainly interfering with VEGF mRNA expression but also with the signalling pathways (Tamura et al. 2009).

It is clearly evident that many anti-angiogenic agents alone or in combination with conventional therapies have shown efficacy and the potential to be used as anti-cancer therapy based on preclinical trials. The common strategies used include interference with angiogenic ligands, receptors or downstream signalling, upregulating endogenous inhibitors or directly targeting tumour vasculature. However, much more research are needed to establish accurately the efficacy of these agents as in the previous preclinical studies, the effect of drug was measured over too short a time and using tumour regression instead of eradication as an endpoint, which permits recurrence. In addition, the site where tumours are grown and the proliferation rate of the cells in most preclinical trials are not typical of human cancers and studies are needed in slow-growing spontaneous tumours. Furthermore, due to the heterogeneity of the expression of vascular markers in tumour vasculature, a cocktail therapy combining multiple anti-angiogenic agents might be needed and the long-term side effects of most of them in normal tissues and physiological angiogenesis are largely unknown (Nagy et al. 2009, 2010).

More recently, there have been interest in the possibility of stimulating angiogenesis and inducing new vascular growth and formation of collateral vessels for treatment of ischaemic heart diseases (Tabibiazar and Rockson 2001). Research in this field is still principally in the animal models and is focusing on the more well-known and direct angiogenic factors such as fibroblast growth factor-1 and VEGF.

References

Bach Leon A, Fu P, Yang Z (2013) Insulin-like growth factor-binding protein-6 and cancer. Clin Sci 124(4):215–229. doi:10.1042/cs20120343

Beattie J, Allan Gordon J, Lochrie Jennifer D, Flint David J (2006) Insulin-like growth factor-binding protein-5 (IGFBP-5): a critical member of the IGF axis. Biochem J 395(Pt 1):1–19. doi:10.1042/BJ20060086

Chen B, Liu S, Xu W, Wang X, Zhao W, Wu J (2009) IGF-I and IGFBP-3 and the risk of lung cancer: a meta-analysis based on nested case-control studies. J Exp Clin Cancer Res 28:89. doi:10.1186/1756-9966-28-89

Chong YM, Subramanian A, Sharma AK, Mokbel K (2007) The potential clinical applications of insulin-like growth factor-1 ligand in human breast cancer. Anticancer Res 27(3b):1617–1624

Clayton PE, Banerjee I, Murray PG, Renehan AG (2011) Growth hormone, the insulin-like growth factor axis, insulin and cancer risk. Nat Rev Endocrinol 7(1):11–24. doi:10.1038/nrendo.2010.171

Cohen BD, Baker DA, Soderstrom C, Tkalcevic G, Rossi AM, Miller PE, Tengowski MW, Wang F, Gualberto A, Beebe JS, Moyer JD (2005) Combination therapy enhances the inhibition of tumor growth with the fully human anti-type 1 insulin-like growth factor receptor monoclonal antibody CP-751,871. Clin Cancer Res 11(5):2063–2073. doi:10.1158/1078-0432.ccr-04-1070

Dobrucki LW, Tsutsumi Y, Kalinowski L, Dean J, Gavin M, Sen S, Mendizabal M, Sinusas AJ, Aikawa R (2010) Analysis of angiogenesis induced by local IGF-1 expression after myocardial infarction using microSPECT-CT imaging. J Mol Cell Cardiol 48(6):1071–1079. doi:10.1016/j.yjmcc.2009.10.008

Firth SM, Baxter RC (2002) Cellular actions of the insulin-like growth factor binding proteins. Endocr Rev 23(6):824–854. doi:10.1210/er.2001-0033

Goya M, Miyamoto S, Nagai K, Ohki Y, Nakamura K, Shitara K, Maeda H, Sangai T, Kodama K, Endoh Y, Ishii G, Hasebe T, Yonou H, Hatano T, Ogawa Y, Ochiai A (2004) Growth inhibition of human prostate cancer cells in human adult bone implanted into nonobese diabetic/severe combined immunodeficient mice by a ligand-specific antibody to human insulin-like growth factors. Cancer Res 64(17):6252–6258. doi:10.1158/0008-5472.can-04-0919

Grulich-Henn J, Ritter J, Mesewinkel S, Heinrich U, Bettendorf M, Preissner KT (2002) Transport of insulin-like growth factor-I across endothelial cell monolayers and its binding to the subendothelial matrix. Exp Clin Endocrinol Diabetes 110(2):67–73. doi:10.1055/s-2002-23488

Haluska P, Carboni JM, Loegering DA, Lee FY, Wittman M, Saulnier MG, Frennesson DB, Kalli KR, Conover CA, Attar RM, Kaufmann SH, Gottardis M, Erlichman C (2006) In vitro and in vivo antitumor effects of the dual insulin-like growth factor-I/insulin receptor inhibitor, BMS-554417. Cancer Res 66(1):362–371. doi:10.1158/0008-5472.can-05-1107

Jameson JL, De Groot LJ (2010) Endocrinology: adult and pediatric, vol 1. Elsevier Health Sciences, Missouri

Ji QS, Mulvihill MJ, Rosenfeld-Franklin M, Cooke A, Feng L, Mak G, O'Connor M, Yao Y, Pirritt C, Buck E, Eyzaguirre A, Arnold LD, Gibson NW, Pachter JA (2007) A novel, potent, and selective insulin-like growth factor-I receptor kinase inhibitor blocks insulin-like growth factor-I receptor signaling in vitro and inhibits insulin-like growth factor-I receptor dependent tumor growth in vivo. Mol Cancer Ther 6(8):2158–2167. doi:10.1158/1535-7163.mct-07-0070

Key TJ, Appleby PN, Reeves GK, Roddam AW (2010) Insulin-like growth factor 1 (IGF1), IGF binding protein 3 (IGFBP3), and breast cancer risk: pooled individual data analysis of 17 prospective studies. Lancet Oncol 11(6):530–542. doi:10.1016/s1470-2045(10)70095-4

Lee OH, Bae SK, Bae MH, Lee YM, Moon EJ, Cha HJ, Kwon YG, Kim KW (2000) Identification of angiogenic properties of insulin-like growth factor II in in vitro angiogenesis models. Br J Cancer 82(2):385–391. doi:10.1054/bjoc.1999.0931

Lopez-Lopez C, LeRoith D, Torres-Aleman I (2004) Insulin-like growth factor I is required for vessel remodeling in the adult brain. Proc Natl Acad Sci 101(26):9833–9838. doi:10.1073/pnas.0400337101

Mira E, Manes S, Lacalle RA, Marquez G, Martinez AC (1999) Insulin-like growth factor I-triggered cell migration and invasion are mediated by matrix metalloproteinase-9. Endocrinology 140(4):1657–1664. doi:10.1210/endo.140.4.6623

Moromisato DY, Moromisato MY, Zanconato S, Roberts J, Brasel JA, Cooper DM (1996) Effect of hypoxia on lung, heart, and liver insulin-like growth factor-I gene and receptor expression in the newborn rat. Crit Care Med 24(6):919–924. doi:10.1097/00003246-199606000-00008

Nagy JA, Chang SH, Dvorak AM, Dvorak HF (2009) Why are tumour blood vessels abnormal and why is it important to know? Br J Cancer 100(6):865–869. doi:10.1038/sj.bjc.6604929

Nagy JA, Chang SH, Shih SC, Dvorak AM, Dvorak HF (2010) Heterogeneity of the tumor vasculature. Semin Thromb Hemost 36(3):321–331. doi:10.1055/s-0030-1253454

Oh JS, Kucab JE, Bushel PR, Martin K, Bennett L, Collins J, DiAugustine RP, Barrett JC, Afshari CA, Dunn SE (2002) Insulin-like growth factor-1 inscribes a gene expression profile for angiogenic factors and cancer progression in breast epithelial cells. Neoplasia 4(3):204–217. doi:10.1038/sj/neo/7900229

Pan W, Kastin AJ (2000) Interactions of IGF-1 with the blood-brain barrier in vivo and in situ. Neuroendocrinology 72(3):171–178. doi:10.1159/000054584

Pescovitz OH, Eugster EA (2004) Pediatric endocrinology: mechanisms, manifestations, and management. Lippincott Williams & Wilkins, Philadelphia

Reinmuth N, Fan F, Liu W, Parikh AA, Stoeltzing O, Jung YD, Bucana CD, Radinsky R, Gallick GE, Ellis LM (2002) Impact of insulin-like growth factor receptor-I function on angiogenesis, growth, and metastasis of colon cancer. Lab Invest 82(10):1377–1389. doi:10.1097/01.lab.0000032411.41603.c2

Rho SB, Dong SM, Kang S, Seo SS, Yoo CW, Lee DO, Woo JS, Park SY (2008) Insulin-like growth factor-binding protein-5 (IGFBP-5) acts as a tumor suppressor by inhibiting angiogenesis. Carcinogenesis 29(11):2106–2111. doi:10.1093/carcin/bgn206

Rinaldi S, Cleveland R, Norat T, Biessy C, Rohrmann S, Linseisen J, Boeing H, Pischon T, Panico S, Agnoli C, Palli D, Tumino R, Vineis P, Peeters PH, van Gils CH, Bueno-de-Mesquita BH, Vrieling A, Allen NE, Roddam A, Bingham S, Khaw KT, Manjer J, Borgquist S, Dumeaux V, Torhild Gram I, Lund E, Trichopoulou A, Makrygiannis G, Benetou V, Molina E, Donate Suarez I, Barricarte Gurrea A, Gonzalez CA, Tormo MJ, Altzibar JM, Olsen A, Tjonneland A, Gronbaek H, Overvad K, Clavel-Chapelon F, Boutron-Ruault MC, Morois S, Slimani N, Boffetta P, Jenab M, Riboli E, Kaaks R (2010) Serum levels of IGF-I, IGFBP-3 and colorectal cancer risk: results from the EPIC cohort, plus a meta-analysis of prospective studies. Int J Cancer 126(7):1702–1715. doi:10.1002/ijc.24927

Roddam AW, Allen NE, Appleby P, Key TJ, Ferrucci L, Carter HB, Metter EJ, Chen C, Weiss NS, Fitzpatrick A, Hsing AW, Lacey JV Jr, Helzlsouer K, Rinaldi S, Riboli E, Kaaks R, Janssen JA, Wildhagen MF, Schroder FH, Platz EA, Pollak M, Giovannucci E, Schaefer C, Quesenberry CP Jr, Vogelman JH, Severi G, English DR, Giles GG, Stattin P, Hallmans G, Johansson M, Chan JM, Gann P, Oliver SE, Holly JM, Donovan J, Meyer F, Bairati I, Galan P (2008) Insulin-like growth factors, their binding proteins, and prostate cancer risk: analysis of individual patient data from 12 prospective studies. Ann Intern Med 149(7):461–471, w483–w468

Rowinsky EK, Youssoufian H, Tonra JR, Solomon P, Burtrum D, Ludwig DL (2007) IMC-A12, a human IgG1 monoclonal antibody to the insulin-like growth factor I receptor. Clin Cancer Res 13(18 Pt 2):5549s–5555s. doi:10.1158/1078-0432.ccr-07-1109

Sachdev D, Yee D (2001) The IGF system and breast cancer. Endocr Relat Cancer 8(3):197–209. doi:10.1677/erc.0.0080197

Samani AA, Yakar S, LeRoith D, Brodt P (2007) The role of the IGF system in cancer growth and metastasis: overview and recent insights. Endocr Rev 28(1):20–47. doi:10.1210/er.2006-0001

Shigematsu S, Yamauchi K, Nakajima K, Iijima S, Aizawa T, Hashizume K (1999) IGF-1 regulates migration and angiogenesis of human endothelial cells. Endocr J 46(Suppl):S59–S62

Stoeltzing O, Liu W, Reinmuth N, Fan F, Parikh AA, Bucana CD, Evans DB, Semenza GL, Ellis LM (2003) Regulation of hypoxia-inducible factor-1α, vascular endothelial growth factor, and angiogenesis by an insulin-like growth factor-I receptor autocrine loop in human pancreatic cancer. Am J Pathol 163(3):1001–1011. doi:10.1016/s0002-9440(10)63460-8

Tabibiazar R, Rockson SG (2001) Angiogenesis and the ischaemic heart. Eur Heart J 22(11):903–918. doi:10.1053/euhj.2000.2372

Tamura K, Hashimoto K, Suzuki K, Yoshie M, Kutsukake M, Sakurai T (2009) Insulin-like growth factor binding protein-7 (IGFBP7) blocks vascular endothelial cell growth factor

(VEGF)-induced angiogenesis in human vascular endothelial cells. Eur J Pharmacol 610(1–3):61–67. doi:10.1016/j.ejphar.2009.01.045

Tolcher AW, Sarantopoulos J, Patnaik A, Papadopoulos K, Lin CC, Rodon J, Murphy B, Roth B, McCaffery I, Gorski KS, Kaiser B, Zhu M, Deng H, Friberg G, Puzanov I (2009) Phase I, pharmacokinetic, and pharmacodynamic study of AMG 479, a fully human monoclonal antibody to insulin-like growth factor receptor 1. J Clin Oncol 27(34):5800–5807. doi:10.1200/jco.2009.23.6745

Tuomisto TT, Rissanen TT, Vajanto I, Korkeela A, Rutanen J, Ylä-Herttuala S (2004) HIF-VEGF-VEGFR-2, TNF-α and IGF pathways are upregulated in critical human skeletal muscle ischemia as studied with DNA array. Atherosclerosis 174(1):111–120. doi:10.1016/j.atherosclerosis.2004.01.015

Welch D, Dawes PJD (2007) Childhood hearing is associated with growth rates in infancy and adolescence. Pediatr Res 62(4):495–498. doi:10.1203/pdr.0b013e3181425869

Wu JD, Haugk K, Coleman I, Woodke L, Vessella R, Nelson P, Montgomery RB, Ludwig DL, Plymate SR (2006) Combined in vivo effect of A12, a type 1 insulin-like growth factor receptor antibody, and docetaxel against prostate cancer tumors. Clin Cancer Res 12(20 Pt 1):6153–6160. doi:10.1158/1078-0432.ccr-06-0443

Yuen JS, Akkaya E, Wang Y, Takiguchi M, Peak S, Sullivan M, Protheroe AS, Macaulay VM (2009) Validation of the type 1 insulin-like growth factor receptor as a therapeutic target in renal cancer. Mol Cancer Ther 8(6):1448–1459. doi:10.1158/1535-7163.mct-09-0101

Zhang D, Bar-Eli M, Meloche S, Brodt P (2004) Dual regulation of MMP-2 expression by the type 1 insulin-like growth factor receptor: the phosphatidylinositol 3-kinase/Akt and Raf/ERK pathways transmit opposing signals. J Biol Chem 279(19):19683–19690. doi:10.1074/jbc.M313145200

Chapter 26
Integrins

Abstract A component of the molecular system controlling angiogenesis is the need for endothelial cells to infiltrate, proliferate and avoid cell death through apoptosis. Integrin are the main receptors that enable endothelial cells to relate with their extracellular surroundings; and relations enabled by integrin have a crucial part in ensuring that cells migrate, proliferate and do not die. Endothelial cells show many integrin on the cell surface according to where they are located and the condition in which they are upregulated. Integrins are heterodimeric transmembrane receptors, which consist of α and β subunit. A variety of different subunit pairings make up a family of integrins. These receptors bind extracellular matrix (ECM) proteins or ligands, and their subunit combination gives specificity of binding to a specific ECM protein, or a range of ECM proteins. The αv integrins, involved in the growth and survival of newly forming vessels from pre-existing vessels (angiogenesis). Integrins αvβ3 and αvβ5 promote distinct pathways of angiogenesis by interacting with various factors and, depending on the interaction, have pro- or anti-angiogenic effects. Integrin α5 leads to the formation of a fibronectin receptor and involved in wound contraction within granulation tissue and in neoplastic situation leads to up-regulation of α5β1 integrin receptors which in turn involves in activation, adhesion and angiogenic cells proliferation of blood vessels. Integrin β5 is activated after activation of VEGF and it encourages the movement of cells and infiltration of tissue. Integrin β8 is part of the integrin beta chain family, which encodes for a single-pass type I membrane protein. Integrin β8 plays a key role in regulating vascular homeostasis and correlates with the process of angiogenesis. The future of therapies targeting integrin could diminish cancer cell progression and their invasiveness in addition to modulation of neovascularization and angiogenesis.

Keywords Angiogenesis • Integrin • Integrin αvβ3 • Integrin αvβ5 • Integrin α5 • Integrin β5 • Integrin β8 • Normal physiology • Disease • Malignancy

A. Salajegheh, *Angiogenesis in Health, Disease and Malignancy*,
DOI 10.1007/978-3-319-28140-7_26

26.1 Integrins

Angiogenesis is influenced by many factors, one of which is integrins. Integrins are heterodimeric transmembrane receptors, which consist of α and β subunit (Robinson and Hodivala-Dilke 2011; Weis and Cheresh 2011). A component of the molecular system controlling angiogenesis is the need for endothelial cells to infiltrate, proliferate and avoid cell death through apoptosis (Stupack and Cheresh 2004). Integrin are the main receptors that enable endothelial cells to relate with their extracellular surroundings; and relations enabled by integrin have a crucial part in ensuring that cells migrate, proliferate and do not die (Stupack and Cheresh 2004). Changes in the action of integrin, the presence of binding molecules and the structures of these binding molecules or ions affect blood vessel cells as they grow or restore the vasculature (Stupack and Cheresh 2004). Endothelial cells show many integrin on the cell surface according to where they are located and the condition in which they are upregulated.

A variety of different subunit pairings make up a family of integrins (Weis and Cheresh 2011). These receptors bind extracellular matrix (ECM) proteins or ligands, and their subunit combination gives specificity of binding to a specific ECM protein, or a range of ECM proteins (Weis and Cheresh 2011). These receptor-ligand interactions mediate the transmission of extracellular signals across the plasma membrane, which links up to the intracellular signalling machinery via their 'cytoplasmic tails' (Robinson and Hodivala-Dilke 2011). This transmission of information across the plasma membrane is bidirectional and these receptors also respond to intracellular stimuli, and can undergo conformational changes that affects how their 'extracellular heads' interact with the environment surrounding the cell (Robinson and Hodivala-Dilke 2011).

The αv integrins, which represent a subgroup of integrins that bind to ECM proteins consisting of the arginine-glycine-aspartic acid (RGD) sequence, have been found to be over-expressed on the surface of endothelial cells (EC) involved in the growth and survival of newly forming vessels from pre-existing vessels (angiogenesis) (Weis and Cheresh 2011). Although their exact role remains unclear despite numerous studies, they have been identified as playing a key role in vascular remodelling associated with angiogenesis in development, wound repair, and in the invasive potential of cancer cells; and blocking the function of these αv integrins by inhibiting the binding of the ligand, can produce an anti-angiogenic effect (van der Flier et al. 2010; Robinson and Hodivala-Dilke 2011).

Integrins αvβ3 and αvβ5 have been observed to promote distinct pathways of angiogenesis by interacting with various factors and, depending on the interaction, have pro- or anti-angiogenic effects (Robinson and Hodivala-Dilke 2011; Weis and Cheresh 2011). Most of the research has been conducted on integrin-αvβ3, which in 1994, was discovered to be over-expressed (the expression is mediated by transcriptional activator, Hox D3) on the vasculature EC associated with angiogenesis in remodelling and cancer (in stark contrast to that of quiescent EC); and thus has become a drug target with the aim to inhibit the ability of tumours to establish their own blood supply via angiogenesis (Robinson and Hodivala-Dilke 2011; Weis and Cheresh 2011).

Another gene in these series is integrin alpha 5 (ITGA5) different from integrin αv and encodes for the integrin alpha 5 (Integrin α5) chain that ultimately leads to the formation of a fibronectin receptor, which plays an important role in endothelial cell adhesion and hence, angiogenesis (Hsia et al. 2014). Once Integrin α5 is translated by a ribosome, posttranslational modifications yield different isoforms of the integrin, or simply different versions of that specific integrin protein (Hsia et al. 2014).

Integrin beta 5 (ITG β5 or ITGB5) is expressed on angiogenic cells. ITGβ5 is a receptor for vitronectin and fibronectin molecules (Stupack and Cheresh 2004). ITG β5 can relate with attachment molecules L1-CAM or angiopoietin-1 on the exterior surface of cells (Stupack and Cheresh 2004).

The integrin beta 8 (Integrin β8 or ITGB8) gene, located on the cytogenetic band 7p21.1 of the human chromosome, is responsible for the coding and synthesis of the integrin beta-8 protein. As its name suggests, this gene is part of the integrin beta chain family, which encodes for a single-pass type I membrane protein. It plays a key role in regulating vascular homeostasis in brain development (Mobley et al. 2009). This directly correlates with the process of angiogenesis (Wang and Olson 2009).

26.2 Integrins in Normal Physiology

Integrin-αvβ3 is a receptor for a range of ligands with a arginine-glycine-aspartic acid, RGD sequence, including but not limited to fibronectin, fibrinogen, vitronectin and collagens, while integrin-αvβ5 appears to preferentially bind vitronectin (Weis and Cheresh 2011).

The integrin-ligand interaction regulates endothelial cell adhesion, proliferation, migration and survival by triggering intra-cellular signalling pathways through the recruitment and activation of specific kinases and signalling intermediates (Weis and Cheresh 2011). Integrin αvβ3 is involved in inducing angiogenesis via tumour necrosis factor α (TNF-α) or fibroblast growth factor (FGF), and more specifically, involves p21-activated kinase (PAK) (Weis and Cheresh 2011). Integrin αvβ5 is involved in inducing angiogenesis via (VEGF) triggering focal adhesion kinase (FAK) and Src kinase, or transforming growth factor α (TGF-α) (Weis and Cheresh 2011). This αvβ5 pathway also leads to activation of Raf on serines 338/339 which results in trans-location of mitochondrial in Raf-1 and MEK1-independent endothelial cell protection from the inherent pathway of programmed cell death triggered by stress or DNA damage (Weis and Cheresh 2011). In contrast, the αvβ3 pathway stimulates Raf on tyrosines 340/341 and MEK1-dependent protection from externally mediated cell death which is triggered by binding of receptor to pro-apoptotic ligands including TNF-α and Fas (Weis and Cheresh 2011). The difference in the signalling pathways which require the function of these two αv integrins, demonstrates the complexity of these receptor-ligand interactions, and gives an indication as to how their expression can drive certain cellular behaviours in angiogenesis (Weis and Cheresh 2011).

In regards to angiogenesis, the isoform of Integrin Alpha 5 (the protein encoded by ITGA5) that is of importance in the Beta 1 isoform (α5β1) (Boudreau and Varner 2004). α5β1 integrin is used in normal cells as a receptor for the ligand fibronectin. One of the roles of α5β1 in normal cells is that it induces cellular changes and protein expression within the cell that increases the adhesion of a cell to the surrounding extracellular matrix.

ITG β5 is activated after activation of VEGF and it encourages the movement of cells and infiltration of tissue (Stupack and Cheresh 2004). This progression is promoted by the cytokine-induced attraction of FAK to the cytoplasm end of ITG β5 (Stupack and Cheresh 2004). Overexpression of ITG β5 has been shown to promote new blood vessel formation in vivo. ITG β5 combines with VEGF to work on endothelial cells involved in angiogenesis in vivo (Stupack and Cheresh 2004). A combination of VEGF and VEGFR2 upregulates ITG β5. Opposition of ITG β5 antagonises angiogenesis, which is activated by VEGF (Stupack and Cheresh 2004). Studies have shown that the excessive expression of ITG β5 in mice results in its phosphorylation and increases in the paracrine properties of cells involved in angiogenesis through Src Kinase mediated activation of signal transducer and activator of transcription 3 (STAT3) (Leifheit-Nestler et al. 2010). Circulating angiogenic cells in vivo and in vitro have greater angiogenic abilities when ITG β5 expression is excessive (Leifheit-Nestler et al. 2010). Through 3D spheroid assays it was observed that ITG β5 overexpression increased the number of circulating angiogenic cells moving through the human umbilical vein endothelial cell (HUVEC) branches (Leifheit-Nestler et al. 2010). Spheroids that had HUVECs and CACs expressing ITG β5 made a greater number and longer branches than the ones consisting of CACs, HUVEC and no ITG β5 (Leifheit-Nestler et al. 2010). The effects of ITG β5 in excessive expression in blood vessel formation include paracrine systems (Leifheit-Nestler et al. 2010).

The function of integrin β8 *in vivo* is the essential development of the vertebrate vascular network required for embryonic survival (Zhu et al. 2002). Expression of the β8 mRNA occurs in the adult brain, kidney and placenta but is mostly localised to the neuroepithelial cells in the brain synapses (Lakhe-Reddy et al. 2014). Experimentation upon integrin β8- deficient mice revealed that the absence of β8 subunit led to fatal defects in vascular development (Zhang et al. 2012). As a result of inadequate vascularisation of the placenta and yolk sac, 65 % of integrin β8-deficient embryos died mid-gestation and those that survived died shortly after birth due to extensive intra-cerebral haemorrhage (Zhu et al. 2002). It has been proposed that despite the presence of vascular endothelial growth factor (VEGF) and its receptors, which are crucial for early endothelial cell differentiation and for angiogenesis, integrin beta-8 is a necessary component as it signals for the appropriate patterning of the embryonic brain vascular network (Zhu et al. 2002). Analysis of brain tissue found within integrin β8 deficient mice shows that capillaries are formed with abnormally patterned hyper-proliferative endothelial cells and the disorganization of neuroepithelial cells (Zhu et al. 2002).

26.3 Integrins in Disease

In physiological or pathological angiogenesis, the progress of vascular remodelling is determined by a balance of pro- and anti-angiogenic factors (Weis and Cheresh 2011). While the integrin-ligand interaction mediates a signalling pathway and promotes cell adhesion, proliferation, migration and survival, this is suppressed when binding to a soluble fragment occurs (Weis and Cheresh 2011).

It has been suggested that αvβ3 and αvβ5 integrins have a differing role in blood vessel formation during development to that in adulthood (Serini et al. 2006; van der Flier et al. 2010). Activation of αvβ3 and αvβ5 integrins by insoluble ligands such as fibronectin and vitronectin respectively, is required for VEGF-dependent neovascularisation in the adult organism, but not in the embryo (Serini et al. 2006). It has been observed that the blocking of αvβ3 and αvβ5 integrins can inhibit angiogenesis in arthritis and ischaemic retinopathy in adults, however mice embryos lacking αvβ3 and/or αvβ5 can not only continue to develop normally in terms of vascular angiogenesis (lethality of embryos did in most cases occur at a certain point) but can show increased pathological angiogenesis (Serini et al. 2006). Another study had similar findings, however it appeared that the embryonic lethality that occurred in the majority around mid-gestation in those lacking αv integrins (and/or α5 integrins), arose from defects in the heart and great vessels (van der Flier et al. 2010).

Therefore, although αv integrins are implicated in both physiological and pathological angiogenesis, their involvement appears to not be essential, and their exact roles remain uncertain in both development and disease (Serini et al. 2006; van der Flier et al. 2010; Robinson and Hodivala-Dilke 2011).

In addition to developmental blood vessel formation, angiogenesis is an important part of wound healing, as it is crucial for the damaged capillaries to be replaced and the supply of oxygen and nutrients to be restored to injured tissue (Greaves et al. 2013).

The structure and composition of ECM alters significantly during wound healing (Greaves et al. 2013). Consequently its effect on angiogenesis is variable depending on the protein constituents at any one time, in addition to other factors (Greaves et al. 2013). Blood vessel 'sprouts' invade the wound together with fibroblast ingrowth in response to pro-angiogenic factors including αvβ3 related FGF and αvβ5 related VEGF, as well as angiogenin, released by infiltrating macrophages and keratinocytes (Greaves et al. 2013). If this fails to occur, a delay in wound healing results (Greaves et al. 2013).

Integrin-αvβ3 has been identified as a key angiogenic marker on the vasculature EC in human granulation tissue, as it is absent in uninjured tissue (Greaves et al. 2013). The blocking of αvβ3 expression in wound granulation tissue has been found to suppress neo-angiogenesis induced by αvβ3 related FGF and TNF-α (Maquart and Monboisse 2014).

Furthermore, in a study investigating the ability of human microvascular EC to form capillary-like structures in vitro, the addition of αvβ3 related FGF and/or αvβ5 related VEGF with αvβ3 related TNF-α, resulted in tubule outgrowth, with addition of all three factors leading to increased effects (Greaves et al. 2013). Furthermore, a hypoxic environment increased the tube formation to an even greater extent (Greaves et al. 2013).

In a non-neoplastic cell, this is of vital importance in wound healing. One of the primary steps in wound healing is the contraction of the wound, once granulation tissue has been formed. Fibroblasts in the granulation tissue area secrete proteases that are enzymes that break down proteins (Rosenberg 2001). Matrix Metalloproteinases are the proteases responsible for the fragmentation of fibronectin. Fragmentation of the fibronectin exposes a site that is complementary to the α5β1 integrin receptor. Once this occurs, the cell synthesises proteins that cause it to become more adhesive. (Rosenberg 2001). This adhesion is what results in the vigorous contraction of the surrounding cells, and hence the overall contraction of the granulation tissue. The exact mechanism of how this occurs is poorly understood, however it is known that the cell produces transmembrane proteins which link up to adjacent cells making the adhesion between successive cells stronger. This process is highly regulated in normal tissue, as well as areas of granulation tissue as proteases required by fibroblasts are required to create the appropriate ligand for the α5β1 integrin receptor, and hence the initiation of this process (Boudreau and Varner 2004; Hsia et al. 2014; Rosenberg 2001).

A study has suggested that different variants of ITG β5 and AGFG1 genes are associated in the modulation of airway hyper responsiveness. This condition is a characteristic of asthma in which there is an increased contraction of the smooth muscles in the airways (Himes et al. 2013). However, it was noted that more studies are required to confirm these findings (Himes et al. 2013). Another study was constructed to investigate the association between different genotypes of ITG β5 and liver fibrosis. It was revealed that this association is considered as a false positive result and the role of ITG β5 in liver fibrosis is not well defined. Moreover, it has been proved that there is no correlation between the histological stages of liver fibrosis and different alleles of ITG β5 (Grünhage et al. 2011). There is a controversy about the role of ITG β5 in causing an increase in the susceptibility to adenovirus infections (Kasuga et al. 2009). Few studies have stated that there is a correlation between ITG β5 and higher adenovirus susceptibility, while others studies shown that there is no significant influence between different genotypes of this gene and the latent adenovirus infection (Kasuga et al. 2009). The impact of ITG β5 on wound healing was suggested in a number of findings (Garlet et al. 2012). One particular study revealed a fivefold or more increase in ITG β5 in periapical granulomas along with the increase of other genes such as SERPINE1, TIMP1, COL1A1, COL5A1, VTN, CTGF, FGF7, TGFB1, TNF, CXCL11 and ITGA4 (Garlet et al. 2012). The study has shown that identification of the genes involved in the progression of these periapical lesions can be used to make a more accurate diagnosis, and to formulate a more appropriate treatment which can lead to an increase in the success rate of endodontic treatment (Garlet et al. 2012).

Integrin β8 prevents oxygen-glucose deprivation-induced neuronal apoptosis (Zhang et al. 2012). Inhibition of integrin β8, established by the use of lentiviral vector-mediated β8 RNAi, results in down regulation of vascular endothelial growth factor (VEGF) (Zhang et al. 2012). VEGF is a signal protein responsible for stimulating angiogenesis. In the absence of β8 subunit in the integrin heterodimeric complex of integrin αvβ8, brain vascular abnormalities results. This is termed hypoxic-ischaemic brain injury, which leads to neurological dysfunction in newborns (Zhang et al. 2012). Integrin β8 is also an essential mediator in brain recovery after such injury. Neurogenesis and neurovascular physiology is essential for the healing process but such processes are inhibited in the absence of integrin β8. Moreover, integrin β8 expressed on glial cells regulate vascular development that is angiogenesis, by indirectly monitoring the release of secreted factors such as VEGF (Zhang et al. 2012).

Integrin β8 is essential in the management of neurovascular physiology (Mobley et al. 2009). Removal of the integrin β8 in the perivascular neural cells results in an inability to regulate the active phases of blood vessel growth, leading to widespread issues in neurovascular homeostasis. A null integrin β8 adult mice exhibits increased numbers of intracerebral blood vessels as well as increased perivascular astrogliosis. The mechanism by which integrin β8 functions occurs through activation of latent forms of TGFβ3 secreted from cells at extracellular matrix (ECM)- bound complexes. Neural vascular cell adhesion and communication is absent in null integrin β8 mice due to diminished integrin-mediated activation of TGFβs (Mobley et al. 2009). The process of angiogenesis terminates by day 30 post-natal. This coincides with lowered neurovascular pathologies of the brain as integrin β8 in adult neural progenitor cells continue to promote interactions in blood vessels of neurovascular niches (Mobley et al. 2009).

26.4 Integrins in Malignancy

In cancer, the integrin-ligand mediated signalling pathways are similar to those in vascular development, physiological angiogenesis, and pathological vascular hyperproliferation; yet are found in tumour-associated fibroblasts and inflammatory cells, and are exploited to allow for the survival and spread of tumour cells by manipulating the host microenvironment (Weis and Cheresh 2011). The integrin-ligand interaction which inhibits cells from proliferating and migrating beyond the normal boundaries can also enable cells to survive and migrate within an abnormal environment on receiving new signals from the extracellular environment, or in response to intracellular reprogramming (Weis and Cheresh 2011).

The expression of αvβ3 integrin by tumour cells has been linked to primary tumour growth and metastatic potential through the promotion of anchorage-independent growth and increased transendothelial migration (Weis and Cheresh

2011). This $\alpha v\beta 3$ expression is seen in approximately half of patients with pancreatic cancer and positively correlated with lymph node metastasis, breast tumours, and to an even greater extent in breast cancer lymph node metastases (Weis and Cheresh 2011). These findings have indicated the therapeutic use of blocking $\alpha v\beta 3$ in the suppression of such tumours (Weis and Cheresh 2011). In $\alpha v\beta 5$ integrin expression, the tumour cells require cytokine or growth factor to stimulate migration and invasion (Weis and Cheresh 2011).

However, the $\alpha v\beta 3$ integrin has been found to have either pro- or anti-angiogenic effects on tumour-associated angiogenesis depending on what it interacts with in the angiogenic pathway, thus suggesting that a multi-target approach may be required in therapeutic management (Robinson and Hodivala-Dilke 2011; Weis and Cheresh 2011). Furthermore, different responses have been seen in different types of tumours (Robinson and Hodivala-Dilke 2011; Weis and Cheresh 2011).

Other suggestion linking integrin-5 and integrin-3 with pathological angiogenesis have been conducted (Reynolds et al. 2002). These studies have illustrated that the inhibition of Integrin 5 and integrin 3 function leads to a decrease in neovascularization and tumour growth (Reynolds et al. 2002). Interestingly it was observed that even when cells were lacking these two genes, pathological angiogenesis and tumour growth continued (Reynolds et al. 2002). However, further studies are required to confirm the effectiveness of integrin antagonists which can be used as anti-angiogenic therapy (Reynolds et al. 2002). In addition, changes in the ITGB-5 gene along with other genes is associated with autosomal dominant osteoporosis type II which is a rare heritable bone disorder characterised by high bone mass and decrease in the osteoclastic activity (Coudert et al. 2014).

In neoplastic cells, $\alpha 5\beta 1$ integrin receptors are not activated by fibronectin fragments as in non-neoplastic cells, however more $\alpha 5\beta 1$ integrin receptors are created in cells that express a neoplastic nature. A transcription factor named Hox D3 in neoplastic cells is implicated to over express $\alpha 5\beta 1$ integrin receptors (Boudreau and Varner 2004). Transcription factors can either block or promote transcription, and hence they can either up-regulate or down-regulate the expression of a specific protein (Walter et al. 2013; Wang et al. 2013; Xuan et al. 2013; Zhang et al. 2008).

Research gathered from studies that examined the relationship between Hox D3 and $\alpha 5\beta 1$ integrin expression clearly demonstrate a link between the two (Boudreau and Varner 2004). What is less understood is why neoplastic cells have an abundant supply of the Hox D3 transcription factor. It is hypothesised that an error in the regulation of the Hox D3 transcription factor results in its over activation. This results in it binding to the promoter region of the $\alpha 5\beta 1$ integrin gene and results in the over expression of the gene's protein product (Boudreau and Varner 2004).

Since the neoplastic cells now have more $\alpha 5\beta 1$ integrin receptors, the mechanism for production of adhesive molecules of the cell are more easily activated. Since $\alpha 5\beta 1$ integrin receptors are usually found in endothelial cells, the resulting adhesion of endothelial cells leads to the proliferation of blood vessels, and therefore an increase in angiogenesis.

It has been shown that ITG $\beta 5$ is overexpressed in colorectal adenocarcinoma and it correlates with the presence of malignant neoplasia (mucinous type) (Denadai et al. 2013). Tahira et al suggested that ITG β5 is overexpressed in pancreatic cancer (Tahira et al. 2011). ITG β5 has also been reported as a potential biomarker in lung cancer and it's currently the target for an inhibitor drug that is undergoing clinical trials (Tahira et al. 2011). A more recent study published in June 2013, investigated the signalling pathways and the role of ITG β5 in breast cancer (Bianchi-Smiraglia et al. 2013). It has been shown that ITG β5 is upregulated in breast carcinoma and it mediates Src FAK (focal adhesion kinase) and MEK ERK signalling pathways (Bianchi-Smiraglia et al. 2013). These two pathways work independently and are involved in the tumorigenic activity of cancer cells (Bianchi-Smiraglia et al. 2013). This same study has shown that ITG β5 has a role in breast cancer angiogenesis and it also facilitates cancer cells' migration (Bianchi-Smiraglia et al. 2013). Moreover, ITG β5 depletion from metastatic breast cancer cells resulted in the hindrance of tumorigenic activity in vitro and in mouse model (Bianchi-Smiraglia et al. 2013). It has been suggested that ITG β5 contributes to tumour angiogenesis and resistance to chemotherapy and radiation (Maubant et al. 2002; Hood et al. 2003). In conclusion, these few publications in the literature about the role of ITG β5 in cancer provide novel insights into its function in cancer. Tumour growth and angiogenesis are facilitated by ITG β5 through matrix adhesions or through its signalling pathways (Bianchi-Smiraglia et al. 2013). Future therapeutic strategies could be achieved by inhibiting the link between ITG β5 and its signalling events or by disabling those two pathways (Bianchi-Smiraglia et al. 2013). The reason why integrins are attractive therapeutic targets is because of their elevated levels in certain aggressive types of tumours (Hood et al. 2003). Thus, ITG β5 can be considered as a potential therapeutic target (Maubant et al. 2002). Further studies are required to investigate the role of integrins as potential biomarkers that could possibly determine the extent of certain types of cancer (Maubant et al. 2002).

Integrin β8 is also responsible for the vascular integrity and regulating angiogenesis in tumour cells. Suppression of integrin β8 expression, prompted by the injection of microRNA, miR-93, promotes tumour growth and angiogenesis (Fang et al. 2011). Comparison of the tumour development and growth within the control group of mice and mice injected with miR-93 shows a larger tumour size and a less abundant amount of fragmented nuclei in the miR-93 tumours. An abundance of condensed and fragmented nuclei leads to severe cell death. Hence, it has been suggested that miR-93, to a certain extent, increases the survival rates of cancer cells. Dissection of the mock and miR-93 tumours revealed an abundance of blood vessels in the miR-93 tumours compared to the mock tumours. Additionally, the blood vessels were more scattered and showed signs of better health, with improved endothelial cell growth. Through the use of computational analysis, there is a direct correlation between integrin- β8 and miR-93 (Fang et al. 2011). The larger number of proteins associated with cell growth and tumour formation are targets of miR-93. Integrin β8 contains 3′-untranslated region (3′UTR), a target sequence for miR-93. Anti-integrin β8 antibody was used to probe the tumour sections, indicating that integrin β8 levels are significantly lower in miR-93 tumours than mock tumours. Hence, integrin β8 expression is suppressed in miR-93 tumour cells. Repression of this gene, visible in the observation of the mock and miR-93 tumours, reveals that integrin β8 protein directly correlates with the dead sections of the tumours. Down regulation of integrin β8 via miR-93 injections promoted tumour cell survival. Integrin β8 is responsible not only for the normal vascular physiology of the brain but also for regulating angiogenesis in tumour cells. Integrin β8 expression is linked with nuclear condensation and promotion of cell death. Hence, integrin β8 is a negative regulator of cell proliferation, where lowered activity of this gene favours endothelial cell growth and angiogenesis of tumour cells (Fang et al. 2011).

In another study the role of integrin β8 in angiogenesis regulation of Glioblastoma multiforme (GBM) has been tested (Tchaicha et al. 2011). Glioblastoma multiforme is an invasive brain tumour that is well known for its highly vascularised and haemorrhagic features. The study revealed that integrin β8 is not only essential for regulation of angiogenesis but it is also moderating the invasiveness of tumour cells. It was concluded that future therapies through integrin β8 could diminish cell progression and invasion in addition to moderation of angiogenesis (Tchaicha et al. 2011).

Therefore, the normal expression of integrin β8 gene is essential for the production of integrin β8 protein which functions as a mediator and regulator of angiogenesis. Mutated individuals that lack the gene or contain abnormalities that prevent normal functioning of the gene, experience vascular pathologies localised in the brain. Suppression of the ITGB8 gene expression can be achieved via administration of other factors such as miR-93, which can be used in the treatment of aggravated stages of tumour growth. Such is possible due to its impact on the activity of integrin β8, which regulates angiogenesis. Repressed integrin β8 results in excessive vascularisation of tumour leading to cell proliferation and malignancy. Further research into this gene is necessary in improving modern scientists' understanding of the neurovascular mechanisms of the human body and can pave way for the establishment of treatment and prevention of pathologies associated with angiogenesis and tumour growth.

References

Bianchi-Smiraglia A, Paesante S, Bakin AV (2013) Integrin beta5 contributes to the tumorigenic potential of breast cancer cells through the Src-FAK and MEK-ERK signaling pathways. Oncogene 32(25):3049–3058. doi:10.1038/onc.2012.320

Boudreau NJ, Varner JA (2004) The homeobox transcription factor Hox D3 promotes integrin alpha5beta1 expression and function during angiogenesis. J Biol Chem 279(6):4862–4868. doi:10.1074/jbc.M305190200

Coudert AE, Del Fattore A, Baulard C, Olaso R, Schiltz C, Collet C, Teti A, de Vernejoul MC (2014) Differentially expressed genes in autosomal dominant osteopetrosis type II osteoclasts reveal known and novel pathways for osteoclast biology. Lab Invest 94(3):275–285. doi:10.1038/labinvest.2013.140

Denadai MVA, Viana LS, Affonso RJ Jr, Silva SR, Oliveira ID, Toledo SR, Matos D (2013) Expression of integrin genes and proteins in progression and dissemination of colorectal adenocarcinoma. BMC Clin Pathol 13:16. doi:10.1186/1472-6890-13-16

Fang L, Deng Z, Shatseva T, Yang J, Peng C, Du WW, Yee AJ, Ang LC, He C, Shan SW, Yang BB (2011) MicroRNA miR-93 promotes tumor growth and angiogenesis by targeting integrin-beta8. Oncogene 30(7):806–821. doi:10.1038/onc.2010.465

Garlet GP, Horwat R, Ray HL Jr, Garlet TP, Silveira EM, Campanelli AP, Trombone AP, Letra A, Silva RM (2012) Expression analysis of wound healing genes in human periapical granulomas of progressive and stable nature. J Endod 38(2):185–190. doi:10.1016/j.joen.2011.09.011

Greaves NS, Ashcroft KJ, Baguneid M, Bayat A (2013) Current understanding of molecular and cellular mechanisms in fibroplasia and angiogenesis during acute wound healing. J Dermatol Sci 72(3):206–217. doi:10.1016/j.jdermsci.2013.07.008

Grünhage F, Mahler M, Lammert F (2011) 1156 evaluation of integrin beta 5 (Itbg5) haplotypes as genetic risk factors for advanced liver fibrosis in a large fibroscan-staged cohort. J Hepatol 54:S457. doi:10.1016/S0168-8278(11)61158-6

Himes BE, Qiu W, Klanderman B, Ziniti J, Senter-Sylvia J, Szefler SJ, Lemanske RF Jr, Zeiger RS, Strunk RC, Martinez FD, Boushey H, Chinchilli VM, Israel E, Mauger D, Koppelman GH, Nieuwenhuis MA, Postma DS, Vonk JM, Rafaels N, Hansel NN, Barnes K, Raby B, Tantisira KG, Weiss ST (2013) ITGB5 and AGFG1 variants are associated with severity of airway responsiveness. BMC Med Genet 14:86. doi:10.1186/1471-2350-14-86

Hood JD, Frausto R, Kiosses WB, Schwartz MA, Cheresh DA (2003) Differential αv integrin–mediated Ras-ERK signaling during two pathways of angiogenesis. J Cell Biol 162(5):933–943. doi:10.1083/jcb.200304105

Hsia HC, Nair MR, Corbett SA (2014) The fate of internalized α5 integrin is regulated by matrix-capable fibronectin. J Surg Res 191(2):268–279. doi:10.1016/j.jss.2014.05.084

Kasuga I, Hogg JC, Pare PD, Hayashi S, Sedgwick EG, Ruan J, Wallace AM, He JQ, Zhang X, Sandford AJ (2009) Role of genetic susceptibility to latent adenoviral infection and decreased lung function. Respir Med 103(11):1672–1680. doi:10.1016/j.rmed.2009.05.008

Lakhe-Reddy S, Li V, Arnold TD, Khan S, Schelling JR (2014) Mesangial cell alphavbeta8-integrin regulates glomerular capillary integrity and repair. Am J Physiol Renal Physiol 306(12):F1400–F1409. doi:10.1152/ajprenal.00624.2013

Leifheit-Nestler M, Conrad G, Heida NM, Limbourg A, Limbourg FP, Seidler T, Schroeter MR, Hasenfuss G, Konstantinides S, Schafer K (2010) Overexpression of integrin beta 5 enhances the paracrine properties of circulating angiogenic cells via Src kinase-mediated activation of STAT3. Arterioscler Thromb Vasc Biol 30(7):1398–1406. doi:10.1161/atvbaha.110.206086

Maquart FX, Monboisse JC (2014) Extracellular matrix and wound healing. Pathol Biol 62(2):91–95. doi:10.1016/j.patbio.2014.02.007

Maubant S, Cruet-Hennequart S, Poulain L, Carreiras F, Sichel F, Luis J, Staedel C, Gauduchon P (2002) Altered adhesion properties and alphav integrin expression in a cisplatin-resistant human ovarian carcinoma cell line. Int J Cancer J Int du Cancer 97(2):186–194

Mobley AK, Tchaicha JH, Shin J, Hossain MG, McCarty JH (2009) Beta8 integrin regulates neurogenesis and neurovascular homeostasis in the adult brain. J Cell Sci 122(Pt 11):1842–1851. doi:10.1242/jcs.043257

Reynolds LE, Wyder L, Lively JC, Taverna D, Robinson SD, Huang X, Sheppard D, Hynes RO, Hodivala-Dilke KM (2002) Enhanced pathological angiogenesis in mice lacking beta3 integrin or beta3 and beta5 integrins. Nat Med 8(1):27–34. doi:10.1038/nm0102-27

Robinson SD, Hodivala-Dilke KM (2011) The role of beta3-integrins in tumor angiogenesis: context is everything. Curr Opin Cell Biol 23(5):630–637. doi:10.1016/j.ceb.2011.03.014

Rosenberg GA (2001) Extracellular matrix-degrading metalloproteinases and neuroinflammation in stroke. In: Inflammation and stroke. Springer. doi:10.1007/978-3-0348-8297-2_22

Serini G, Valdembri D, Bussolino F (2006) Integrins and angiogenesis: a sticky business. Exp Cell Res 312(5):651–658. doi:10.1016/j.yexcr.2005.10.020

Stupack DG, Cheresh DA (2004) Integrins and angiogenesis. Curr Top Dev Biol 64:207–238. doi:10.1016/s0070-2153(04)64009-9

Tahira AC, Kubrusly MS, Faria MF, Dazzani B, Fonseca RS, Maracaja-Coutinho V, Verjovski-Almeida S, Machado MC, Reis EM (2011) Long noncoding intronic RNAs are differentially expressed in primary and metastatic pancreatic cancer. Mol Cancer 10:141. doi:10.1186/1476-4598-10-141

Tchaicha JH, Reyes SB, Shin J, Hossain MG, Lang FF, McCarty JH (2011) Glioblastoma angiogenesis and tumor cell invasiveness are differentially regulated by β8 integrin. Cancer Res 71(20):6371–6381. doi:10.1158/0008-5472.CAN-11-0991

van der Flier A, Badu-Nkansah K, Whittaker CA, Crowley D, Bronson RT, Lacy-Hulbert A, Hynes RO (2010) Endothelial alpha5 and alphav integrins cooperate in remodeling of the vasculature during development. Development 137(14):2439–2449. doi:10.1242/dev.049551

Walter RB, Laszlo GS, Alonzo TA, Gerbing RB, Levy S, Fitzgibbon MP, Gudgeon CJ, Ries RE, Harrington KH, Raimondi SC, Hirsch BA, Gamis AS, McIntosh MW, Meshinchi S (2013) Significance of expression of ITGA5 and its splice variants in acute myeloid leukemia: a report from the children's oncology group. Am J Hematol 88(8):694–702. doi:10.1002/ajh.23486

Wang S, Olson EN (2009) AngiomiRs–key regulators of angiogenesis. Curr Opin Genet Dev 19(3):205–211. doi:10.1016/j.gde.2009.04.002

Wang HY, Chen Z, Wang ZH, Wang H, Huang LM (2013) Prognostic significance of alpha5beta1-integrin expression in cervical cancer. Asian Pac J Cancer Prev 14(6):3891–3895

Weis SM, Cheresh DA (2011) αv integrins in angiogenesis and cancer. Cold Spring Harbor Perspect Med 1(1):a006478. doi:10.1101/cshperspect.a006478

Xuan SH, Zhou YG, Pan JQ, Zhu W, Xu P (2013) Overexpression of integrin alphav in the human nasopharyngeal carcinoma associated with metastasis and progression. Cancer Biomark 13(5):323–328. doi:10.3233/cbm-130361

Zhang S, Yue J-X, Yang J-H, Cai P-C, Kong W-J (2008) Overexpression of transketolase protein TKTL1 is associated with occurrence and progression in nasopharyngeal carcinoma: a potential therapeutic target in nasopharyngeal carcinoma. Cancer Biol Ther 7(4):517–522. doi:10.4161/cbt.7.4.5479

Zhang L, Qu Y, Tang B, Zhao F, Xiong T, Ferriero D, Mu D (2012) Integrin beta8 signaling in neonatal hypoxic-ischemic brain injury. Neurotox Res 22(4):280–291. doi:10.1007/s12640-012-9312-0

Zhu J, Motejlek K, Wang D, Zang K, Schmidt A, Reichardt LF (2002) beta8 integrins are required for vascular morphogenesis in mouse embryos. Development 129(12):2891–2903

Chapter 27
Interleukins

Abstract Interleukin are integral part of the immune system and are secreted by leukocytes to clear wound of infection and foreign bodies in order to initiate healing process. Some examples of Interleukins are IL-1, 6, 8, 15 and 17. These Interleukins are responsible for signalling to attract phagocytes and other repair cells to the site of infection or injury through a number of processes but one of their key roles is in angiogenesis. Interleukins can either promote or inhibit angiogenesis. The Interleukin-1 (IL-1) family contains pro-inflammatory cytokines that function to protect inflammatory responses. Interleukin-1 has an influence on angiogenesis under normal and malignant conditions. Interleukin-6 (IL-6) has been demonstrated to induce VEGF expression and thus promote angiogenesis. Interleukin-8 (IL-8) plays a role in stimulating angiogenesis leading to wound repair. It has also been suggested to directly correlate with the extent of local growth and distal metastasis of cancer cells. Interleukin-15 (IL-15) may be inherently involved in the immune system and the anti-tumour response, emerging evidences are suggesting that IL-15 may also have pro-tumour properties through stimulation of the production of pro-angiogenic factors such as VEGF and interleukin 17 (IL-17). IL-17 has been linked to both stimulation and inhibition of angiogenesis. However, a stronger correlation to the stimulation of angiogenesis under malignant conditions has been shown. It is noteworthy to know the role of cytokines in angiogenesis during normal situation or tumour progression as the ability to manipulate vascular formation is an area of interest for therapeutic considerations.

Keywords Interleukin • IL-1 • IL-6 • IL-8 • IL-15 • IL-17 • Angiogenesis • Normal physiology • Disease • Malignancy

27.1 Interleukins

Several factors and biochemical agents have been implicated in the development and regulation of angiogenesis. Oxygen has been considered a master signal molecule in the growth regulation of vasculature (Ferrara 1999). Because it cannot be stored in cells in large quantity like other metabolites so cells/tissues with high metabolic rates have higher O_2 demand. Therefore, it is tempting to speculate that

A. Salajegheh, *Angiogenesis in Health, Disease and Malignancy*,
DOI 10.1007/978-3-319-28140-7_27

actively dividing tumour cells and heavily exercising muscle tissues may face hypoxic/ischaemic conditions triggering development and proliferation of blood vessels. Several studies have demonstrated that hypoxia/ischemia regulates pro-angiogenic factors such as vascular endothelial growth factor (VEGF-A), VEGFR2, Angipoietin-1 (Ang-1), Angipoietin-2 (Ang-2) and their Tie 2 receptor (Ferrara 1999; Rissanen et al. 2002). Recent study has shown that O_2 starved cells produce a transcription factor, hypoxia inducible factor (HIF1) which in turn induces VEGF expression in hypoxic tissues (Semenza 2009). VEGF expression is also stimulated by cytokines such as interleukin1 (IL1) providing an evidence for involvement of immune system in the regulation of angiogenesis during trauma, wound healing after surgery and tumour growth (Salven 2002).

Interleukin are integral part of the immune system and are secreted by leukocytes to clear wound of infection and foreign bodies in order to initiate healing process. Some examples of Interleukins are IL-1, 6, 8, 15 and 17. These Interleukins are responsible for signalling to attract phagocytes and other repair cells to the site of infection or injury through a number of processes but one of their key roles is in angiogenesis. Interleukins can either promote or inhibit angiogenesis.

27.2 Interleukin-1 in Normal Physiology, Disease and Malignancy

There are eleven members of the interleukin-1 (IL-1) family, and the genes encoding this group are clustered in a 400 kb region of chromosome 2 (Sims and Smith 2010). IL-1 members mediate biological responses through binding to receptors with specific sub-units (i.e. IL-1 receptor type 1) and a Toll/IL-1 domain, which then activates the nuclear-factor κB and mitogen-activated protein kinase (MAPK) pathway (Sims and Smith 2010).

The IL-1 family contains pro-inflammatory cytokines that function to protect inflammatory responses (Matsuo et al. 2009). There are two agonistic IL-1 proteins, IL-1α and IL-1β, which may play important roles in tumour metastasis and angiogenesis (Voronov et al. 2003). IL-1β is thought to be more important for the ability of tumours to initiate and complete angiogenesis, as it operates in the tumour micro-environment to activate nearby cells in the stroma (Voronov et al. 2003). IL-1β may stimulate angiogenesis through causing an up-regulation of vascular endothelial growth factor (VEGF), as is the case when nucleus pulposus cells are exposed to IL-1β (Lee et al. 2011). IL-1α is largely associated within the cell, and has been suggested to be not as important in angiogenesis (Voronov et al. 2003). Further studies indicate that it is the IL-1β that induce the production of VEGF by the endothelial cells which results in inflammation and thus angiogenesis (Carmi et al. 2013). It is also observed in this study that through the inhibition of IL-1β, by recombinant VEGF through rIL-1β, the inflammatory response is reduced thereby, reducing tumour growth. However, in the study, inhibition of tumour growth is only observed for a short period as angiogenesis leads to the recurrence of the tumour.

However, research has demonstrated that IL-1α may be produced by tumour cells (Ma et al. 2008), IL-1α RNA may be expressed more greatly in cancerous tissue at risk of metastasis (Matsuo et al. 2009), and VEGF secretion may be regulated by IL-1α (Ma et al. 2008). Interleukin-1 has an influence on angiogenesis under normal and malignant conditions. This is achieved through interaction with other cytokines and inflammatory cells migration to the site of infection/injury by altering the adhesion molecule expressed on the endothelial cells and inflammatory cells (Voronov et al. 2003). As indicated IL-1β can induce the production of VEGF by the endothelial cells which results in inflammation and thus angiogenesis (Carmi et al. 2013). It is also observed in this study that through the inhibition of IL-1β, by recombinant VEGF through rIL-1β, the inflammatory response is reduced thereby, reducing tumour growth. However, in the study, inhibition of tumour growth is only observed for a short period as angiogenesis leads to the recurrence of the tumour (Carmi et al. 2013).

27.3 Interleukin 6 in Normal Physiology, Disease and Malignancy

The nine members of the interleukin-6 (IL-6) family of cytokines bind to a receptor complex, which includes IL-6 receptor-α and gp130 receptor subunits and signal transducers (Lee et al. 2011), to mediate their biological activities in homeostasis and the immune response. Under physiological conditions, the levels of IL-6 protein are practically undetectable due to tightly regulated IL-6 expression (Chalaris et al. 2011).

IL-6 has also been demonstrated to induce VEGF expression and thus promote angiogenesis (Cohen et al. 1996). IL-6 has purported roles in inflammatory processes seen in myocardial infarction, autoimmune disorders and cancer (Hsu et al. 2011). This may be because the protein has been shown to be synthesised by macrophages, neutrophils, fibroblasts and endothelial cells, particularly when stimulated by inflammation (Nibali et al. 2012). IL-6 could potentially be a key regulator in metastatic spread of cancer through its positive effects on angiogenesis. A study by Hsu and authors demonstrated that IL-6 in colorectal cancer promoted tumour angiogenesis and invasiveness through the PI3K /AKT and MAPK pathways, as well as enhancing the expression of MMP-2 and MMP-9 (Hsu et al. 2011).

Interleukin-6 (Il-6) is also linked to the inhibition of angiogenesis where the endothelial cell proliferation and VEGF are inhibited (Hatzi et al. 2002). This is shown through xenograft of tumours expressing IL-6 in mice resulting in reduced vascularisation. In this study, it is suggested that a correlation exits between oncogenes, such as N-Myc, and IL-6 where angiogenesis is enhanced through the down regulation of IL-6 and the endothelial cells become insensitive to IL-6. This enhances the malignancy of the tumour caused by the oncogene N-Myc.

27.4 Interleukin 8 in Normal Physiology, Disease and Malignancy

Interleukin 8 (IL-8) is a small protein that is secreted from a range of nucleated cells, with monocytes and macrophages being the principal producers (Apostolakis et al. 2009). IL-8 is primarily responsible for the recruitment of monocytes and neutrophils to an inflammatory site by establishing a chemokine gradient (Apostolakis et al. 2009). IL-8 binds to two cell surface G-protein-coupled receptors, CXCR1 and CXCR2, to mediate its biological effects (Apostolakis et al. 2009).

IL-8 may also be inherently involved in the angiogenic process through its action on endothelial cells to induce their division and replication (Ning et al. 2011). IL-8 expression has been found in a vast number of cancers, including acute myelogenous leukaemia, B-cell chronic lymphocytic leukaemia, brain tumours, breast cancer, colon cancer, cervical cancer, gastric cancer, Hodgkin's disease, lung cancer, melanoma, mesothelioma, ovarian cancer, pituitary adenomas, prostate cancer, renal cell carcinoma, and thyroid tumours (Xie 2001). Many of these cancer types express IL-8 at a level much higher than in normal tissue (Weidner et al. 1991), which is enhanced further by hypoxic conditions (Xie 2001). The overexpression of IL-8 has been previously shown to induce an increased, VEGF-independent, angiogenic response in tumour xenograph models, where tumours were of an increased size with a much greater microvascular density (Ning et al. 2011). IL-8 expression has also been suggested to directly correlate with the extent of local growth and distal metastasis of cancer cells (Xie 2001). Indeed, these results indicate that IL-8 is strongly associated with angiogenesis and subsequent metastatic spread of cancer, and that a future avenue of treatment may be to block IL-8 receptors or decrease the production of this cytokine through mRNA manipulation.

27.5 Interleukin 15 in Normal Physiology, Disease and Malignancy

Interleukin 15 (IL-15) has been mapped to the chromosome 4 region q25–35, and the mature protein is largely produced by dendritic cells, monocytes, macrophages and stromal cells (Jakobisiak et al. 2011). IL-15 seems to stimulate biological activity mostly through juxtacrine signalling (Jakobisiak et al. 2011). IL-15 has been purported to stimulate the development and activity of memory CD8+ T cells and Natural Killer cells (Jakobisiak et al. 2011). Interleukin-15 (IL-15) is also another cytokine that induces angiogenesis in both healthy and malignant subjects. IL-15 has been shown to promote angiogenesis when introduced subcutaneously into mice where it results in vascularisation of Matrigel plugs (Angiolillo et al. 1997). Despite the angiogenic effect *in vivo* the mechanisms are unknown as experiments to stimulate endothelial cell proliferation *in vitro* did not occur as expected. However, further studies are able to determine the link between angiogenesis and IL-15 through immunostaining specimens of colon cancer from a large sample of patients (Kuniyasu et al. 2005). Additionally, the introduction of human colon cancer cells in nude mice resulted in the hyperplasia of the surrounding mucosa and also the expression of IL-15 receptors in the rat's intestinal epithelial cells. On treatment with IL-15, these cells show an increase in VEGF and other growth factors that are linked to angiogenesis. This confirms the angiogenic effect *in vivo* where IL-15 can be linked to increasing tumour growth and progression of colon cancer (Kuniyasu et al. 2005).

IL-15 may be inherently involved in the immune system and the anti-tumour response (Jakobisiak et al. 2011). It has been suggested that treatment aimed at enhancing the activity of IL-15 may assist the immune system in overcoming carcinoma invasion (Jakobisiak et al. 2011). However, evidence has emerged recently that IL-15 may also have pro-tumour properties. In a study by Badoual and colleagues, there was a correlation between high intra-tumoural IL-15 concentrations and poor clinical outcomes in head and neck cancer patients (Badoual et al. 2008). It has been hypothesised that this is because IL-15 may stimulate the production of pro-angiogenic factors such as VEGF and interleukin 17 (IL-17) (Tartour et al. 2011).

27.6 Interleukin 17 in Normal Physiology, Disease and Malignancy

IL-17 is produced by Th17 CD4+ helper T cells (Pickens et al. 2010), and induces the induction of TNFα, IL-1, and IL-6, as well as chemokines IL-8 and monocyte chemoattractant protein1 (MCP1) in various cell types (Maione et al. 2009).

Interleukin-17 (IL-17) has been linked to both stimulation and inhibition of angiogenesis. However, a stronger correlation to the stimulation of angiogenesis under malignant conditions has been shown (Yang et al. 2014). IL-17 has been shown to not only aid in angiogenesis and metastasis through the stimulation of VEGF but also to be capable of negating the effects of anti-vascular endothelial growth factor (anti-VEGF) chemotherapy. This is supported by the finding that by inhibiting IL-17, anti-VEGF chemotherapy is successful in reducing malignant tumour growth. Further studies have also shown that IL-17 plays an important role in chronic inflammation through stimulating pro-inflammatory cytokines that maintain the inflammatory environment. This effect is particularly important as pro-inflammatory cytokines have been shown to play a role in angiogenesis and thus stimulate growth of tumours (Voronov et al. 2003).

Extensive documents demonstrated the major role of IL-17 in angiogenesis in a number of settings. IL-17 producing cells have been demonstrated in prostate cancer, non-small-cell lung carcinoma, hepatocellular carcinoma, breast cancer, ovarian cancer, and colorectal cancer (Liu et al. 2011). Angiogenesis has been shown to proceed following the delivery of IL-17 to tumour vascular beds (Pasche et al. 2012) and when levels of IL-17 increase in a model of arthritis (Pickens et al. 2010). A study by Wakita and authors also echoed these findings, showing that the growth of tumour cells was significantly reduced in IL-17 negative mutants, presumably due to a lack of blood supply from reduced angiogenesis (Wakita et al. 2010). IL-17 may promote angiogenesis indirectly through stimulating the production of VEGF (Liu et al. 2011), and encouraging endothelial cell migration and cord formation by the actions of VEGF (Numasaki et al. 2003).

References

Angiolillo AL, Kanegane H, Sgadari C, Reaman GH, Tosato G (1997) Interleukin-15 promotes angiogenesisin vivo. Biochem Biophys Res Commun 233(1):231–237. doi:10.1006/bbrc.1997.6435

Apostolakis S, Vogiatzi K, Amanatidou V, Spandidos DA (2009) Interleukin 8 and cardiovascular disease. Cardiovasc Res 84(3):353–360. doi:10.1093/cvr/cvp241

Badoual C, Bouchaud G, Agueznay Nel H, Mortier E, Hans S, Gey A, Fernani F, Peyrard S, Puig PL, Bruneval P, Sastre X, Plet A, Garrigue-Antar L, Quintin-Colonna F, Fridman WH, Brasnu D, Jacques Y, Tartour E (2008) The soluble alpha chain of interleukin-15 receptor: a proinflammatory molecule associated with tumor progression in head and neck cancer. Cancer Res 68(10):3907–3914. doi:10.1158/0008-5472.can-07-6842

Carmi Y, Dotan S, Rider P, Kaplanov I, White MR, Baron R, Abutbul S, Huszar M, Dinarello CA, Apte RN, Voronov E (2013) The role of IL-1beta in the early tumor cell-induced angiogenic response. J Immunol 190(7):3500–3509. doi:10.4049/jimmunol.1202769

Chalaris A, Garbers C, Rabe B, Rose-John S, Scheller J (2011) The soluble Interleukin 6 receptor: generation and role in inflammation and cancer. Eur J Cell Biol 90(6–7):484–494. doi:10.1016/j.ejcb.2010.10.007

Cohen T, Nahari D, Cerem LW, Neufeld G, Levi BZ (1996) Interleukin 6 induces the expression of vascular endothelial growth factor. J Biol Chem 271(2):736–741

Ferrara N (1999) Role of vascular endothelial growth factor in regulation of angiogenesis. In: Antiangiogenic agents in cancer therapy. Springer. doi:10.1007/978-1-59259-453-5_8

Hatzi E, Murphy C, Zoephel A, Rasmussen H, Morbidelli L, Ahorn H, Kunisada K, Tontsch U, Klenk M, Yamauchi-Takihara K, Ziche M, Rofstad EK, Schweigerer L, Fotsis T (2002) N-myc oncogene overexpression down-regulates IL-6; evidence that IL-6 inhibits angiogenesis and suppresses neuroblastoma tumor growth. Oncogene 21(22):3552–3561. doi:10.1038/sj.onc.1205440

Hsu CP, Chen YL, Huang CC, Chou CC, Liu CL, Hung CH, Kao TY, Chung YC (2011) Anti-interleukin-6 receptor antibody inhibits the progression in human colon carcinoma cells. Eur J Clin Invest 41(3):277–284. doi:10.1111/j.1365-2362.2010.02405.x

Jakobisiak M, Golab J, Lasek W (2011) Interleukin 15 as a promising candidate for tumor immunotherapy. Cytokine Growth Factor Rev 22(2):99–108. doi:10.1016/j.cytogfr.2011.04.001

Kuniyasu H, Yoshida K, Sasaki T, Sasahira T, Fujii K, Ohmori H (2005) Conjugated linoleic acid inhibits peritoneal metastasis in human gastrointestinal cancer cells. Int J Cancer 118(3):571–576. doi:10.1002/ijc.21368

Lee JM, Song JY, Baek M, Jung HY, Kang H, Han IB, Kwon YD, Shin DE (2011) Interleukin-1beta induces angiogenesis and innervation in human intervertebral disc degeneration. J Orthop Res 29(2):265–269. doi:10.1002/jor.21210

Liu J, Duan Y, Cheng X, Chen X, Xie W, Long H, Lin Z, Zhu B (2011) IL-17 is associated with poor prognosis and promotes angiogenesis via stimulating VEGF production of cancer cells in colorectal carcinoma. Biochem Biophys Res Commun 407(2):348–354. doi:10.1016/j.bbrc.2011.03.021

Ma J, Sawai H, Matsuo Y, Ochi N, Yasuda A, Takahashi H, Wakasugi T, Funahashi H, Sato M, Okada Y, Takeyama H, Manabe T (2008) Interleukin-1alpha enhances angiogenesis and is associated with liver metastatic potential in human gastric cancer cell lines. J Surg Res 148(2):197–204. doi:10.1016/j.jss.2007.08.014

Maione F, Paschalidis N, Mascolo N, Dufton N, Perretti M, D'Acquisto F (2009) Interleukin 17 sustains rather than induces inflammation. Biochem Pharmacol 77(5):878–887. doi:10.1016/j.bcp.2008.11.011

Matsuo Y, Sawai H, Ochi N, Yasuda A, Takahashi H, Funahashi H, Takeyama H, Guha S (2009) Interleukin-1alpha secreted by pancreatic cancer cells promotes angiogenesis and its therapeutic implications. J Surg Res 153(2):274–281. doi:10.1016/j.jss.2008.04.040

Nibali L, Fedele S, D'Aiuto F, Donos N (2012) Interleukin-6 in oral diseases: a review. Oral Dis 18(3):236–243. doi:10.1111/j.1601-0825.2011.01867.x

Ning Y, Manegold PC, Hong YK, Zhang W, Pohl A, Lurje G, Winder T, Yang D, LaBonte MJ, Wilson PM, Ladner RD, Lenz HJ (2011) Interleukin-8 is associated with proliferation, migration, angiogenesis and chemosensitivity in vitro and in vivo in colon cancer cell line models. Int J Cancer 128(9):2038–2049. doi:10.1002/ijc.25562

Numasaki M, Fukushi J, Ono M, Narula SK, Zavodny PJ, Kudo T, Robbins PD, Tahara H, Lotze MT (2003) Interleukin-17 promotes angiogenesis and tumor growth. Blood 101(7):2620–2627. doi:10.1182/blood-2002-05-1461

Pasche N, Frey K, Neri D (2012) The targeted delivery of IL17 to the mouse tumor neo-vasculature enhances angiogenesis but does not reduce tumor growth rate. Angiogenesis 15(1):165–169. doi:10.1007/s10456-011-9239-8

Pickens SR, Volin MV, Mandelin AM 2nd, Kolls JK, Pope RM, Shahrara S (2010) IL-17 contributes to angiogenesis in rheumatoid arthritis. J Immunol 184(6):3233–3241. doi:10.4049/jimmunol.0903271

Rissanen TT, Vajanto I, Hiltunen MO, Rutanen J, Kettunen MI, Niemi M, Leppänen P, Turunen MP, Markkanen JE, Arve K, Alhava E, Kauppinen RA, Ylä-Herttuala S (2002) Expression of vascular endothelial growth factor and vascular endothelial growth factor receptor-2 (KDR/Flk-1) in ischemic skeletal muscle and its regeneration. Am J Pathol 160(4):1393–1403. doi:10.1016/s0002-9440(10)62566-7

Salven P (2002) Interleukin-1a (IL-1a) promotes angiogenesis in vivo via VEGFR-2 pathway by inducing inflammatory cell VEGF synthesis and secretion. FASEB J 16(11):1471–1473. doi:10.1096/fj.02-0134fje

Semenza GL (2009) Defining the role of hypoxia-inducible factor 1 in cancer biology and therapeutics. Oncogene 29(5):625–634. doi:10.1038/onc.2009.441

Sims JE, Smith DE (2010) The IL-1 family: regulators of immunity. Nat Rev Immunol 10(2):89–102. doi:10.1038/nri2691

Tartour E, Pere H, Maillere B, Terme M, Merillon N, Taieb J, Sandoval F, Quintin-Colonna F, Lacerda K, Karadimou A, Badoual C, Tedgui A, Fridman WH, Oudard S (2011) Angiogenesis and immunity: a bidirectional link potentially relevant for the monitoring of antiangiogenic therapy and the development of novel therapeutic combination with immunotherapy. Cancer Metastasis Rev 30(1):83–95. doi:10.1007/s10555-011-9281-4

Voronov E, Shouval DS, Krelin Y, Cagnano E, Benharroch D, Iwakura Y, Dinarello CA, Apte RN (2003) IL-1 is required for tumor invasiveness and angiogenesis. Proc Natl Acad Sci U S A 100(5):2645–2650. doi:10.1073/pnas.0437939100

Wakita D, Sumida K, Iwakura Y, Nishikawa H, Ohkuri T, Chamoto K, Kitamura H, Nishimura T (2010) Tumor-infiltrating IL-17-producing gammadelta T cells support the progression of tumor by promoting angiogenesis. Eur J Immunol 40(7):1927–1937. doi:10.1002/eji.200940157

Weidner N, Semple JP, Welch WR, Folkman J (1991) Tumor angiogenesis and metastasis–correlation in invasive breast carcinoma. N Engl J Med 324(1):1–8. doi:10.1056/nejm199101033240101

Xie K (2001) Interleukin-8 and human cancer biology. Cytokine Growth Factor Rev 12(4):375–391

Yang B, Kang H, Fung A, Zhao H, Wang T, Ma D (2014) The role of interleukin 17 in tumour proliferation, angiogenesis, and metastasis. Mediators Inflamm 2014:1–12. doi:10.1155/2014/623759

Chapter 28
Insulin Receptor Substrate (IRS-1)

Abstract Insulin receptor substrate (IRS-1) is an adapter signalling protein which lacks intrinsic kinase activity. IRS-1 was isolated as an insulin receptor substrate but over time various studies have shown that it is capable of working in a range of growth hormone and cytokine receptor signalling cascades. The process that involves IRS-1 consists of recruiting various proteins to its receptors and then activating intracellular signalling cascades that facilitate responses at a cellular level to insulin. Studies have shown that IGF-I is capable of expressing a range of genes, including the encoding of VEGF and activates the formation of capillaries. Considering the described role of IRS-1 in angiogenesis, it seems that numerous papers have argued that IRS-1 is important in angiogenesis. Variety of methods showed that inhibition of IRS-1 results in down regulation or inhibition of angiogenesis. There are also findings with regards to the IRS-1 role in lymphangiogenesis. Future research question about angiogenic role of IRS-1 is whether lymphatic endothelial cells are more vulnerable to IRS-1 than blood endothelial cells.

Keywords Insulin receptor substrate • IRS-1 • Angiogenesis • Normal physiology • Disease • Malignancy

28.1 Insulin Receptor Substrate (IRS-1)

Through energy storage and consumption insulin regulates metabolic pathways and controls cell growth. Once ligand binding has occurred the insulin receptor kinase is stimulated, this leads to the phosphorylation of intracellular proteins in particular IRS-1 (Cheatham and Kahn 1995). Insulin receptor substrate (IRS-1) is an adapter signalling protein which lacks intrinsic kinase activity. IRS-1 was isolated as an insulin receptor substrate but over time various studies have shown that it is capable of working in a range of growth hormone and cytokine receptor signalling cascades. The process that involves IRS-1 consists of recruiting various proteins to its receptors and then activating intracellular signalling cascades that facilitate responses at a cellular level to insulin. IRS-1 was the first insulin receptor substrate to be cloned and therefore serves as an example standard of docking proteins (Saad et al. 1993).

A. Salajegheh, *Angiogenesis in Health, Disease and Malignancy*,
DOI 10.1007/978-3-319-28140-7_28

28.2 Insulin Receptor Substrate (IRS-1) in Normal Physiology

The role of IRS-1 in angiogenesis comes from the previous evidence that insulin and IGF systems have been associated with one another in a range of vascular diseases. This includes angiogenesis (Delafontaine 2000). In endothelial cells IRS-1 is overexpressed when placed in neo-vascular conditions (Jiang et al. 2003). In addition, numerous studies have hypothesised the possibility that IRS-1 can be studied as a target in regulation angiogenesis. This particularly included mediations by hypoxia, insulin and inflammation (Andrieu-Soler et al. 2005).

Studies conducted by Zelzer et al. (1998) have shown that IGF-I is capable of expressing a range of genes, including the encoding of VEGF (Vascular endothelial growth factor). VEGF is an angiogenic factor which activates the formation of capillaries. This results in a mitogenic cascade that is isolated to endothelial cells. In addition to this IGF-I is a potent mitogen that causes tumour growth and endorses the altered phenotype (Macaulay 1992; Baserga 1995).

28.3 Insulin Receptor Substrate (IRS-1) in Disease and Malignancy

A number of studies have explored the role of IRS-1 in hemangiogenesis. A study conducted by Jiang et al. (2003) found that IRS-1 decreased the amount of corneal neovascularization. This therefore created new pathways for treatment of ocular neo-vascular processes. In this study it was observed that down regulation of IRS-1 would cause the suppression of IL-1β expression in treated corneas. The importance of using IL-1β in this study was that it is a vital factor in the expression of inflammatory angiogenesis. This study paves way for the possibility of using Aganirsen in neovascularization treatment. This is because Aganirsen activity stops the production of IRS-1 and therefore causes an upstream inhibition of the neovascularization pathway (BenEzra et al. 1997; Moore et al. 2002; Voronov et al. 2003). Another study in relation to the role of IRS-1 in hemangiogenesis showed that in human endothelial cells IRS-1 expression is up regulated with regards to capillary formation (Al-Mahmood et al. 2009). Furthermore this study established that inhibition of IRS-1 by an Antisense Oligonucleotide will cause the inhibition of angiogenesis. This once again supports the previous evidence that IRS-1 plays a vital role in angiogenesis. The study conducted by Jiang et al. (2003) found that IRS-1 as part of angiogenesis shows hyperoxia induced retinal angiogenesis in mice. This therefore means that IRS-1 can play a role in dealing with vascular problems that are related to diabetes in particular through the hemangiogenesis pathway (Jiang et al. 2003).

With regards to lymphangiogenesis, there has been little research looking into the role IRS-1 plays in this process. Lymphangiogenesis is the formation of lymphatic vessels from pre-existing lymphatic vessels. However one recent study has concluded that IRS-1 is able to interact with integrins, multifunctional proteins also involved in lymphangiogenesis (Avraamides et al. 2008). In addition to this study, a study conducted by Hos et al. (2011) explored the role of IRS-1 in corneal lymphangiogenesis. Hos et al. (2011) concluded that IRS-1 signalling plays a role in developing new lymphatic vessels. Furthermore they found that the blockage of IRS-1 expression through the use of Aganirsen inhibits lymphangiogenesis and corneal hemangiogenesis. To understand the significance of IRSA-1 signalling, in this study the authors used corneal neovascularization assay as well as an established model for analysis of inflammatory corneal hemangiogenesis and lymphangiogenesis (Cursiefen et al. 2004). Once the suture was placed the authors saw a reduction in inflammatory corneal neovascularization. Another interesting finding from this study was that with IRS-1 blockage, this incidentally impairs lymphangiogenesis. This occurred due to a reduction in growth factor expression. Therefore this study concluded that IRS-1 is part of the molecular pathway that leads to lymphangiogenesis. This study concluded that it was not only hemangiogenesis that IRS-1 plays a role in but also in lymphangiogenesis.

To summarise the role of IRS-1 with regards to angiogenesis, it seems that IRS-1 does indeed play a role in angiogenesis. Numerous papers have argued that IRS-1 is important in angiogenesis. This has been done through a variety of methods but the majority of conclusions have all been consistent. The inhibition of IRS-1 results in down regulation or inhibition of angiogenesis. As IRS-1's role in lymphangiogenesis is under researched more studies will need to be done to establish a scientifically accepted opinion. Therefore more investigation is required to analyse whether there is a possibility that not only does IRS-1 play a role in lymphangiogenesis; but also whether lymphatic endothelial cells are more vulnerable to IRS-1 blocking than blood endothelial cells.

References

Al-Mahmood S, Colin S, Farhat N, Thorin E, Steverlynck C, Chemtob S (2009) Potent in vivo antiangiogenic effects of GS-101 (5′-TATCCGGAGGGCTCGCCATGCTGCT-3′), an antisense oligonucleotide preventing the expression of insulin receptor substrate-1. J Pharmacol Exp Ther 329(2):496–504. doi:10.1124/jpet.108.147496

Andrieu-Soler C, Berdugo M, Doat M, Courtois Y, BenEzra D, Behar-Cohen F (2005) Downregulation of IRS-1 expression causes inhibition of corneal angiogenesis. Invest Ophthalmol Vis Sci 46(11):4072–4078. doi:10.1167/iovs.05-0105

Avraamides CJ, Garmy-Susini B, Varner JA (2008) Integrins in angiogenesis and lymphangiogenesis. Nat Rev Cancer 8(8):604–617. doi:10.1038/nrc2353

Baserga R (1995) The insulin-like growth factor I receptor: a key to tumor growth? Cancer Res 55(2):249–252

BenEzra D, Griffin BW, Maftzir G, Sharif NA, Clark AF (1997) Topical formulations of novel angiostatic steroids inhibit rabbit corneal neovascularization. Invest Ophthalmol Vis Sci 38(10):1954–1962

Cheatham B, Kahn CR (1995) Insulin action and the insulin signaling network. Endocr Rev 16(2):117–142. doi:10.1210/edrv-16-2-117

Cursiefen C, Chen L, Borges LP, Jackson D, Cao J, Radziejewski C, D'Amore PA, Dana MR, Wiegand SJ, Streilein JW (2004) VEGF-A stimulates lymphangiogenesis and hemangiogenesis in inflammatory neovascularization via macrophage recruitment. J Clin Invest 113(7):1040–1050. doi:10.1172/jci20465

Delafontaine P (2000) IRS-1 variant: a new risk factor for coronary artery disease? Arterioscler Thromb Vasc Biol 20(2):283–284

Hos D, Regenfuss B, Bock F, Onderka J, Cursiefen C (2011) Blockade of insulin receptor substrate-1 inhibits corneal lymphangiogenesis. Invest Ophthalmol Vis Sci 52(8):5778–5785. doi:10.1167/iovs.10-6816

Jiang ZY, He Z, King BL, Kuroki T, Opland DM, Suzuma K, Suzuma I, Ueki K, Kulkarni RN, Kahn CR, King GL (2003) Characterization of multiple signaling pathways of insulin in the regulation of vascular endothelial growth factor expression in vascular cells and angiogenesis. J Biol Chem 278(34):31964–31971. doi:10.1074/jbc.M303314200

Macaulay VM (1992) Insulin-like growth factors and cancer. Br J Cancer 65(3):311–320

Moore JE, McMullen TC, Campbell IL, Rohan R, Kaji Y, Afshari NA, Usui T, Archer DB, Adamis AP (2002) The inflammatory milieu associated with conjunctivalized cornea and its alteration with IL-1 RA gene therapy. Invest Ophthalmol Vis Sci 43(9):2905–2915

Saad MJ, Folli F, Kahn JA, Kahn CR (1993) Modulation of insulin receptor, insulin receptor substrate-1, and phosphatidylinositol 3-kinase in liver and muscle of dexamethasone-treated rats. J Clin Invest 92(4):2065–2072. doi:10.1172/jci116803

Voronov E, Shouval DS, Krelin Y, Cagnano E, Benharroch D, Iwakura Y, Dinarello CA, Apte RN (2003) IL-1 is required for tumor invasiveness and angiogenesis. Proc Natl Acad Sci U S A 100(5):2645–2650. doi:10.1073/pnas.0437939100

Zelzer E, Levy Y, Kahana C, Shilo BZ, Rubinstein M, Cohen B (1998) Insulin induces transcription of target genes through the hypoxia-inducible factor HIF-1α/ARNT. EMBO J 17(17):5085–5094

Chapter 29
Iron-Sulfur Clusters (ISCU)

Abstract Iron-sulphur cluster assembly enzyme, ISCU, is a protein coding gene which is engaged in repression of mitochondrial metabolism. In human, the two isoforms of ISCU exist as iron-sulphur (Fe-S) cluster scaffold protein, ISCU1 and ISCU2. ISCU1 is located in cytosol while ISCU2 presents in mitochondria. ISCU1 and ISCU2 promote the assembly of iron-sulphur clusters, which is involved in redox reactions, electron transport, and other cellular processes such as catalysis, transcription, respiration, DNA repair, iron metabolism and heme biosynthesis. In normoxia, formation of iron-sulfur clusters result in the respiration of mitochondria. ISCU potential role in therapy selection of cancer is explored as it was discovered that miR-210 is closely linked to tumour growth initiation. This is evidenced by the higher amounts of miR-210 found in cancer stricken individuals that may reveal the metabolic state of their primary cancer. Thus, ISCU down regulation pathway may be the 'next therapeutic approach of synthetic lethality' for individuals found with high miR-210 levels. So drugs that effect ISCU's DNA damage repaired composing of helicases Rad3 or Fanconi anaemia can be used together with PARP or glycolysis inhibitors and glutaminolysis.

Keywords Iron-sulphur cluster assembly enzyme • ISCU • Angiogenesis • Normal physiology • Disease • Malignancy

29.1 Iron-Sulfur Clusters (ISCU)

Iron-sulphur cluster assembly enzyme, ISCU, is a protein coding gene which is engaged in repression of mitochondrial metabolism (Huang et al. 2010). In human, the two isoforms of ISCU exist as iron-sulphur (Fe-S) cluster scaffold protein, ISCU1 and ISCU2. ISCU1 is located in cytosol while ISCU2 presents in mitochondria. ISCU1 and ISCU2 promote the assembly of iron-sulphur clusters, which is involved in redox reactions, electron transport, and other cellular processes such as catalysis, transcription, respiration, DNA repair, iron metabolism and heme biosynthesis. In particular, mitochondrial aconitase in tricarboxylic acid (TCA) cycle and the mitochondrial respiratory complexes (Complexes 1, 2, and 3) demand iron-sulphur clusters for energy production and the respiration of mitochondria.,

A. Salajegheh, *Angiogenesis in Health, Disease and Malignancy*,
DOI 10.1007/978-3-319-28140-7_29

29.2 ISCU in Normal Physiology

ISCU are included into enzymes controlling mitochondrial respiration and the manufacture of energy. Examples of these enzymes include aconitase (key player in tricarboxylic acid-TCA cycle) and the mitochondrial respiratory complexes (important in electron transport) (Chan et al. 2009). In normoxia, ISCU1 and ISCU2 stimulate iron-sulfur cluster formation, which promotes aconitase activity of TCA cycle and mitochondrial electron transport, resulting in the respiration of mitochondria (Chan et al. 2009). ISCU1 are necessary scaffolding proteins that work together with a variety of evolutionarily conserved assembly and carry factors to manufacture [4Fe-4S] and [2Fe-2S] iron-sulfur clusters. These prosthetic groups encourage electron transport and mitochondrial oxidation-reduction activities necessary for various cellular activities, varying from ribosome biogenesis, purine catabolism, heme biosynthesis, DNA repair, to iron metabolism.

The role of ISCU 1 is closely related to mir-210 microRNA (miR-210), which regulates the expression levels of ISCU. Several events happen at the miR-210/ISCU1 pathway, where the up regulation of miR-210 leads to down regulation of ISCU.

29.3 ISCU in Disease

Mutation of iron-sulphur cluster has significant effects on the activities of iron-sulphur enzymes, including aconitase and succinate dehydrogenase as well as controlling the cellular iron homeostasis, resulting in deficiency of irons in cytosol or unnecessary accumulation of iron in mitochondria. Defective iron-sulphur cluster biosynthesis can not only develop Friedreich's ataxia where the myocytes and sensory neurons are damaged, causing heart problems and afferent impairments, but also cause glutaredoxin 5 deficiency which shows clinical symptoms of anaemia (Ye and Rouault 2010). One of the well known diseases caused by deficiency of the ISCU gene is myopathy (Kollberg et al. 2011; Crooks et al. 2014). The disease is generally inherited recessively and develops in childhood (Kollberg et al. 2009). Specifically, intronic IVS5+382 G>C mutation in ISCU affects decreased iron-sulphur cluster protein formation and myopathy with exercise intolerance. This mutation reduces ISCU protein level in muscle tissue by formation of truncated protein due to abnormally spliced mRNA (Larsson 2012). Clinical symptoms of myopathy are dyspnoea, myoglobinuria, muscle fatigue, increase in heart rate, and cardiac palpitations. Myopathy patients demonstrate overloads of iron in muscle fibres and the deficiency of histochemical succinate dehydrogenase. Although myopathy is not progressive, severe metabolic acidosis, rhabdomolysis and widespread of muscle weakness can occur. Even though effective iron-sulphur cluster deficiency myopathy treatment has not developed, yet, successful in vitro studies have conducted on patients' fibroblast samples. RNA regulating therapy with antisense oligonucleotides which blocks the site of splicing, inducing normal ISCU mRNA splice, was performed. In conclusion, stimulation of normal ISCU mRNA splicing by blocking the site of splice indicates a therapeutic probability. Another research on Duchenne muscular dystrophy treatment with antisense oligonucleotides also supports the accessibility of antisense-therapy on muscle tissue (Kollberg et al. 2011). VEGF proteins are indicative of angiogenesis related gene expression, as it is associated with new capillary formation. Since it is found in higher amounts in vastus lateralis and gastrocnemius biopsy protein extracts in ISCU myopathy suffers. Thus, it is obvious that ISCU myopathy, resulting from ISCU deficiency involves angiogenesis mechanism (Crooks et al. 2014).

Hypoxia inducible factor 1α (HIF-1α) induces hypoxia, causing biological processes including angiogenesis, glycolysis, and pH regulation. Reactive oxygen species (ROS) are produced by mitochondria in hypoxia, arbitrated by electron transport continuing in hypoxic condition. This free radical which promotes HIF and enhanced growth in vivo is toxic. In specific, to protect from the toxic effect, HIF-1α induces miR-210 in hypoxia. miR-210 is one of the microRNAs which control gene expression by stimulating its degradation or inhibiting target transcription (Favaro et al. 2010). During hypoxia and after the release of HIF1α, ISCU down regulation were observed (Favaro et al. 2010).

In hypoxic conditions, repression of ISCU1 would lead to various events. It was also shown that ISCU2 is a predicted target of miR-210, regulating ROS production level in hypoxia. With suppression of ISCU, it inhibits the activity of Krebs cycle and the function of mitochondrial, providing an increased free radical production. This escalated ROS increase HIF-1α expression, which lead to transcription of HIF response element, those element encodes proteins that increase chances of cell survival under hypoxia, inducing VEGF-dependant angiogenesis, glycolysis switch and stimulation of the iron uptake needed for the growth of the cells (Favaro et al. 2010). This is also believed to mediate cancer cell growth in vivo through the mechanism of the HIF dependent (Devlin et al. 2011).

29.4 ISCU in Malignancy

Hypoxia is a symptom of low oxygen level in body due to inadequate blood supply. Many cancer cells are hypoxic because of the growth of tumour, depriving blood supply from other cells and consuming oxygen for angiogenesis. Therefore, hypoxia is a physiological factor between normal and cancer cells. The initiation of the Pasteur effect: stability distortion of Fe-S cluster (evidenced by electron paramagnetic resonance spectroscopy) leads to repression in the electron transport chain, which corrects the mismatch with reduced oxygen tension, leading to an increase in ATP and induction of mitochondrial biogenesis repression through induction of most glycolytic enzymes (pyruvate dehydrogenase kinase) (Devlin et al. 2011). This ultimately leads to a from mitochondrial oxidative phosphorylation to high glycolytic metabolic shift in tumours phenotype. And this would be favourable to tumor growth and angiogenesis (Devlin et al. 2011).

During normoxia, repression of ISCU1, ISCU2 and the biogenesis of the iron clusters causes increased toxic ROS and apoptosis due to imbalance between the concentration of oxygen and electron transport. On the other hand, miR-210 reduces the gap between the concentration of oxygen and electron transport by decreasing ROS production as a homeostatic mechanism during hypoxia. This represents that miR-210 is maladaptive in normoxia, resulting in homeostatic imbalance between oxygen concentration and electron transport, yet, it is adaptive during hypoxia because of the mechanism of the homeostatic correction of reducing the imbalance (Chan et al. 2009). Given miR-210 adaptive role, repression of ISCU would lead to the homeostatic reaction mechanism to the hypoxic conditions. With the release of HIF-1α would be a primary rise in mitochondrial reactive oxygen species (ROS). This ROS released acts as O2 sensors and initiate HIF to kick-start the Pasteur effect. When Pasteur effect has corrected the inconsistency, there will be a secondary decrease in ROS. This reduces the likelihood of apoptosis while maximizing the manufacture of energy in hypoxic cells and enables effective tumor growth in vivo through a HIF dependent process, ultimately extending the expectancy of tumour cells. Thus, down regulation of ISCU encourages angiogenesis in hyper-proliferative disorders and cancer by providing conditions advantageous to tumor growth (Favaro et al. 2010).

Down regulation of ISCU would also lead to a decreased level of the iron-sulfur-containing protein succinate dehydrogenase B (SDHB). Head and neck paragangliomas (HNPGLs) are rare tumours of the parasympathetic nervous systems and are related to paragangliomas of the sympathetic nervous system. Where there is 30–45 % familial due to mutations in genes encoding subunits such as SDHB. Thus ISCU plays a role in malignancies such as HNPLGLs through the HIF-1α/miRNA-210/ISCU mechanism (Merlo et al. 2012). The down regulation of ISCU would lead to a decreased aconitase enzyme activity as it converts to iron-responsive element-binding protein (IRP1) that controls the stability and translation of mRNA for transferrin and ferritin. This results in modification of iron metabolism and more absorption, which are necessary for tumour growth. This relationship between ISCU/miRNA-210 and tumour growth is evidence by the positive correlation between miRNA-210 overexpression in various hypoxic related diseases such as solid tumours and to adverse clinical outcomes in breast cancer, renal clear cell carcinomas, pancreatic cancer patients and in chronic hypoxia that is a key player in pulmonary hypertension (Ivan et al. 2014).

ISCU role in mitochondrial respiration is further explored in the area of cell therapy. Today it is widely known that bone marrow mesenchymal stem cells (MSCs) in stem cell therapy hold future possibilities in improving cardiac function in cardiac diseases. Thus, better understanding of ISCU's role in mitochondrial respiration would contribute in helping to resolve the current impediment of poor survival of implanted MSCs. This is because the expression miRNA-210 induced by hypoxia/SD may be involved in protecting MSCs against apoptosis. The up regulation of miRNA results in a shift to glycolysis and decrease FLASH/capase-8 associated protein-2 expression, thereby enhancing MSC survival without protecting mitochondria. However, miR-210 overexpression was still unable to rescue mitochondrial membrane potential under hypoxia/SD treatment. This was due to ISCU down regulation that resulted in mitochondrial dysfunction (Nie et al. 2011).

Moreover, HIF-1α/miR-210 pathway can cause glycolytic pathway instead of respiration and assist the survival of the cancerous cells. It is expected that miR-210 supports tumour cells to adapt to hypoxic condition. In addition, the elevated expression of miR-210 in breast cancer, diffuse large B cell lymphoma, and pancreatic cancer represents that miR-210 helps tumour cells to grow (Chen et al. 2010). Furthermore, reduced Fe-S clusters due to ISCU inhibition causes the loss of aconitase, increasing uptake of iron, which is the crucial condition of cancer growth (Favaro et al. 2010).

Suppression of ISCU indicates possible therapeutic approach of synthetic lethality could be utilised in combination with PARP inhibitors, or glycolysis and glutaminolysis suppressors for the high miR-210 level patients (Chan et al. 2009; Favaro et al. 2010). Thus better understanding of the role and mechanism of ISCU may enable us to find solutions or alternatives to curb the effects of ISCU effects in normal physiology.

References

Chan SY, Zhang YY, Hemann C, Mahoney CE, Zweier JL, Loscalzo J (2009) MicroRNA-210 controls mitochondrial metabolism during hypoxia by repressing the iron-sulfur cluster assembly proteins ISCU1/2. Cell Metab 10(4):273–284. doi:10.1016/j.cmet.2009.08.015

Chen Z, Li Y, Zhang H, Huang P, Luthra R (2010) Hypoxia-regulated microRNA-210 modulates mitochondrial function and decreases ISCU and COX10 expression. Oncogene 29(30):4362–4368. doi:10.1038/onc.2010.193

Crooks DR, Natarajan TG, Jeong SY, Chen C, Park SY, Huang H, Ghosh MC, Tong WH, Haller RG, Wu C, Rouault TA (2014) Elevated FGF21 secretion, PGC-1alpha and ketogenic enzyme expression are hallmarks of iron-sulfur cluster depletion in human skeletal muscle. Hum Mol Genet 23(1):24–39. doi:10.1093/hmg/ddt393

Devlin C, Greco S, Martelli F, Ivan M (2011) miR-210: more than a silent player in hypoxia. IUBMB Life 63(2):94–100. doi:10.1002/iub.427

Favaro E, Ramachandran A, McCormick R, Gee H, Blancher C, Crosby M, Devlin C, Blick C, Buffa F, Li JL, Vojnovic B, Pires das Neves R, Glazer P, Iborra F, Ivan M, Ragoussis J, Harris AL (2010) MicroRNA-210 regulates mitochondrial free radical response to hypoxia and krebs cycle in cancer cells by targeting iron sulfur cluster protein ISCU. PLoS One 5(4):e10345. doi:10.1371/journal.pone.0010345

Huang X, Le Q-T, Giaccia AJ (2010) MiR-210 – micromanager of the hypoxia pathway. Trends Mol Med 16(5):230–237. doi:10.1016/j.molmed.2010.03.004

Ivan M, Zhong X, Greco S, Martelli F (2014) Emerging roles of non-coding RNAs in the hypoxic response. In: Hypoxia and cancer. Springer, New York, pp 43–64

Kollberg G, Tulinius M, Melberg A, Darin N, Andersen O, Holmgren D, Oldfors A, Holme E (2009) Clinical manifestation and a new ISCU mutation in iron-sulphur cluster deficiency myopathy. Brain 132(8):2170–2179. doi:10.1093/brain/awp152

Kollberg G, Melberg A, Holme E, Oldfors A (2011) Transient restoration of succinate dehydrogenase activity after rhabdomyolysis in iron–sulphur cluster deficiency myopathy. Neuromuscul Disord 21(2):115–120. doi:10.1016/j.nmd.2010.11.010

Larsson E (2012) The use of monogenic disease to study basal and disease associated mechanisms with focus on NGF dependent pain insensitivity and ISCU myopathy. Print & Media, Umeå

Merlo A, de Quiros SB, Secades P, Zambrano I, Balbin M, Astudillo A, Scola B, Aristegui M, Suarez C, Chiara MD (2012) Identification of a signaling axis HIF-1alpha/microRNA-210/ISCU independent of SDH mutation that defines a subgroup of head and neck paragangliomas. J Clin Endocrinol Metab 97(11):E2194–E2200. doi:10.1210/jc.2012-2410

Nie Y, Han BM, Liu XB, Yang JJ, Wang F, Cong XF, Chen X (2011) Identification of MicroRNAs involved in hypoxia- and serum deprivation-induced apoptosis in mesenchymal stem cells. Int J Biol Sci 7(6):762–768

Ye H, Rouault TA (2010) Human iron–sulfur cluster assembly, cellular iron homeostasis, and disease. Biochemistry 49(24):4945–4956. doi:10.1021/bi1004798

Chapter 30
MDM4

Abstract The role that MDM4 plays in angiogenesis in tumor growth and metastatic spread is an important area of medical research and has major implications clinically. The MDM4 gene and its corresponding MDM4 protein play an important role in tumor angiogenesis primarily through its role in down regulating the P53 protein. MDM4 is a nuclear protein-encoding gene, resulting in a protein that is 490 amino acids in length. The protein encoded by MDM4 has a ring finger domain at the C-terminus and possesses a p53-binding domain at the N-terminus. MDM4's p53 binding domain plays an important role in the proliferation of some tumours and cancerous growths. MDM4 protein acts as an oncogene, inhibiting p53 activity by binding its transcriptional activation domain. MDM4 inhibits and regulates the transcriptional activity of P53. It has been shown that when healthy cells experience DNA damage, MDM4 contributes significantly to P53 stabilization and activation. In acute hypoxia it has been shown that P53 promotes the expression of VEGF. However, in tissues undergoing sustained hypoxia, as would occur in the cells of a rapidly growing tumor, it has been shown that P53 down regulates the VEGF pathway. MDM4 in conjunction with MDM2, therefore, are likely to promote tumor angiogenesis by inhibiting P53. The vital role of MDM4 in tumor development has become important area in medical research for its interaction with P53.

Keywords MDM4 • Angiogenesis • Normal physiology • Disease • Malignancy

30.1 MDM4

The role that MDM4 plays in angiogenesis in tumor growth and metastatic spread is an important area of medical research and has major implications clinically (Salvi et al. 2014). The MDM4 gene and its corresponding MDM4 protein play an important role in tumor angiogenesis primarily through its role in down regulating the P53 protein. MDM4 is a nuclear protein-encoding gene, resulting in a protein that is 490 amino acids in length (Zhao et al. 2014). The protein encoded by MDM4 has a ring finger domain at the C-terminus and possesses a p53-binding domain at the N-terminus. MDM4's p53 binding domain plays an important role in the

A. Salajegheh, *Angiogenesis in Health, Disease and Malignancy*,
DOI 10.1007/978-3-319-28140-7_30

proliferation of some tumours and cancerous growths. MDM4 protein acts as an oncogene, inhibiting p53 activity by binding its transcriptional activation domain hence named also as p53 regulator. (Zhao et al. 2014).

30.2 MDM4 in Normal Physiology

MDM4 and MDM2 both inhibit the transcription activity of P53 (Perry 2010). However MDM4 and MDM2 both exhibit distinct but complimentary inhibition of P53 with MDM4 mainly regulating P53 activity while MDM2 is involved in the regulation of P53 stability (Toledo and Wahl 2007). MDM2 acts as an E3 ubiquitin ligase in which the ubiqitination of P53 is catalysed. Conversely, MDM4 does not cause degradation to P53. P53 is inhibited by MDM4 through MDM4 binding and masking its transcriptional activation domain (Pei et al. 2012). Additionally MDM4 and MDM2 can work in combination by the formation of heterodimers inducing the ubiquitination of P53 (Perry 2010). It has been shown that when healthy cells experience DNA damage, MDM2 is involved in the degradation of itself and MDM4 thereby contributing significantly to P53 stabilization and activation (Toledo and Wahl 2007).

30.3 MDM4 in Malignancy

The importance of p53 in tumour suppression is unequivocal (Bieging et al. 2014). P53 is a powerful tumor suppressor which helps regulate the cell cycle, apoptosis, DNA repair, cellular metabolism, innate immunity as well as angiogenesis (Perry 2010). P53 is also important in most cellular stress responses such as hypoxia, where it positively or negatively regulates the expression of numerous target genes involved in vital cellular processes including cell proliferation and angiogenesis (Farhang Ghahremani et al. 2013).

P53 is the most frequently inactivated tumor suppressor gene in human cancer (Assadian et al. 2012). P53 inactivation plays a vital role in almost all human cancers. In approximately 50 % of neoplasia the tumor suppressor function of P53 is inactivated by mutation or deletion of its geneTP53 (Perry 2010). In the other 50 % of cancers the TP53 gene remains active but the tumor suppression function of P53 is compromised. Multiple mechanisms play a role in this down-regulation but MDM4 and its close analogue MDM2 both play vital roles (Perry 2010).

P53 has been shown in a number of studies to have the ability to limit angiogenesis and that its actions can actually induce tumor dormancy (Assadian et al. 2012). Furthermore P53 inactivation increases microvessel density in tumours (Assadian et al. 2012). P53 has also been shown to induce production of antiangiogenic factors derived from collagen such as endostatin and tumstatin and arrestin (Assadian et al. 2012).

MDM4 plays an important role in modulating angiogenesis of tumours through its role in inhibiting P53 activity. Studies have shown that p53 inhibition enables tumours to be more vascularised and hence more aggressive (Bieging et al. 2014). P53 has been shown in a number of studies to have the ability to limit angiogenesis and that its actions can actually induce tumor dormancy (Assadian et al. 2012). Furthermore P53 inactivation increases microvessel density in tumours (Assadian et al. 2012). P53 controls angiogenesis through a number of pathways; inducing the production of anti-angiogenic factors such as arrestin, endostatin and tumstatin (Assadian et al. 2012). P53 also inhibits the production of angiogenic factors such as VEGF.

VEGF is essential for solid tumor development (Biderman et al. 2012). It is the key mediator of angiogenesis in tumours but the exact action of tumor suppressors such as P53, and hence MDM4 proteins in inhibiting VEGF are not fully understood (Zdravkovic et al. 2014; Farhang Ghahremani et al. 2013). VEGF plays a number of roles in the various steps of tumor angiogenesis thereby providing tumours with increased oxygen and nutrients. This may play a key role in driving tumor metastasis (Farhang Ghahremani et al. 2013). Farhang Ghahremani et al. described VEGF as 'an essential modulator of angiogenesis during both physiological and pathological hypoxia'. P53 is also a key component of cellular response to hypoxia (Farhang Ghahremani et al. 2013). P53 can positively or negatively regulate the expression of a number of genes within the cell (Farhang Ghahremani et al. 2013).

In acute hypoxia it has been shown that P53 actually promotes the expression of VEGF. However, in tissues undergoing sustained hypoxia, as would occur in the cells of a rapidly growing tumor, it has been shown that P53 down regulates the VEGF pathway via the retinoblastoma pathway in a P21 dependant manner (Farhang Ghahremani et al. 2013). MDM4 in conjunction with MDM2, therefore, are likely to promote tumor angiogenesis by inhibiting P53. P53 inhibits angiogenesis directly by the stimulating the production of arresten by the tumor cells. This significantly inhibited angiogenesis and limited tumor growth in vivo. Arrestin levels in human prostate tumors directly correlates with P53 levels (Assadian et al. 2012). MDM4 prevents the production of arrestin through its inhibition of P53 (Assadian et al. 2012).

Because of its vital role in tumor development, MDM4 has become important in medical research aimed at blocking its interaction with P53. There are a number of MDM2-P53 inhibitors with six molecules undergoing current clinical trials (Perry 2010). MDM4 inhibitors are proving more difficult to develop (Zak et al. 2013). Because MDM4 and MDM2 work in conjunction future molecules may work by targeting both proteins (Zak et al. 2013). More research still needs to be done on how to limit MDM4's binding with p53, which ultimately leads to angiogenesis and tumor growth.

It is clear that angiogenesis plays a crucial role in tumor development and growth. Although in some cases not fully understood, certain genes play a role in the regulation of angiogenesis. MDM4 directly or indirectly stimulates angiogenesis in certain tumor types. For this reason, ground-breaking research into gene modulation is taking place in the hopes of limiting or in some cases ceasing tumor development.

References

Assadian S, El-Assaad W, Wang XQ, Gannon PO, Barres V, Latour M, Mes-Masson AM, Saad F, Sado Y, Dostie J, Teodoro JG (2012) p53 inhibits angiogenesis by inducing the production of Arresten. Cancer Res 72(5):1270–1279. doi:10.1158/0008-5472.can-11-2348

Biderman L, Poyurovsky MV, Assia Y, Manley JL, Prives C (2012) MdmX is required for p53 interaction with and full induction of the Mdm2 promoter after cellular stress. Mol Cell Biol 32(7):1214–1225. doi:10.1128/mcb.06150-11

Bieging KT, Mello SS, Attardi LD (2014) Unravelling mechanisms of p53-mediated tumour suppression. Nat Rev Cancer 14(5):359–370. doi:10.1038/nrc3711

Farhang Ghahremani M, Goossens S, Nittner D, Bisteau X, Bartunkova S, Zwolinska A, Hulpiau P, Haigh K, Haenebalcke L, Drogat B, Jochemsen A, Roger PP, Marine JC, Haigh JJ (2013) p53 promotes VEGF expression and angiogenesis in the absence of an intact p21-Rb pathway. Cell Death Differ 20(7):888–897. doi:10.1038/cdd.2013.12

Pei D, Zhang Y, Zheng J (2012) Regulation of p53: a collaboration between Mdm2 and Mdmx. Oncotarget 3(3):228–235

Perry ME (2010) The regulation of the p53-mediated stress response by MDM2 and MDM4. Cold Spring Harb Perspect Biol 2(1):a000968. doi:10.1101/cshperspect.a000968

Salvi S, Calistri D, Gurioli G, Carretta E, Serra L, Gunelli R, Zoli W, Casadio V (2014) Copy number analysis of 24 oncogenes: MDM4 identified as a putative marker for low recurrence risk in non muscle invasive bladder cancer. Int J Mol Sci 15(7):12458–12468. doi:10.3390/ijms150712458

Toledo F, Wahl GM (2007) MDM2 and MDM4: p53 regulators as targets in anticancer therapy. Int J Biochem Cell Biol 39(7–8):1476–1482. doi:10.1016/j.biocel.2007.03.022

Zak K, Pecak A, Rys B, Wladyka B, Domling A, Weber L, Holak TA, Dubin G (2013) Mdm2 and MdmX inhibitors for the treatment of cancer: a patent review (2011-present). Expert Opin Ther Pat 23(4):425–448. doi:10.1517/13543776.2013.765405

Zdravkovic ND, Jovanovic IP, Radosavljevic GD, Arsenijevic AN, Zdravkovic ND, Mitrovic S, Arsenijevic NN (2014) Potential dual immunomodulatory role of VEGF in ulcerative colitis and colorectal carcinoma. Int J Med Sci 11(9):936–947. doi:10.7150/ijms.8277

Zhao Y, Aguilar A, Bernard D, Wang S (2014) Small-molecule inhibitors of the MDM2-p53 protein-protein interaction (MDM2 inhibitors) in clinical trials for cancer treatment. J Med Chem 58:1038–1052. doi:10.1021/jm501092z

Chapter 31
Matrix Metalloproteinase 2 (MMP2)

Abstract Matrix Metalloproteinases (MMPs) are a family of glycoprotein enzymes, which are zinc-dependent endoproteinases. These degradative enzymes digest the blood-vessel walls to allow endothelial cells to escape and migrate toward the site of the angiogenic stimuli. MMPs play a role in angiogenesis in many different ways; endothelial cell migration caused by surrounding tissues by interrupting ECM barriers, release of angiogenic factors, such as fibroblast growth factor-2 or vascular endothelial growth factor. MMP-2, also known as Gelatinase-A, is type IV collagenase which can not only break down type IV collagen of the basal laminae but also other areas of non-helical collagen and proteins such as Fibronectin, Laminin, Natural Insoluble Elastin, Aggregan, and Vitronectin. Many different cell types express MMP-2, such as fibroblasts, keratinocytes, endothelial cells, chondrocytes and monocytes. MMP-2 is known to contribute in various diseases, such as atherosclerosis and skeletal disorder. MMP2 can regulate pathological growth in vasculature and plays a role in angiogenesis of variety of diseases by releasing angiogenic factors and assisting in cell migration and cell adhesion. Further, MMP-2 can be a potential cause in initial tumor formation by altering ECM, which provides optimal environment for tumor growth. Hence, understanding the pathological and physiological aspects of MMP-2 remains crucial in order to develop therapeutic interventions.

Keywords Matrix metalloproteinase 2 • MMP2 • Angiogenesis • Normal physiology • Disease • Malignancy

31.1 Matrix Metalloproteinase 2 (MMP2)

Under normal physiological conditions, angiogenesis is a slow process. It affords diffusion exchange of nutrients and metabolites within tissues. Angiogenesis primarily occurs during embryonic development, wound healing and in response to ovulation (Carmeliet 2003). It is well described that vascular system and capillaries grow and relapse in healthy tissues on functional demands (Carmeliet 2003; Folkman 2007). In shortage of oxygen, the cell releases molecules, which induce proliferation of endothelial and inflammatory cells. During their migration,

A. Salajegheh, *Angiogenesis in Health, Disease and Malignancy*,
DOI 10.1007/978-3-319-28140-7_31

inflammatory cells secrete proangiogenic factors such as Vascular Endothelial Growth Factors (VEGF) (Rundhaug 2005). These factors bind to endothelial cell surface receptors of local pre-existing blood vessels, which cause them to differentiate and secrete Matrix Metalloproteases (MMP). These degradative enzymes digest the blood-vessel walls to allow endothelial cells to escape and migrate toward the site of the angiogenic stimuli. At the same time, protein fragments released through blood vessel wall digestion, enhance the migratory and proliferative activity of endothelial cells (Folkman 2007). Endothelial cells continue to sprout and anastomose until stable blood vessels are formed. These newly formed vessels are reinforced through the action of specialized muscle cells called pericytes, at which point blood flow is initiated. This type of angiogenesis is known as 'Sprouting angiogenesis' as blood vessels are formed in portions of tissues previously devoid of any vasculature (Kurz et al. 2003).

31.2 MMP-2 in Normal Physiology

Matrix Metalloproteinases (MMPs) are a family of glycoprotein enzymes, which are zinc-dependent endoproteinases (Bellafiore et al. 2013). In human MMP family, there are at least 24 different vertebrae MMPs found so far which are able to break down various extracellular matrix (ECM) components (Clements et al. 2003). This property enables MMPs not only to affect cellular events, such as cell proliferation, migration and adhesion but also many significant physiological processes like angiogenesis, bone development and wound healing (Clements et al. 2003). MMPs are synthesized as zymogens and activated via cleavage of signal peptide (Clements et al. 2003). These enzymes are then led to secretory pathway from the cell and this restricts their catalytic activity to the extracellular space (Clements et al. 2003).

It is well known that there is a significant link between MMPs and angiogenesis, in both pathological and physiological conditions, such as tumours and ovarian cycle (Corry 2004). MMPs play a role in angiogenesis in many different ways; endothelial cell migration caused by surrounding tissues by interrupting ECM barriers, release of angiogenic factors, such as fibroblast growth factor-2 or vascular endothelial growth factor (Corry 2004). The breakdown products of ECM create integrin-binding sites, which stimulate integrin intracellular signalling (Fanjul-Fernández et al. 2010). MMP-2, also known as Gelatinase-A, is type IV collagenase which can not only break down type IV collagen of the basal laminae but also other areas of non-helical collagen and proteins such as Fibronectin, Laminin, Natural Insoluble Elastin, Aggregan, and Vitronectin (Fanjul-Fernández et al. 2010). Many different cell types express MMP-2, such as fibroblasts, keratinocytes, endothelial cells, chondrocytes and monocytes (Bellafiore et al. 2013). MMP-2 can induce integrin signalling which enables survival and proliferation of endothelial cells (Fanjul-Fernández et al. 2010). However, it can also disturb angiogenesis by releasing angiogenesis inhibitors, such as Endostatin and Tumstatin (Fanjul-Fernández et al. 2010).

MMPs can modify the cells ability to proliferate, survive or differentiate changes by acting on the ECM. MMP-2 or Gelatinase A plays a significant role in the breakdown of ECM in normal physiological events.

MMP-2 contributes in activation of mitotic cell division, activity of vascular smooth muscle cells and airway smooth muscle cells (Gormus et al. 2011). Furthermore, MMP-2 plays a role in tissue morphogenesis by affecting cell organisation. MMP-2, which is secreted by adipocytes, also plays a role in islet morphogenesis by migration and association into islets (Gormus et al. 2011). It was discussed that activation of MMP-2 differentiated embryonic pancreatic epithelial cells and established clusters that have characteristics of islets of Langerhans. When the activity of MMP-2 was inhibited, islet cells differentiated, however did not migrate and associated into islets, obliterating islet morphogenesis (Gormus et al. 2011).

In addition, MMP-2 is involved in inflammatory responses. MMP-2 can act either as pro-inflammatory or anti-inflammatory by splitting inflammatory mediator (Clements et al. 2003). This eventually leads to a regulated inflammatory response. One experiment (Inoue et al. 2006) showed that in mice with asthma, eosinophils ended up accumulating in the interstitium, instead of migrating into the airways, which put animals at risk of suffocation. Furthermore, MMP-2 interrupts lethal asphyxiation by building chemostatic gradient that is needed to clear inflammatory cells in lungs (Kuzuya 2006). MMPs may also contribute in modulation of biologically active molecules. For example, MMP-2 can cleave and activate insulin-like growth factors and latent TGF-β, which is a protein that regulate proliferation, differentiation of cells, and thereby, activate them (Clements et al. 2003). Furthermore, it is also suggested that essential steps in the epithelial-mesenchymal transformation are triggered by MMP-2 (Murillo et al. 2009).

31.3 MMP-2 in Disease

MMP-2 is known to contribute in various diseases, such as atherosclerosis and skeletal disorder. It has been proposed that MMP-2 plays a critical role in destabilisation of atherosclerotic plaque (Bellafiore et al. 2013). A few studies have revealed that there was an overexpression and enzymatically activation of MMP-2 and MMP-9 in human atheroma. Moreover, increased number of MMP-2 has also been detected in plaques with extensive remodelling, which increases the risk of plaque rupture (Page-McCaw et al. 2007). Indeed, MMP-2 has been associated with coronary heart disease, which might have been caused by activity of MMP-2 on fibrous cap thickness and the initial plaque formation (Shah and Galis 2001). In fact, it has been shown that removal of MMP-2 or MMP-9 tends to reduce the plaque formation and decrease incidence of cardiac fibrosis after experimental myocardial infarction (Raffetto et al. 2007).

Mutations in MMP-2 have been found in three different skeletal disorders, such as Osteolysis Syndrome which is caused by progressive resorption of bone

(Bellafiore et al. 2013). The absence of MMP-2 activity seems to be the common cause in osteolysis, since no MMP-2 activity could be detected in serum and fibroblasts from patients with mutations in this gene (Bellafiore et al. 2013). Additionally Inoue et al. suggested that poor MMP-2 activity interrupts osteocyte networks and decreased bone density (Rundhaug 2005). In fact, mice with a deficiency in MMP-2 tended to develop more severe arthritis than control animals with normal levels (Tutton et al. 2003).

According to Gormus et al., MMP-2 seems to play a role in acceleration of matrix turnover in venous leg ulcers (Vu 2000). In another study, it was stated that venous dilation, chronic venous insufficiency and varicose veins can be caused by MMP-2. In fact, higher number of MMP-2 mRNA expression was found in patients with varicose veins and coronary artery disease (Vu 2000).

31.4 MMP-2 in Malignancy

Up-regulation of several MMPs is found in higher levels in many different types of human cancer than in normal tissue or benign adenomas (Fanjul-Fernández et al. 2010). Early regulation of MMPs which might have been caused by the tumor cells or surrounding stromal cells, assist in alteration of ECM and releasing ECM and membrane growth factors, which establish an optimal environment for the initial tumor formation (Fanjul-Fernández et al. 2010). With time, as tumor grows, proangiogenic factors eventually become dominant over angiogenic inhibitors (Zheng et al. 2000). MMP-2 has been associated with the stimulation of angiogenic shift in different model systems. Further regulation of MMP expression, particularly MMP-2 and MMP-9, which can break down basement membrane components, permit the invasion of tumor cells into the adjacent stroma and to degrade the basement membranes in capillaries and lymphatic vessels, and this leads to the invasion of tumor cells into the circulation (Fanjul-Fernández et al. 2010). Furthermore, it has been proposed that MMPs in body fluids tend to contribute to the malignant tumor invasion (Zheng et al. 2000). A few studies indicate that MMP-2 expression is able to predict invasion, metastasis, and relapse in urothelial and prostate cancer patients. For this reason, MMP-2 could be used in diagnosis of malignant tumor and metastasis that contribute to the breakdown of type IV collagen of extracellular matrix and basal membranes (Zheng et al. 2000). This explains the role of MMP-2 in the migration of malignant cells during pathological progression. It has been assumed that MMP-2 is a potential indicator in the progression of tumor and diagnosis of a few different kinds of malignancies, such as lung cancer and gastric cancer (Zheng et al. 2000).

Angiogenesis is initially the growth and remodelling process of the primitive vascular system to form a complex vascular network in foetal development. Reparative angiogenesis or pathological angiogenesis proceeds after birth as in the case of wound healing. Pathological angiogenesis, as suggested by the name, leads to a negative effect on the body as it assists in the growth of tumours and drives metastasis.

Various proteins including MMP2 can regulate this pathological growth in vasculature. It was shown that MMP-2 plays a role in angiogenesis and a variety of diseases by releasing angiogenic factors and assisting in cell migration and cell adhesion. Further, MMP-2 can be a potential cause in initial tumor formation by altering ECM, which provides optimal environment for tumor growth. Hence, understanding the pathological and physiological aspects of MMP-2 remains crucial in order to develop therapeutic interventions.

References

Bellafiore M, Battaglia G, Bianco A, Farina F, Palma A, Paoli A (2013) The involvement of MMP-2 and MMP-9 in heart exercise-related angiogenesis. J Transl Med 11(1):283. doi:10.1186/1479-5876-11-283

Carmeliet P (2003) Angiogenesis in health and disease. Nat Med 9(6):653–660. doi:10.1038/nm0603-653 nm0603-653 [pii]

Clements KM, Price JS, Chambers MG, Visco DM, Poole AR, Mason RM (2003) Gene deletion of either interleukin-1?, interleukin-1?-converting enzyme, inducible nitric oxide synthase, or stromelysin 1 accelerates the development of knee osteoarthritis in mice after surgical transection of the medial collateral ligament and partial medial meniscectomy. Arthritis Rheum 48(12):3452–3463. doi:10.1002/art.11355

Corry DB (2004) Overlapping and independent contributions of MMP2 and MMP9 to lung allergic inflammatory cell egression through decreased CC chemokines. FASEB J 18:995–997. doi:10.1096/fj.03-1412fje

Fanjul-Fernández M, Folgueras AR, Cabrera S, López-Otín C (2010) Matrix metalloproteinases: evolution, gene regulation and functional analysis in mouse models. Biochim Biophys Acta Mol Cell Res 1803(1):3–19. doi:10.1016/j.bbamcr.2009.07.004

Folkman J (2007) Angiogenesis: an organizing principle for drug discovery? Nat Rev Drug Discov 6(4):273–286. doi:10.1038/nrd2115

Gormus U, Kahraman OT, Isbir S, Tekeli A, Isbir T (2011) MMP2 gene polymorphisms and MMP2 mRNA levels in patients with superficial varices of lower extremities. In Vivo (Athens, Greece) 25(3):387–391

Inoue K, Mikuni-Takagaki Y, Oikawa K, Itoh T, Inada M, Noguchi T, Park JS, Onodera T, Krane SM, Noda M, Itohara S (2006) A crucial role for matrix metalloproteinase 2 in osteocytic canalicular formation and bone metabolism. J Biol Chem 281(44):33814–33824. doi:10.1074/jbc.m607290200

Kurz H, Burri PH, Djonov VG (2003) Angiogenesis and vascular remodeling by intussusception: from form to function. Physiology 18(2):65–70. doi:10.1152/nips.01417.2002

Kuzuya M (2006) Effect of MMP-2 deficiency on atherosclerotic lesion formation in ApoE-deficient mice. Arterioscler Thromb Vasc Biol 26(5):1120–1125. doi:10.1161/01.atv.0000218496.60097.e0

Murillo CA, Woodside KJ, Guo Q, Zhang S, O'Connor KL, Hunter GC (2009) Integrin and matrix metalloproteinase expression in human carotid plaque. J Surg Res 155(1):157–164. doi:10.1016/j.jss.2008.08.005

Page-McCaw A, Ewald AJ, Werb Z (2007) Matrix metalloproteinases and the regulation of tissue remodelling. Nat Rev Mol Cell Biol 8(3):221–233. doi:10.1038/nrm2125

Raffetto JD, Ross RL, Khalil RA (2007) Matrix metalloproteinase 2–induced venous dilation via hyperpolarization and activation of K+ channels: relevance to varicose vein formation. J Vasc Surg 45(2):373–380. doi:10.1016/j.jvs.2006.10.041

Rundhaug JE (2005) Matrix metalloproteinases and angiogenesis. J Cell Mol Med 9(2):267–285. doi:10.1111/j.1582-4934.2005.tb00355.x
Shah PK, Galis ZS (2001) Matrix metalloproteinase hypothesis of plaque rupture: players keep piling up but questions remain. Circulation 104(16):1878–1880
Tutton MG, George ML, Eccles SA, Burton S, Swift RI, Abulafi AM (2003) Use of plasma MMP-2 and MMP-9 levels as a surrogate for tumour expression in colorectal cancer patients. Int J Cancer 107(4):541–550. doi:10.1002/ijc.11436
Vu TH (2000) Matrix metalloproteinases: effectors of development and normal physiology. Genes Dev 14(17):2123–2133. doi:10.1101/gad.815400
Zheng T, Zhu Z, Wang Z, Homer RJ, Ma B, Riese RJ, Chapman HA, Shapiro SD, Elias JA (2000) Inducible targeting of IL-13 to the adult lung causes matrix metalloproteinase– and cathepsin-dependent emphysema. J Clin Investig 106(9):1081–1093. doi:10.1172/jci10458

Chapter 32
Monobutyrin

Abstract Monobutyrin, also known as 1-butyrylglycerol, is a monoglyceride of butyric acid. The chiral, non-protein, small lipid molecule was first recognised as a regulator of angiogenesis in 1990 when a chicken embryo chorioallantoic membrane assay was used to demonstrate its ability to stimulate angiogenesis. The results from the study strongly indicated that monobutyrin was a major regulatory molecule of angiogenesis involved in the normal development of cells and tissues. The expansion of adipose tissue in adults is similar to the growth of a tumour or neoplasm, where rapid growth gives rise to hypoxia, which induces angiogenesis, and consequently promotes further tissue growth. Adipose tissue that has been newly formed will require continued angiogenesis for additional growth. Monobutyrin acts as a messenger for the paracrine interactions between adipocytes and endothelial cells that mediate this coordinated growth. Although angiogenesis in adipose tissue is carefully balanced during a state of tissue homeostasis, a neoplasm or lesion beginning in adipose tissue, such as in the mammary gland, will exploit the microvasculature for its own local growth and metastasis. The therapeutic benefits of monobutyrin have been examined in nutritional studies. As the knowledge surrounding monobutyrin develops, so will the information on the potential angiogenic inhibitors as a therapeutic option for treatment and prevention of obesity and its associated disorders.

Keywords Monobutyrin • Angiogenesis • Normal physiology • Disease • Malignancy

32.1 Monobutyrin

Monobutyrin, also known as 1-butyrylglycerol, is a monoglyceride of butyric acid. The chiral, non-protein, small lipid molecule was first recognised as a regulator of angiogenesis in 1990 when a chicken embryo chorioallantoic membrane assay was used to demonstrate its ability to stimulate angiogenesis at doses as low as 20 pg. The results from the study strongly indicated that monobutyrin was a major regulatory molecule of angiogenesis involved in the normal development of cells and tissues (Dobson et al. 1990).

A. Salajegheh, *Angiogenesis in Health, Disease and Malignancy*,
DOI 10.1007/978-3-319-28140-7_32

32.2 Monobutyrin in Normal Physiology

Monobutyrin is synthesised and secreted by adipocytes after differentiation from 3T3-F442A preadipocytes. This differentiation-dependent characteristic of monobutyrin synthesis was discovered when adipocytes were compared with preadipocytes in early labelling studies that employed the use of various radiochemical precursors, such as [14C]acetate (Wilkison and Spiegelman 1993). Whilst the specific biosynthetic pathway of monobutyrin has been difficult to ascertain, it has been shown that diacylglycerol can stimulate the synthesis of the short-chain monoacylglycerol from an adipocyte particulate fraction by acting as a substrate. Diacylglycerol undergoes acylation with butyryl-CoA to form a triacylglycerol intermediate that is then hydrolysed by long-chain fatty acyl-specific enzymes called lipases. During this hydrolysis reaction, the lipase enzyme selectively cleaves off the two remaining long-chain fatty acyl groups to produce monobutyrin. This acyltransferase activity is present in several cell types, although at much lower degrees than in adipocytes. The enzymatic activity responsible for the synthesis of monobutyrin have been found to occur at high levels in adipocytes and at either indiscernible or relatively low levels in cells of other tissue types. In adipocytes, it occurs in two distinct steps that are highly dependent on the differentiation process (Wilkison et al. 1991). During differentiation, levels of monobutyrin multiply by approximately 200-fold and are responsible for a large portion of the overall angiogenic activity (Dobson et al. 1990). A study conducted by the American Society for Biochemistry and Microbiology also suggested monobutyrin production may also be dependent on hormones to a certain extent, as the addition of epinephrine to adipose tissue increased the amount of monobutyrin secreted by tenfold (Wilkison and Spiegelman 1993).

32.3 Monobutyrin in Disease

The secretion of monobutyrin, which is tightly coupled to the differentiation process, has shown to be detectable only from adipocytes from studies involving analysis of various cultured cell types, including tumour cells (Wilkison and Spiegelman 1993). Studies comparing liver, muscle and adipose tissue showed that only adipose was able to induce its own neovascularisation (Dobson et al. 1990). This is evident in situations where wound healing and revascularisation of ischemic myocardium are promoted by fat tissue and the ability of its cells to stimulate angiogenesis (Liu and Meydani 2003). The capacity of adipose tissue to grow throughout an individual's life is dependent on this ability to secrete angiogenic stimulators, such as monobutyrin, that promote the formation of new blood vessels. Monobutyrin acts as a messenger for the paracrine interactions between adipocytes and endothelial cells that mediate this coordinated growth (Cao 2007). The relatively low levels of monobutyrin that were able to stimulate angiogenesis in the chicken embryo chorioallantoic membrane assay suggest that the molecule works by means of an intracellular or cell surface receptor-mediated mechanism (Dobson et al. 1990).

The expansion of adipose tissue in adults is similar to the growth of a tumour or neoplasm, where rapid growth gives rise to hypoxia, which induces angiogenesis, and consequently promotes further tissue growth (Rutkowski et al. 2009). In recent studies, obesity, a result of excessive expansion of adipose tissue, has been regarded as an endocrine tumour because of the ability of adipose tissue to secrete angiogenic factors, such as monobutyrin, that facilitate the tissue expansion. Angiogenesis is a necessary event in adipose tissue expansion as it counteracts the hypoxia that arises (Dizdar and Alyamac 2004). The expansion of adipose tissue can result in drastic consequences on health, including cancer, diabetes and cardiovascular disease (Rutkowski et al. 2009). Adipose secretory products, such as monobutyrin, contribute to this expansion process, ensuring that the adipose tissue is highly vascularised with each adipocyte receiving nourishment from an extensive capillary network (Cao 2007). The extensive vasculature of adipose tissue allows for efficient circulation of blood that fuels adipose tissue function and transports monobutyrin, along with other factors secreted by adipocytes, for further angiogenesis and tissue growth (Rutkowski et al. 2009). This is especially significant during embryonic development, where adipogenesis and angiogenesis are temporally and spatially coupled (Cao 2007). Adipose tissue that has been newly formed will require continued angiogenesis for additional growth (Rutkowski et al. 2009).

32.4 Monobutyrin in Cancer

Although angiogenesis in adipose tissue is carefully balanced during a state of tissue homeostasis, a neoplasm or lesion beginning in adipose tissue, such as in the mammary gland, will exploit the microvasculature for its own local growth and metastasis. Breast cancer is one example of a neoplasm infiltrating a region that is rich in adipose tissue. Once transformed ductal epithelial cells have invaded the adipose-rich mammary stromal compartment, angiogenic stimulators such as monobutyrin will operate in conjunction with tumour-derived factors to fulfil the circulatory needs of the invading lesion (Rutkowski et al. 2009).

The understanding surrounding monobutyrin and its use is still very much in the early stages. This is due to improve as increasing studies regarding the regulation of adipose tissue mass via vasculature are undertaken with the purified synthetic compound, which has shown to have a biological activity profile identical to its adipocyte-derived counterpart in terms of its ability to stimulate vascular endothelial cell motility in vitro and angiogenesis in vivo without posing a direct effect on endothelial cell growth (Dobson et al. 1990). The therapeutic benefits of monobutyrin have also been examined in nutritional studies, which have demonstrated its effectiveness as a dietary energy source, food preservative and solubilising agent (Dobson et al. 1990). As the knowledge surrounding monobutyrin develops, so will the information on the potential angiogenic inhibitors as a therapeutic option for treatment and prevention of obesity and its associated disorders (Cao 2007). So far, studies exploring the concept of systemic treatment with angiogenic inhibitors have managed to achieve weight reduction and adipose tissue reduction in obese mouse (Dizdar and Alyamac 2004).

References

Cao Y (2007) Angiogenesis modulates adipogenesis and obesity. J Clin Invest 117(9):2362–2368. doi:10.1172/jci32239

Dizdar O, Alyamac E (2004) Obesity: an endocrine tumor? Med Hypotheses 63(5):790–792. doi:10.1016/j.mehy.2004.01.046

Dobson DE, Kambe A, Block E, Dion T, Lu H, Castellot JJ Jr, Spiegelman BM (1990) 1-Butyryl-glycerol: a novel angiogenesis factor secreted by differentiating adipocytes. Cell 61(2):223–230

Liu L, Meydani M (2003) Angiogenesis inhibitors may regulate adiposity. Nutr Rev 61(11):384–387

Rutkowski JM, Davis KE, Scherer PE (2009) Mechanisms of obesity and related pathologies: the macro- and microcirculation of adipose tissue. FEBS J 276(20):5738–5746. doi:10.1111/j.1742-4658.2009.07303.x

Wilkison WO, Spiegelman BM (1993) Biosynthesis of the vasoactive lipid monobutyrin. Central role of diacylglycerol. J Biol Chem 268(4):2844–2849

Wilkison WO, Choy L, Spiegelman BM (1991) Biosynthetic regulation of monobutyrin, an adipocyte-secreted lipid with angiogenic activity. J Biol Chem 266(25):16886–16891

Chapter 33
Neutrophil Activating Protein-2 (NAP-2)

Abstract Neutrophil activating protein-2 (NAP-2) is a potent chemoattractant and activator of neutrophils, stimulating degranulation, and release of inflammatory mediators and degradative enzymes, such as serine proteases and lysozymes, however, NAP-2 has also been shown to stimulate a number of other cellular processes, including mitosis, extracellular matrix production, and angiogenesis, making it a novel target for cancer treatment strategies due to its angiogenic and mitotic roles. NAP-2 has been shown to induce endothelial cell proliferation and chemotaxis (required for angiogenesis) *in vitro*, and angiogenesis *in vivo*, in the absence of any preceding inflammation, indicating that its role in neutrophil chemoattraction and activation is distinct from its potential physiological angiogenic role. Whilst neovascularisation is important for the growth of the tumour, the formation of new blood vessels is also critical to the process of invasion and metastasis of tumour cells, as the new blood vessel provide a way for the cancer cells to enter the circulation and travel around the body. Majority of the pro-cancerous effects of NAP-2 have been well documented to occur as a result of its interaction with its primary receptor. It is clear that NAP-2 has a role in the growth, survival, invasion and metastasis of various cancer subtypes, but the exact function of NAP-2 in cancer still remains broad and indistinct, however, angiogenesis has been consistently associated with NAP-2 in tumorigenesis.

Keywords Neutrophil activating protein-2 • NAP-2 • Angiogenesis • Normal physiology • Disease • Malignancy

33.1 Neutrophil Activating Protein-2 (NAP-2)

Neutrophil activating protein-2 (NAP-2), also known as CXCL7, belongs to the CXC family of chemokines (Grepin et al. 2014). CXCL7 is produced when a protein secreted by platelets, named platelet basic (PBP) protein, is cleaved by cathepsin-G at a tyrosine-alanine bond situated 24 amino acid residues from the N-terminus, however, other studies have shown that NAP-2 may be produced via cleavage of connective tissue- activating peptide III (CTAP-III), which is also a product of differential cleavage of PBP (Cohen et al. 1992; Belperio et al. 2000).

A. Salajegheh, *Angiogenesis in Health, Disease and Malignancy*,
DOI 10.1007/978-3-319-28140-7_33

Being of the chemokine family, NAP-2 is a potent chemoattractant and activator of neutrophils, stimulating degranulation, and release of inflammatory mediators and degradative enzymes, such as serine proteases and lysozymes, however, NAP-2 has also been shown to stimulate a number of other cellular processes, including mitosis, extracellular matrix production, and angiogenesis, making it a novel target for cancer treatment strategies due to its angiogenic and mitotic roles (Cohen et al. 1992; Witko-Sarsat et al. 2000; Russo et al. 2014). NAP-2 stimulates these processes via binding to the chemokine receptor CXCR2, which is present on a number of cell types, including polymorphonuclear leukocytes, epithelial cells, endothelial cells, and fibroblasts (Russo et al. 2014).

33.2 Neutrophil Activating Protein-2 (NAP-2) in Normal Physiology

NAP-2 is produced by a variety of cells, including platelets, activated endothelial cells and by leukocytes (Belperio et al. 2000). Interestingly, research has shown that the production of NAP-2 occurs to a large degree in areas of high neutrophil numbers due to the presence of a surface bound cathepsin-G-like enzyme, which can generate a positive feedback cycle by cleaving more CTAP-III to NAP-2, attracting more neutrophils to the site, and so on (Cohen et al. 1992; Belperio et al. 2000).

Many of the known physiological and pathological processes (including tumorigenesis) involving angiogenesis tend to show leukocytic infiltrates, that is, involving the presence of neutrophils amongst other inflammatory cells (Tazzyman et al. 2009). Wound healing, inflammatory conditions (such as rheumatoid arthritis) and malignant cancers are all inherently intertwined with neovascularisation, and tend to exhibit chemokine profiles that attract various types of leukocytes, the presence of which may lead to the production of high amounts of NAP-2, which in in turn may be a key player in the induction of angiogenesis associated with these conditions (Tazzyman et al. 2009; Witko-Sarsat et al. 2000).

Angiogenesis is a critical process essential to various physiological processes, of which numerous chemokines, including NAP-2, stimulate and promote (Gillitzer and Goebeler 2001). NAP-2 has been shown to induce endothelial cell proliferation and chemotaxis (required for angiogenesis) *in vitro*, and angiogenesis *in vivo*, in the absence of any preceding inflammation, indicating that its role in neutrophil chemoattraction and activation is distinct from its potential physiological angiogenic role, however, very little published data implicates NAP-2 in physiological angiogenesis, focusing more on its role in pathological angiogenesis (Belperio et al. 2000). Despite this, it is entirely possible that NAP-2 plays a role in angiogenesis within a normal physiological context.

33.3 Neutrophil Activating Protein-2 (NAP-2) in Disease

Angiogenesis is a tightly regulated process, seldom seen in normal adult tissues, with little evidence showing it to be frequently involved with inflammation, however, some research has shown that angiogenesis accompanies some chronic inflammatory conditions, such as inflammatory bowel disease, prolonging and intensifying the inflammatory process (Granger and Senchenkova 2010). One well characterised process known to involve an intimate link between angiogenesis and inflammation is the physiological process of cutaneous wound healing (Gillitzer and Goebeler 2001).

When there is a break in the skin, formation of a platelet plug and blood clot forms, acting to stop the bleeding, which is followed by an acute inflammatory response characterised by a neutrophil rich exudate. Aggregated platelets in the blood clot release a variety of growth factors to initiate the process of healing, including PBP, which is subsequently cleaved to CTAP-III, and then to NAP-2 by neutrophils around the clot, leading to attraction of more neutrophils via its stimulation of CXCR2 (Gillitzer and Goebeler 2001). Angiogenesis and the formation of new blood vessels is a hallmark process involved in the formation of granulation tissue during the process of wound repair, and despite the known proangiogenic properties of NAP-2, very little published data is available to show a direct involvement of NAP-2 in the process of angiogenesis in wound healing, despite its known role in the acute inflammatory response of wound healing described prior (Gillitzer and Goebeler 2001; Granger and Senchenkova 2010). It is fairly reasonable to assume that NAP-2 has a role amongst the various other proangiogenic factors in the initiation of angiogenesis during wound healing, as it has been shown to induce a potent angiogenic response in a chick embryo membrane which was saturated in NAP-2 (Powell and Mousa 2007). Thus further research into NAP-2's role in physiological angiogenesis is required before any concrete conclusions can be made as to its potential role in this scope, but despite this, the knowledge of NAP-2's proangiogenic activity means there is promise of finding a more intricate involvement of the chemokine in physiological angiogenic processes.

Angiogenesis has been linked to a variety of pathological conditions, in particular, those involving chronic inflammatory responses, with studies suggesting that angiogenesis and chronic inflammation may even be co-dependent in certain circumstances (Jackson et al. 1997). The basis of this pairing is due to the fact that various inflammatory mediators can both directly and indirectly promote angiogenesis, with angiogenesis thus contributing to and maintaining the pathological inflammatory state, delivering more inflammatory cells to the diseased tissue (Jackson et al. 1997; Szekanecz et al. 2009).

Rheumatoid arthritis is a chronic inflammatory disease characterised by extensive angiogenesis in the inflamed Synovium (Szekanecz et al. 2009). The chemokine profile of serum derived from joints affected by rheumatoid arthritis contains a large variety of proinflammatory and proangiogenic cytokines, including NAP-2, which is consistently recognised in such profiles (Jackson et al. 1997; Szekanecz et al. 2009). NAP-2 has become a potential novel target for rheumatoid arthritis therapy due to its consistent presence in rheumatoid arthritis and known proangiogenic function (Szekanecz et al. 2009; Jorgensen 2000; Nataraj et al. 2013).

Chemokine induced angiogenesis has also been linked to fibroproliferative disorders (Belperio et al. 2005; Strieter et al. 2004b). Various studies have described the presence of high levels of NAP-2 in tissue samples from lung transplant patients suffering from a chronic inflammatory fibroproliferative complication of the surgery, called bronchiolitis obliterans, in comparison to tissue samples from healthy lung transplant patients. In addition, the corresponding receptor of NAP-2, CXCR2, was found to be expressed in higher levels on endothelium surrounding sites of bronchiolitis obliterans. In the same study, the use of an anti-CXCR2 antibody significantly reduced angiogenic activity in bronchiolitis obliterans, whilst another antibody raised against the receptor of the powerful angiogenic factor, vascular endothelial growth factor (VEGF), showed little significant effects at reducing angiogenesis in bronchiolitis obliterans, indicating a critical role of NAP-2 and the CXCR2/CXCR ligand axis in angiogenesis and the pathogenesis of this fibroproliferative lung disorder (Belperio et al. 2005).

There is undeniable evidence of a role of NAP-2, amongst other chemokines, in the pathogenesis of various chronic inflammatory conditions, which are intimately intertwined with the process of angiogenesis. Based on this knowledge and the results of studies such as the aforementioned, there is huge potential for the use of novel treatments targeting NAP-2 in an effort to treat these chronic debilitating conditions. However, whilst the literature shows a direct link between angiogenesis and NAP-2 in such inflammatory diseases, the degree to which NAP-2 induces angiogenesis in these conditions, amongst its various other inflammatory roles, is still vague, as these chemokine receptors are highly promiscuous, meaning definite localisation and determination of particular ligand driven responses with CXCR2 activation is not as simple as showing a decrease in angiogenic activity when there is high NAP-2 levels in inflamed tissues along with CXCR2 inhibition (Belperio et al. 2005). Rather, this tends to only show that inhibition of a highly promiscuous chemokine receptor acts to reduce angiogenesis in these diseases, however, that receptor could be activated by a number of proangiogenic factors known to interact with CXCR2 in the process of angiogenesis (for example, IL-8, CXCL2, CXCL6, CXCL5 etc.), which are also found in high levels in the same conditions (Tazzyman et al. 2009; Nataraj et al. 2013; Belperio et al. 2000, 2005). The wide ligand range of these chemokine receptors thus make the determination of the degree to which NAP-2 induces angiogenesis in the aforementioned conditions difficult, however, results are promising, but further research and more conclusive evidence is required if NAP-2 is to become a reliable drug target for the treatment of such disorders in the future.

33.4 Neutrophil Activating Protein-2 (NAP-2) in Malignancy

Contrary to the role of NAP-2 in normal physiological angiogenesis and pathology based angiogenesis, there is a large amount of clear experimental evidence that implicates various CXC chemokines, including NAP-2, in angiogenic events that are critical for the growth, invasion and metastasis of neoplastic masses (Strieter et al. 2004b). Studies have even shown that tumours are unable to grow beyond 2–3 mm (Witko-Sarsat et al. 2000) without developing a consistent bloody supply through angiogenesis (Tazzyman et al. 2009). Furthermore, studies have shown that tumour cells may promote the formation of neutrophilic infiltrates, which may promote angiogenesis via a mechanism associated with the release of a proangiogenic cytokine like NAP-2 (Tazzyman et al. 2009; Cohen et al. 1992).

The importance of angiogenesis in the growth of tumours is well documented, with various studies showing inhibition of angiogenesis reduces tumour cell mass, with preclinical trials using the endogenous anti-angiogenic molecule endostatin to treat certain tumours showing a generalised reduction in both the mass of the tumours treated, as well as a reduction in neovascularisation associated with the tumour (Folkman 2002). Conversely, NAP-2 has been shown to play a key role in the growth of tumours, particularly in the growth and survival of clear cell renal cell carcinoma (ccRCC), most likely due to its proangiogenic role. Overexpression of NAP-2 acted to accelerate ccRCC development, whilst the use of anti-NAP-2 antibodies greatly reduced the growth of ccRCC in nude mice models. Interestingly, in the same study, the ccRCC were shown to normally secrete low amounts of NAP-2, however, production of IL-1β by the ccRCC acted to positively feedback and increase the amount of NAP-2 production, further promoting its own growth (Grepin et al. 2014).

Whilst neovascularisation is important for the growth of the tumour, the formation of new blood vessels is also critical to the process of invasion and metastasis of tumour cells, as the new blood vessel provide a way for the cancer cells to enter the circulation and travel around the body, where they deposit at distant sites forming secondary tumours (Folkman 2002; Tang et al. 2008). The receptor for NAP-2, CXCR2, has been shown to be the primary receptor that drives angiogenesis and subsequent metastatic potential, with both *in-vitro* and *in-vivo* studies using CXCR2−/− mice lung cancer models showing a reduction in angiogenesis and metastasis in comparison to CXCR2+/+ models. However, as mentioned previously, CXCR2 also interacts with various other proangiogenic CXC chemokines, thus it cannot be undeniably assumed that it is NAP-2 which is promoting metastasis in these models by inducing tumour angiogenesis, but more likely synergistic effect of various proangiogeneic cytokines (Strieter et al. 2004a).

Interestingly, NAP-2's angiogenic role may not be the only process by which it could promote tumour invasion and metastasis, with some studies in breast cancer cell lines showing that NAP-2 has heparanase activity, allowing it to degrade heparin sulphate in the extracellular matrix (ECM) which allows tumour cells to move through the ECM and basement membrane to invade local tissues, as well as to intravasate into blood vessels and metastasise around the body (Tang et al. 2008; Hoogewerf et al. 1995). Along with this, premalignant cells transfected with NAP-2 also showed higher levels of NAP-2 secretion and invasiveness than non-transfected counterparts (Hoogewerf et al. 1995). This shows that NAP-2 indeed plays a role in tumour growth and metastasis, however, whether its pro-cancerous effect is due to increased angiogenesis or ECM destruction is still not clear, but it is more than likely a mixture of both would account for NAP-2's pro-cancerous effects, which makes inhibition of NAP-2 function an interesting molecular target for future breast cancer treatment (Hoogewerf et al. 1995).

There has also been evidence showing that NAP-2 levels may actually decrease in some cancers (Matsubara et al. 2011). One study showed that there was an overall reduction in serum NAP-2 levels in patients with pancreatic adenocarcinoma compared to normal healthy subjects, however, histologically, pancreatic adenocarcinoma is consistently characterised by hypovascularisation, explaining the decreased circulating NAP-2 levels, however, this also shows that the proangiogenic role of NAP-2 is not universally seen across all cancer subtypes, and hence future use of anti-NAP-2 drugs to treat cancer may only be suitable in certain cases (Matsubara et al. 2011; Miura et al. 2006). From results of studies like this, it has been suggested that NAP-2 may actually be able to be used diagnostically as a biomarker for early detection of pancreatic adenocarcinoma, promoting a better prognosis (Matsubara et al. 2011).

Majority of the pro-cancerous effects of NAP-2 have been well documented to occur as a result of its interaction with its primary receptor, CXCR2 (Nataraj et al. 2013; Belperio et al. 2005; Belperio et al. 2000). However, a very recent study investigated the molecular mechanisms by which NAP-2 induces breast cancer cell growth and metastasis, and found co-localisation of NAP-2 and the growth factor receptor EGFR, and it was subsequently suggested that interaction of NAP-2 with EGFR, rather than CXCR2, promoted angiogenesis and induced cell cycle progression in breast cancer cells, leading to their growth and subsequent metastasis, making inhibition of not only the NAP-2/CXCR2 axis, but the NAP-2/EGFR axis as well, a promising novel target for the treatment of breast cancer cells (Salazar et al. 2014).

It is undeniably clear that NAP-2 has a role in the growth, survival, invasion and metastasis of various cancer subtypes, but the exact function of NAP-2 in cancer still remains broad and indistinct, however, angiogenesis has been consistently associated with NAP-2 in tumorigenesis (Grepin et al. 2014; Tazzyman et al. 2009; Strieter et al. 2004a; Folkman 2002). With continuing research into the molecular mechanisms and effects of NAP-2 in malignancy, there is almost certainly the potential for NAP-2 to become an effective drug target to reduce the incidence and morbidity associated with cancer.

References

Belperio JA, Keane MP, Arenberg DA, Addison CL, Ehlert JE, Burdick MD, Strieter RM (2000) CXC chemokines in angiogenesis. J Leukoc Biol 68(1):1–8

Belperio JA, Keane MP, Burdick MD, Gomperts B, Xue YY, Hong K, Mestas J, Ardehali A, Mehrad B, Saggar R, Lynch JP, Ross DJ, Strieter RM (2005) Role of CXCR2/CXCR2 ligands in vascular remodeling during bronchiolitis obliterans syndrome. J Clin Invest 115(5):1150–1162. doi:10.1172/jci24233

Cohen AB, Stevens MD, Miller EJ, Atkinson MA, Mullenbach G (1992) Generation of the neutrophil-activating peptide-2 by cathepsin G and cathepsin G-treated human platelets. Am J Physiol 263(2 Pt 1):L249–L256

Folkman J (2002) Role of angiogenesis in tumor growth and metastasis. Semin Oncol 29(6 Suppl 16):15–18. doi:10.1053/sonc.2002.37263

Gillitzer R, Goebeler M (2001) Chemokines in cutaneous wound healing. J Leukoc Biol 69(4):513–521

Granger DN, Senchenkova E (2010) Integrated systems physiology–from cell to function. In: Inflammation and the microcirculation. Morgan & Claypool Life Sciences. Copyright (c) 2010 by Morgan & Claypool Life Sciences, San Rafael

Grepin R, Guyot M, Giuliano S, Boncompagni M, Ambrosetti D, Chamorey E, Scoazec JY, Negrier S, Simonnet H, Pages G (2014) The CXCL7/CXCR1/2 axis is a key driver in the growth of clear cell renal cell carcinoma. Cancer Res 74(3):873–883. doi:10.1158/0008-5472.can-13-1267

Hoogewerf AJ, Leone JW, Reardon IM, Howe WJ, Asa D, Heinrikson RL, Ledbetter SR (1995) CXC chemokines connective tissue activating peptide-III and neutrophil activating peptide-2 are heparin/heparan sulfate-degrading enzymes. J Biol Chem 270(7):3268–3277

Jackson JR, Seed MP, Kircher CH, Willoughby DA, Winkler JD (1997) The codependence of angiogenesis and chronic inflammation. FASEB J 11(6):457–465

Jorgensen C (2000) New therapeutic targets in rheumatoid arthritis. Joint Bone Spine revue du rhumatisme 67(4):277–282

Matsubara J, Honda K, Ono M, Tanaka Y, Kobayashi M, Jung G, Yanagisawa K, Sakuma T, Nakamori S, Sata N, Nagai H, Ioka T, Okusaka T, Kosuge T, Tsuchida A, Shimahara M, Yasunami Y, Chiba T, Hirohashi S, Yamada T (2011) Reduced plasma level of CXC chemokine ligand 7 in patients with pancreatic cancer. Cancer Epidemiol Biomarkers Prev 20(1):160–171. doi:10.1158/1055-9965.epi-10-0397

Miura F, Takada T, Amano H, Yoshida M, Furui S, Takeshita K (2006) Diagnosis of pancreatic cancer. HPB Off J Int Hepato Pancreato Biliary Assoc 8(5):337–342. doi:10.1080/13651820500540949

Nataraj NB, Krishnamurthy J, Salimath BP (2013) Treatment with anti-NAP monoclonal antibody reduces disease severity in murine model of novel angiogenic protein-induced or ovalbumin-induced arthritis. Clin Exp Immunol 171(2):155–163. doi:10.1111/cei.12009

Powell JA Jr, Mousa SA (2007) Neutrophil-activating protein-2- and interleukin-8-mediated angiogenesis. J Cell Biochem 102(2):412–420. doi:10.1002/jcb.21302

Russo RC, Garcia CC, Teixeira MM, Amaral FA (2014) The CXCL8/IL-8 chemokine family and its receptors in inflammatory diseases. Expert Rev Clin Immunol 10(5):593–619. doi:10.1586/1744666x.2014.894886

Salazar N, Munoz D, Kallifatidis G, Singh RK, Jorda M, Lokeshwar BL (2014) The chemokine receptor CXCR7 interacts with EGFR to promote breast cancer cell proliferation. Mol Cancer 13:198. doi:10.1186/1476-4598-13-198

Strieter RM, Belperio JA, Burdick MD, Sharma S, Dubinett SM, Keane MP (2004a) CXC chemokines: angiogenesis, immunoangiostasis, and metastases in lung cancer. Ann N Y Acad Sci 1028:351–360. doi:1028/1/351 [pii]10.1196/annals.1322.041

Strieter RM, Belperio JA, Phillips RJ, Keane MP (2004b) CXC chemokines in angiogenesis of cancer. Semin Cancer Biol 14(3):195–200. doi:10.1016/j.semcancer.2003.10.006

Szekanecz Z, Besenyei T, Paragh G, Koch AE (2009) Angiogenesis in rheumatoid arthritis. Autoimmunity 42(7):563–573

Tang Z, Yu M, Miller F, Berk RS, Tromp G, Kosir MA (2008) Increased invasion through basement membrane by CXCL7-transfected breast cells. Am J Surg 196(5):690–696. doi:10.1016/j.amjsurg.2008.08.001

Tazzyman S, Lewis CE, Murdoch C (2009) Neutrophils: key mediators of tumour angiogenesis. Int J Exp Pathol 90(3):222–231. doi:10.1111/j.1365-2613.2009.00641.x

Witko-Sarsat V, Rieu P, Descamps-Latscha B, Lesavre P, Halbwachs-Mecarelli L (2000) Neutrophils: molecules, functions and pathophysiological aspects. Lab Invest 80(5):617–653

Chapter 34
p27kip1 and p57Kip2 (CDKN1B and CDKN1C)

Abstract Under normal circumstances, the progression of the cell cycle is controlled by the activation of cyclin-dependent kinases (CDKs) and the regulatory subunits called cyclins in eukaryotic cells. p27Kip1 encodes the cyclin-dependent kinase inhibitors (CDKIs) in Cip/Kip family. These CDKIs bind to CDKs, thus terminating cell cycle. nuclear p27kip1 protein content is high in tissues that are not expected to or have already undergone mitotic divisions and expression is lower in tissues that are expected to undergo proliferation such as basal layer of oral mucosa and skin or germinal centres of normal lymph nodes. p57Kip2 protein together with CDK interacting proteins (CIP) also belongs to the CIP/KIP family of cyclin dependent kinase inhibitors. p57Kip2 is the least studied protein within the family, but research has shown that it plays a key role in the regulation of the cell cycle. p57Kip2 believed to have an anti-oncogenic role involved in the regulation of several hallmarks for tumor growth including apoptosis, cell differentiation and proliferation, control of tissue invasion and metastasis and angiogenesis. Down regulation of p57Kip2 is commonly associated and inversely correlated with the proliferation of many malignancies. To date, there are few studies linking p57Kip2 and angiogenesis. Reports have noted an increase in the expression of vascular endothelial growth factor (VEGF), an angiogenesis promoting factor in the absence of p57Kip2.

Keywords p27kip1 • p57Kip2 • CDKN1B • CDKN1C • Angiogenesis • Normal physiology • Disease • Tumour • Malignancy

34.1 p27kip1 and p57Kip2 (CDKN1B and CDKN1C)

Under normal circumstances, the progression of the cell cycle is controlled by the activation of cyclin-dependent kinases (CDKs) and the regulatory subunits called cyclins in eukaryotic cells (Besson et al. 2004).

The p27kip1 gene, formally CDKN1B, is located on chromosome 12p13 (Ponce-Castaneda et al. 1995) and exhibits characteristics of "natively unfolded," "intrinsically disordered" or "intrinsically unstructured" proteins (Ponce-Castaneda et al. 1995). This characteristic allows for a large number of possible conformations and therefore interactions with many targets comprising both nucleic acids and proteins.

A. Salajegheh, *Angiogenesis in Health, Disease and Malignancy*,
DOI 10.1007/978-3-319-28140-7_34

The aforementioned functionality is enhanced as a result of p27kip1's ability to transform from its' disordered state to a well-defined conformations with high affinity for that target. In short, due to the absence of a pre-existing folding, p27kip1 can interact with a number of targets (Borriello et al. 2007).

p57Kip2, also known as CDKN1C, is a maternally imprinted gene located on the chromosomal locus 11p15.5 (Herceg and Ushijima 2009). This protein together with CDK interacting proteins (CIP) belongs to the CIP/KIP family of cyclin dependent kinase inhibitors. It is the least studied protein within the family (Pateras et al. 2009), but research has shown that it plays a key role in the regulation of the cell cycle. The unique role of this protein in embryonic development has been extensively studied and is commonly implicated in Beckwith-Wiedemann syndrome (BWS) and sporadic cancers (Pateras et al. 2009). The down-regulation of this protein is frequently connected with many common human malignancies illustrating its antioncogeneic role. The tumor suppressor capabilities of this protein are now being explored where its reactivation is seen to have therapeutic potential.

34.2 p27kip1 and p57Kip2 in Normal Physiology

p27Kip1 encodes the cyclin-dependent kinase inhibitors (CDKIs) in Cip/Kip family. This CDKIs binds to CDKs, thus terminating cell cycle. To understand p27Kip1 role in cellular proliferation we require a few details about mitotic cell division. In order for the endothelial cells to proliferate mitotically they must leave their idle state (G0) and re-enter the cell cycle. When p27 is expressed in cell, CDKIs bind to CDKs and prevent the transition between G1 and S phases of the cell cycle (Besson et al. 2006). Once the progression of the cell cycle is interrupted, cells can no longer undergo proliferation. This ultimately stops angiogenesis since endothelial cell proliferation is a prerequisite of angiogenesis (Folkman 1971). Therefore, tissues can control the degree of angiogenesis by regulating the expression of p27 gene. As is to be expected, nuclear p27kip1 protein content is high in tissues that are not expected to or have already undergone mitotic divisions. Examples of such include epithelial tissues of the breast, lung, prostate, ovaries, and lymphocytes of the blood and lymph nodes (Slingerland and Pagano 2000). As an extension, p27kip1 expression is also lower in tissues that are expected to undergo proliferation: basal layer of oral mucosa and skin, germinal centres of normal lymph nodes and the tonsils (Slingerland and Pagano 2000).

CKIs are found to be involved in the negative regulation of the G1 phase of the cell cycle. Progression of the cell cycle is governed by a group of cyclins and cyclin-dependent kinase (CDK) complexes. CDK complexes are regulated at a number of different points by various numbers of guiding processes, including CKI binding. CKI binding will negatively regulate the progression forward through the cell cycle. This negative regulation by CKIs (including p57Kip2) illustrates their antioncogenic role.

34.3 p27kip1 and p57Kip2 in Disease and Malignancy

With respect to cancer, p27kip1 is known primarily as a tumor-suppressor gene through its inhibition of the progression of cell cycle. However, it is not a typical tumour-suppressor gene; The absence of normal tumor-suppressor gene in a cell is responsible for tumour formation or progression (Besson et al. 2007). Hence its deletion from the genome would terminate tumourigenesis and cancer would not be developed. Also, studies have never been found that p27 gene to be mutated in tumours, for reasons unknown (Besson et al. 2004). Instead, it is always either the expression of p27 are either down-regulated, or the gene is excluded from the nucleus of the tumor cells (Besson et al. 2004). It is therefore not surprising that deficiency in p27Kip1 is often linked to cancer cells frequently, and this is in close association with elevated tumour aggression and a poor prognosis (Slingerland and Pagano 2000). Cases of these malignancies can be seen in brain tumours, adenocarcinomas, and lymphoproliferative neoplasms. Study done by Chien et al. (2007) have shown that the p27Kip1 knockout mice displayed spleen and thymus hyperplasia (Chien et al. 2007). They also exhibited spontaneous tumours in their pars intermedia, the barrier between the posterior and anterior lobes of the pituitary gland. The presence of haemorrhage in these pituitary tumours means that the tumours could maintain its vascular density. The newly capillaries formed may have abnormal structure and function, hence the haemorrhage (Chien et al. 2007). This proves that in the absence of p27Kip1 protein, angiogenesis was stimulated in these tumours at a rate that is proportional to its growing mass (Chien et al. 2007). The decrease in nuclear expression of p27kip1 in tumor cells may result from the combination of accelerated rates of degradation and nuclear export. However, little is known about the specific pathways of degradation and nuclear export (Besson et al. 2006).

As mentioned earlier, p27Kip1 modifies its folding upon binding to its target. Concordantly it is not surprising to see different properties when p27Kip1 enters a different cellular environment. Once p27Kip1 leaves the nucleus and enters the cytoplasm, p27Kip1 exerts other functions. p27Kip1 has the ability to control cell motility and cell migration by inhibition of Ras homolog gene family member A (RhoA) activity (Kimura et al. 1996). RhoA stimulates the formation of actin stress fibers and focal adhesions. It is also responsible for cell motility by controlling actin-myosin contractility to drive the translocation and the retraction of the cell (Kimura et al. 1996). Studies have shown that cells with engineered deletion of the p27kip1 gene have no reaction in reorganizing their actin cytoskeleton and focal adhesion in response to growth factors, whereas their control group did (Besson et al. 2004). The presence of p27Kip1 is necessary in cell migration and once it leaves its nuclear environment, it promotes angiogenesis and therefore tumor progression and invasiveness.

This contradiction of p27Kip1 being a tumor-suppressor gene or oncogene is shown in many studies in regards to p27Kip1 activity in human patients and genetically engineered animals. A low level of p27Kip1 is often found in human cancers, but rarely the homozygous loss of p27Kip1 (Sicinski et al. 2007). Also, mice with p27kip1 haplo-insufficiency were showed increased susceptibility to malignant tumour induction by chemical carcinogens and radiation when compared to p27kip1 knockout mice (Slingerland and Pagano 2000). The results suggest that the existence of p27Kip1 allele alone contributes tumourigenesis, even when the gene is not expressed. The protein acts as a tumour suppressor in the nucleus whereas in the dominant cytosolic oncogene with possibly metastatic potential (Sicinski et al. 2007). This explains the unusual role of p27Kip1 in regulation in cancer.

Studies performed with embryonic mice have illustrated the key role played by p57Kip2 protein in the regulation of growth, particularly in early organogenesis (Beretta et al. 2005). Its ubiquitous distribution through the entire embryo has been noted, with decreasing concentrations as the growth progresses. When protein dosages were manipulated there were notable changes to the phenotypes expressed, hence highlighting the importance of the proteins expression.

Consequently, a twofold expression lead to embryonic growth retardation due to premature cell cycle arrest causing 10–30 % reductions in the weight of the embryo and embryonic lethality (Andrews et al. 2007). Those mice lacking p57Kip2 presented with severe developmental abnormalities, often leading to post-natal mortality. Targeted disruptions showed "limb shortening due to incomplete endochrondral ossification, cleft palates, gastrointestinal tract defects, adrenal cortex enlargements, renal medullary dysplasia and increased body weight" (Pateras et al. 2009). Many of these phenotypes have been implicated in patients with Beckwith-Weidemann syndrome, suggesting the loss of the protein plays a role in the disorders progression (Zhang et al. 1997). Current results suggest normal embryonic growth is highly sensitivity to the expression of p57Kip2 at appropriate dosages.

p57Kip2 believed to have an anti-oncogenic role involved in the regulation of several hallmarks for tumor growth including apoptosis, cell differentiation and proliferation, control of tissue invasion and metastasis and angiogenesis (Kavanagh and Joseph 2011). The regulatory role of this protein in controlling cell differentiation is well established and its down regulation has been associated with many common human malignancies including "adrenal, head and neck, gastrointestinal, urothelial and lung carcinomas" (Larson et al. 2008). Associations have also been made with gestational trophoblastic diseases as well as Wilms tumour and Beckwith-Wiedemann syndrome (Larson et al. 2008). The reduction of the protein in cancers is believed to be caused by the methylation of the CDKN1C gene and lowered expression, however post-transcriptional adaptations have been described in some instances. A second mechanism leading to the diminished protein concentrations is due to increase in protein metabolism by Skp2 (S-phase kinas-associated protein 2) (Kavanagh and Joseph 2011). Overexpression of this protein has been implicated in several cancers, however the importance p57Kip2 degradation mediated by Skp2 is not fully understood and is still being investigated (Kavanagh and Joseph 2011).

As mentioned, the down-regulation of p57Kip2 is often associated with Beckwith-Wiedemann syndrome (BWS). BWS is a developmental disorder commonly implicated with the down regulation of p57Kip2. It is a heterogeneous disease where most patients present with "macrosomia, macroglossia, omphalocoele and a predisposition for paediatric tumours" (Pateras et al. 2009). The lack of p57Kip2 in combination with overexpression of insulin-like growth factors-II (IGF2) produces phenotypes indicative of the overgrowth syndrome. Data has suggested that the effect of increased IGF2 in BWS is, in part, mediated by decreased expression of p57Kip2 (Grandjean et al. 2000). This highlights the significant role of p57Kip2 as an indirect suppressant of some pathological conditions.

Studies have explored the effect of p57Kip2 in Wilms tumor, where reduced expression of the protein is often implicated. Research has shown that the transcription of the protein comes predominantly from the maternal allele and loss of the maternal allele or abnormal imprinting reduces its expression (Hatada et al. 1996).

Interestingly, other studies have explored the increases in immunostaining of p57Kip2 in oesophageal squamous cell carcinomas and positive relationships with Ki-67 (a protein that is strictly related to cell proliferation) (Scholzen and Gerdes 2000) and cyclin E (forms complex with CDK2, playing a critical role in transition from G1 to S phase of cell cycle). These relationships suggest limitation to increased proliferation via the activation of feedback mechanisms. In the same form, it was found that p57Kip2 levels increased in colorectal carcinogenesis when progressing from normal mucosa to adenomas and ultimately the levels were down regulated in primary carcinomas. Poor prognosis was often shown in patients presenting with complete loss of p57Kip2 (Li et al. 2003). Furthermore, down regulation is commonly associated and inversely correlated with the proliferation of many malignancies including "non-small cell lung carcinoma, pancreatic adenocarcinoma, hepatocellular carcinoma, extrahepatic bile duct carcinoma and intrahepatic cholangiocellular carcinoma" (Kavanagh and Joseph 2011).

Although correlations have been made between p57Kip2 down regulation and the aggressiveness of tumor progression, little has been done to explore the direct mechanisms of action, particularly in the field of angiogenesis. Indirectly, this protein can be associated with tumor vascularisation. The stimulatory effect of the proteins down-regulation on tumorigenesis ultimately leads to vascular proliferation as the tumour proceeds in its malignancy.

Most of the studies relating angiogenesis and p27Kip1 are done on its effects on tumourigenesis and thus leads to cancer development. Little has shown how p27Kip1 expression causes diseases in relation to angiogenesis. However, the role of p27Kip1 in disease can be demonstrated in atherosclerosis. Studies have shown that p27Kip1 expression is inversely correlated with vascular cell proliferation in human atherosclerotic tissue. The down-regulation of p27Kip1 at the injury site contributes to atherosclerosis and restenosis after coronary intervention (Breitenstein et al. 2013) through the CDK activation mechanism mentioned above. CDKs are not inactivated by the binding of CKDIs, leading to the progression of cell cycle, which in turn increases proliferation and migration of endothelial and vascular smooth muscle cells. The resulting angiogenesis stimulates the development of atherosclerotic lesions that ultimately leads to atherosclerosis (Papetti and Herman 2002).

Studies suggest the key to treatment of cancer and atherosclerosis is to restore the function of p27kip1, possibly by gene therapy. However, one should not overlook the possibility of overexpressing a therapeutic gene and causing undesirable systemic effects. More studies and development are needed to perfect the current vectors and/or gene delivery strategies (Goukassian et al. 2001).

To date, there are few studies linking p57Kip2 and angiogenesis. Reports have noted an increase in the expression of vascular endothelial growth factor (VEGF), an angiogenesis promoting factor, in the placenta of mice carrying p57Kip2 null embryos as compared to those carrying wild type (Matsuura et al. 2002). A relationship has been suggested between the expression of VEGF and the decreased expression of p57Kip2, however this has not been largely explored. Of the few studies performed, null expression of p57Kip2 showed a significant increase in VEGF protein levels, particularly the VEGF164 isoform that has shown more potent induction of tumorigenesis than other isoforms (Matsuura et al. 2002). However, the relationship was unrelated to tumor growth. The loss of p57Kip2 expression caused insufficient trophoblastic proliferation and invasion into the placental bed leading to poor perfusion of the placenta and fetoplacental unit VEGF expression within the placenta increased in concentration due to the hypoxic conditions (Matsuura et al. 2002).

Considering the relationships found between the down-regulation of p57Kip2 and common human malignancies, the protein offers great potential as a tumour suppressor candidate. Explorations into p57Kip2 reactivation as a potential therapeutic strategy for cancer are being made (Kavanagh and Joseph 2011). De-methylating agents are currently being examined to treat those patients with CDKN1C silencing by promotor methylations. These forms of silencing are commonly reported and de-methylation of the protein offers an avenue for re-expression or even over-expression to re-harness the proteins tumour suppressor capabilities (Kavanagh and Joseph 2011). Other mechanisms to stimulate the up-regulation of this protein based on the types of mutation or alterations causing the deregulation are also being examined.

References

Andrews SC, Wood MD, Tunster SJ, Barton SC, Surani MA, John RM (2007) Cdkn1c (p57Kip2) is the major regulator of embryonic growth within its imprinted domain on mouse distal chromosome 7. BMC Dev Biol 7:53. doi:10.1186/1471-213x-7-53

Beretta C, Chiarelli A, Testoni B, Mantovani R, Guerrini L (2005) Regulation of the cyclin-dependent kinase inhibitor p57Kip2 expression by p63. Cell Cycle 4(11):1625–1631. Retrieved from http://www.tandfonline.com/doi/pdf/10.4161/cc.4.11.2135

Besson A, Gurian-West M, Schmidt A, Hall A, Roberts JM (2004) p27Kip1 modulates cell migration through the regulation of RhoA activation. Genes Dev 18(8):862–876. doi:10.1101/gad.1185504

Besson A, Gurian-West M, Chen X, Kelly-Spratt KS, Kemp CJ, Roberts JM (2006) A pathway in quiescent cells that controls p27Kip1 stability, subcellular localization, and tumor suppression. Genes Dev 20(1):47–64. doi:10.1101/gad.1384406

Besson A, Hwang HC, Cicero S, Donovan SL, Gurian-West M, Johnson D, … Roberts JM (2007) Discovery of an oncogenic activity in p27Kip1 that causes stem cell expansion and a multiple tumor phenotype. Genes Dev 21(14):1731–1746. doi:10.1101/gad.1556607

Borriello A, Cucciolla V, Oliva A, Zappia V, Della Ragione F (2007) p27Kip1 metabolism: a fascinating labyrinth. Cell Cycle 6(9):1053–1061. Retrieved from http://www.tandfonline.com/doi/pdf/10.4161/cc.6.9.4142

Breitenstein A, Akhmedov A, Camici GG, Luscher TF, Tanner FC (2013) p27(Kip1) inhibits tissue factor expression. Biochem Biophys Res Commun 439(4):559–563. doi:10.1016/j.bbrc.2013.09.002

Chien WM, Garrison K, Caufield E, Orthel J, Dill J, Fero ML (2007) Differential gene expression of p27Kip1 and Rb knockout pituitary tumors associated with altered growth and angiogenesis. Cell Cycle 6(6):750–757. Retrieved from http://www.tandfonline.com/doi/pdf/10.4161/cc.6.6.3986

Folkman J (1971) Tumor angiogenesis: therapeutic implications. N Engl J Med 285(21):1182–1186. doi:10.1056/nejm197111182852108

Goukassian D, Diez-Juan A, Asahara T, Schratzberger P, Silver M, Murayama T, . . . Andres V (2001) Overexpression of p27(Kip1) by doxycycline-regulated adenoviral vectors inhibits endothelial cell proliferation and migration and impairs angiogenesis. Faseb J 15(11):1877–1885. doi:10.1096/fj.01-0065com

Grandjean V, Smith J, Schofield PN, Ferguson-Smith AC (2000) Increased IGF-II protein affects p57kip2 expression in vivo and in vitro: implications for Beckwith-Wiedemann syndrome. Proc Natl Acad Sci U S A 97(10):5279–5284. doi:10.1073/pnas.080409297

Hatada I, Inazawa J, Abe T, Nakayama M, Kaneko Y, Jinno Y, … Mukai T (1996) Genomic imprinting of human p57KIP2 and its reduced expression in Wilms' tumors. Hum Mol Genet 5(6):783–788. Retrieved from http://hmg.oxfordjournals.org/content/5/6/783.full.pdf

Herceg Z, Ushijima T (2009) Introduction: epigenetics and cancer. Adv Genet 70:1–23

Kavanagh E, Joseph B (2011) The hallmarks of CDKN1C (p57, KIP2) in cancer. Biochim Biophys Acta 1816(1):50–56. doi:10.1016/j.bbcan.2011.03.002

Kimura K, Ito M, Amano M, Chihara K, Fukata Y, Nakafuku M, … Kaibuchi K (1996) Regulation of myosin phosphatase by Rho and Rho-associated kinase (Rho-kinase). Science 273(5272):245–248

Larson PS, Schlechter BL, King CL, Yang Q, Glass CN, Mack C, … Rosenberg CL (2008) CDKN1C/p57kip2 is a candidate tumor suppressor gene in human breast cancer. BMC Cancer 8:68. doi:10.1186/1471-2407-8-68

Li JQ, Wu F, Usuki H, Kubo A, Masaki T, Fujita J, . . . Imaida K (2003) Loss of p57KIP2 is associated with colorectal carcinogenesis. Int J Oncol 23(6):1537–1543. Retrieved from http://www.spandidos-publications.com/ijo/23/6/1537

Matsuura T, Takahashi K, Nakayama K, Kobayashi T, Choi-Miura NH, Tomita M, Kanayama N (2002) Increased expression of vascular endothelial growth factor in placentas of p57(Kip2) null embryos. FEBS Lett 532(3):283–288. Retrieved from http://www.febsletters.org/article/S0014-5793(02)03681-5/pdf

Papetti M, Herman IM (2002) Mechanisms of normal and tumor-derived angiogenesis. Am J Physiol Cell Physiol 282(5):C947–C970. doi:10.1152/ajpcell.00389.2001

Pateras IS, Apostolopoulou K, Niforou K, Kotsinas A, Gorgoulis VG (2009) p57KIP2: "Kip"ing the cell under control. Mol Cancer Res 7(12):1902–1919. doi:10.1158/1541-7786.mcr-09-0317

Ponce-Castaneda MV, Lee MH, Latres E, Polyak K, Lacombe L, Montgomery K, … et al (1995) p27Kip1: chromosomal mapping to 12p12-12p13.1 and absence of mutations in human tumors. Cancer Res 55(6):1211–1214. Retrieved from http://cancerres.aacrjournals.org/content/55/6/1211.full.pdf

Scholzen T, Gerdes J (2000) The Ki-67 protein: from the known and the unknown. J Cell Physiol 182(3):311–322. doi:10.1002/(sici)1097-4652(200003)182:3<311::aid-jcp1>3.0.co;2-9

Sicinski P, Zacharek S, Kim C (2007) Duality of p27Kip1 function in tumorigenesis. Genes Dev 21(14):1703–1706. doi:10.1101/gad.1583207

Slingerland J, Pagano M (2000) Regulation of the cdk inhibitor p27 and its deregulation in cancer. J Cell Physiol 183(1):10–17. doi:10.1002/(sici)1097-4652(200004)183:1<10::aid-jcp2>3.0.co;2-i

Zhang P, Liegeois NJ, Wong C, Finegold M, Hou H, Thompson JC, … Elledge SJ (1997) Altered cell differentiation and proliferation in mice lacking p57KIP2 indicates a role in Beckwith-Wiedemann syndrome. Nature 387(6629):151–158. doi:10.1038/387151a0

Chapter 35
Platelet-Derived Endothelial Cell Growth Factor (PDGF)

Abstract Platelet-derived endothelial cell growth factor (PDGF) is a large family of polypeptide chain and can be categorized in 4 groups (PDGF-A, B, C and D). PDGF-A, B and C isoforms were found as early as 1970s whereas PDGF-D has firstly been found and identified as highly related to VEGF. PDGF and VEGF may not only have similar functions but also have its own specific purposes. PDGF is a potent mitogen for mesenchymal origin cells, including glial cells and smooth muscle cells. PDGF-A and -D were localized within the endothelial cells, immune cells and myofibroblasts following MI. These significant findings suggest pivotal role of PDGFs during the repair process. The synergistic ability of PDGFs with VEGFs has been discussed and in fact there are presence of other "angiogenic synergism" discovered in other diseases. The main mechanism of PDGF involved in tumor angiogenesis can be mainly summarized in matrix metalloproteinases (MMPs) and VEGF. Currently, the rationale behind clinical trials are using combined treatments, which would involve targeting the formation of new vessels via the inhibition of PDGFR and VEGFR signaling, hence suppressing the tumor cell proliferation, survival and migration. Future researches on the mechanism of PDGF gene expression and larger clinical trials will be beneficial to all specialties and provide more promising outcome.

Keywords Platelet-derived endothelial cell growth factor • PDGF • Angiogenesis • Normal physiology • Disease • Malignancy

35.1 Platelet-Derived Endothelial Cell Growth Factor (PDGF)

Platelet-derived endothelial cell growth factor (PDGF) is a large family of polypeptide chain and can be categorized in four groups (PDGF-A, B, C and D). It is presented in disulphate-bonded dimer (4 homodimer and 1 heterodimer types) (Wang et al. 2010).

Dr. Moein Amin contributed equally with Dr. Ali Salajegheh.

A. Salajegheh, *Angiogenesis in Health, Disease and Malignancy*,
DOI 10.1007/978-3-319-28140-7_35

PDGF is one of the angiogenic growth factors released from human dentine matrix for wound healing and tissue regeneration activity in both hard and soft tissue (Roberts-Clark and Smith 2000; Derringer and Linden 2004; Rasubala et al. 2003; Tran-Hung et al. 2008). In the 1980s, Lynch et al. (1987) was the first to investigate the PDGF expression with tissue regeneration and found PDGF promotes tissue regeneration. During the wound healing process for bone tissue, PDGF expression was only detected in the early and mid stages of the healing process, and its expression was responsible for the primitive mesenchymal cell proliferation and migration (Rasubala et al. 2003). During the process of mechanically injured dental pulp tissue healing, significant amount of PDGF was secreted by pulp cells and shown to promote angiogenesis before mineralisation of reparative dentine occurs (Tran-Hung et al. 2008). Several researches regarding osseointegration and wound healing were conducted after understanding the mechanisms of tissue regeneration by PDGF expression.

35.2 PDGF in Normal Physiology

PDGF-A, B and C isoforms were found as early as 1970s in the past 10 years. PDGF-D has firstly been found and identified as highly related to VEGF (Andrae et al. 2008). The most possible explanation is the two proteins may not only have similar functions but also have its own specific purposes. PDGF is a potent mitogen for mesenchymal origin cells, including glial cells and smooth muscle cells. In humans, the PDGF signaling network consists of four ligands, PDGFA-δ, and two receptors of PDGFR- α, and PDGFR- β (Matsui et al. 1988).

35.3 PDGF in Disease

The interaction between the isoforms of PDGF and their receptors has been identified and many studies reveal crucial roles of PDGFRs for down-stream signaling pathways, which are important for vascular remodeling and angiogenesis after injuries (Wang et al. 2012; Andrae et al. 2008; Li et al. 2003; Moriya et al. 2014).

In Myocardial Infarction (MI), cardiac remodeling process occurs in the infarcted myocardium (Khurana et al. 2005). During this repair, angiogenesis plays a huge role in the formation of new blood vessels (Raines 2004). Zhao et al. (2011) showed that in both early and late stages of MI, significant increase in the expression levels of PDGFR isoforms (PDGFR-α and -β) were detected in infarcted myocardium of rats in comparison to control. Furthermore, increased expression levels of PDGF-A and –D were localized within the endothelial cells, immune cells and myofibroblasts following MI. These significant findings suggest pivotal role of PDGFs during the repair process of MI and serve as potent inducers of angiogenesis. In addition to this, another study by Wang et al. (2012) investigated the effects of over-expressing

PDGFR-β signaling following vascular injury using mice spleen-derived EPCs. The results demonstrated strong correlation between PDGF-β over-expression increased activity of EPCs via PI3K/Akt pathway, which is a down stream of PDGF-β signaling responsible for endothelial cell activation. Consequently, PDGFs are essential for promoting angiogenesis in response to injuries (Wang et al. 2012; Khurana et al. 2005; Raines 2004; Andrae et al. 2008).

However, this repair process can be a negative mechanism in other diseases. Such pathological angiogenesis is prominent in chronic liver diseases such as Cirrhosis (Fernandez et al. 2009). Intrahepatic vascular remodeling driven by PDGF and other growth factors such as VEGF, FGFs and TGF-β1 have been shown cause increased hepatic pressure, also known as portal hypertension, hence leading to permanent liver failure (Floege et al. 2008; Fernandez et al. 2009; Carmeliet 2000). Studies have indicated that VEGF may be a potent angiogenic factor leading to portal hypertension seen in animal models (Floege et al. 2008; Cao et al. 2010; Fernandez et al. 2004, 2005). Furthermore, it has been suggested PDGF may have a synergistic role together with VEGF in promoting angiogenesis. This postulation was confirmed by Fernandez et al. (2007), who showed significant reduction in the neovascularization activity in portal hypertensive rats following co-administration of rapamycin (VEGF signaling inhibitor) and Gleevec (PDGF signaling inhibitor). Moreover, human clinical trials have shown dramatic decrease in the risk of portal hypertension and other complications under combined drug treatments (Fernandez et al. 2009). Therefore, these findings suggest possible interplay between pro-angiogenic factors and that process of vessel growth can cause adverse effects as seen in Cirrhosis.

The synergistic ability of PDGFs with VEGFs has been discussed and in fact there are presence of other "angiogenic synergism" discovered in other diseases. Cao et al. (2003) reported that the combination of PDGF-B with FGF-2 synergistically induced angiogenesis in ischaemic hind limb of rabbits that improved stability and functioning of limbs. Similarly, as a follow up study to this, Li et al. (2010) demonstrated synergistic effects of PDGF-B with FGF-2 by intramuscular injection of DNA plasmid, which resulted in greater capillary growth and also muscle repair in ischaemic hind limb of rabbits. Thus, these findings provide advancement in the future therapeutics.

35.4 PDGF in Malignancy

The main mechanism of PDGF involved in tumor angiogenesis can be mainly summarized in matrix metalloproteinases (MMPs) and VEGF. The inhibition of PDGF-D can simultaneously down-regulate the angiogenesis via MMP9 and MMP2, and also partially inhibit the VEGF-regulated angiogenesis in gastric cancer (Zhao et al. 2010). Hsu and his colleagues summarized the current clinical trial in the head and neck squamous cell carcinoma (HNSCC) (Hsu et al. 2014). Currently, the rationale behind clinical trials are using combined treatments, which would involve targeting the formation of new vessels via the inhibition of PDGFR and VEGFR signaling, hence suppressing the tumor cell proliferation, survival and migration.

In the RNA level, miR-29b has been reported to regulate the microenvironment genes-PDGF, VEGF and MMP9 that are highly related to the angiogenesis and promoted the metastasis in the later process (Chou et al. 2013). In 2011, Zhang and colleagues proposed that miR-9 interacted with the 3′-UTR of PDGFR-β and this led to the overexpression of miR-9 and caused the GBM cell line-U87 cell and cardiomyocytes (Zhang et al. 2011). Furthermore, miR-146a was found to promote the angiogenic activities on HUVECs cell and hepatocellular carcinoma (HCC) cell co-culture via breast cancer 1 (BRCA1) and the PDGFR-A signalling (Zhu et al. 2013). Although Shao et al. tried to use the microRNA (miR) array to screen the PDGF induction in the glioblastoma and ovarian cancer, they found that PDGF can induce miR-146b to regulate not only the activation of Akt via PDGF-AA and Mitogen-Activated Protein Kinase (MAPK) via PDGF-BB, but also regulate epidermal growth factor receptor (EGFR) signalling. Despite this finding, there is still no evidence suggesting direct link to angiogenesis (Shao et al. 2011).

During early embryonic development, PDGF acts as a key regulator for tooth cusp and palate developments. Decreased levels of PDGF expression can disturb greatly in the extracellular matrix formation and the organisation of odontoblasts can ultimately induce defects in tooth development, and cleft palate in craniofacial development (Xu et al. 2005).

In the review by Kiagler et al. (2011) it was suggested that the use of growth factors such as recombinant human platelet-derived growth factor-BB (rhPDGF) promotes periodontal and pre-implant healing with more predictable outcome. The use of PDGF-BB in re-implantation of avulsed tooth can reduce occurrence of ankylosis and root resorption (Noda et al. 2012). Furthermore, the use of PDGF and insulin-like growth factor-1 (IGF-1) has shown improved osseointegration and bone remodeling on the titanium surface of implants (Ortolani et al. 2014). As a result, future researches on the mechanism of PDGF gene expression and larger clinical trials will be beneficial to all specialties and provide more promising outcome.

References

Andrae J, Gallini R, Betsholtz C (2008) Role of platelet-derived growth factors in physiology and medicine. Genes Dev 22(10):1276–1312. doi:10.1101/gad.1653708

Cao R, Brakenhielm E, Pawliuk R, Wariaro D, Post MJ, Wahlberg E, Leboulch P, Cao Y (2003) Angiogenic synergism, vascular stability and improvement of hind-limb ischemia by a combination of PDGF-BB and FGF-2. Nat Med 9(5):604–613. doi:10.1038/nm848

Cao S, Yaqoob U, Das A, Shergill U, Jagavelu K, Huebert RC, Routray C, Abdelmoneim S, Vasdev M, Leof E, Charlton M, Watts RJ, Mukhopadhyay D, Shah VH (2010) Neuropilin-1 promotes cirrhosis of the rodent and human liver by enhancing PDGF/TGF-beta signaling in hepatic stellate cells. J Clin Invest 120(7):2379–2394. doi:10.1172/jci41203

Carmeliet P (2000) Mechanisms of angiogenesis and arteriogenesis. Nat Med 6(4):389–395. doi:10.1038/74651

Chou J, Lin JH, Brenot A, Kim JW, Provot S, Werb Z (2013) GATA3 suppresses metastasis and modulates the tumour microenvironment by regulating microRNA-29b expression. Nat Cell Biol 15(2):201–213. doi:10.1038/ncb2672

Derringer KA, Linden RW (2004) Vascular endothelial growth factor, fibroblast growth factor 2, platelet derived growth factor and transforming growth factor beta released in human dental pulp following orthodontic force. Arch Oral Biol 49(8):631–641. doi:10.1016/j.archoralbio.2004.02.011

Fernandez M, Vizzutti F, Garcia-Pagan JC, Rodes J, Bosch J (2004) Anti-VEGF receptor-2 monoclonal antibody prevents portal systemic collateral vessel formation in portal hypertensive mice. Gastroenterology 126:886–894

Fernandez M, Mejias M, Angermayr B, Garcia-Pagan JC, Rodes J, Bosch J (2005) Inhibition of VEGF receptor-2 decreases the development of hyperdynamic splanchnic circuation and portal-systemic collateral vessels in portal hyperensive rats. J Hepatol 43:98–103

Fernandez M, Mejias M, Garcia-Pras E, Mendez R, Garcia-Pagan JC, Bosch J (2007) Reversal of portal hypertension and hyperdynamic splanchnic circulation by combined vascular endothelial growth factor and platelet-derived growth factor blockade in rats. Hepatology 46:1208–1217

Fernandez M, Semela D, Bruix J, Colle I, Pinzani M, Bosch J (2009) Angiogenesis in liver disease. J Hepatol 50(3):604–620. doi:10.1016/j.jhep.2008.12.011

Floege J, Eitner F, Alpers CE (2008) A new look at platelet-derived growth factor in renal disease. J Am Soc Nephrol 19:12–23

Hsu HW, Wall NR, Hsueh CT, Kim S, Ferris RL, Chen CS, Mirshahidi S (2014) Combination antiangiogenic therapy and radiation in head and neck cancers. Oral Oncol 50(1):19–26. doi:10.1016/j.oraloncology.2013.10.003

Kaigler D, Avila G, Wisner-Lynch L, Nevins ML, Nevins M, Rasperini G, Lynch SE, Giannobile WV (2011) Platelet-derived growth factor applications in periodontal and peri-implant bone regeneration. Expert Opin Biol Ther 11(3):375–385. doi:10.1517/14712598.2011.554814

Khurana R, Simons M, Martin JF, Zachary IC (2005) Role of angiogenesis in cardiovascular disease: a critical appraisal. Circulation 112(12):1813–1824. doi:10.1161/circulationaha.105.535294

Li H, Fredriksson L, Li X, Eriksson U (2003) PDGF-D is a potent transforming and angiogenic growth factor. Oncogene 22(10):1501–1510. doi:10.1038/sj.onc.1206223

Li J, Wei Y, Liu K, Yuan C, Tang Y, Quan Q, Chen P, Wang W, Hu H, Yang L (2010) Synergistic effects of FGF-2 and PDGF-BB on angiogenesis and muscle regeneration in rabbit hindlimb ischemia model. Microvasc Res 80(1):10–17. doi:10.1016/j.mvr.2009.12.002

Lynch SE, Nixon JC, Colvin RB, Antoniades HN (1987) Role of platelet-derived growth factor in wound healing: synergistic effects with other growth factors. Proc Natl Acad Sci U S A 84(21):7696–7700

Matsui T, Heidaran M, Miki T, Popescu N, La Rochelle W, Kraus M, Jacalyn P, Aaronson S (1988) Isolation of a novel receptor cDNA establishes the existence of two PDGF receptor genes. Science 243:4

Moriya J, Wu X, Zavala-Solorio J, Ross J, Liang XH, Ferrara N (2014) Platelet-derived growth factor C promotes revascularization in ischemic limbs of diabetic mice. J Vasc Surg 59(5):1402–1409.e1401–1404. doi:10.1016/j.jvs.2013.04.053

Noda K, Seshima F, Okubo N, Ishii Y, Ota M, Yamada S, Saito A (2012) Effect of platelet-derived growth factor-BB on root resorption after reimplantation of partially denuded tooth in dog. Dent Traumatol 28(3):217–225. doi:10.1111/j.1600-9657.2011.01070.x

Ortolani E, Guerriero M, Coli A, Di Giannuario A, Minniti G, Polimeni A (2014) Effect of PDGF, IGF-1 and PRP on the implant osseointegration. An histological and immunohistochemical study in rabbits. Ann Stomatol 5(2):66–68

Raines EW (2004) PDGF and cardiovascular disease. Cytokine Growth Factor Rev 15(4):237–254. doi:10.1016/j.cytogfr.2004.03.004

Rasubala L, Yoshikawa H, Nagata K, Iijima T, Ohishi M (2003) Platelet-derived growth factor and bone morphogenetic protein in the healing of mandibular fractures in rats. Br J Oral Maxillofac Surg 41(3):173–178

Roberts-Clark DJ, Smith AJ (2000) Angiogenic growth factors in human dentine matrix. Arch Oral Biol 45(11):1013–1016

Shao M, Rossi S, Chelladurai B, Shimizu M, Ntukogu O, Ivan M, Calin GA, Matei D (2011) PDGF induced microRNA alterations in cancer cells. Nucleic Acids Res 39(10):4035–4047. doi:10.1093/nar/gkq1305

Tran-Hung L, Laurent P, Camps J, About I (2008) Quantification of angiogenic growth factors released by human dental cells after injury. Arch Oral Biol 53(1):9–13. doi:10.1016/j.archoralbio.2007.07.001

Wang Z, Ahmad A, Li Y, Kong D, Azmi AS, Banerjee S, Sarkar FH (2010) Emerging roles of PDGF-D signaling pathway in tumor development and progression. Biochim Biophys Acta 1806(1):122–130. doi:10.1016/j.bbcan.2010.04.003

Wang H, Yin Y, Li W, Zhao X, Yu Y, Zhu J, Qin Z, Wang Q, Wang K, Lu W, Liu J, Huang L (2012) Over-expression of PDGFR-B promotes PDGF-induced proliferation, migration and angiogenesis of EPCs through PI3K/Akt signaling pathway. PLoS One 7(2):1–14

Xu X, Bringas P Jr, Soriano P, Chai Y (2005) PDGFR-alpha signaling is critical for tooth cusp and palate morphogenesis. Dev Dyn 232(1):75–84. doi:10.1002/dvdy.20197

Zhang J, Chintalgattu V, Shih T, Ai D, Xia Y, Khakoo AY (2011) MicroRNA-9 is an activation-induced regulator of PDGFR-beta expression in cardiomyocytes. J Mol Cell Cardiol 51(3):337–346. doi:10.1016/j.yjmcc.2011.05.019

Zhao L, Zhang C, Liao G, Long J (2010) RNAi-mediated inhibition of PDGF-D leads to decreased cell growth, invasion and angiogenesis in the SGC-7901 gastric cancer xenograft model. Cancer Biol Ther 9(1):42–48

Zhao W, Zhao T, Huang V, Chen Y, Ahokas RA, Sun Y (2011) Platelet-derived growth factor involvement in myocardial remodeling following infarction. J Mol Cell Cardiol 51(5):830–838. doi:10.1016/j.yjmcc.2011.06.023

Zhu K, Pan Q, Zhang X, Kong LQ, Fan J, Dai Z, Wang L, Yang XR, Hu J, Wan JL, Zhao YM, Tao ZH, Chai ZT, Zeng HY, Tang ZY, Sun HC, Zhou J (2013) MiR-146a enhances angiogenic activity of endothelial cells in hepatocellular carcinoma by promoting PDGFRA expression. Carcinogenesis 34(9):2071–2079. doi:10.1093/carcin/bgt160

Chapter 36
Prolyl Hydroxylase Domain-2 (PHD-2)

Abstract Angiogenesis has long been considered as an important aspect for cancer therapy. Tumor angiogenesis, the pathological process that induced by solid tumours includes one important process: neovascularization. This process forms new blood vessels by the mobilization and recruitment of bone marrow-derived cells. Previous studies have shown that Hypoxia-inducible factors (HIF) positively correlated with tumor growth and HIF are key regulators of oxygen homeostasis and are composed of an oxygen-labile α subunit and a constitutive β subunit. Prolyl hydroxylase domain (PHD) proteins, which control oxygen-dependent degradation HIF, include PHD3, PHD2, and PHD1. PHD enzymes are inhibited during hypoxic conditions, allowing for HIF accumulation and subsequent induction of angiogenesis. PHD-2 is believed to be the key propyl hydroxylase in controlling HIF-1α during hypoxia (Kant et al. 2013). Under normoxic conditions, molecular oxygen, 2-oxoglutarate, iron ions (Fe2+) and ascorbic acid are required to fully activate these enzymes. Additionally, HIF-1 inactivates PHD-2 in a negative feedback manner. Although PHD-2 is inactive during hypoxia, PHD-2 levels are also increased by hypoxia, which provides a HIF-1-dependent auto-regulating mechanism. Suppression of HIF-1α by Methylselenocysteine (MSC) can be achieved through stabilization of PHD2 and PHD3 although uncertainty about their side effects of this method demands more work before therapeutic interventions can be applied in clinical cases.

Keywords Prolyl hydroxylase domain • PHD-2 • Angiogenesis • Normal physiology • Disease • Malignancy

36.1 Prolyl Hydroxylase Domain-2 (PHD-2)

Angiogenesis has long been considered as an important aspect for cancer therapy. Tumor angiogenesis, the pathological process that induced by solid tumours includes one important process: neovascularization. This process forms new blood vessels by the mobilization and recruitment of bone marrow-derived cells. Previous studies have shown that Hypoxia-inducible factors (HIF) positively correlated with tumor growth and HIF are key regulators of oxygen homeostasis and are composed

A. Salajegheh, *Angiogenesis in Health, Disease and Malignancy*,
DOI 10.1007/978-3-319-28140-7_36

of an oxygen-labile α subunit and a constitutive β subunit (Berra et al. 2003). Prolyl hydroxylase domain (PHD) proteins, which control oxygen-dependent degradation HIF, include PHD3, PHD2, and PHD1 (Duan et al. 2011). The aim of this chapter is to discuss the relationship among HIF, PHD-2 and angiogenesis in malignant situation as well as the strategies to treat the tumor.

36.2 Prolyl Hydroxylase Domain-2 (PHD-2) in Normal Physiology

Exposure of a cell or tissue to inadequate oxygen levels causes hypoxia and results in cellular changes in gene expression. Hypoxia is also associated with pathological events such as solid tumours and ischemic disease (Madanecki et al. 2013). Prolyl-4-hydroxylase domain (PHDs) family is one of the most important regulatory factors in hypoxic stress. PHD-2 plays a critical role in making cells and tissues adapt to the low oxygen environment and PHD-2 also acts as an important factor in oxygen homeostasis. PHD-2 is one of three prolyl hydroxylases originally identified as negative regulators of HIF. It is the key oxygen sensor because PHD-2 results in stabilization of HIF-1α (Duan et al. 2011).

36.3 Prolyl Hydroxylase Domain-2 (PHD-2) in Disease

PHD-2 regulates the stability and transcriptional activity of the hypoxia-inducible factor 1 (HIF-1), which is the key factor in response to hypoxic stress. Moreover, through the regulation of PHD-2 on HIF-1, it plays an important role in the post-ischemic neovascularization (Jia et al. 2014). Furthermore, PHD-2 also regulates other ways that positively regulate angiogenesis factors HIF-1 under hypoxic condition. Based on these facts, PHD-2 has been considered as a potential therapeutic target in treating tumours (Bottsford-Miller et al. 2012).

PHD enzymes are inhibited during hypoxic conditions, allowing for HIF accumulation and subsequent induction of angiogenesis. PHD-2 is believed to be the key propyl hydroxylase in controlling HIF-1α during hypoxia (Kant et al. 2013). Under normoxic conditions, molecular oxygen, 2-oxoglutarate, iron ions ($Fe2+$) and ascorbic acid are required to fully activate these enzymes. Additionally, HIF-1 inactivates PHD-2 in a negative feedback manner. Although PHD-2 is inactive during hypoxia, PHD-2 levels are also increased by hypoxia, which provides a HIF-1-dependent auto-regulating mechanism (Bottsford-Miller et al. 2012). PHD-2 siRNA treatment not only improves the fibroblast proliferation, but also enhances the expression of HIF-1. These effects may affect the diabetic wound healing in positive way (Jia et al. 2014).

36.4 Prolyl Hydroxylase Domain-2 (PHD-2) in Malignancy

More than 80 % of renal cell carcinomas are influenced by clear cell renal cell carcinoma (ccRCC). When ccRCC deactivates Von-Hippel-Lindau (VHL), expression of HIF-α, which is involved in tumor Angiogenesis, is affected (Bottsford-Miller et al. 2012). Researchers hypothesized that deregulation of PHD-2 might be used for treating several kinds of tumor that have high level of HIF, which accelerates the growth of tumor angiogenesis. They firstly investigated the mRNA and protein level of PHD2; they found it decreased in several types of tumor. Secondly, they used medication to limit the level of PHD-2 and found that tumor progression in these cases was increased. Finally, they used the same medication to inhibit the level of HIF and found that the tumor growth did not reduce, thus the researchers thought tumor suppressor function of PHD2 is HIF-independent. They suggest that low incidence of PHD-2 (32 %), undetectable PHD-3 and high incidence of HIF-α (92 %) are expressed by ccRCC in head & neck and colon cancers (Madanecki et al. 2013). This study highlights the importance of controlling tumor angiogenesis by limiting the level of HIF with the help of PHD-2.

Moreover, reports have also demonstrated that suppression of HIF-1α by Methylselenocysteine (MSC) can be achieved through stabilization of PHD2/3 and chemotherapy when combined with therapeutic synergy. This research started by investigating the level of PHDs and HIF-α in selected human cancers, the mechanism of HIF-α limitation by MSC, and documenting antitumor action of MSC against ccRCC xenografts (Kant et al. 2013). And it is the first time to discover that MSC is a highly efficient suppressor for controlling HIF-α in ccRCC tumours. MSC does not limit the synthesis of HIF-1α protein, but accelerates the degradation of HIF. The treatment of MSC on HIF-1α is connected with important antitumor action which is against ccRCC xenograft (Madanecki et al. 2013).

The researchers have proposed different molecular mechanisms for increasing the level of the HIF-1α system in tumorigenesis and they suggest that its opposite regulation may be useful for tumor treatment. Bordoli demonstrates that the lowering the level of PHD2 will result in accelerating the growth of tumor in a carcinoma and there are clinical examples of breast cancer showing that the time which patient can survive with low-level PHD2 will be much shorter (Kant et al. 2013). It is reported that if PHD2 is absent, the number in amphiregulin will be increased and its levels will go down when level of PHD2 is normal. Amphiregulin is regulated specifically by HIF-2, not HIF-1, which is an angiogenesis-related antibody (Chintala et al. 2012). Accordingly, it has been suggested that in breast cancer and other related cancers, PHD2 is a potential tumor inhibitor (Vogel et al. 2010). The researchers have also demonstrated the influence which PHDs have on human colorectal cancer (CRC) as tumor suppressors (Chintala et al. 2012). A clinical research has shown that the low level of PHD2 in CRC will result in poor survival of HIF-1α especially in early stage tumours (Madanecki et al. 2013). The progression of these tumours is inhibited for the reason that they grow their vascular supply too fast and then they will become hypoxic. Therefore, it is suggested by researchers that PHD2 is good target for therapeutic control as lacking PHD2 will cause increase of HIF-1α, which promotes the growth of angiogenesis in these tumours in mutant mice (Robinson et al. 2008).

Additionally, PHD-2 also is related to m*iR-200b, miR-200c* and *miR-429, which* increase during ischemic preconditioning. These miRNAs target PHD-2 in order to lead to the accumulation of HIF-1α and induction of angiogenesis (Madanecki et al. 2013). However, a recent study demonstrates that the angiogenic response is suppressed by miR-200b, which overexpresses in human microvascular endothelial cells (HMECs). In HMECs, miR-200b levels are suppressed by hypoxia, and the direct target for this miRNA is Ets-1mRNA, which is a critical angiogenesis related transcription factor. Thus, hypoxia-induced miR-200b inhibition allows Ets-1 accumulation to promote angiogenesis (Verma 2012).

Results show that how HIF-α in human ccRCC is affected by PHD2/3 in stable expression. What is more, lowering HIF-1α by MSC is fulfilled by regulating the level of PHD2 and there is another pathway called "VHL independent" which is unique for controlling HIF-α. The results present us with the basic information for connect currently used agents with MSC for treating cancer. However, because of uncertain side effects of the treating methods, more work is needed before therapeutic interventions can be applied in clinical cases.

References

Berra E, Benizri E, Ginouves A, Volmat V, Roux D, Pouyssegur J (2003) HIF prolyl-hydroxylase 2 is the key oxygen sensor setting low steady-state levels of HIF-1alpha in normoxia. Embo J 22(16):4082–4090. doi:10.1093/emboj/cdg392

Bottsford-Miller JN, Coleman RL, Sood AK (2012) Resistance and escape from antiangiogenesis therapy: clinical implications and future strategies. J Clin Oncol 30(32):4026–4034. doi:10.1200/jco.2012.41.9242

Chintala S, Najrana T, Toth K, Cao S, Durrani FA, Pili R, Rustum YM (2012) Prolyl hydroxylase 2 dependent and Von-Hippel-Lindau independent degradation of Hypoxia-inducible factor 1 and 2 alpha by selenium in clear cell renal cell carcinoma leads to tumor growth inhibition. BMC Cancer 12:293. doi:10.1186/1471-2407-12-293

Duan LJ, Takeda K, Fong GH (2011) Prolyl hydroxylase domain protein 2 (PHD2) mediates oxygen-induced retinopathy in neonatal mice. Am J Pathol 178(4):1881–1890. doi:10.1016/j.ajpath.2010.12.016

Jia HZ, Kasim V, Xu ZL, Yang L, Wu SR (2014) Hypoxia-responsive factor PHD2 and angiogenic diseases. Yao Xue Xue Bao 49(2):151–157

Kant R, Bali A, Singh N, Jaggi AS (2013) Prolyl 4 hydroxylase: a critical target in the pathophysiology of diseases. Korean J Physiol Pharmacol Off J Kor Physiol Soc Kor Soc Pharmacol 17(2):111–120. doi:10.4196/kjpp.2013.17.2.111

Madanecki P, Kapoor N, Bebok Z, Ochocka R, Collawn JF, Bartoszewski R (2013) Regulation of angiogenesis by hypoxia: the role of microRNA. Cell Mol Biol Lett 18(1):47–57. doi:10.2478/s11658-012-0037-0

Robinson A, Keely S, Karhausen J, Gerich ME, Furuta GT, Colgan SP (2008) Mucosal protection by hypoxia-inducible factor prolyl hydroxylase inhibition. Gastroenterology 134(1):145–155. doi:10.1053/j.gastro.2007.09.033

Verma M (2012) Cancer control and prevention by nutrition and epigenetic approaches. Antioxid Redox Signal 17(2):355–364. doi:10.1089/ars.2011.4388

Vogel S, Wottawa M, Farhat K, Zieseniss A, Schnelle M, Le-Huu S, von Ahlen M, Malz C, Camenisch G, Katschinski DM (2010) Prolyl hydroxylase domain (PHD) 2 affects cell migration and F-actin formation via RhoA/rho-associated kinase-dependent cofilin phosphorylation. J Biol Chem 285(44):33756–33763. doi:10.1074/jbc.M110.132985

Chapter 37
Phosphatidylinositol-4, 5-Bisphosphate 3-Kinase (PIK3Ca)

Abstract PIK3Ca (phosphatidylinositol-4, 5-bisphosphate 3-kinase) gene belongs to the family of genes, Phosphoinositide 3-kinase. PIK3Ca is composed of a 110 kDa catalytic subunit and an 85 kDa regulatory subunit, located in the cytoplasm and the catalytic subunit uses ATP to phosphorylate phosphatidylinositols. PIK3Ca is responsible for the action of PI3K. The mutation in the PIK3Ca gene causes the megalencephaly-capillary malformation syndrome (MCAP) characterized by overgrowth of brain tissue (megalencephaly) and abnormalities caused by enlarged small blood vessels in the skin. PIK3Ca mutations have been also strongly linked to a congenital lipoma known as CLOVES (congenital lipomatous overgrowth, vascular malformations, epidermal nevi and skeletal/spinal abnormalities) syndrome. It has been found that PI3KCa is oncogenic and is involved particularly in cervical, breast and colon cancers. Additionally, it was indicated that oncogenic PI3KCa is involved in about 30 % of all breast and colon cancers and less frequent in cancers of the stomach, brain, liver and ovaries. Expression of PIK3Ca is positively correlated with the expression of vascular endothelial growth factor (VEGF). It has been shown that PIK3Ca inhibitor decreased both inducible and constitutive expression of the hypoxia-inducible factor-1α (HIF1α) and revoked VEGF upregulation by starvation of glucose. PIK3Ca does promote angiogenesis and endorse angiogenesis; perhaps the importance of this gene in cell proliferation and survival is part of the future of anti-cancer treatments.

Keywords Phosphatidylinositol-4 • 5-bisphosphate 3-kinase • PIK3Ca • Angiogenesis • Normal physiology • Disease • Malignancy

37.1 Phosphatidylinositol-4, 5-Bisphosphate 3-Kinase (PIK3Ca)

PIK3Ca (phosphatidylinositol-4, 5-bisphosphate 3-kinase) gene belongs to the family of genes, Phosphoinositide 3-kinase (Karakas et al. 2006). PIK3Ca gene is responsible for the production of p110α protein which is a subunit of an enzyme called phosphatidylinositol 3-kinase (PI3K). PIK3Ca is composed of a 110 kDa catalytic subunit and an 85 kDa regulatory subunit. The catalytic subunit uses ATP

A. Salajegheh, *Angiogenesis in Health, Disease and Malignancy*,
DOI 10.1007/978-3-319-28140-7_37

to phosphorylate phosphatidylinositols (PtdIns), PTdIns4P and PtdIns (4,5) P2 (Karakas et al. 2006). p110α, located in the cytoplasm, is responsible for the action of PI3K. PIK3Ca's cytogenic location is 3q26.3 (chromosome 3, (q) arm). PIK3Ca is widely expressed (German et al. 2013).

37.2 PIK3Ca in Normal Physiology

Being a kinase, PI3K adds a phosphate groups to other proteins (phosphorylation). Specifically, PI3K phosphorylates certain signaling molecules responsible for the transmission of additional reactions that occur within particular cells (German et al. 2013). In particular, PI3K plays an important role in cell activities such as cell proliferation (division and growth), migration, protein production, transport and survival. Furthermore, the gene may be implicated in the regulation of different hormones and play a role in adipocyte maturation (Ma et al. 2000). Studies have correlated the involvement of PIK3Ca to certain diseases and even cancers (German et al. 2013).

37.3 PIK3Ca in Disease

A disease of note that is highly interrelated to the gene PIK3Ca is megalencephaly-capillary malformation syndrome (MCAP) (Mirzaa et al. 1993). It is the mutation in the PIK3Ca gene, which causes the most interest with this disease, of which, there have been 15 mutations linked to MCAP (Docker et al. 2014). MCAP is characterized by overgrowth of brain tissue (megalencephaly) and abnormalities caused by enlarged small blood vessels in the skin (malformed capillaries – pink/red spots on the skin, particularly on the face). Other abnormalities associated with MCAP included excess fluid within the brain, brain structure abnormalities, intellectual disabilities, seizures, weak muscle tone, speech delays and difficulty eating (Mirzaa et al. 2013). The majority of mutations of the PIK3Ca gene implicated in MCAP involve a change in an amino acid of the protein p110α (Mirzaa et al. 2013). Again, an altered p110α is produced that makes PI3K unusually active. Just like in the case of cancer, unregulated signals are now produced causing the cells affected, to continuously divide and growth. This uncontrolled growth is a characteristic of MCAP. Despite PI3K's strong link to cancer development, those with MCAP don't appear to have an increased chance of developing cancer (Docker et al. 2014).

PIK3Ca mutations have been strongly linked to CLOVES (congenital lipomatous overgrowth, vascular malformations, epidermal nevi and skeletal/spinal abnormalities) syndrome. CLOVES syndrome has recently been depicted as an overgrowth syndrome with complex vascular anomalies, which very much links to PIK3Ca (responsible often for overgrowth and excessive proliferation) (Mirzaa et al. 1993).

Other such diseases that are a result of mosaic mutations of the PIK3Ca gene include hemimegalencephaly and fibroapidose hyperplasia. Hemimegalencephaly characterized by the overgrowth of one side of the brain and is known to cause seizures and intellectual disability. Fibroapidose Hyperplasia also is characterized by overgrowth but of fibrous and fatty tissues in various regions of the body (Mirzaa et al. 1993). Despite the involvement of PIK3Ca in these diseases, it is unknown whether these disorders have an elevated risk of developing cancer.

37.4 PIK3Ca in Malignancy

PI3KCa mutations are often involved in various types of cancer (notably breast, colon and cervical but also involved in brain, lung, stomach) (German et al. 2013). The PI3KCa mutations correlated to cancer are somatic, heterozygous point mutations – acquired during one's lifetime (not inherited) and present only in tumour cells (Mirzaa et al. 1993). PIK3Ca-associated overgrowth is established in a person with a mutation in allele of PIK3Ca – usually in affected tissues. PIK3Ca mutations are known to appear postzygotic (therefore, mosaic), meaning more than one tissue may need to be tested (Mirzaa et al. 1993). The mutations of the PIK3Ca gene implicated in cancer development involve a change in an amino acid of the protein p110α. Altered p110α is produced that makes PI3K unusually active causing unregulated signals leading to continuous cell division and growth. This uncontrolled proliferation of cells is a characteristic of cancer development (Mirzaa et al. 1993). Furthermore, when expressed in normal cells, these mutations allow anchorage-independent growth, which is a characteristic commonly associated with cancer development. Of note recently, it has been found that PI3KCa is oncogenic and is involved particularly in cervical, breast and colon cancers (Mirzaa et al. 1993). A recent study indicated that that oncogenic PI3KCa is involved in about 30 % of all breast and colon cancers and less frequent in cancers of the stomach, brain, liver and ovaries (Bader et al. 2006). As studies have conclusively correlated the involvement of the gene PIK3Ca and its implication in cancer, efforts are now directed toward the production of molecules that successfully inhibit the activity of PI3K (Bader et al. 2006; German et al. 2013; Karakas et al. 2006). Of note, PTEN, a well-known tumour suppressor that counteracts the action of PI3K by focusing on dephosphorylation of phosphoinositide-3, 4,5-triphosphate (PIP3). Therefore, treatment with drugs that act to inhibit p110α's activity would possibly be efficient in patients whose tumors carry genetic alterations at PTEN (Chow and Baker 2006).

In relation to the gene's angiogenic influences, an *in vitro* study, positively correlated PIK3Ca with the expression of Vascular Endothelial Growth Factor (VEGF) in ovarian cancer cells, at both mRNA and protein levels (Zhang et al. 2003). PIK3Ca amplification in ovarian cells leads to increased PI3k activity, which affects the expression of HIF-1α (an upstream regulator of VEGF). Another study, where staining was involved showed microvascular development was much more pronounced in tumours with high p110α compared to tumours with low p110α levels.

Furthermore, an in vitro study positively correlated the expression of PIK3Ca with the expression of VEGF in ovarian cancer cells, whereas the phosphatidylinositol 3-kinase inhibitor, Ly294002, decreased both inducible and constitutive expression of the hypoxia-inducible factor-1α (HIF1α) at both protein and mRNA levels and revoked VEGF up-regulation by starvation of glucose (Zhang et al. 2003). Moreover, Ly294002 inhibited cell proliferation and induced marked apoptosis in ovarian cancer cells (Zhang et al. 2003). Collectively, the statements provided strongly specifies that PIK3Ca supports ovarian cancer cell growth through pathways affecting proliferation, angiogenesis and apoptosis. Therefore, it can be concluded that PIK3Ca does promote angiogenesis. As PIK3Ca has been linked and successfully determined that it does endorse angiogenesis, perhaps this gene is important in cell proliferation and survival (Zhang et al. 2003).

In conclusion, PIK3Ca has been seen to cause many diseases including various cancers and diseases, specifically those that involve overgrowth of various body tissues. PIK3Ca has also been successfully linked with angiogenesis (Jiang et al. 2000).

References

Bader AG, Kang S, Vogt PK (2006) Cancer-specific mutations in PIK3CA are oncogenic in vivo. Proc Natl Acad Sci U S A 103(5):1475–1479. doi:10.1073/pnas.0510857103

Chow LM, Baker SJ (2006) PTEN function in normal and neoplastic growth. Cancer Lett 241(2):184–196. doi:10.1016/j.canlet.2005.11.042

Docker D, Schubach M, Menzel M, Spaich C, Gabriel HD, Zenker M, Bartholdi D, Biskup S (2014) Germline PTPN11 and somatic PIK3CA variant in a boy with megalencephaly-capillary malformation syndrome (MCAP) – pure coincidence? Eur J Hum Genet. doi:10.1038/ejhg.2014.118

German S, Aslam HM, Saleem S, Raees A, Anum T, Alvi AA, Haseeb A (2013) Carcinogenesis of PIK3CA. Hered Cancer Clin Pract 11(1):5. doi:10.1186/1897-4287-11-5

Jiang BH, Zheng JZ, Aoki M, Vogt PK (2000) Phosphatidylinositol 3-kinase signaling mediates angiogenesis and expression of vascular endothelial growth factor in endothelial cells. Proc Natl Acad Sci U S A 97(4):1749–1753. doi:10.1073/pnas.040560897

Karakas B, Bachman KE, Park BH (2006) Mutation of the PIK3CA oncogene in human cancers. Br J Cancer 94(4):455–459

Ma YY, Wei SJ, Lin YC, Lung JC, Chang TC, Whang-Peng J, Liu JM, Yang DM, Yang WK, Shen CY (2000) PIK3CA as an oncogene in cervical cancer. Oncogene 19(23):2739–2744. doi:10.1038/sj.onc.1203597

Mirzaa G, Conway R, Graham JM, Dobyns WB et al (1993) PIK3CA-related segmental overgrowth. In: Pagon RA, Adam MP, Ardinger HH (eds) GeneReviews(R). University of Washington, Seattle University of Washington, Seattle

Mirzaa GM, RiviÈRe J-B, Dobyns WB (2013) Megalencephaly syndromes and activating mutations in the PI3K-AKT pathway: MPPH and MCAP. Am J Med Genet C Semin Med Genet 163(2):122–130. doi:10.1002/ajmg.c.31361

Zhang L, Yang N, Katsaros D, Huang W, Park JW, Fracchioli S, Vezzani C, Rigault de la Longrais IA, Yao W, Rubin SC, Coukos G (2003) The oncogene phosphatidylinositol 3′-kinase catalytic subunit alpha promotes angiogenesis via vascular endothelial growth factor in ovarian carcinoma. Cancer Res 63(14):4225–4231

Chapter 38
PIK3R2 (p85β) – Phosphatidylinositol 3-Kinase β-Subunit

Abstract PIK3R2 is the gene that codes for the β-subunit (85 kDa) of the lipid kinase enzyme, Phosphatidylinositol 3-kinase, also known as PI3K, in the human body. More commonly referred to as p85β, it forms a regulatory subunit component of the heterodimeric PI3K enzyme complex with a larger p110 α-catalytic subunit. Phosphorylation of phosphatidylinositol directed by the PI3K enzyme gives rise to average levels of PIP3, or 3-polyphosphoinositides, at cell membrane surfaces. In particular, the p85β-subunit interacts with Macrophage-colony stimulating factors, PIK3CD, Cbl genes, Epidermal growth factors, FYN, and HER2/neu. PIK3R2 have been primarily studied for its impact on various cellular signalling pathways, including Insulin-signalling, PI3K-AKT signalling Pathway, EGF/PDGF, and also the mTOR signaling pathways. PI3K/AKT pathway hyperactivation in cancer reduces PIK3R2-mediated apoptosis and stimulates increased VEGF release and consequently initiate cell migration and angiogenesis. Mutation of PIK3R2 and its isoforms in the PIK3 family has been associated with anti-cancer drug resistance development in recent studies. miRNA-126 represses PIK3R2 which is acting as negative-regulators in the pathway of angiogenesis. Therefore, PIK3R2 and the role of miR-126 is potentially useful therapeutic tool that could be applied in addition to conventional anticancer treatment in different malignancies.

Keywords PIK3R2 • p85β • Phosphatidylinositol 3-kinase β-subunit • Angiogenesis • Normal physiology • Disease • Malignancy

38.1 PIK3R2 (p85β) – Phosphatidylinositol 3-Kinase β-Subunit

PIK3R2 is the gene that codes for the β-subunit (85 kDa) of the lipid kinase enzyme, Phosphatidylinositol 3-kinase, also known as PI3K, in the human body. More commonly referred to as p85β, it forms a regulatory subunit component of the heterodimeric PI3K enzyme complex with a larger p110 α-catalytic subunit (Kok et al. 2009; Fruman 2010). Phosphorylation of phosphatidylinositol directed by the PI3K enzyme gives rise to average levels of PIP3, or 3-polyphosphoinositides, at cell membrane surfaces, leading to generation of 2nd messengers which are essential in

A. Salajegheh, *Angiogenesis in Health, Disease and Malignancy*,
DOI 10.1007/978-3-319-28140-7_38

interaction with several cellular growth factors and receptors. This ultimately enhances cell migration and eventual cell death, or apoptosis, via the PI3K/Akt pathway (Fruman 2010; Kok et al. 2009). Akt is a protein kinase B that intricately interacts with PI3K enzymes. This pathway is essential in normal development of blood vessels and intubation during embryogenic development, as embryos that have dead kinase p110A subunits in the PI3K complex demonstrate defects in vasculature (Graupera et al. 2008).

38.2 PIK3R2 (p85β) – Phosphatidylinositol 3-Kinase β-Subunit in Normal Physiology

In particular, the p85β-subunit interacts with Macrophage-colony stimulating factors, PIK3CD, Cbl genes, Epidermal growth factors, FYN, and HER2/neu (Graupera et al. 2008). It is located on the long q-arm of chromosome #19 in between positions 13.2 & 13.4, and gene expression patterns for PIK3R2 have been till date primarily studied in murigenic models, rats, and humans (Volinia et al. 1992) for its impact on various cellular signalling pathways, including Insulin-signalling, PI3K-AKT signalling Pathway, EGF/PDGF, and also the mTOR (Mammalian target of rapamycin) signaling pathway. However, in the majority of tissues, p85α-subunit, an isoform of p85β was found expressed at a higher level, than the latter (Volinia et al. 1992).

38.3 PIK3R2 (p85β) – Phosphatidylinositol 3-Kinase β-Subunit in Disease

The p85β-subunit has also been studied previously for its potential to attenuate the influenza A virus and form the basis for a useful new strategy against viruses (Shin et al. 2007). PCR profiling revealed that the human p85β-subunit has Src-homology 3, SH3, and an iSH2 coiled-coil kind domain structure with conformational interhelical region plasticity, both of which are able to bind with the influenza A virus, through the nonstructural protein 1, NS1, receptor that is present on their surfaces. Heterologous over-expression of the SH3-domain showed successfully reduced Influenza A replication and reduced phosphorylation of Akt in infected cells, via down-regulation of the PI3K/Akt pathway (Shin et al. 2007).

Use of baculovirus vectors in insects allowed expression and isolation of the two PI3K subunits. In addition, mutated expression of the p85 protein in embryonic fibroblasts of chickens was found to first induce oncogenic proliferation and transformations. P85-mediated tumor cell proliferation has also been demonstrated to be inhibited by in vivo delivery of P-1257, phosphopeptide 1257 (Graupera et al. 2008).

38.4 PIK3R2 (p85β) – Phosphatidylinositol 3-Kinase β-Subunit in Malignancy

In colon and breast carcinomas, genetically altered and decreased p85β ubiquitous expression has been directly correlated with decreased activation of the PI3K pathway, aiding in cell growth, proliferation and tumor progression by inhibiting apoptosis (in vivo). Hence, suppressed expression of the PIK3R2 gene and p85 β-subunit via mutation represents key molecular strategies of cancers for progression (Zhou et al. 2014). Besides PIK3R2, cancer pathogenesis is linked to activation via growth factor receptors of the majority of Class I – PI3K p85-type subunits that are regulatory in function, including PIK3R1, PIK3R2, and even PIK3R3. These genes code for p85α, p55α, p50α, p85B, and p55γ respectively (Zhou et al. 2014).

The p85 β -subunit and its isoforms have been indicated for function in the pathogenesis & progression of cancers of the human breast too (Zhou et al. 2014). In a recent study, immune-histochemical stains to p85β of more than a 100 breast cancer samples revealed high expression of the mutated p85 protein (non-apoptosis-inducing) to be significantly correlated with advanced tumor grade, metastases, vascular invasion, and even recurrence, leading to shorter survival times. However, p85β significance as one of the prognostic markers in breast cancer protein expression is yet to be fully understood. In the PI3K signaling pathway, other components have also been shown to effect in tumor progression, including those of the thyroid, lung, breast, ovarian, and glioblastoma (Zhou et al. 2014).

Furthermore, megaloencephaly, molymicrogyria, polydactyly, hydrocephalus syndrome, or MPPH, is a syndrome characterized by the above four conditions occurring in unison in the affected patient. It has been found to be linked to higher PIK3R2 expression (Riviere et al. 2012). MPPH is a disorder concerned with sporadic overgrowth and significant expansion in brain size, together with recurrent germline mutations of PIK3R2. To be noted is the remarkable phenotypical similarity between this condition and the capillary-malformation syndrome that is associated with megalo-encephaly, demonstrating the pro-angiogenic effects of PIK3R2 (Riviere et al. 2012).

Hence, several recent studies have distinctly identified the mutated PIK3R2 gene's overexpression of the p85B regulatory subunit as being closely linked to multiple kinds of tumor progression, determining it also to be a potential therapeutic target for restriction of tumoregenic developments in the human body (Kok et al. 2009; Fruman 2010).

The primary apoptotic (cell-death inducing) function of PIK3R2 encoded p85B is via the PKI3/Akt pathway. PI3K/AKT/MTOR is a pathway of intrinsic intracellular signalling that is central to apoptosis induction. In this pathway, essentially PI3K is responsible for activating AKT (protein B kinase specific to serine/threonine), which in turn activates mTOR (mammalian target of rapamycin, a protein kinase family member, and a serine/threonine protein kinase) (Riviere et al. 2012). In this pathway, a number of other signalling and control factors are involved, particularly PDK1 (Phosphoinositide dependent kinase 1) and PTEN (Bhatt and

Damania 2013). Activation of this pathway can lead to hypertrophy in the muscles and cancers when apoptosis is inhibited (via PIK3R2 targeting), hence inducing cell proliferation (Riviere et al. 2012).

The PIP3 produced at plasma membranes activates other seine/threonine type kinases, such as *PDKI* and also AKT. Simultaneously, PI3K's apoptotic action is endogenously opposed by a phosphatase which is encoded for by the PTEN gene. This reduces activated AKT levels (Romano 2013). AKT is responsible for control of protein synthesis and also cell growth, leading to eventual mTOR phosphorylation. The PI3K/Akt also partly mediates (via downregulation) the effect of VEGF-induced (vascular-endothelial growth-factor) angiogenesis on endothelial cells by binding with receptor tyrosine-kinases (Romano 2013).

In spite of its isoform p85α being characterized as a colon and ovarian tumor causing oncogene more than a decade ago (Romano 2013), potential for cancer due to mutations in p85β are only recently coming into focus (Cheung et al. 2011). In a study on endometrial cancer cases, PI3K pathway mutations leading to hyper-activation of the PI3K pathway occurred in more than 80 % of cases, with PIK3R1 (coding for p85α) occurring at high-frequency, and PIK3R2 (p85B) also frequently being mutated. This comes in spite of PIK3R2 not being officially recognized as a cancer-causing gene till this point (Cheung et al. 2011). Furthermore, the PI3K/Akt pathway has also been proven to play a central role in promoting angiogenesis and cancer progression when hyper-activated (Romano 2013). Activation of the PI3K pathway will typically occur via the PTEN molecule (phosphatase-and-tensin homolog) leading to protein loss, increased expression of growth factor-receptors (such as epidermal growth-factor EGF receptor), or via mutations in the RAS pathway. Loss of PTEN may be caused by promoter methylation, genetic mutation, and even protein breakdown. This has been found in at least 20 % of hyperplasias in endometrial regions. PTEN signaling inactivation is able to negatively regulate PI3K pathways and provide sufficient drive to murigenic models in tumorigenesis (Cheung et al. 2011).

PI3K/AKT pathway hyperactivation in cells of cancer reduced PIK3R2-mediated apoptosis and also stimulated increased VEGF release, which is quite important for cell migration and angiogenesis. Continued activation of endothelial AKT1 (13) has also been shown to stimulate structural angiopoeitins (Cortes et al. 2012) and release of nitric oxide, both of which are stimulatory angiogenic factor with up-modulated expression in the hyperactive PI3K/AKt pathway (Romano 2013).

Associated PI3K pathway protein-coding genes found mutated in endometrial cancers in high frequencies were AKT1, fibroblast growth-factor receptor 2, and also PIK3CA. The PI3K pathway has also been found to interact in a bidirectional manner with the MAPK (mitogen-activated protein-kinase) and the Ras pathway, both of which have been profiled for their central roles in carcinogenesis when dysfunctional (Cortes et al. 2012). This interaction suggests the PI3K pathway is helped or inhibited by the RAS/MAPK system in determining final functional outcomes in terms of angiogenesis. Hence, optimal cancer therapy targeting the PI3K pathway may require special classification of those patients likely to be benefited from combinatorial & rational therapy, which lies in targeting both the RAS and the PI3K/Akt signaling pathways simultaneously (Cortes et al. 2012).

MiRNA-126 is a variant microRNA highly expressed in vascular endothelial cells that has been shown to have a pro-angiogenic effect in both cultured (in vitro) endothelial cells and in vivo angiogenesis (ischemia-induced) (Zhu et al. 2011). Endothelial cells are key apparatus in the vascular intima, where growth modulation plays a pivotal role in angiogenesis, tumor growth, and ultimate vasculogenesis. MicroRNAs, 21–23 nucleotides in length, are non-protein coding RNAs with high levels of conservation, and they are responsible for targeting the 3′-UTR of mRNAs in order to regulate gene expression at a post-transcriptional level via inhibition or stimulation of mRNA degradation (Cortes et al. 2012). miRNA-126 also promotes angiogenesis by aiding the transition of endothelial cells in the vascular intima into mesenchymal cells (Zhu et al. 2011).

MiRNA-126 represses PIK3R2 as well as SPRED1 (sprout-related protein 1), both of which are acting as negative-regulators in the pathway of angiogenesis. This endothelial-specific (intron-derived) microRNA is coded for by the EGF-like domain 7, EGFL-7 (Cortes et al. 2012). It has also been demonstrated via Luciferase reporter assays, Western blots, and gene-knockout experiments conducted in bone marrows (mice) to directly target endothelial responses with VEGFA, hence promoting angiogenesis (Zhu et al. 2011). The above study hence establishes that miRNA-126 plays a central role in tumourogenesis through regulation of various components in the VEGF/PI3K/AKT signaling pathways, especially the PIK3 enzyme that is partially coded for by the gene of interest, PIK3R2 (Zhu et al. 2011).

Low expression of miRNA-126 was associated with poor prognosis for cases of colorectal cancer, involving distant metastases and poor survival periods (Liu et al. 2014). miRNA-126 functions as a tumor-suppressor by targeting the CXCR4 protein, central in colorectal cancer. Also, experiments that restored MiR-126 expression to normal levels in lung cancer cells were found to down-regulate VEGF, inhibiting the growth of cancer cell lines both *in vivo* and *in vitro*, demonstrating that the restoration of miR-126 expression is a potential therapeutic target in anti-cancer drugs positively regulating the p85β-subunit activity and, hence, inhibiting tumor progression via increased apoptosis (Liu et al. 2014).

Mutation of PIK3R2 and its isoforms in the PIK3 family has been associated with anti-cancer drug resistance development in recent studies (Courtney et al. 2010). The effect is suspected to be mediated via drug-induced reprogramming of the tyrosine kinase receptor as an after effect of targeted combination therapy or mono-targeted anti-cancer therapy. For example, resistance to Tamoxifen, a commonly used breast cancer therapy drug, was demonstrated in TamR cells along with high levels of expression of MAPK1 (tyrosine-phosphorylated), PIK3R2 and also SHC1, demonstrating involvement of the PI3K/AKT signaling and Ras/MAPK pathways in Tamoxifen-resistant phenotypes (Browne et al. 2013). Integrative analysis (genomic) of the PI3K family of lipid kinases has also established PIK3R1, a Class I sister of the PIK3R2 gene, as a potential target in therapy of epithelial ovarian cancers, with PIK3R3 being significantly up-regulated in liver, breast, ovarian, and even prostate cancers (Browne et al. 2013).

Several components of the PI3K/Akt signaling pathway are currently the target of anti-angiogenic therapeutic drug development initiative, including PIK3R2. For example, a PI3K/mTOR inhibitor currently being studied is XL765 (SAR245409). Dose-dependently, this drug was successful in demonstrating anti-angiogenic properties, along with pro-apoptotic and anti-proliferative effects in nude mice models (Yu et al. 2014). Agents of this nature typically aim to decrease VEGF levels in secretion and repress angiogenesis. Upregulation of PIK3R2's non-mutated form can also trigger increased apoptosis and prevent tumourogenesis. Hence, the PI3K pathway plays a central role in both normal tissue and targeted therapeutics to the angiogenic pathways. In particular, drug therapy compounds for breast cancer and endometrial endometreoid cancer (EEC) that may target the down-regulation of PIK3R2 gene's expression via up-regulation of miRNA-126 are of interest (Graupera et al. 2008). EEC is one of the most prevalent kinds of malignancies of the gynaecological system and also the 4th most common type of cancer found occurring amongst females in Western countries. It has limited treatment choices in metastatic EEC with poor outcomes at present, presenting an urgent need for development of an effective and novel therapy compound, possibly targeting non-mutated PIK3R2 upregulation in the PI3K/Akt and PTEN loss pathway as a primary option (Yu et al. 2014).

References

Bhatt AP, Damania B (2013) AKTivation of the PI3K/AKT/mTOR signaling pathway by KSHV. Front Immunol 3. doi:10.3389/fimmu.2012.00401

Browne BC, Hochgrafe F, Wu J, Millar EK, Barraclough J, Stone A, McCloy RA, Lee CS, Roberts C, Ali NA, Boulghourjian A, Schmich F, Linding R, Farrow L, Gee JM, Nicholson RI, O'Toole SA, Sutherland RL, Musgrove EA, Butt AJ, Daly RJ (2013) Global characterization of signalling networks associated with tamoxifen resistance in breast cancer. FEBS J 280(21):5237–5257. doi:10.1111/febs.12441

Cheung LW, Hennessy BT, Li J, Yu S, Myers AP, Djordjevic B, Lu Y, Stemke-Hale K, Dyer MD, Zhang F, Ju Z, Cantley LC, Scherer SE, Liang H, Lu KH, Broaddus RR, Mills GB (2011) High frequency of PIK3R1 and PIK3R2 mutations in endometrial cancer elucidates a novel mechanism for regulation of PTEN protein stability. Cancer Discov 1(2):170–185. doi:10.1158/2159-8290.cd-11-0039

Cortes I, Sanchez-Ruiz J, Zuluaga S, Calvanese V, Marques M, Hernandez C, Rivera T, Kremer L, Gonzalez-Garcia A, Carrera AC (2012) p85beta phosphoinositide 3-kinase subunit regulates tumor progression. Proc Natl Acad Sci U S A 109(28):11318–11323. doi:10.1073/pnas.1118138109

Courtney KD, Corcoran RB, Engelman JA (2010) The PI3K pathway as drug target in human cancer. J Clin Oncol 28(6):1075–1083. doi:10.1200/jco.2009.25.3641

Fruman DA (2010) Regulatory subunits of class IA PI3K. Curr Top Microbiol Immunol 346:225–244. doi:10.1007/82_2010_39

Graupera M, Guillermet-Guibert J, Foukas LC, Phng LK, Cain RJ, Salpekar A, Pearce W, Meek S, Millan J, Cutillas PR, Smith AJ, Ridley AJ, Ruhrberg C, Gerhardt H, Vanhaesebroeck B (2008) Angiogenesis selectively requires the p110alpha isoform of PI3K to control endothelial cell migration. Nature 453(7195):662–666. doi:10.1038/nature06892

Kok K, Geering B, Vanhaesebroeck B (2009) Regulation of phosphoinositide 3-kinase expression in health and disease. Trends Biochem Sci 34(3):115–127. doi:10.1016/j.tibs.2009.01.003

Liu Y, Zhou Y, Feng X, Yang P, Yang J, An P, Wang H, Ye S, Yu C, He Y, Luo H (2014) Low expression of microRNA-126 is associated with poor prognosis in colorectal cancer. Genes Chromosomes Cancer 53(4):358–365. doi:10.1002/gcc.22146

Riviere JB, Mirzaa GM, O'Roak BJ, Beddaoui M, Alcantara D, Conway RL, St-Onge J, Schwartzentruber JA, Gripp KW, Nikkel SM, Worthylake T, Sullivan CT, Ward TR, Butler HE, Kramer NA, Albrecht B, Armour CM, Armstrong L, Caluseriu O, Cytrynbaum C, Drolet BA, Innes AM, Lauzon JL, Lin AE, Mancini GM, Meschino WS, Reggin JD, Saggar AK, Lerman-Sagie T, Uyanik G, Weksberg R, Zirn B, Beaulieu CL, Majewski J, Bulman DE, O'Driscoll M, Shendure J, Graham JM Jr, Boycott KM, Dobyns WB (2012) De novo germline and postzygotic mutations in AKT3, PIK3R2 and PIK3CA cause a spectrum of related megalencephaly syndromes. Nat Genet 44(8):934–940. doi:10.1038/ng.2331

Romano G (2013) The role of the dysfunctional akt-related pathway in cancer: establishment and maintenance of a malignant cell phenotype, resistance to therapy, and future strategies for drug development. Scientifica 2013:317186. doi:10.1155/2013/317186

Shin YK, Liu Q, Tikoo SK, Babiuk LA, Zhou Y (2007) Influenza A virus NS1 protein activates the phosphatidylinositol 3-kinase (PI3K)/Akt pathway by direct interaction with the p85 subunit of PI3K. J Gen Virol 88(Pt 1):13–18. doi:10.1099/vir.0.82419-0

Volinia S, Patracchini P, Otsu M, Hiles I, Gout I, Calzolari E, Bernardi F, Rooke L, Waterfield MD (1992) Chromosomal localization of human p85 alpha, a subunit of phosphatidylinositol 3-kinase, and its homologue p85 beta. Oncogene 7(4):789–793

Yu P, Laird AD, Du X, Wu J, Won KA, Yamaguchi K, Hsu PP, Qian F, Jaeger CT, Zhang W, Buhr CA, Shen P, Abulafia W, Chen J, Young J, Plonowski A, Yakes FM, Chu F, Lee M, Bentzien F, Lam ST, Dale S, Matthews DJ, Lamb P, Foster P (2014) Characterization of the activity of the PI3K/mTOR inhibitor XL765 (SAR245409) in tumor models with diverse genetic alterations affecting the PI3K pathway. Mol Cancer Ther 13(5):1078–1091. doi:10.1158/1535-7163.mct-13-0709

Zhou W, An G, Wei P, Chen W (2014) Significance of p85 expression as a prognostic factor for patients with breast cancer. Oncol Lett 8(4):1657–1661. doi:10.3892/ol.2014.2359

Zhu N, Zhang D, Xie H, Zhou Z, Chen H, Hu T, Bai Y, Shen Y, Yuan W, Jing Q, Qin Y (2011) Endothelial-specific intron-derived miR-126 is down-regulated in human breast cancer and targets both VEGFA and PIK3R2. Mol Cell Biochem 351(1–2):157–164. doi:10.1007/s11010-011-0723-7

Chapter 39
Platelet-Activating Factor

Abstract Platelet-activating factor (PAF) is a phospholipid produced by a variety of cells including platelets, endothelial cells, neutrophils, monocytes and macrophages. It has a wide range of potent biological activities including wound healing, physiological inflammation, apoptosis, angiogenesis, reproduction and long-term potentiation. This factor through modulation of intracellular pathways, can mediate various disease and malignancy processes. A wide range of pathological processes are affected by PAF including multiple sclerosis, CNS ischemia, meningitis, encephalitis, epilepsy and panic disorders, cardiac anaphylaxis, reperfusion injury and myocardial infarction, bronchial asthma, allergic rhinitis and pulmonary oedema, ulcers, acute pancreatitis and inflammatory bowel disease and many others. PAF can affect tumour development in several ways. Activated endothelium or cancer cells themselves may be the source of PAF production in the tumour microenvironment as well as expressing PAF receptor on their membranes. Several PAF-specific antagonists have been shown to be able to inhibit the malignant processes of cancer cells and further research in this field could develop novel cancer therapeutics.

Keywords Platelet-activating factor • PAF • Angiogenesis • Normal physiology • Disease • Malignancy

39.1 Platelet-Activating Factor

Platelet-activating factor (PAF) is the trivial name for a phospholipid, 1-O-alkyl-2-acetyl-sn-glycero-3-phosphocholine. PAF is produced by a variety of cells including platelets, endothelial cells, neutrophils, monocytes and macrophages. It has a wide range of potent biological activities including wound healing, physiological inflammation, apoptosis, angiogenesis, reproduction and long-term potentiation.

Dr. Moein Amin contributed equally with Dr. Ali Salajegheh.

A. Salajegheh, *Angiogenesis in Health, Disease and Malignancy*,
DOI 10.1007/978-3-319-28140-7_39

39.2 Platelet-Activating Factor in Normal Physiology

The platelet-activating factor (PAF) can be synthesized through two different pathways. The remodelling pathway (the main pathway) involves substitution of an acetyl residue for the long-chain fatty acyl residue of cellular phospholipids. The de novo pathway parallels phospholipid synthesis, in which a phosphocholine function is transferred to alkyl acetyl glycerol (Prescott et al. 2000). PAF is synthesised on demand and in response to specific stimuli. The enzymatic production of PAF seems to be regulated by changes in levels of intracellular Ca2+ (Baker and Chang 1997).

The primary role of PAF in normal physiology seems to be to mediate intercellular interactions. It binds to receptors on the plasma membranes of other cells, activating them and causing a change in their phenotypes (Prescott et al. 2000). Various mechanisms regulate the PAF intercellular-signalling system, including tightly controlled synthesis, biologic availability of PAF, expression of the PAFR on specific cells, desensitization of the receptor and rapid degradation of PAF by extracellular and intracellular acetyl-hydrolases (Arai et al. 2002).

PAF and PAF-like lipids (PAF-LL) have the ability to induce PAF signalling cascade by binding to a 7 transmembrane G-protein coupled receptor, the PAF receptor (PAFR) on the target cells of many physiological systems including inflammatory, immune and haemostatic systems (Yost et al. 2010). Through a Gq-linked mechanism, activation of PAFR initiates hydrolysis of PIP2 to produce IP3 and DAG, leading to transient elevation of cytosolic Ca2+ released from intracellular stores and activation of PKC (Honda et al. 2002). There is also evidence that part of the response to PAF is sensitive to pertussis toxin, suggesting that PAFR may also be Gi-linked. Signalling through Gi inhibits the conversion of ATP to cAMP by adenylate cyclase, preventing the activation of PKA and downstream signalling events (Lad et al. 1985).

The PAFR is constitutively present on platelets, leukocytes and endothelial cells and further expression may be induced by inflammatory cytokines. Ligand binding promotes the activation and aggregation of platelets and leukocytes, promotes leukocyte adherence, motility, chemotaxis, invasion, and ROS generation (Brock 2012). Many of these effects involve activation of p38 and ERK1/2 MAP kinases. PAF induces the expression of numerous genes, including COX-2, iNOS, IL-6, MT1-MMP, MMP-9, and TIMP-2 (Brock 2012). Downstream effects of PAF include activation of endothelial cells, promotion of endothelial cell migration, and stimulation of wound healing and angiogenesis (Stafforini et al. 2003).

Role of PAF in angiogenesis has been shown in various studies. PAF directly stimulates the in vitro migration of endothelial cells, enhances vascular permeability, and promotes in vivo angiogenesis (Camussi et al. 1995; Montrucchio et al. 1994). PAF also contributes to angiogenesis by promoting the synthesis of angiogenic cytokines such as VEGF, tumour necrosis factor-alpha, and hepatocyte growth factor (Montrucchio et al. 1994; Camussi et al. 1997; He et al. 2009).

39.3 Platelet-Activating Factor in Disease

Inflammation is a necessary bodily process needed for various physiological responses. However excessive inflammation can often result in injury to tissues. Since PAF is one of the most potent inflammatory mediators (see earlier discussion), it is not surprising that it has been implicated in many pathologies.

A wide range of pathological processes are affected by PAF. Some of the pathologies in various bodily systems include nervous system (e.g. multiple sclerosis, CNS ischemia, meningitis, encephalitis, epilepsy and panic disorders), cardiovascular system (e.g. cardiac anaphylaxis, reperfusion injury and myocardial infarction), respiratory system (e.g. bronchial asthma, allergic rhinitis and pulmonary edema), gastrointestinal system (e.g. ulcers, acute pancreatitis and inflammatory bowel disease), urogenital system (e.g. renal ischemia) and many others (e.g. rheumatoid arthritis, sepsis, thrombosis and osteoporosis) (Nitoda et al. 2012). The role of PAF in regards to some of these pathologies will be discussed below.

In bronchial asthma PAF contributes to airway hypersensitivity (Ishii and Shimizu 2000). PAF is capable of stimulating eosinophil transmigration across endothelial barriers (Casale et al. 1993) and can activate eosinophils and neutrophils to release granule constituents and to generate superoxide anions (Zoratti et al. 1991). Furthermore it induces production of LTC4 (Bruynzeel et al. 1986) leading to bronchial spasm and increases the expression of matrix metalloproteinase-9 which may be actively involved in PAF-induced remodelling in human bronchial epithelial cells (Shan et al. 1999).

PAF has been associated with acute pancreatitis. It promotes leukocyte adhesion, chemotaxis and degranulation which prolongs the inflammatory response. Furthermore PAF upregulates adhesion molecules and leukocyte-endothelial cell interactions which can eventually lead to endothelial barrier dysfunction and bacterial translocation. Finally PAF has been associated with acinar cell necrosis and apoptosis by prolonging the inflammatory response and delaying neutrophil apoptosis (Liu and Xia 2006).

Shock syndrome is one of the cardiovascular pathologies studied involving PAF. PAF has been shown to be involved in the hemoconcentration and the increased systemic vascular permeability seen in immune complex-induced shock. Furthermore PAF has been suggested as the most proximal mediator in the cytokine cascade seen in septic shock. Downstream effects may include promotion of factors such as TNF-alpha which has been shown to have negative ionotropic effect. PAF has also been shown to be involved in mediating the effects of endotoxins leading to injury in various tissues such as cardiac, pulmonary and renal tissues (Montrucchio et al. 2000).

The role of PAF in modulating angiogenesis can be seen in a range of pathological conditions. For example PAF has been associated with corneal neovascularization and its downstream destructive effects (Ma et al. 2004). PAF has been shown to stimulate the expression of VEGF mRNA and protein as well as increased expression of its receptors flt-1 and KDR in endothelial cells of vascularized corneas

(Philipp et al. 2000). Furthermore PAF has been shown to be involved in the pathological neovascularization, exudate accumulation and subretinal fibrosis leading to age related macular degeneration (Zhang et al. 2013).

Therapeutic effects of PAF antagonists further support the role of this mediator in pathophysiology of disease. PAF antagonists have shown beneficial effects on the manifestations of acute pancreatitis (Liu and Xia 2006), promote corneal wound healing and preventing corneal ulcerations (Nitoda et al. 2012), delaying effects of age related macular disease (Nitoda et al. 2012; Zhang et al. 2013) and reduce the autoimmune damage in conditions such as multiple sclerosis, rheumatoid arthritis and inflammatory bowel disease (Edwards and Constantinescu 2009).

39.4 Platelet-Activating Factor in Malignancy

In regard to malignancy it has been shown that PAF can affect tumour development in several ways. Several studies have indicated that activated endothelium and/or cancer cells themselves may be the source of PAF production in the tumour microenvironment as well as expressing PAF receptor on their membranes (Montrucchio et al. 1998; Ishii et al. 1997; Camussi et al. 1995; Sobhani et al. 1996). Production of PAF in these cells can be induced by various cytokines such as VEGF, FGF and TNF-α (Bussolati et al. 2000; Fallani et al. 2002). In human leukocytes interaction of CD40 with soluble CD154 or with CD154 expressed on the membrane of leukocytes (CD154-transfected J558 cells) or of activated platelets, may stimulate the synthesis of PAF. This mechanism of PAF production may be involved in some cancers which have elevated levels of CD154 (Bussolati et al. 2002).

Recent studies have shown that PAF expression correlates with malignancy of various tumours. In breast cancer cells, cells that express higher ability to synthesise and release PAF where found to be more malignant with higher metastatic ability than cells that did not. These cells were also found to be expressing more PAF receptor on their membranes (Bussolati et al. 2000). In colon cancer, analysing blood samples from patients revealed elevated levels of both PAF-AH and PLA2 (precursors of PAF) compared to healthy individuals (Denizot et al. 2004). Also a significant increase in PAF levels were observed in colon cancer metastases samples as compared to non-metastatic cells (Denizot et al. 2005; Denizot et al. 2003). In hepatocellular carcinoma, elevated levels of PAF and PAFR transcripts 1 and 2 were found in carcinoma specimens as compared with non-tumor tissues (Mathonnet et al. 2006). In oesophageal squamous cell carcinoma (ESCC), tumor cell lines expressed more PAFR, compared with controls. Furthermore PAFR levels were positively correlated with ESCC clinical stages with PAFR promoting the malignant development of ESCC in vitro and in vivo (Chen et al. 2015).

PAF stimulates the production of inflammatory cytokines TNF-α, VEGF, FGF and IL-1 which possess metastatic-enhancing properties through activation of NF-kB pathway (Heon Seo et al. 2006; Ko et al. 2002). Furthermore it upregulates the production of adhesion factors (e.g. VCAM-1 and E-Selectin) and thereby increasing the ability of cancer cells to attach to endothelium and invade (Im et al. 1996). Therefore it can be concluded that presence of PAF in tumor microenvironment promotes processes such as angiogenesis and metastasis through enhancing the role of cytokines. The ability of cancer cells to produce PAF and to express PAF-receptor on their membrane provides them with the autonomy to trigger such processes associated with higher malignancy.

It is important to note that in some cancers, PAF has been shown to have beneficial effects. In these studies the presence of PAF-receptor enhance cancer cell apoptosis by the activation of NF-kB (Li et al. 2003; Darst et al. 2004). This is due to the dual action of NF-kB pathway implicated in both malignancy and apoptosis through immune response (Shishodia and Aggarwal 2004).

In terms of therapeutics, several PAF-specific antagonists have been shown to be able to inhibit the above malignant processes of cancer cells *in vitro* and *in vivo* (Bussolati et al. 2000; Im et al. 1996; Mannori et al. 1997, 2000; Brizzi et al. 1999). Further research is needed in this field to develop novel cancer therapeutics.

Platelet-activating factor (PAF), a bioactive phospholipid synthesized and secreted by a variety of cells. It mediates many physiological responses such as microcirculatory disturbance and inflammation through intracellular pathways acting on PAF receptor which is a G protein coupled receptor. PAF is catabolized by PAF-acetylhydrolase and disturbances in its synthesis or degradation can lead to pathology. PAF has been shown to be involved in many disease states involving cardiovascular, respiratory, gastrointestinal, inflammatory and central nervous systems through its central role in regulating inflammation and angiogenesis. Angiogenic role of PAF has also been widely implicated in various cancers modulating the ability of the cancer cells to invade and metastasise. It is therefore evident that PAF antagonists may be used as a possible therapeutic agent in an attempt to suppress the aforementioned adverse effects.

References

Arai H, Koizumi H, Aoki J, Inoue K (2002) Platelet-activating factor acetylhydrolase (PAF-AH). J Biochem 131(5):635–640

Baker RR, Chang HY (1997) Neuronal nuclear acetyltransferases involved in the synthesis of platelet-activating factor are located in the nuclear envelope and show differential losses in activity. Biochim Biophys Acta 1345(2):197–206

Brizzi MF, Battaglia E, Montrucchio G, Dentelli P, Del Sorbo L, Garbarino G, Pegoraro L, Camussi G (1999) Thrombopoietin stimulates endothelial cell motility and neoangiogenesis by a platelet-activating factor-dependent mechanism. Circ Res 84(7):785–796

Brock T (2012) Platelet-activating factor: activator of inflammation, angiogenesis, and metastasis. Available via Cayman Chemical. https://www.caymanchem.com/article/2135: Platelet-Activating Factor: Activator of Inflammation, Angiogenesis, and Metastasis. Accessed 15 May 2015

Bruynzeel PL, Koenderman L, Kok PT, Hameling ML, Verhagen J (1986) Platelet-activating factor (PAF-acether) induced leukotriene C4 formation and luminol dependent chemiluminescence by human eosinophils. Pharmacol Res Commun 18(Suppl):61–69

Bussolati B, Biancone L, Cassoni P, Russo S, Rola-Pleszczynski M, Montrucchio G, Camussi G (2000) PAF produced by human breast cancer cells promotes migration and proliferation of tumor cells and neo-angiogenesis. Am J Pathol 157(5):1713–1725. doi:10.1016/s0002-9440(10)64808-0

Bussolati B, Russo S, Deambrosis I, Cantaluppi V, Volpe A, Ferrando U, Camussi G (2002) Expression of CD154 on renal cell carcinomas and effect on cell proliferation, motility and platelet-activating factor synthesis. Int J Cancer/Journal international du cancer 100(6):654–661. doi:10.1002/ijc.10545

Camussi G, Montrucchio G, Lupia E, De Martino A, Perona L, Arese M, Vercellone A, Toniolo A, Bussolino F (1995) Platelet-activating factor directly stimulates in vitro migration of endothelial cells and promotes in vivo angiogenesis by a heparin-dependent mechanism. J Immunol (Baltimore, MD: 1950) 154(12):6492–6501

Camussi G, Montrucchio G, Lupia E, Soldi R, Comoglio PM, Bussolino F (1997) Angiogenesis induced in vivo by hepatocyte growth factor is mediated by platelet-activating factor synthesis from macrophages. J Immunol (Baltimore, MD: 1950) 158(3):1302–1309

Casale TB, Erger RA, Little MM (1993) Platelet-activating factor-induced human eosinophil transendothelial migration: evidence for a dynamic role of the endothelium. Am J Respir Cell Mol Biol 8(1):77–82. doi:10.1165/ajrcmb/8.1.77

Chen J, Lan T, Zhang W, Dong L, Kang N, Zhang S, Fu M, Liu B, Liu K, Zhang C, Hou J, Zhan Q (2015) Platelet-activating factor receptor-mediated PI3K/AKT activation contributes to the malignant development of esophageal squamous cell carcinoma. Oncogene 34(40):5114–5127. doi:10.1038/onc.2014.434

Darst M, Al-Hassani M, Li T, Yi Q, Travers JM, Lewis DA, Travers JB (2004) Augmentation of chemotherapy-induced cytokine production by expression of the platelet-activating factor receptor in a human epithelial carcinoma cell line. J Immunol (Baltimore, MD: 1950) 172(10):6330–6335

Denizot Y, Gainant A, Guglielmi L, Bouvier S, Cubertafond P, Mathonnet M (2003) Tissue concentrations of platelet-activating factor in colorectal carcinoma: inverse relationships with Dukes' stage of patients. Oncogene 22(46):7222–7224. doi:10.1038/sj.onc.1207032

Denizot Y, Truffinet V, Bouvier S, Gainant A, Cubertafond P, Mathonnet M (2004) Elevated plasma phospholipase A2 and platelet-activating factor acetylhydrolase activity in colorectal cancer. Mediators Inflamm 13(1):53–54. doi:10.1080/09629350410001664824

Denizot Y, Descottes B, Truffinet V, Valleix D, Labrousse F, Mathonnet M (2005) Platelet-activating factor and liver metastasis of colorectal cancer. Int J Cancer/Journal international du cancer 113(3):503–505. doi:10.1002/ijc.20585

Edwards LJ, Constantinescu CS (2009) Platelet activating factor/platelet activating factor receptor pathway as a potential therapeutic target in autoimmune diseases. Inflamm Allergy Drug Targets 8(3):182–190

Fallani A, Grieco B, Barletta E, Mugnai G, Giorgi G, Salvini L, Ruggieri S (2002) Synthesis of platelet-activating factor (PAF) in transformed cell lines of a different origin. Prostaglandins Other Lipid Mediat 70(1–2):209–226

He YG, Wang H, Zhao B, Hsu M, Liao D, Owens J, Johnston J (2009) Functional analysis of platelet-activating factor in the retinal pigment epithelial cells and choroidal endothelial cells. Curr Eye Res 34(11):957–965. doi:10.3109/02713680903231135

Heon Seo K, Ko HM, Kim HA, Choi JH, Jun Park S, Kim KJ, Lee HK, Im SY (2006) Platelet-activating factor induces up-regulation of antiapoptotic factors in a melanoma cell line through nuclear factor-kappaB activation. Cancer Res 66(9):4681–4686. doi:10.1158/0008-5472.can-05-3186

Honda Z-I, Ishii S, Shimizu T (2002) Platelet-activating factor receptor. J Biochem 131(6):773–779

Im SY, Ko HM, Kim JW, Lee HK, Ha TY, Lee HB, Oh SJ, Bai S, Chung KC, Lee YB, Kang HS, Chun SB (1996) Augmentation of tumor metastasis by platelet-activating factor. Cancer Res 56(11):2662–2665

Ishii S, Shimizu T (2000) Platelet-activating factor (PAF) receptor and genetically engineered PAF receptor mutant mice. Prog Lipid Res 39(1):41–82

Ishii S, Nagase T, Tashiro F, Ikuta K, Sato S, Waga I, Kume K, Miyazaki J, Shimizu T (1997) Bronchial hyperreactivity, increased endotoxin lethality and melanocytic tumorigenesis in transgenic mice overexpressing platelet-activating factor receptor. EMBO J 16(1):133–142. doi:10.1093/emboj/16.1.133

Ko HM, Seo KH, Han SJ, Ahn KY, Choi IH, Koh GY, Lee HK, Ra MS, Im SY (2002) Nuclear factor kappaB dependency of platelet-activating factor-induced angiogenesis. Cancer Res 62(6):1809–1814

Lad PM, Olson CV, Grewal IS (1985) Platelet-activating factor mediated effects on human neutrophil function are inhibited by pertussis toxin. Biochem Biophys Res Commun 129(3):632–638

Li T, Southall MD, Yi Q, Pei Y, Lewis D, Al-Hassani M, Spandau D, Travers JB (2003) The epidermal platelet-activating factor receptor augments chemotherapy-induced apoptosis in human carcinoma cell lines. J Biol Chem 278(19):16614–16621. doi:10.1074/jbc.M211287200

Liu LR, Xia SH (2006) Role of platelet-activating factor in the pathogenesis of acute pancreatitis. World J Gastroenterol 12(4):539–545

Ma X, Ottino P, Bazan HE, Bazan NG (2004) Platelet-activating factor (PAF) induces corneal neovascularization and upregulates VEGF expression in endothelial cells. Invest Ophthalmol Vis Sci 45(9):2915–2921. doi:10.1167/iovs.04-0128

Mannori G, Santoro D, Carter L, Corless C, Nelson RM, Bevilacqua MP (1997) Inhibition of colon carcinoma cell lung colony formation by a soluble form of E-selectin. Am J Pathol 151(1):233–243

Mannori G, Barletta E, Mugnai G, Ruggieri S (2000) Interaction of tumor cells with vascular endothelia: role of platelet-activating factor. Clin Exp Metastasis 18(1):89–96

Mathonnet M, Descottes B, Valleix D, Truffinet V, Labrousse F, Denizot Y (2006) Platelet-activating factor in cirrhotic liver and hepatocellular carcinoma. World J Gastroenterol 12(17):2773–2778

Montrucchio G, Lupia E, Battaglia E, Passerini G, Bussolino F, Emanuelli G, Camussi G (1994) Tumor necrosis factor alpha-induced angiogenesis depends on in situ platelet-activating factor biosynthesis. J Exp Med 180(1):377–382

Montrucchio G, Sapino A, Bussolati B, Ghisolfi G, Rizea-Savu S, Silvestro L, Lupia E, Camussi G (1998) Potential angiogenic role of platelet-activating factor in human breast cancer. Am J Pathol 153(5):1589–1596. doi:10.1016/s0002-9440(10)65747-1

Montrucchio G, Alloatti G, Camussi G (2000) Role of platelet-activating factor in cardiovascular pathophysiology. Physiol Rev 80(4):1669–1699

Nitoda E, Moschos MM, Mavragani CP, Koutsilieris M (2012) Ocular actions of platelet-activating factor: clinical implications. Expert Opin Ther Targets 16(10):1027–1039. doi:10.1517/14728222.2012.712961

Philipp W, Speicher L, Humpel C (2000) Expression of vascular endothelial growth factor and its receptors in inflamed and vascularized human corneas. Invest Ophthalmol Vis Sci 41(9):2514–2522

Prescott SM, Zimmerman GA, Stafforini DM, McIntyre TM (2000) Platelet-activating factor and related lipid mediators. Annu Rev Biochem 69:419–445. doi:10.1146/annurev.biochem.69.1.419

Shan L, Nishimura Y, Kotani Y, Yokoyama M (1999) Platelet-activating factor increases the expression of metalloproteinase-9 in human bronchial epithelial cells. Eur J Pharmacol 374(1):147–156

Shishodia S, Aggarwal BB (2004) Nuclear factor-kappaB: a friend or a foe in cancer? Biochem Pharmacol 68(6):1071–1080. doi:10.1016/j.bcp.2004.04.026

Sobhani I, Denizot Y, Moizo L, Laigneau JP, Bado A, Laboisse C, Benveniste J, Lewin MJ (1996) Regulation of platelet-activating factor production in gastric epithelial cells. Eur J Clin Invest 26(1):53–58

Stafforini DM, McIntyre TM, Zimmerman GA, Prescott SM (2003) Platelet-activating factor, a pleiotrophic mediator of physiological and pathological processes. Crit Rev Clin Lab Sci 40(6):643–672. doi:10.1080/714037693

Yost CC, Weyrich AS, Zimmerman GA (2010) The platelet activating factor (PAF) signaling cascade in systemic inflammatory responses. Biochimie 92(6):692–697. doi:10.1016/j.biochi.2010.02.011

Zhang H, Yang Y, Takeda A, Yoshimura T, Oshima Y, Sonoda KH, Ishibashi T (2013) A novel platelet-activating factor receptor antagonist inhibits choroidal neovascularization and subretinal fibrosis. PLoS One 8(6):e68173. doi:10.1371/journal.pone.0068173

Zoratti EM, Sedgwick JB, Vrtis RR, Busse WW (1991) The effect of platelet-activating factor on the generation of superoxide anion in human eosinophils and neutrophils. J Allergy Clin Immunol 88(5):749–758

Chapter 40
Placenta Growth Factor (PlGF)

Abstract PlGF, also called PGF, is a gene encoding placenta growth factor (PlGF) protein, which belongs to VEGF family. This protein was first discovered from human placenta where it is highly expressed to promote placenta growth, thus providing its name –placenta growth factor and later identified to be expressed in many other organs like the heart, brain, thyroid, and skeletal muscle. There are four isoforms of PlGF protein made from alternative splicing of PlGF gene from chromosome 14. PlGF protein may bind to vascular endothelial receptor-1 (VEGFR-1) on monocytes and endothelial cells and stimulates migration of these cells, and thus improves the activity of VEGF, an important protein for human angiogenesis for enhancing vascular permeability, which is one of the main steps for angiogenesis. Because of PlGF potential for promoting angiogenesis, it is suggested that it may be used in skin repair. PlGF is involved in angiogenesis and growth of several kinds of tumours and its blockage inhibited the formation of new vessels at healthy cells around tumours although it did not necessarily result in prevention of primary tumor growth. A treatment with anti-PlGF and anti-VEGF-A antibodies showed similar efficiency against tumor angiogenesis with that with only anti-VEGF-A antibodies. This suggests that the roles of PlGF in tumor angiogenesis might not be as significant as it is believed.

Keywords Placenta growth factor • PlGF • Angiogenesis • Normal physiology • Disease • Malignancy

40.1 Placenta Growth Factor (PlGF)

PlGF, also called PGF, is a gene which encodes for placenta growth factor (PlGF) protein, which belongs to VEGF family. This protein was first discovered from human placenta where it is highly expressed to promote placenta growth, thus providing its name –placenta growth factor (Maglione et al. 1991). Later, this gene was found to be expressed in many other organs like the heart, brain, thyroid, and skeletal muscle. As a result, it was suggested that PlGF also has roles other than promoting placenta growth (Persico et al. 1999). There are four isoforms of PlGF protein made from alternative splicing of PlGF gene from chromosome 14. PlGF proteins

A. Salajegheh, *Angiogenesis in Health, Disease and Malignancy*,
DOI 10.1007/978-3-319-28140-7_40

form two disulfide bonds with each other to form homodimers, but according to in vitro study, their efficiency for improving vascular permeability and cell division is lower than monodimers of vascular endothelial growth factor (VEGF) (Park et al. 1994).

40.2 Placenta Growth Factor (PlGF) in Normal Physiology

PlGF protein may bind to vascular endothelial receptor-1 (VEGFR-1) on monocytes and endothelial cells (De Falco et al. 2002). This stimulates migration of these cells, and thus improves the activity of VEGF, an important protein for human angiogenesis for enhancing vascular permeability, which is one of the main steps for angiogenesis. Also, PlGF may bind to VEGF-A to form PlGF/VEGF-A heterodimers. Just like VEGF-A homodimers, PlGF/VEGF heterodimers bind to VEGFR-2 and enhances cell division of vascular endothelial cells, thus promoting angiogenesis. The efficiency of PlGF/VEGF-A heterodimers for angiogenesis is similar to that of VEGF-A homodimers (DiSalvo et al. 1995).

40.3 Placenta Growth Factor (PlGF) in Disease

Like many other genes, PlGF plays important roles in development of diseases – especially in angiogenesis of diseases because of its roles mentioned above. In rheumatoid arthritis, an autoimmune disease which primarily damages synovial joints, it is suggested that new blood formation is associated with synovial cells hyperplasia, vessel proliferation and mononuclear cell infiltration (Rooney et al. 1988). Therefore, angiogenesis is highly likely to be correlated with synovial inflammation in rheumatoid arthritis. Recently, it has been suggested that PlGF is upregulated in lining of hyperplastic synovium and synovial fluid of rheumatoid arthritis patients (Yoo et al. 2009), and PlGF promotes angiogenesis there. As a result, many new sub-lining blood vessels are found in pannus –an abnormal fibrovascular tissue in rheumatoid arthritis. These blood vessels are used for carrying inflammatory cells to inflammatory sites and providing nutrients and oxygen to inflamed tissues, thus maintaining chronic inflammation during rheumatoid arthritis. Also, angiogenesis increases endothelial surface area and this enhances production of cytokines, adhesion molecules, and other inflammatory stimuli, which contribute to maintenance of inflammation as well. Furthermore, angiogenesis in synovial membrane makes it easier for the tissue helping the active infiltration of synovial membrane to invade into cartilage, thus helping erosion and destruction of cartilage (Marrelli et al. 2011).

PlGF also promotes angiogenesis in eye disease, which is proven by the result that PlGF-deficient mice and those treated with 5D11D4, anti-PlGF antibody, showed less total vessel area, vessel density and macrophage infiltration in eyes

(Van de Veire et al. 2010). Ocular angiogenesis is one of the main causes for severe loss of vision because increased vascular permeability from ocular angiogenesis may lead to retina oedema and vascular fragility, which may cause haemorrhage (Rajappa et al. 2010).

PlGF also works in development of atherosclerosis –a chronic inflammatory disease where arteries of various diameters are blocked by lipid plaques. In early stage of atherosclerosis, PlGF is upregulated in atherosclerotic lesions and usually found at the shoulders of atherosclerotic plaque (Roncal et al. 2010), implying its involvement in development of atherosclerosis. In fact, PlGF is important in atherosclerotic intimal thickening and macrophage infiltration in early atherosclerotic lesions (Khurana et al. 2005). Macrophages play several roles in development of atherosclerosis, and in relation to angiogenesis, they produce many potent angiogenic cytokines and growth factors like VEGF, which are known to induce angiogenesis (Ono et al. 1999). Also, it is suggested from the experiment with ApoE-deficient mice that PlGF knockout reduced the number of macrophages in atherosclerotic plaques in ApoE-deficient mice, compared to those without PlGF knockout (Khurana et al. 2005). Apolipoprotein E (ApoE) deficiency causes lipoprotein accumulation, and therefore increases plasma cholesterol level, but is not involved in atherosclerotic angiogenesis or macrophage infiltration (Curtiss 2000). Therefore, it is shown that PlGF contributes to angiogenesis of early atherosclerosis –and thus its development– indirectly by promoting macrophage infiltration, rather than by inducing angiogenesis directly.

Angiogenesis promoted by PlGF, however, is not important just in development of diseases; on the contrary, it may contribute to treatment of diseases as well –for example, of atherosclerosis. In the late stage of this disease, angiogenesis plays opposite roles in relation to the disease, compared to its earlier stage, where angiogenesis contributes to its development. Arteries are blocked by thickened plaque during late atherosclerosis, so oxygenated blood flow is interrupted, and therefore vessel walls suffer from hypoxia –lack of oxygen supply, and hypoxia enhances PlGF expression in fibroblasts (Green et al. 2001). As a result, more blood vessels are built, so artery blockage is reduced and blood flow is recovered. In other words, PlGF may contribute to development or treatment of atherosclerosis, depending on the progress of the disease. Therefore, it may be necessary to check the progress level of disease first, when deciding whether to use PlGF for its treatment or not.

In myocardial ischemia, PlGF contributes to the treatment in a similar way as in atherosclerosis. In this disease, coronary arteries are partially or completely blocked, and therefore blood flow and oxygen supply to heart muscle is reduced. During myocardial ischemia, myocardium suffers from hypoxia, which enhances PlGF expression in myocardium (Torry et al. 2009). As a result, more coronary arteries are created, so blood flow to heart is recovered.

Because of potential for promoting angiogenesis, it is suggested that PlGF may be used in skin repair. Mice whose PlGF was over-expressed due to keratin 14 promoter, which promotes PlGF activity, expressed hyperaemic skin with vessels of increased number and permeability. Wound healing rate was enhanced for these mice, compared to the control group (Odorisio et al. 2002). This is because new blood vessels enhance delivery of oxygen and nutrients necessary for cell metabolism, and fibroblasts which can now migrate through larger pathway produce new extracellular matrix necessary for cell growth (Tonnesen et al. 2000).

40.4 Placenta Growth Factor (PlGF) in Malignancy

PlGF is involved in angiogenesis and growth of several kinds of tumours. PlGF blockage delayed formation of skin papillomas, which are benign epithelial tumours, and their associated neovessels. Also, PlGF blockage inhibited the formation of neovessels at healthy cells around tumours. Likewise, in hepatocellular carcinoma (HCC), PlGF blockage decreased tumor vessel abnormalization (Van de Veire et al. 2010). PlGF induces angiogenesis in HCC and excessive production of angiogenic molecules render tumor vessels abnormal (De Bock et al. 2011). Then, vessel abnormalization blocks oxygen delivery to tissues and causes hypoxia, and this results in invasion, metastasis and malignancy (Van de Veire et al. 2010). PlGF is involved in angiogenesis, metastasis and tumor growth in colorectal cancer as well (Wei et al. 2005). Also, small-cell lung cancer (SCLC) patients showed increased PlGF level, indicating that PlGF level is related to brain metastasis as well (Li et al. 2013).

However, PlGF blockage does not necessarily result in prevention of primary tumor growth. On the contrary, the treatment with anti-PlGF and anti-VEGF-A antibodies showed similar efficiency against tumor angiogenesis with that with only anti-VEGF-A antibodies (Bais et al. 2010). Also, inhibition of PlGF did not affect growth of several kinds of tumours like pancreatic tumor (Van de Veire et al. 2010).

In terms of tumor, PlGF may have an advantage over VEGF. The over-expression of PlGF gene in tumours increased the number of PlGF homodimers and PlGF/VEGF-A heterodimers but decreased VEGF-A homodimers. Tumours with PlGF over-expression showed similar vessel permeability but lower vessel density and larger vessel diameter. PlGF over-expression inhibited tumor growth in orthotopic tumor of mice (Xu et al. 2006), probably because its influence on tumor angiogenesis was not as great as that of VEGF. This shows that the roles of PlGF in tumor angiogenesis may not be as significant as many scientists have believed.

References

Bais C, Wu X, Yao J, Yang S, Crawford Y, McCutcheon K, Tan C, Kolumam G, Vernes JM, Eastham-Anderson J, Haughney P, Kowanetz M, Hagenbeek T, Kasman I, Reslan HB, Ross J, Van Bruggen N, Carano RA, Meng YJ, Hongo JA, Stephan JP, Shibuya M, Ferrara N (2010) PlGF blockade does not inhibit angiogenesis during primary tumor growth. Cell 141(1):166–177. doi:10.1016/j.cell.2010.01.033

Curtiss LK (2000) ApoE in atherosclerosis: a protein with multiple hats. Arterioscler Thromb Vasc Biol 20(8):1852–1853

De Bock K, Cauwenberghs S, Carmeliet P (2011) Vessel abnormalization: another hallmark of cancer? Molecular mechanisms and therapeutic implications. Curr Opin Genet Dev 21(1):73–79. doi:10.1016/j.gde.2010.10.008

De Falco S, Gigante B, Persico MG (2002) Structure and function of placental growth factor. Trends Cardiovasc Med 12(6):241–246

DiSalvo J, Bayne ML, Conn G, Kwok PW, Trivedi PG, Soderman DD, Palisi TM, Sullivan KA, Thomas KA (1995) Purification and characterization of a naturally occurring vascular endothelial growth factor: placenta growth factor heterodimer. J Biol Chem 270(13):7717–7723

Green CJ, Lichtlen P, Huynh NT, Yanovsky M, Laderoute KR, Schaffner W, Murphy BJ (2001) Placenta growth factor gene expression is induced by hypoxia in fibroblasts: a central role for metal transcription factor-1. Cancer Res 61(6):2696–2703

Khurana R, Moons L, Shafi S, Luttun A, Collen D, Martin JF, Carmeliet P, Zachary IC (2005) Placental growth factor promotes atherosclerotic intimal thickening and macrophage accumulation. Circulation 111(21):2828–2836. doi:10.1161/circulationaha.104.495887

Li B, Wang C, Zhang Y, Zhao XY, Huang B, Wu PF, Li Q, Li H, Liu YS, Cao LY, Dai WM, Fang WG, Shang DS, Cao L, Zhao WD, Chen YH (2013) Elevated PLGF contributes to small-cell lung cancer brain metastasis. Oncogene 32(24):2952–2962. doi:10.1038/onc.2012.313

Maglione D, Guerriero V, Viglietto G, Delli-Bovi P, Persico MG (1991) Isolation of a human placenta cDNA coding for a protein related to the vascular permeability factor. Proc Natl Acad Sci U S A 88(20):9267–9271

Marrelli A, Cipriani P, Liakouli V, Carubbi F, Perricone C, Perricone R, Giacomelli R (2011) Angiogenesis in rheumatoid arthritis: a disease specific process or a common response to chronic inflammation? Autoimmun Rev 10(10):595–598. doi:10.1016/j.autrev.2011.04.020

Odorisio T, Schietroma C, Zaccaria ML, Cianfarani F, Tiveron C, Tatangelo L, Failla CM, Zambruno G (2002) Mice overexpressing placenta growth factor exhibit increased vascularization and vessel permeability. J Cell Sci 115(Pt 12):2559–2567

Ono M, Torisu H, Fukushi J, Nishie A, Kuwano M (1999) Biological implications of macrophage infiltration in human tumor angiogenesis. Cancer Chemother Pharmacol 43(Suppl):S69–S71

Park JE, Chen HH, Winer J, Houck KA, Ferrara N (1994) Placenta growth factor. Potentiation of vascular endothelial growth factor bioactivity, in vitro and in vivo, and high affinity binding to Flt-1 but not to Flk-1/KDR. J Biol Chem 269(41):25646–25654

Persico MG, Vincenti V, DiPalma T (1999) Structure, expression and receptor-binding properties of placenta growth factor (PlGF). Curr Top Microbiol Immunol 237:31–40

Rajappa M, Saxena P, Kaur J (2010) Ocular angiogenesis: mechanisms and recent advances in therapy. Adv Clin Chem 50:103–121

Roncal C, Buysschaert I, Gerdes N, Georgiadou M, Ovchinnikova O, Fischer C, Stassen JM, Moons L, Collen D, De Bock K, Hansson GK, Carmeliet P (2010) Short-term delivery of anti-PlGF antibody delays progression of atherosclerotic plaques to vulnerable lesions. Cardiovasc Res 86(1):29–36. doi:10.1093/cvr/cvp380

Rooney M, Condell D, Quinlan W, Daly L, Whelan A, Feighery C, Bresnihan B (1988) Analysis of the histologic variation of synovitis in rheumatoid arthritis. Arthritis Rheum 31(8):956–963

Tonnesen MG, Feng X, Clark RA (2000) Angiogenesis in wound healing. J Invest Dermatol Symp Proc/Soc Invest Dermatol Inc [and] Eur Soc Dermatol Res 5(1):40–46. doi:10.1046/j.1087-0024.2000.00014.x

Torry RJ, Tomanek RJ, Zheng W, Miller SJ, Labarrere CA, Torry DS (2009) Hypoxia increases placenta growth factor expression in human myocardium and cultured neonatal rat cardiomyocytes. J Heart Lung Transplant 28(2):183–190. doi:10.1016/j.healun.2008.11.917

Van de Veire S, Stalmans I, Heindryckx F, Oura H, Tijeras-Raballand A, Schmidt T, Loges S, Albrecht I, Jonckx B, Vinckier S, Van Steenkiste C, Tugues S, Rolny C, De Mol M, Dettori D, Hainaud P, Coenegrachts L, Contreres JO, Van Bergen T, Cuervo H, Xiao WH, Le Henaff C, Buysschaert I, Kharabi Masouleh B, Geerts A, Schomber T, Bonnin P, Lambert V, Haustraete J, Zacchigna S, Rakic JM, Jimenez W, Noel A, Giacca M, Colle I, Foidart JM, Tobelem G, Morales-Ruiz M, Vilar J, Maxwell P, Vinores SA, Carmeliet G, Dewerchin M, Claesson-Welsh L, Dupuy E, Van Vlierberghe H, Christofori G, Mazzone M, Detmar M, Collen D, Carmeliet P

(2010) Further pharmacological and genetic evidence for the efficacy of PlGF inhibition in cancer and eye disease. Cell 141(1):178–190. doi:10.1016/j.cell.2010.02.039

Wei SC, Tsao PN, Yu SC, Shun CT, Tsai-Wu JJ, Wu CH, Su YN, Hsieh FJ, Wong JM (2005) Placenta growth factor expression is correlated with survival of patients with colorectal cancer. Gut 54(5):666–672. doi:10.1136/gut.2004.050831

Xu L, Cochran DM, Tong RT, Winkler F, Kashiwagi S, Jain RK, Fukumura D (2006) Placenta growth factor overexpression inhibits tumor growth, angiogenesis, and metastasis by depleting vascular endothelial growth factor homodimers in orthotopic mouse models. Cancer Res 66(8):3971–3977. doi:10.1158/0008-5472.can-04-3085

Yoo SA, Yoon HJ, Kim HS, Chae CB, De Falco S, Cho CS, Kim WU (2009) Role of placenta growth factor and its receptor flt-1 in rheumatoid inflammation: a link between angiogenesis and inflammation. Arthritis Rheum 60(2):345–354. doi:10.1002/art.24289

Chapter 41
Preproendothelin-1 (PreproET-1)

Abstract The Preproendothelin-1 (PreproET-1) gene is located on chromosome 6. Physiologically it is induced by thrombin, which then further is synthesized into ET-1 by endothelin-converting enzyme (ECE)-1. After synthesize, the secreted ET-1 from the endothelial cell disperses to the vascular smooth muscle and plays a paracrine or autocrine action. Smooth muscle tissues in the stomach, duodenum, colon, trachea, and uterus are affected by ET-1. The endothelin has also been shown to stimulate the proliferation of vascular smooth muscle cells along with Fibroblasts. With about 100 times more effect on vasculature than noradrenaline, endothelins are considered one of the most potent and long-lasting vasoconstrictor in human body. Endothelin-1 has been shown to directly act on endothelial cells receptor, which then modulates neovascularization, proliferation, migration, invasion, protease production and morphogenesis, and also stimulates neovascularization in vivo, thus promoting angiogenesis and metastasis through tumor growth.

Keywords Preproendothelin-1 • PreproET-1 • Endothelin • ET-1 • Angiogenesis • Normal physiology • Disease • Malignancy

41.1 Preproendothelin-1 (PreproET-1)

The Preproendothelin-1 also known as the preproET-1 gene is located on chromosome 6 and in its mature peptide form it is coded by the second exon (Levin and Nadler 1998). Physiologically preproET-1 gene is induced by thrombin, which then further is synthesized into ET-1. This synthesis happens through the enzyme endothelin-converting enzyme (ECE)-1 (Marsen et al. 1995). This synthesis occurs on the level of gene transcription which product is referred to as preproET-1 with 212 amino acid (Kohan and Barton 2014). PreproET-1 cleaves to ET-1, which then further divides into a biologically active ET-1 and C-terminal fragment. After synthesize the secreted ET-1 from the endothelial cell disperses to the vascular smooth muscle, which suggests a paracrine or autocrine mechanism of action (Kohan and Barton 2014).

A. Salajegheh, *Angiogenesis in Health, Disease and Malignancy*,
DOI 10.1007/978-3-319-28140-7_41

41.2 Preproendothelin-1 (PreproET-1) in Normal Physiology

There are a number of actions that endothelin has in the body. There are smooth muscle tissues that are affected which include: stomach, duodenum, colon, trachea, and uterus, which contracts in response to treatment with ET. In human myometrial smooth muscle cells, ET-1 causes an increase in the intracellular calcium and the myosin light chain phosphorylation. The ET has also been shown to stimulate the proliferation of vascular smooth muscle cells along with Fibroblasts. In the sympathetic nerve terminals the ET elicits norepinephrine release (Casey et al. 1991). Furthermore Endothelins are involved in the functioning of nervous, cardiovascular, renal, respiratory, endocrine and gastrointestinal systems within the body (Mayes 2003). Endothelins are one of the most potent and long-lasting vasoconstrictor known and is about 100 times more potent than noradrenaline (Barton 2010).

41.3 Preproendothelin-1 (PreproET-1) in Disease

ET-1 is very significant in wound healing in fact endothelin helps to control the immune function and inflammation (Barton 2010; Li et al. 2012). In hepatic injury for example a wound healing response of the ET-1 causes the Stellate Cells to proliferate, contract, and collagen production to occur (Rockey et al. 1998). In regards to the inflammatory response these processes are not fully understood but there are studies that have shown that ET-1 is involved in the stress-triggered ACS (Acute Coronary Syndrome) that provides evidence with regards to this pathway (Fernandez et al. 2010).

As a cognate receptor to ET-1, preproET-1 has been attributing to many diseases like bronchoconstriction, carcinogenesis, fibrosis, pulmonary hypertension and heart failure (Mayes 2003).

41.4 Preproendothelin-1 (PreproET-1) in Malignancy

One of the important controllers of vascular growth, inflammation and vascular tone is endothelin family and their associated receptors (Irani et al. 2014). Numerous controlling roles in different malignancies have been described for this family of endothelins including cancer development, progression, angiogenesis, stromal reaction, epithelial mesenchymal transitions, apoptosis, invasion, metastases and drug resistance (Irani et al. 2014).

Research has shown that Endothelin-1 directly acts on endothelial cells receptor, which modulates neovascularization, proliferation, migration, invasion, protease production and morphogenesis, and also stimulates neovascularization in vivo, thus promoting angiogenesis and metastasis through tumor growth (Bagnato and Spinella

2003; Grant et al. 2003; Wu et al. 2014). This is further confirmed as there have been studies showing increased plasma levels of ET-1 in patients with solid tumours, including hepatocellular, gastric and prostate cancer (Wu et al. 2014; Loizidou 2008). As Endothelin mediates pathways of cancer growth and progression it can also be a possible target for cancer therapeutics as well.

Further dissemination of ET-1 according to scientists as it relates to angiogenesis is that there are small molecule inhibitors of the Endothelin receptor that exist which the Endothelin axis which could be blocked therefore improving the antitumor treatment and inhibit neovascularization and tumor cell growth (Salani et al. 2000; Bagnato and Spinella 2003; Zubar 2005).

As stated previously the ET-1 activates the autocrine and paracrine signaling which intern modulates cell proliferation, apoptosis, migration, epithelial-t-mesenchyme transition, chemo-resistance and neovascularization, which provides good logic into targeting ET1 receptors in cancer treatment (Rosano et al. 2013).

Despite promising outcome in inhibition of endothelin axis in cell lines and xenografts have been demonstrated, limited information are available in its clinical trial application and specifically any significant survival benefit (Irani et al. 2014). The potential of targeting both cancer cells and vasculature around them through endothelins is an important and promising approach for therapeutic strategies and demands further scientific investigation and exploration.

References

Bagnato A, Spinella F (2003) Emerging role of endothelin-1 in tumor angiogenesis. Trends Endocrinol Metab 14(1):44–50. doi:S1043276002000103 [pii]

Barton M (2010) Therapeutic potential of endothelin receptor antagonists for chronic proteinuric renal disease in humans. Biochim Biophys Acta 1802(12):1203–1213. doi:10.1016/j.bbadis.2010.03.012

Casey ML, Word RA, MacDonald PC (1991) Endothelin-1 gene expression and regulation of endothelin mRNA and protein biosynthesis in avascular human amnion. Potential source of amniotic fluid endothelin. J Biol Chem 266(9):5762–5768

Fernandez AB, Soufer R, Collins D, Soufer A, Ranjbaran H, Burg MM (2010) Tendency to angry rumination predicts stress-provoked endothelin-1 increase in patients with coronary artery disease. Psychosom Med 72(4):348–353. doi:10.1097/PSY.0b013e3181d71982

Grant K, Loizidou M, Taylor I (2003) Endothelin-1: a multifunctional molecule in cancer. Br J Cancer 88(2):163–166. doi:10.1038/sj.eye.6700750

Irani S, Salajegheh A, Smith RA, Lam AK (2014) A review of the profile of endothelin axis in cancer and its management. Crit Rev Oncol Hematol 89(2):314–321. doi:10.1016/j.critrevonc.2013.08.011

Kohan DE, Barton M (2014) Endothelin and endothelin antagonists in chronic kidney disease. Kidney Int 86(5):896–904. doi:10.1038/ki.2014.143

Levin ER, Nadler JL (1998) Endocrinology of cardiovascular function, vol 1. Springer Science & Business Media, New York

Li T, Shi Z, Rockey DC (2012) Preproendothelin-1 expression is negatively regulated by IFNgamma during hepatic stellate cell activation. Am J Physiol Gastrointest Liver Physiol 302(9):G948–G957. doi:10.1152/ajpgi.00359.2011

Loizidou M (2008) Endothelin-1-promoting actions in the growth and angiogenesis of solid cancers. In: Vascular complications in human disease. Springer, London, pp 187–196
Marsen TA, Simonson MS, Dunn MJ (1995) Thrombin induces the preproendothelin-1 gene in endothelial cells by a protein tyrosine kinase-linked mechanism. Circ Res 76(6):987–995
Mayes MD (2003) Endothelin and endothelin receptor antagonists in systemic rheumatic disease. Arthritis Rheum 48(5):1190–1199. doi:10.1002/art.10895
Rockey DC, Fouassier L, Chung JJ, Carayon A, Vallee P, Rey C, Housset C (1998) Cellular localization of endothelin-1 and increased production in liver injury in the rat: potential for autocrine and paracrine effects on stellate cells. Hepatology (Baltimore, MD) 27(2):472–480. doi:10.1002/hep.510270222
Rosano L, Spinella F, Bagnato A (2013) Endothelin 1 in cancer: biological implications and therapeutic opportunities. Nat Rev Cancer 13(9):637–651. doi:10.1038/nrc3546
Salani D, Taraboletti G, Rosano L, Di Castro V, Borsotti P, Giavazzi R, Bagnato A (2000) Endothelin-1 induces an angiogenic phenotype in cultured endothelial cells and stimulates neovascularization in vivo. Am J Pathol 157(5):1703–1711. doi:10.1016/s0002-9440(10)64807-9
Wu MH, Huang CY, Lin JA, Wang SW, Peng CY, Cheng HC, Tang CH (2014) Endothelin-1 promotes vascular endothelial growth factor-dependent angiogenesis in human chondrosarcoma cells. Oncogene 33(13):1725–1735. doi:10.1038/onc.2013.109
Zubar RV (2005) Trends in angiogenesis research. Nova Science Publishers, New York

Chapter 42
Proliferin

Abstract The formation of new blood vessels (angiogenesis/neovascularisation) is essential during growth and development of mammalian organisms to provide nutrients and regulatory signals for the said growth and development of the foetus. In many mammalian species, the placenta is the site of synthesis of growth hormone related proteins. Analysis of two such proteins, proliferin (PLF) and proliferin-related protein reveal their potent regulation of mammalian angiogenesis or neovascularisation. The embryonic functions currently known of this family include stimulation of uterine cell proliferation and development of placental capillary network thus vascularisation of the placenta. Researchers demonstrated the profound effect of PLF's in wound healing and the hair follicle cycle as a growth factor and/or angiogenesis inducer. Thus it would seem that the PLF gene, which previously had only been seen to be expressed by the placenta and damaged tissue, can be reactivated by tumour cells to secure a nutrients source and a mode of metastasis. The discovery that tumour cells use PLF to regulate angiogenesis provides an attractive anti-tumour therapy pathway. By disrupting angiogenesis by interfering with PLF secretion, tumour metastasis may be prevented.

Keywords Proliferin • Angiogenesis • Normal physiology • Disease • Malignancy

42.1 Proliferin

The formation of new blood vessels (angiogenesis/neovascularisation) is essential during growth and development of mammalian organisms to provide nutrients and regulatory signals for the said growth and development of the foetus. In many mammalian species, the placenta is the site of synthesis of growth hormone related proteins. Analysis of two such proteins, proliferin (PLF) and proliferin-related protein reveal their potent regulation of mammalian angiogenesis or neovascularisation. Proliferin belongs to the prolactin/growth hormone/placental lactogen family of polypeptide hormones. It is synthesised in placental trophoblast giant cells and found specifically on chromosome 13 and 23 of the genomes of species that express PLF (Yang et al. 2011).

A. Salajegheh, *Angiogenesis in Health, Disease and Malignancy*,
DOI 10.1007/978-3-319-28140-7_42

42.2 Proliferin in Normal Physiology

The embryonic functions currently known of this family include stimulation of uterine cell proliferation and development of placental capillary network thus vascularisation of the placenta. PLF's have been identified to bind to the mannose-6-phosphate (IGF II/M6P) receptor of endothelial cells and stimulate their migration via the mitogen-activated protein kinase-dependent pathway, thus neovascularisation in vivo (Jackson and Linzer 1997; Fang et al. 1999). Studies by Fang et al. have shown a direct correlation between the amount of PLF detected in gestational mice and the degree of vascularisation of the placenta. Specifically, this study identifies MRP4, a sub type of the proliferin family as being located within the placenta for the longest duration during gestation and thus this protein is considered as the main activator of angiogenesis (Fang et al. 1999). Another study by Xinhai Yang et al. evaluated not only the specific effects of PLF on vascularisation of the placenta but also its interactions with various proteins and molecules. The study identified that FGF2, a proangiogenic stimulator, along with others activate the STAT5 gene which in turn activates the expression of PLF which promotes endothelial cell migration, invasion and tube formation. Interestingly, the same researchers discovered that one of the roles of PLF was to activate STAT5. Thus it was concluded that STAT5 was involved in both the upstream and downstream interactions of PLF establishing a positive feedback loop that amplified and sustained proangiogenic signals in the placenta. This discovered 'redundancy' of pathways to stimulate neovascularisation portrayed a biologically economic mechanism for integrating signals from several proangiogenic factors in the interest of rapidly and efficiently forming new blood vessels (Yang et al. 2011). Besides its obvious role in vascularisation of the placenta during foetal development, PLF's also contribute to other physiological functions in the mammalian body. For example, Tsuruta et al. have investigated and contributed to the vast publications on the effects of PLF's on spermatogenesis (Tsuruta 2000).

42.3 Proliferin in Disease

Fasset et al. demonstrated the profound effect of PLF's in wound healing and the hair follicle cycle as a growth factor and/or angiogenesis inducer. The results of this study indicated that the nutritional and structural components required by repairing tissue induce the wounded tissue to re-activate expression of the PLF gene and produce the PLF protein. PLF in turn stimulated angiogenesis and thus provided the circulatory route for nutrients, structural components and immune cells to infiltrate the region of damage (Fassett 2001). In addition, research by Dowdy Jackson et al. has identified it to play a key role in directly stimulating angiogenesis in the rat cornea (Jackson et al. 1994). At this point it is appropriate to distinguish proliferin from proliferin-related protein, which, as the aforementioned research has identified, has been found to be implemented in the inhibition of angiogenesis.

Proliferin-related protein (PRP) is a potent placental anti-angiogenesis hormone. It belongs to the same family as proliferin with only minor amino acid sequence changes. The effects of proliferin-related protein are however completely opposite to proliferin. Research shows that proliferin-related protein inhibits angiogenesis as depicted in the picture above in which proliferin-related protein impregnated pellets were introduced into an endothelial cell medium and prevented the growth of new blood vessels (Jackson et al. 1994).

Another study shows that physiological placental expression of PRP occurs after complete angiogenesis has taken place (through the expression of PLF) in order to prevent the further migration of endothelial cells and the subsequent resealing of the newly created vessels (Bengtson and Linzer 2000).

42.4 Proliferin and Malignancy

As of yet, only the physiologic roles of proliferin and proliferin related protein have been discussed. Of specific relevance to this collection is the current research into the effects of proliferin and proliferin-related protein on tumorigenesis and metastasis. Major contributors in this field are Toft et al. who discovered, through their experimentation on fibrosarcaomas, the direct correlation between the degree of cell transformation and tumorigenicity and the level of secreted PLF (Toft et al. 2001). Through isolation of RNA from mild fibromatosis, aggressive fibromatosis and fibrosarcoma-derived cell lines, Toft et al. identified a direct correlation between the degree of transformation of the cells and the expression of the PLF gene. Research indicates that the secretion of PLF is greatest at the Fibrosarcoma stage, at which stage the tumour becomes highly angiogenic. In addition, researchers discovered that tumour cells that were cultured in an immunodepleted PLF medium showed a dramatic reduction in angiogenic behaviour (Toft et al. 2001).

Thus it would seem that the PLF gene, which previously had only been seen to be expressed by the placenta and damaged tissue, can be reactivated by tumour cells to secure a nutrients source and a mode of metastasis. The placenta has many properties similar to a tumour: both grow rapidly and generate a blood supply by secreting PLF angiogenic factors, both escape immune responses, and both are invasive. It thus seems a logical mechanism for tumours to reactivate the genes that evolved to promote placental vascularisation. The discovery that tumour cells use PLF to regulate angiogenesis provides an attractive anti-tumour therapy pathway. By disrupting angiogenesis by interfering with PLF secretion, tumour metastasis may be prevented.

In addition, a study was conducted to test the anti-angiogenic potential of proliferin-related protein (Toft et al. 2001). The cultured tumour cells were engineered to express PRP. The results revealed that all cells expressing the PRP gene had markedly reduced tumour growth rates and were largely avascular. This contrasts to the control tumours grown, all of which were well developed and rich in blood vessels.

These results were consistent with the theory that PRP prevented neovascularization of tumors. The discovery of the anti-angiogenic potential of PRP poses interesting questions for future research into the cancer-therapy potential of PRP in inhibiting angiogenesis by tumors (Bengtson and Linzer 2000).

References

Bengtson NW, Linzer DIH (2000) Inhibition of tumor growth by the antiangiogenic placental hormone, proliferin-related protein. Mol Endocrinol 14(12):1934–1943. doi:10.1210/mend.14.12.0573

Fang Y, Lepont P, Fassett JT, Ford SP, Mubaidin A, Hamilton RT, Nilsen-Hamilton M (1999) Signaling between the placenta and the uterus involving the mitogen-regulated protein/proliferins 1. Endocrinology 140(11):5239–5249. doi:10.1210/endo.140.11.7142

Fassett JT (2001) Mrp3, a mitogen-regulated protein/proliferin gene expressed in wound healing and in hair follicles. Endocrinology 142(5):2129–2137. doi:10.1210/en.142.5.2129

Jackson D, Linzer DIH (1997) Proliferin transport and binding in the mouse fetus 1. Endocrinology 138(1):149–155. doi:10.1210/endo.138.1.4828

Jackson D, Volpert O, Bouck N, Linzer D (1994) Stimulation and inhibition of angiogenesis by placental proliferin and proliferin-related protein. Science 266(5190):1581–1584. doi:10.1126/science.7527157

Toft DJ, Rosenberg SB, Bergers G, Volpert O, Linzer DIH (2001) Reactivation of proliferin gene expression is associated with increased angiogenesis in a cell culture model of fibrosarcoma tumor progression. Proc Natl Acad Sci 98(23):13055–13059. doi:10.1073/pnas.231364798

Tsuruta JK (2000) Insulin-like growth factor-II/cation-independent mannose 6-phosphate receptor mediates paracrine interactions during spermatogonial development. Biol Reprod 63(4):1006–1013. doi:10.1095/biolreprod63.4.1006

Yang X, Qiao D, Meyer K, Pier T, Keles S, Friedl A (2011) Angiogenesis induced by signal transducer and activator of transcription 5A (STAT5A) is dependent on autocrine activity of proliferin. J Biol Chem 287(9):6490–6502. doi:10.1074/jbc.m111.254631

Chapter 43
Prostaglandins

Abstract Prostaglandins are lipid autacoid derived from arachidonic acid and helping in sustaining homeostatic functions and mediating pathogenic mechanisms, such as inflammatory responses. PGE1 inhibits platelet aggression, it is a potent vasodilator, stimulates the intestinal and uterine smooth muscles, and is also found to be linked with the induction and enhancement of angiogenesis. In addition to PGE1 anti-inflammatory effects with regards to the inhibition of neutrophil dependent immune complex, PGE1 has a stimulatory interaction with VEGFs to promote angiogenesis. Prostaglandin E2 (PGE2) is one of the most commonly produced PGs in the body. It acts as a local messenger molecule; exerting itself both in an autocrine and paracrine fashion, and synthesized in the cancer epithelial cells, blood vessel endothelial, and immune cells. Prostaglandins play important role in pro-inflammatory mediation; promoting tumor growth by inhibiting apoptosis, stimulating angiogenesis, cell invasion, and cell growth. However, the exact mechanism of just how they promote angiogenesis still remains largely unclear, as it is part of a multifactorial disease process.

Keywords Prostaglandin E1 • PGE1 • Prostaglandin E1 • PGE2 • Angiogenesis • Normal physiology • Disease • Malignancy

43.1 Prostaglandins

Prostaglandins are naturally occurring lipids that are derived from the breakdown of arachidonic acid through the action of cyclooxygenase (COX) isoenzymes (Zhang and Daaka 2011; Ricciotti and FitzGerald 2011). Their primary function is to maintain homeostasis within the body, and mediate pathogenic mechanisms such as the promotion and resolution of inflammatory responses (Zhang and Daaka 2011; Ricciotti and FitzGerald 2011). Such examples include prostaglandin E1 (PGE1) and prostaglandin E2 (PGE2).

PGE1 inhibits platelet aggression, it is a potent vasodilator, stimulates the intestinal and uterine smooth muscles, and is also found to be linked with the induction and enhancement of angiogenesis (Fan and Chapkin 1998). PGE1 is derived from dihomo-γ-linolenic acid (DGLA) via a cyclo-oxygenated reaction mediated by

A. Salajegheh, *Angiogenesis in Health, Disease and Malignancy*,
DOI 10.1007/978-3-319-28140-7_43

COX-1/2 isoenzymes (Fan and Chapkin 1998). DGLA is a byproduct of gamma-linolenic acid (GLA), which is attainable from the diet.

Prostaglandin E2 (PGE2) is one of the most commonly produced PGs in the body (Ricciotti and FitzGerald 2011). It acts as a local messenger molecule; exerting itself both in an autocrine and paracrine fashion, and synthesized in the cancer epithelial cells, blood vessel endothelial, and immune cells (Zhang and Daaka 2011). Under normal physiological conditions, PGE2 mediates many biological functions, such as immune response regulation, gastrointestinal systems, blood pressure and parts of the reproductive system. However, available evidence has suggested the involvement of PGE2 in a wide range of pathological conditions, such as inflammation (Zhang and Daaka 2011). Further studies have also shown associations between pro-inflammatory PGE2 and angiogenesis; whereby it has an effect on cancer cell growth, survival, and migration as well as immune cell responses (Zhang and Daaka 2011).

43.2 Prostaglandins in Normal Physiology

PGE1 is capable of binding onto prostacyclin receptors on platelets, via the Gs-protein coupled receptor; this enhances the activity of adenyl cyclase, leading to an increase in the intracellular concentration of cAMP. The elevated levels of cAMP in turn inhibit phospholipase C, causing a reduction of intracellular calcium mobilization (from storage). As a result, inhibiting platelet aggregation induced by the P2Y1 receptor activation; which would normally increase calcium release resulting in platelet activation and aggregation (Ricciotti and FitzGerald 2011).

PGE1 is able to carry out the relaxation of smooth muscle cells (SMC) and cause vasodilation due to its interaction with a G-Protein coupled receptor (GPCR) which affects the activity of adenyl cyclase (AC) (Majed and Khalil 2012). PGE1 interacts with the G_s – protein receptor, stimulating the activity of AC, causing an increase in the intracellular cAMP levels. The increase in cAMP level leads to the expression of various genes via the protein kinase A (PKA) – mediated phosphorylation (Li et al. 2001). Examples of genes expressed include the genes for the enzymes lipase, glycogen synthase, and phosphorylase kinase. The increase in cAMP leads to the relaxation of the SMC, and thus vasodilation in smooth muscles.

PGE2 affects target cells by activating four G protein–coupled receptors (EP1, EP2, EP3, and EP4). The activation of each receptor leads to the production of GTP on the Gα subunits and subsequently, the dissociation of the Gα-GTP from the GβϒY subunits (Zhang and Daaka 2011). With each EP subtype showing a distinct cellular localization within tissues (Ricciotti and FitzGerald 2011). The binding of PGE2 to the EP1 receptor induces the activation of PKC through the mobilization of intracellular Ca^{2+} ions. While the EP3 receptor inhibits the synthesis of camp, which thereby decreases activity of the cAMP-dependent protein kinase (PKA). Both EP2 and EP4 receptors act by increasing the synthesis of cAMP and consequently, the activation of PKA. PGE2 converts the receptor subtype-specific signaling events, via the stimulation of these EP receptors, causing the transactivation of epidermal growth factor receptor and its multiple downstream effectors (Zhang and Daaka 2011).

43.3 Prostaglandins in Disease

PGE1 produces anti-inflammatory effects, it addresses the clinical signs of inflammatory which are caused by changes in vascular changes and cell recruitment, such as vasodilation and vascular permeability. Since PGE1 is a vasodilator, it helps to lower the temperature of the body by allowing for more heat to escape (release). Another anti-inflammatory mechanism of PGE1 is that it is able to inhibit neutrophil dependent immune complex injuries by preventing formyl-methionyl-leucyl-phenylalanine- (FMLP) induced lysosomal enzyme release from polymorphonuclear leukocytes (PMNs) – resulting in the down regulation and modulation of cells related to the inflammatory response, such as cytokines and interleukins related to the FMLP family (Arbour et al. 1996; Fantone et al. 1983).

Chronic inflammation is a well-known risk factor for cancer, and is recognized as a critical mediator of angiogenesis (Wang et al. 2006; Zhang and Daaka 2011). During the process of inflammation, PGE2 plays an important role in all processes leading to the classic signs of inflammation such as heat, swelling, redness, and pain (Zhang and Daaka 2011). Pro-inflammatory PGE2 plays a vital role in the process of angiogenesis through the activation of EP1 to EP4 receptors (Zhang and Daaka 2011). It is able to impact many steps involved in the inflammatory process; such as inflammatory mediators that can be both pro-inflammatory and anti-inflammatory. In some instances, due to a number of cascading events related to the presence of PGE2, this can promote tumor growth by inhibiting apoptosis, stimulating angiogenesis, cell invasion, and cell growth (Wang et al. 2006).

Another cellular interaction PGE1 has is with vascular endothelial growth factor (VEGF), where it is found to induce its expression (Mehrabi et al. 2002). Research done by Mehrabi et al. (2002) provided evidence where PGE1 was able to induce the production of VEGF in human coronary artery SMC in vitro; where it has been shown that VEGF plays a role in angiogenesis (Reinmuth 2001). Mehrabi et al. (2002) came to the conclusion that patients treated with PGE1 for chronic ischemic heart displayed enhanced levels of VEGF-capillaries compared to those without the PGE1 treatment. This meant that the treated patients displayed a high rate cardiac vascular proliferation than those who did not receive the treatment. It is due to these mechanisms and results that it is believed that PGE1 may be capable of having beneficial effects for patients with chronic ischemic heart disease.

Alprostadil is the drug form of PGE1, and can be used to relax the ductus arteriosus. This mechanism is very beneficial for infants with congenital defects which restrict the systemic or pulmonary blood flow and who depend on a patent ductus arteriosus for sufficient blood flow and oxygen perfusion. The down side of this drug is that it is rapidly metabolized by the body, approximately 80 % through the first pass. Thus to treat infants with congenital cardiac defects, the drug should be administered intravenously to maximize the effects and minimize the dose of the drug. PGE1 is also able to treat erectile dysfunction by stimulating vasodilation of the cavernosal arteries and relaxation of the trabecular smooth muscles. This allows for the enlargement of the lacunar space and collection of blood in the area hence stimulating an erection (Ricciotti and FitzGerald 2011).

43.4 Prostaglandins in Malignancy

Angiogenesis plays an important role in the growth, survival, migration, and formation of endothelial cells. In Angiogenesis, the regulatory mechanisms that are involved are mainly derived from the angiogenic inducer known as vascular endothelial growth factor (VEGF) (Zhang and Daaka 2011). Evidence suggests that in angiogenesis, PGE2 may act on cancer cells, thereby prompting them to produce pro-angiogenic factors such as VEGF, basic fibroblast growth factor (BFGF), and the chemokine CXCL1 that, in turn, activate endothelial cells to promote angiogenesis (Zhang and Daaka 2011). PGE2 may also have a direct action on endothelial cells, and can promote their angiogenic response by increasing the expression of CXCR4, therefore increasing endothelial cell migration.

PGE2 and its receptors play a major role in progression of cancers. In cancer, it has been found that the over-expression of the enzyme in a variety of cell types present in diseased organs, including tumors correlates with poor prognosis of the disease (Ricciotti and FitzGerald 2011). In particular, it was shown that PGE2 produced by the COX2 enzyme occurs in much higher concentrations when it is present in tumor cells than when compared with normal tissues (Greenhough et al. 2009). PGE2 has also been shown to inhibit the mechanism of apoptosis and induce cell proliferation by increasing epithelial cell migration and motility. This has been shown to correlate with the survival of tumour cells. Furthermore, via its effect on the immune system and inflammation, it has adverse effects in relation to the elimination of tumors (Greenhough et al. 2009).

Quite a large body of evidence suggests that there is a link between PGE2 and the promotion of tumour growth. This is largely due to the stimulation of EP receptor signalling, leading to a cascade of downstream factors. Evidence suggests that it plays an important role in pro-inflammatory mediation; promoting tumor growth by inhibiting apoptosis, stimulating angiogenesis, cell invasion, and cell growth (Zhang and Daaka 2011). PGE2 has also been shown to exert pro-oncogenic effects in colorectal neoplasia through the action of autocrine and paracrine growth factors (Finetti et al. 2008). An over-expression of the enzyme in a variety of cell types was also presented in diseased organs, including breast and colon tumors (Greenhough et al. 2009). The exact mechanism of just how PGE2 promotes angiogenesis however, still remains largely unclear, as it is a multifactorial disease. Further research is still required to further explore the area of cancer so that we may better understand the disease.

Prostaglandins play important role in pro-inflammatory mediation; promoting tumor growth by inhibiting apoptosis, stimulating angiogenesis, cell invasion, and cell growth. However, the exact mechanism of just how they promote angiogenesis still remains largely unclear, as it is part of a multifactorial disease process.

References

Arbour N, Tremblay P, Oth D (1996) N-formyl-methionyl-leucyl- phenylalanine induces and modulates Il-1 and Il-6 in human Pbmc. Cytokine 8(6):468–475. doi:10.1006/cyto.1996.0063

Fan YY, Chapkin RS (1998) Importance of dietary gamma-linolenic acid in human health and nutrition. J Nutr 128(9):1411–1414

Fantone JC, Marasco WA, Elgas LJ, Ward PA (1983) Anti-inflammatory effects of prostaglandin E1: in vivo modulation of the formyl peptide chemotactic receptor on the rat neutrophil. J Immunol 130(4):1495–1497

Finetti F, Solito R, Morbidelli L, Giachetti A, Ziche M, Donnini S (2008) Prostaglandin E2 regulates angiogenesis via activation of fibroblast growth factor receptor-1. J Biol Chem 283(4):2139–2146. doi:10.1074/jbc.M703090200

Greenhough A, Smartt HJ, Moore AE, Roberts HR, Williams AC, Paraskeva C, Kaidi A (2009) The COX-2/PGE2 pathway: key roles in the hallmarks of cancer and adaptation to the tumour microenvironment. Carcinogenesis 30(3):377–386. doi:10.1093/carcin/bgp014

Li G, Harton JA, Zhu X, Ting JPY (2001) Downregulation of CIITA function by protein kinase a (PKA)-mediated phosphorylation: mechanism of prostaglandin E, cyclic AMP, and PKA inhibition of class II major histocompatibility complex expression in monocytic lines. Mol Cell Biol 21(14):4626–4635. doi:10.1128/mcb.21.14.4626-4635.2001

Majed BH, Khalil RA (2012) Molecular mechanisms regulating the vascular prostacyclin pathways and their adaptation during pregnancy and in the newborn. Pharmacol Rev 64(3):540–582. doi:10.1124/pr.111.004770

Mehrabi M, Serbecic N, Tamaddon F, Kaun C, Huber K, Pacher R, Wild T, Mall G, Wojta J, Glogar H (2002) Clinical and experimental evidence of prostaglandin E1-induced angiogenesis in the myocardium of patients with ischemic heart disease. Cardiovasc Res 56(2):214–224. doi:10.1016/s0008-6363(02)00591-6

Reinmuth N (2001) Induction of VEGF in perivascular cells defines a potential paracrine mechanism for endothelial cell survival. FASEB J 15(7):1239–1241. doi:10.1096/fj.00-0693fje

Ricciotti E, FitzGerald GA (2011) Prostaglandins and inflammation. Arterioscler Thromb Vasc Biol 31(5):986–1000. doi:10.1161/ATVBAHA.110.207449

Wang D, Wang H, Brown J, Daikoku T, Ning W, Shi Q, Richmond A, Strieter R, Dey SK, DuBois RN (2006) CXCL1 induced by prostaglandin E2 promotes angiogenesis in colorectal cancer. J Exp Med 203(4):941–951. doi:10.1084/jem.20052124

Zhang Y, Daaka Y (2011) PGE2 promotes angiogenesis through EP4 and PKA Cgamma pathway. Blood 118(19):5355–5364. doi:10.1182/blood-2011-04-350587

Chapter 44
ROS 1

Abstract Receptor tyrosine kinases (RTKs) play an important role in the process of signal transduction and cellular communication. A receptor tyrosine kinase, *ROS 1* has been shown as one of the last two remaining orphan receptor tyrosine kinases whose ligand has not completely identified. In addition to being as a role player in epithelial cell differentiation, *ROS 1* can activate number of downstream signaling pathways related to cell survival, proliferation and angiogenesis. The aberrant expression of this gene as well as its mutant forms has been reported in the initiation and progression of variety of human cancers, including non–small cell lung cancer (NSCLC), glioblastoma, ovarian cancer and colorectal cancer. However, it has not been fully found the normal function of ROS kinase in different human body tissues so far. Recent insights have shed light onto ROS 1 signal transduction and its importance in cell development, diseases, clinical applications and treatment of substantial cancers.

Keywords ROS 1 • Angiogenesis • Normal physiology • Disease • Malignancy

44.1 ROS1

ROS 1 gene is found on human chromosome 6 region 6q22 and it encodes for proto-oncogene tyrosine protein kinase ROS, which is an orphan kinase with unconfirmed ligand and unclarified functions (El-Deeb et al. 2011). Having a highly similar structure to receptor tyrosine kinase (RTK), ROS tyrosine kinase is believed to be closely related to RTK. Since ROS tyrosine kinase is similar to RTK, one believes that their functions are similar as well. Hence, it can be concluded the ROS tyrosine kinase behaves like a cell surface receptor for growth factors and hormones, regulates cell to cell communication, and is involved in signal transduction (El-Deeb et al. 2011).

ROS tyrosine kinase has binding sites that are similar to epidermal growth factor receptors, macrophage colony stimulating factor receptors, and HIR genes, indicating that ROS tyrosine kinase have comparable functions to those receptors. Hence, specifically, tyrosine kinase ROS is responsible for growth and maturation of cells (El-Deeb et al. 2011).

Dr. Haleh Vosgha contributed equally with Dr. Ali Salajegheh.

A. Salajegheh, *Angiogenesis in Health, Disease and Malignancy*,
DOI 10.1007/978-3-319-28140-7_44

Its gene expression pattern during embryonic development and post-development suggests its involvement in the maturation of organs. Chromosome 6 specifies non-random chromosomal rearrangement in specific neoplasia in which over expression of mutated *ROS 1* gene leads to the development of malignancy that are commonly found in glioblastoma (El-Deeb et al. 2011).

44.2 ROS 1 in Normal Physiology

Although the ligand for ROS1 tyrosine kinase receptor is unknown, the present discovery of activated ROS protein kinase shows downstream signaling pathways:

(a) Signal Transducers and Activators of Transcription 3 (STAT3)
(b) PI3/KAKT/m-TOR cascade
(c) RAS/MAPK/ERK pathways
(d) VAV 3 proteins
(e) SHP1 and 2

From the above pathways, over activation of STAT 3 and PI3/KAKT/m-TOR signaling cascade are heavily involved in angiogenesis related to oncogenesis.

Tyrosine-phosphorylated STAT3 protein is responsible for cytokine and growth factor induced gene expression which contributes to the survival, proliferation, and differentiation of cells. This includes oncogenesis as well (Chen and Han 2008). Cytoplasmic STAT3 protein undergoes transformation to homo- or hetero-dimer based transcription factors through extracellular ligand induced phosphorylation, allowing its translocation to nucleus followed by transcription of the targeted genes (Chen and Han 2008). Overall, ROS tyrosine kinase activates STAT3, increasing targeted DNA transcription. Activation of STAT3 causes immune-suppression and angiogenic promoting mechanism, both of which are commonly expressed in primary tumour development as they promote tumour survival (Chen and Han 2008).

44.3 ROS 1 in Disease

STAT3, with its target gene transcription, is a regulator of the multistage and multifactorial stimulated angiogenesis in both pathological and physiological conditions (El-Deeb et al. 2011). Through the activation of angiogenic factors- including VEGF, bFGF, matrix metalloproteinase 2, and matrix metalloproteinase 9- angiogenesis is controlled. This comprises the activation of endothelial cell proliferation and differential induction of endothelial stem cell tissue (Karar and Maity 2011). Mutated ROS 1 gene produces a constantly active ROS tyrosine kinase. Therefore, leading to a constant stimulation of STAT 3 pathways and thus, the upregulation of angiogenic factors.

Angiogenesis supports tumour development by ensuring a constant supply of blood to the malignant cells. Furthermore, it can lead to metastasis as oncogenic cells can be transported to other parts of the body through the circulatory system (Yang et al. 2013). Apart from aiding tumour development through angiogenesis, STAT3 signaling pathway also enhances tumour development through the dysregulation of Bcl-xL, Mcl-1, and Bcl 2 anti-apoptotic genes; activation of Cyclin D1 and Myc proteins which induce cellular proliferation; and suppression of both innate and adaptive immune expression (Karar and Maity 2011).

Uncontrolled STAT 3 signaling is due to three main reasons. Firstly, a mutated ROS1 gene will lead to the production of a mutated ROS tyrosine kinase, which will be constantly active and hence, will trigger STAT 3 signaling pathway continuously. Secondly, overstimulation of cytokines and growth factors. Lastly, the silencing of STAT3- inhibiting regulator. Over simulation of STAT3 pathway promotes tumour development (Zong et al. 1998; Nguyen et al. 2002).

PI3K is the key growth regulating cellular factor where its signaling cascade involves AKT kinase and mTOR; Both plays a part in controlling proliferation and differentiation of the cells and inhibition of physiological apoptosis. Recent studies identified PI3K involvement in vasculature development during embryonic stage where defects in PI3K exhibit production of functionless vessels (El-Deeb et al. 2011).

PI3 pathways contribute to tumour development by its regulations on cell growth mainly via angiogenic control of AKT kinase. Angiogenesis in PI3K pathways is achieved through both direct and indirect hypoxic dependent mechanisms which mainly comprises of nitric oxide pathways, HIF 1, angiopoietin, and VEGF induction. Being the key regulator of angiogenesis, VEGF cascades its effect in the cell through PI3K pathway with mTOR as the mediator. Its intracellular effects include a positive feedback mechanism whereby more angiogenic factors such as VEGF and angiopoietin are induced (Yang et al. 2013). Thus, uncontrolled stimulation from PI3K enhances VEGF positive feedback without requiring extracellular VEGF signaling. Of all the ROS1 originated mutagenic proteins, FIG-ROS is specific for PI3K pathway activation which contributes to tumour formation (El-Deeb et al. 2011). This leads to the development of PI3K pathways inhibitor drugs which is meant to halt tumour development by inhibiting angiogenic effect from VEGF secretion on PI3K pathway (Karar and Maity 2011).

44.4 ROS 1 and Malignancy

ROS1 gene is vulnerable to chromosomal mutation, bringing about the production of mutagenic fusion proteins. Current clinical investigation has identified five mutagenic proteins: v-ROS, Mcf3, FIG-ROS, SLC 34A2-ROS and CD74, all of which induce uncontrolled frequent activation of its downstream signaling pathways (El-Deeb et al. 2011). Examples of ROS1 related cancer is glioblastoma and non-small cell lung cancer (Jun et al. 2012).

Glioblastoma is a malignant brain tumour with high expression of mutated ROS1 genes. It is characterized by chromosomal deletion of ROS1 at 6q21, which causes production of GOPC-ROS1 protein. High expression of GOPC-ROS1 protein correlates to glioblastoma tumour formation (Ou et al. 2012). This mutation is expressed in 56 % of glioblastoma cell lines with contrasting minimal expression of ROS gene in functional brain cells (El-Deeb et al. 2011). Recent clinical investigations identified a link between FIG-ROS1, a mutated ROS1 gene, with glioblastoma cancer. FIG-ROS tumorigenicity is explored in an experiment and it was concluded that its tumorigenicity was credited to its promotion of anchorage independent growth of glioblastoma cells. In the experiment, the expression of FIG ROS1 in basal ganglia of mice led to the formation of astrocytomas. Ectopic ROS1 expression pattern exhibits low level in astrocytomas and a higher level in glioblastoma tumour. Hence, it is concluded that ROS1's role in tumour advancement is through proliferation and not so much in initiation (El-Deeb et al. 2011). Researchers are currently focused on discovering the cause of FIG-ROS1 fusion and way to inhibit glioblastoma.

Overall, mutation of ROS1 gene leads to production of mutated ROS tyrosine kinase which constantly stimulates the pathways including STAT3 and PI3K pathways. As mentioned above, over stimulation of these pathways promote angiogenesis and thus, enhance tumour formation. Currently, Crizotinib is a drug given to treat ROS1 gene mutation. Being an anaplastic lymphoma kinase (ALK) inhibitor, it inhibits the proliferation of cells and arrest mitosis cell cycle at G1 phase by preventing downstream signaling cascades. ROS tyrosine kinase shares some pathways with ALK. Hence, but blocking ALK pathways, a few pathways of ROS tyrosine kinase are also blocked, thereby inhibiting ROS tyrosine kinase's effect on tumour development. Despite not knowing its mechanism of action fully, Crizotinib does induce tumour shrinkage in patients with ROS 1 gene rearrangements (Doebele and Camidge 2012). Further research includes determining Crizotinib's mechanism of action and through this knowledge; more drugs can be developed to treat mutated ROS1 induced cancer.

References

Chen Z, Han ZC (2008) STAT3: a critical transcription activator in angiogenesis. Med Res Rev 28(2):185–200. doi:10.1002/med.20101

Doebele RC, Camidge DR (2012) Targeting ALK, ROS1, and BRAF kinases. J Thorac Oncol 7(16 Suppl 5):S375–S376. doi:10.1097/JTO.0b013e31826df05e

El-Deeb IM, Yoo KH, Lee SH (2011) ROS receptor tyrosine kinase: a new potential target for anticancer drugs. Med Res Rev 31(5):794–818. doi:10.1002/med.20206

Jun HJ, Johnson H, Bronson RT, de Feraudy S, White F, Charest A (2012) The oncogenic lung cancer fusion kinase CD74-ROS activates a novel invasiveness pathway through E-Syt1 phosphorylation. Cancer Res 72(15):3764–3774. doi:10.1158/0008-5472.CAN-11-3990

Karar J, Maity A (2011) PI3K/AKT/mTOR pathway in angiogenesis. Front Mol Neurosci 4:51. doi:10.3389/fnmol.2011.00051

Nguyen KT, Zong CS, Uttamsingh S, Sachdev P, Bhanot M, Le MT, Chan JL, Wang LH (2002) The role of phosphatidylinositol 3-kinase, rho family GTPases, and STAT3 in Ros-induced cell transformation. J Biol Chem 277(13):11107–11115. doi:10.1074/jbc.M108166200

Ou SH, Tan J, Yen Y, Soo RA (2012) ROS1 as a 'druggable' receptor tyrosine kinase: lessons learned from inhibiting the ALK pathway. Expert Rev Anticancer Ther 12(4):447–456. doi:10.1586/era.12.17

Yang F, Zhang W, Li D, Zhan Q (2013) Gadd45a suppresses tumor angiogenesis via inhibition of the mTOR/STAT3 protein pathway. J Biol Chem 288(9):6552–6560. doi:10.1074/jbc.M112.418335

Zong CS, Zeng L, Jiang Y, Sadowski HB, Wang LH (1998) Stat3 plays an important role in oncogenic Ros- and insulin-like growth factor I receptor-induced anchorage-independent growth. J Biol Chem 273(43):28065–28072

Chapter 45
SIRT1

Abstract SIRT1 belongs to the family of proteins called the Sirtuins and the mammalian equivalent of the yeast silent information regulator, sir2. SIRT1 has been implicated in a variety of cellular processes such as aging, transcription, neurodegeneration and apoptosis as it has been found to deacetylate the proteins involved these processes. The Sirtuin group of proteins help cells adapt against environmental stressors, such as DNA damage and replication stress which may arise from tissue ischemia. SIRT1 deacetylates key signalling molecules that respond to these stress signals and are consequently involved in many cellular processes such as metabolism, survival, proliferation and apoptosis. Studies have shown that SIRT1 gene was significantly up regulated in leukaemia (AML), prostate, colon and skin cancers. Activation of SIRT1 by resveratrol that can be found in red wine was able to reduce signs of ageing. These signs include decreased inflammation and apoptosis of the vascular system and increased aortic elasticity. SIRT1 has shown a promising role in the modulation of vascular disease as a regulator in vascular growth and homeostasis.

Keywords SIRT1 • Angiogenesis • Normal physiology • Disease • Malignancy

45.1 SIRT1

SIRT1 belongs to the family of proteins called the Sirtuins and the mammalian equivalent of the yeast silent information regulator, sir2. SIRT1 has been implicated in a variety of cellular processes such as aging, transcription, neurodegeneration and apoptosis as it has been found to deacetylate the proteins involved these processes (Guarani and Potente 2010a). It is a nicotinamide adenine dinucleotide (NAD+)-dependent deacetylase and requires an NAD+ group when removing each acetyl (Ac) group from its protein substrate (Stünkel and Campbell 2011). As a result, NAD+ and its reduced form NADH are inhibitors of SIRT1 (Stünkel and Campbell 2011).

A. Salajegheh, *Angiogenesis in Health, Disease and Malignancy*,
DOI 10.1007/978-3-319-28140-7_45

45.2 SIRT1 in Normal Physiology

The Sirtuin group of proteins help cells adapt against environmental stressors, such as DNA damage and replication stress which may arise from tissue ischemia. Within this group, the SIRT1 deacetylates key signalling molecules that respond to these stress signals and are consequently involved in many cellular processes such as metabolism, survival, proliferation and apoptosis. With regards to angiogenesis, SIRT1 acts as a metabolic sensor that adapts the growth of blood vessels to metabolic tissue requirements (Guarani and Potente 2010b).

In a cell and tissue-dependent context, SIRT1 is found in the nucleus and can shuttle to the cytoplasm (Guarani and Potente 2010b; Stünkel and Campbell 2011). SIRT1 deacetylates by transferring the acetyl group of an acetylated protein substrate to the ADP-ribose moiety of NAD+, which results in a deacetylated protein, 2′-acetyl-ADP-ribose, and nicotinamide (NAM) (Guarani and Potente 2010b). The deacetylase activity of SIRT1 is activated via NAD+, whereas NAM and NADH, the reduced form of NAD+, inhibits its catalytic activity. The physiological functions of SIRT1 are result from the adaptation of gene expression to the energetic state of the cell. This energetic state is monitored by the cell through its NAD+ levels (Guarani and Potente 2010b). SIRT1 deacetylates histones, transcription factors and cofactors and is consequently involved in the coordination of gene expression required for a range of processes such as cell survival, metabolism, proliferation and differentiation. SIRT1 can also deacetylate non-histone substrates, such as p53 and forkhead box class O (FOXO) which act as important downregulating effectors in the SIRT1 signaling circuitry that controls metabolism and response to stress (Guarani and Potente 2010b; Stünkel and Campbell 2011). Evidence shows that vascular endothelium are important tissue targets for the biological functions of SIRT1 where it is prominently expressed in endothelial cells of growing blood vessel sprouts (Guarani and Potente 2010b). Repressed SIRT1 activity in endothelial cells has been found to impair the ability of the cells to form new vessels in response to ischemic stress. SIRT1 binds and deacetylates FOXO1, an important negative regulator of angiogenesis resulting in the activation of angiogenic activity (Guarani and Potente 2010b).

45.3 SIRT1 in Disease

Several studies have linked SIRT1 with metabolism. Opposed to sir2, the yeast homolog of SIRT1, SIRT1 acts as a positive regulator of insulin. SIRT1 is downregulated in cells that have high insulin resistance and its activation increases and improves insulin sensitivity patients with type II diabetes mellitus (Sun et al. 2007; Kong et al. 2009). Pancreatic β cells are known to secrete insulin to promote glucose uptake and react to increased glucose levels in peripheral tissues. SIRT1 is expressed in pancreatic β cells and suppresses uncoupling protein 2 (UCP2), which in turn uncouples respiration from ATP production in the mitochondria. SIRT1 indirectly activates two insulin transcription factors, NeuroD and MafA by forming a complex with the promyelocytic leukaemia protein (PML) and FOXO1; this may protect the pancreatic β cell pathway from oxidative damage. Therefore it can be indicated that SIRT1 acts as a positive regulator in the secretion of insulin by pancreatic β cells (Kong et al. 2009).

SIRT1 has been associated with DNA repair by deacetylation of Ku70, a repair protein. Overexpression of SIRT1 increases the DNA repair capacity that may have resulted from radiation. SIRT1 siRNA may also inhibit endogenous SIRT1 expression, thus decreasing its repair ability and regulating the overall process of DNA repair (Jeong et al. 2007).

SIRT1 has been shown to protect against cardiac hypertrophy, which is a major cause of heart failure. Cardiac hypertrophy decreases the ability of the heart to pump blood effectively and results in heart failure. Eventually, this leads to organ failure due to increased apoptosis. SIRT1 protects the primary cultured myocytes from apoptosis stimulated by activation of PARP-1 in a p35-dependent pathway. However, SIRT1 deacetylates the histone variant H2A.Z, which is a factor that induces cardiac hypertrophy (Chen et al. 2006).

SIRT1 also has protective factors in kidney diseases. SIRT1 responds to oxidative stress via inhibition of p53 and reduces mesangial cell apoptosis as it prevents glomerular apoptosis in kidney disease (Kume et al. 2006, 2007).

SIRT1 also regulates transactivator of transcription (Tat) activity, a HIV-1 related protein that is required for the transcriptional activation of the provirus. Cessation of Tat activity compromises the replication of the HIV-1 virus. SIRT1 deacetylates Tat protein, thus activating it and HIV-1 viral transcription is consequently increased (Pagans et al. 2005).

45.4 SIRT1 in Malignancy

A number of cancers have been associated with increased age. Genes that code for proteins that support tumour growth are known as "tumour promoters" whereas genes that code for proteins that inhibit tumour growth are known as "tumour suppressors" (Fang and Nicholl 2011). The pathways of SIRT1 that serve to protect normal cells may be the cause that prevents apoptosis of cancer cells. Contradicting studies have associated SIRT1 as both a tumour suppressor and tumour promoter. This is discussed in the paragraphs below.

Studies have shown the SIRT1 gene to be significantly up regulated in cancers such as acute myeloid leukaemia (AML), prostate, colon and skin cancer (Carafa et al. 2012). Overexpression of SIRT1 has also been shown in cutaneous squamous and basal cell carcinomas (Carafa et al. 2012).

Studies show evidence that SIRT1 may activate a tumour promoter and inactivate tumour suppressors. Hypermethylated in cancer 1 (HIC1) is a tumour suppressor that can be inactivated or deleted in some cancers (Fang and Nicholl 2011). HIC1 forms a regulatory feedback loop with the p53 and SIRT1. In the absence of HIC1, SIRT is activated and deacetylates p53, which acts as a tumour suppressor (Stünkel and Campbell 2011).

SIRT1 may also act as a tumour promoter via the stabilisation of the proto-oncogene protein, N-Myc. This protein induces SIRT1 transcription in neuroblastoma, thus enhancing the stability of N-Myc in a positive feedback loop (Fang and Nicholl 2011). SIRT1 induces the inactivation of MKP3 phosphatase resulting in increased ERK phosphorylation and subsequently N-Myc Ser62 phosphorylation. This has the effect of preventing the degradation of N-Myc (Fang and Nicholl 2011).

Overexpression of SIRT1 can inhibit the apoptosis of tumour cells and result in the ineffectiveness of some anti-cancer drugs (Pearson et al. 2008). Additionally, the overexpression of SIRT1 can inhibit p53-dependent cell-cycle arrest and apoptosis (Carafa et al. 2012). DNA damage causes SIRT1 to deacetylate p53, thereby inhibiting its functions and disrupting p53-dependent pathways. This has the effect of decreasing the cell's ability to respond effectively to the damage and oxidative stress (Carafa et al. 2012). SIRT1 can also deacetylate forkhead box protein O members, inhibiting FOXO-dependent transcription or apoptosis (Pearson et al. 2008). SIRT1 promotes angiogenesis as a result of deacetylation of FOXO1, via inhibition of senescence which promotes endothelial cell growth and vascular development (Pearson et al. 2008).

Tumour growth reduction is an effect of SIRT1's ability to deacetylate β-caternin, a tumour promoting protein, and enhance cytoplasmic localization of the nuclear-localised oncogenic form of β-catenin (Altieri 2008). Overexpression of SIRT1 also decreases cell proliferation in colon cancer, which is promoted by active β-catenin (Fang and Nicholl 2011).

Studies show SIRT1 is down regulated in BRCA1-associated breast cancer, a mutant variant, and acts as a tumour suppressor by inhibiting Survivin, which is an important anti-apoptotic protein that is overexpressed in many cancers (Carafa et al. 2012; Fang and Nicholl 2011).

The equilibrium between SIRT1 as a tumour promoter and tumour suppressor greatly influences its role in oncogenesis. Therefore further studies should investigate the development of anti-cancer drugs, which can simultaneously upregulate and downregulate SIRT1 activity.

SIRT1 is a nuclear protein that is primarily expressed in neurons. SIRT1 binds to and deacetylates a number of essential transcription factors (Donmez and Outeiro 2013). SIRT1 is involved indirectly via the FOXO transcription factors that are involved in different molecular pathways such as neuronal protection, stress resistance and glucose production where NAD+, NADH and NAM regulate SIRT1 function (Donmez and Outeiro 2013). Energy depletion results in an increase of NAD levels and consequently activates SIRT1 deacetylation activity, which is NAD-dependent (Donmez and Outeiro 2013). The AMPK phosphorylates whilst SIRT1 deacetylates thus activating PGC-1α, to stimulate fatty acid oxidation and mitochondrial biogenesis. SIRT1 activates AMPK activity, which then increases NAD levels (Donmez and Outeiro 2013).

It has been shown that SIRT1 has an association with neurodegenerative diseases due to its involvement with gene expression (Donmez and Outeiro 2013). In cases where certain genes are repressed this could result in conditions that are age-related, such as cancer, heart disease, diabetes neurodegenerative diseases such as Alzheimer's disease (Donmez and Outeiro 2013).

Neurodegenerative diseases have prevalence amongst the aging population. The aging process is associated with telomere shortening. SIRT1 is involved with telomere function and maintenance in aging process (Carafa et al. 2012). SIRT1 promotes replicative senescence in reaction to chronic oxidative stress through the overexpression of p19ARF which is a senescence regulator. This has the effect of positively regulating the tumour suppressor, p53, by inhibiting MDM2 mediates p53 degradation (Carafa et al. 2012).

Over expression of SIRT1 protects against Alzeheimer's disease, Huntington's disease and amyotrophic lateral sclerosis (ALS). Activation of SIRT1 also inhibits NFκB signalling and reduces cell death (Carafa et al. 2012). In Alzheimer's disease studies, overexpression of SIRT1 in the brain has been found to reduce β-amyloid production and the formation of aggregates. On the other hand, inhibition of SIRT1 activity results in the formation of β-amyloid aggregates in the brains of Alzheimer's patients (Carafa et al. 2012). Additionally, calorie restriction results in an increase in NAD in cells. NAD converts glucose from food into energy. The decrease of calories increases NAD levels, which is required for SIRT1 activation and has been shown to reduce β-amyloid plaques (Donmez and Outeiro 2013).

SIRT1 has shown a promising role in the modulation of vascular disease. It is seen as a regulator in vascular growth and homeostasis. Current research shows SIRT1 to provide a therapeutic insight in the treatment of ischemic-induced vascular disease and blood vessel growth (Guarani and Potente 2010b). Much of the pharmaceutical research has been invested into using small molecule activators of the SIRT1 to facilitate angiogenesis activity of atherosclerotic-disease vessels after ischemic damage via the activation of PARP-1 (Chen et al. 2006). A key study by Pearson et al. (2008) on mice showed an activator of SIRT1 called resveratrol (found in red wine) was able to reduce signs of ageing. These signs include decreased inflammation and apoptosis of the vascular system and increased aortic elasticity (Pearson et al. 2008). Current studies indicate SIRT1 and its small molecule activators are possible targets for the treatment of the cardiovascular disease (Guarani and Potente 2010b).

References

Altieri DC (2008) Survivin, cancer networks and pathway-directed drug discovery. Nat Rev Cancer 8(1):61–70. doi:http://www.nature.com/nrc/journal/v8/n1/suppinfo/nrc2293_S1.html

Carafa V, Nebbioso A, Altucci L (2012) Sirtuins and disease: the road ahead. Front Pharmacol 3:4. doi:10.3389/fphar.2012.00004

Chen I-Y, Lypowy J, Pain J, Sayed D, Grinberg S, Alcendor RR, Sadoshima J, Abdellatif M (2006) Histone H2A.z is essential for cardiac myocyte hypertrophy but opposed by silent information regulator 2α. J Biol Chem 281(28):19369–19377. doi:10.1074/jbc.M601443200

Donmez G, Outeiro TF (2013) SIRT1 and SIRT2: emerging targets in neurodegeneration. EMBO Mol Med 5(3):344–352. doi:10.1002/emmm.201302451

Fang Y, Nicholl MB (2011) Sirtuin 1 in malignant transformation: friend or foe? Cancer Lett 306(1):10–14. doi:10.1016/j.canlet.2011.02.019

Guarani V, Potente M (2010a) SIRT1 – a metabolic sensor that controls blood vessel growth. Curr Opin Pharmacol 10(2):139–145. doi:10.1016/j.coph.2010.01.001

Guarani V, Potente M (2010b) SIRT1- a metabolic sensor that controls blood vessel growth. Curr Opin Pharmacol 10(2):139–145. doi:http://dx.doi.org/10.1016/j.coph.2010.01.001

Jeong J, Juhn K, Lee H, Kim SH, Min BH, Lee KM, Cho MH, Park GH, Lee KH (2007) SIRT1 promotes DNA repair activity and deacetylation of Ku70. Exp Mol Med 39(1):8–13. doi:10.1038/emm.2007.2

Kong X-X, Wang R, Liu X-J, Zhu L-L, Shao D, Chang Y-S, Fang F-D (2009) Function of SIRT1 in physiology. Biochem Mosc 74(7):703–708. doi:10.1134/S0006297909070013

Kume S, Haneda M, Kanasaki K, Sugimoto T, Araki S, Isono M, Isshiki K, Uzu T, Kashiwagi A, Koya D (2006) Silent information regulator 2 (SIRT1) attenuates oxidative stress-induced mesangial cell apoptosis via p53 deacetylation. Free Radic Biol Med 40(12):2175–2182. doi:10.1016/j.freeradbiomed.2006.02.014

Kume S, Haneda M, Kanasaki K, Sugimoto T, Araki S, Isshiki K, Isono M, Uzu T, Guarente L, Kashiwagi A, Koya D (2007) SIRT1 inhibits transforming growth factor beta-induced apoptosis in glomerular mesangial cells via Smad7 deacetylation. J Biol Chem 282(1):151–158. doi:10.1074/jbc.M605904200

Pagans S, Pedal A, North BJ, Kaehlcke K, Marshall BL, Dorr A, Hetzer-Egger C, Henklein P, Frye R, McBurney MW, Hruby H, Jung M, Verdin E, Ott M (2005) SIRT1 regulates HIV transcription via Tat deacetylation. PLoS Biol 3(2):e41. doi:10.1371/journal.pbio.0030041

Pearson KJ, Baur JA, Lewis KN, Peshkin L, Price NL, Labinskyy N, Swindell WR, Kamara D, Minor RK, Perez E, Jamieson HA, Zhang Y, Dunn SR, Sharma K, Pleshko N, Woollett LA, Csiszar A, Ikeno Y, Le Couteur D, Elliott PJ, Becker KG, Navas P, Ingram DK, Wolf NS, Ungvari Z, Sinclair DA, de Cabo R (2008) Resveratrol delays age-related deterioration and mimics transcriptional aspects of dietary restriction without extending life span. Cell Metab 8(2):157–168. doi:10.1016/j.cmet.2008.06.011

Stünkel W, Campbell RM (2011) Sirtuin 1 (SIRT1): the misunderstood HDAC. J Biomol Screen 16(10):1153–1169. doi:10.1177/1087057111422103

Sun C, Zhang F, Ge X, Yan T, Chen X, Shi X, Zhai Q (2007) SIRT1 improves insulin sensitivity under insulin-resistant conditions by repressing PTP1B. Cell Metab 6(4):307–319. doi:http://dx.doi.org/10.1016/j.cmet.2007.08.014

Chapter 46
SMAD4 (Mothers Against Decepentaplegic Homolog 4)

Abstract The significance of the SMAD4 gene is considerable in various aspects related to angiogenesis activation in normal physiology, cancer and disease and its influences have been extensively researched. The SMAD4 protein is coded by the SMAD4 gene and plays an essential role in the Transforming Growth Factor- β (TGF-β) signalling pathway. TGF-β genes have multitudinous effects on a wide range of biological activities including cell proliferation termination, encouraging cell differentiation and pattern forming during early vertebrate development. The SMAD4/TGF-β signaling pathway induces angiogenesis through inducing the expression of Vascular Endothelial Growth Factor (VEGF) and other proangiogenic factors. Upregulation of these proangiogenic factors hence will target the phosphatase and tensin homolog (PTEN) in endothelial cells which promotes endothelial cell migration (angiogenesis essentiality) and tube formation. The mechanisms and consequences involved in a mutation of SMAD4 mainly relates to its loss of function. Mutation in SMAD4 gene has been shown leading familial juvenile polyposis. Previous researches also showed that tumor growth and vascularization density could be markedly impaired by restoration of SMAD4. It is hypothesised that SMAD4, to a certain extent, controls an angiogenic switch with VEGF (pro-angiogenic) and TSP-1 (anti-angiogenic) as the main targets. Yet to confirm, the mechanism of action in controling VEGF and TSP-1, this potential demands future research to develop a greater understanding of the SMAD4/TGF-β pathway.

Keywords SMAD4 • Angiogenesis • Normal physiology • Disease • Malignancy

46.1 SMAD4 (Mothers Against Decepentaplegic Homolog 4)

The significance of the SMAD4 gene is considerable in various aspects related to angiogenesis activation in normal physiology, cancer and disease and its influences have been extensively researched. The SMAD4 protein is coded by the SMAD4 gene and plays an essential role in the Transforming Growth Factor- β (TGF-β) signalling pathway (Blobe et al. 2000). Binding of TGF-β onto TGF-β receptors leads to SMAD2 or SMAD3 combining with SMAD4 to form a transcription unit, which ultimately allows the external environment of the cell to control the

A. Salajegheh, *Angiogenesis in Health, Disease and Malignancy*,
DOI 10.1007/978-3-319-28140-7_46

transcription of genes intracellularly (Blobe et al. 2000). Furthermore, TGF-β genes have multitudinous effects on a wide range of biological activities including cell proliferation termination, encouraging cell differentiation and pattern forming during early vertebrate development (Blobe et al. 2000). In particular, the role of the SMAD4/TGF-β signaling pathway in the regulation of angiogenesis is of paramount interest when considering the effect of the SMAD4 gene in normal physiology, cancer and disease.

46.2 SMAD4 in Normal Physiology

The SMAD4/TGF-β signaling pathway induces angiogenesis through inducing the expression of Vascular Endothelial Growth Factor (VEGF) and other pro-angiogenic factors. Specifically, the SMAD4/TGF-β pathway up- regulates pro-angiogenic miRNA-29a which targets the phosphatase and tensin homolog (PTEN) in endothelial cells (Ferrari et al. 2009). This leads to the activation of the AKT pathway which promotes endothelial cell migration (angiogenesis essentiality) and tube formation. Published articles have indicated that these mechanisms are involved *in vitro* and *in vivo* (Ferrari et al. 2009). Furthermore, the SMAD4/TGF-β signaling pathway encourages deposition of the extracellular matrix and induces apoptosis. The inducement of apoptosis is especially seen in angiogenesis whereby apoptosis of the endothelial cells lining the blood vessels occur, this process especially important for "pruning the forming of vascular network" (Geng et al. 2013). This is contrasted with the effects of VEGF which antagonizes apoptosis and protects the endothelial cells. A significant imbalance or lack of SMAD4 proteins potentially results in the crippling of the apoptosis mechanism, therefore causing an increase in the development of abnormal blood vessels (Geng et al. 2013). This was further substantiated in a study of SMAD4 absent mutant mice where results showed an unaffected, functioning vasculogenesis mechanism and an affected, abnormal angiogenesis mechanism where primary blood vessels failed to undergo remodeling or efficient sprouting *in vivo* (Geng et al. 2013).

46.3 SMAD4 in Disease

An example of a SMAD4 gene mutation, in this case leading to disease, is familial juvenile polyposis (O'Malley et al. 2012). The mutation involved in this disease is a germ line mutation, hence its classification as a familial/inherited disease (autosomal dominant). In familial juvenile polyposis, the SMAD4 gene mutation directly affects the coding of a protein that allows the environment of the cell to control the regulation of genes, in particular, genes that are involved in the growth and division of cells (O'Malley et al. 2012). This protein pathway is then disrupted and cells are then allowed to grow in an uncontrolled manner leading to polyp formation (O'Malley et al. 2012). These benign polyps develop in the gastrointestinal tract, especially the large colon and can range from a few to hundreds of polyps. Furthermore, polyp formation can lead to gastrointestinal bleeding, anemia, abdominal pain and diarrhoea. Approximately 10–50 % of individuals diagnosed with juvenile polyposis syndrome develop polyps that can become malignant and cancerous. However, if juvenile polyps develop during infancy, children can develop protein-losing enteropathy which results in severe diarrhea, failure to gain weight or grow at a normal rate and general wasting of the muscles. Additionally, a study has found that mutations occurring in SMAD4 before codon 423 resulted in vascular malformations and was identified within the stromal components of the polyps (O'Malley et al. 2012).

Most patients diagnosed with polyposis also present with hereditary hemorrhage telangiectasia; a disorder in the blood vessels characterized by high pressure blood in arteries moving directly into the veins rather than the capillaries (O'Malley et al. 2012). The SMAD4 protein is involved in making proteins that are found in blood vessel linings and interact with growth factors that control angiogenesis. A mutation results in a decrease in the amount of functional protein and this disrupts the development of blood vessels, causing arteriovenous malformation. When this occurs near the skin, it is clinically observed as red markings known as telangiectases. The thinner- walled veins cannot adapt to the high blood pressure, leading to straining and expansion of the veins and compression or irritation of adjacent tissues and frequent episodes of haemorrhage. Thus, a range of diseases can manifest ranging from nose bleeds to hemorrhages in the brain, lover, lungs or other organs (O'Malley et al. 2012).

46.4 SMAD4 in Malignancy

Other research indicates that the tumor suppressive effects of the SMAD4/TGF-β pathway to be context dependant (Zhang et al. 2010). For example, when the TGF-β pathway was found to be functional in the early stages of tumorigenesis (i.e. SMAD4 not mutated), tumor growth was delayed. However, when SMAD4 was lost or mutated, the tumor cells became resistant to the SMAD4/TGF-β pathway and tumor growth promotion and metastasis was observed (Zhang et al. 2010). The mechanisms and consequences involved in a mutation of SMAD4 mainly relates to its loss of function. In particular, this mutation leads to communication breakdown between the external and internal environment (nucleus) of the cell (Miyaki and Kuroki 2003). Therefore, "STOP" signals are not sent and received and the aforementioned functions of this pathway become severely affected. The effects of such mutations are evident in 55 % of all pancreatic adenocarcinomas where a mutation resulting in the interference of the SMAD4 homo-oligomer formation leads to an interference with the hetero-oligomer formation between SMAD2 and SMAD4 (Miyaki and Kuroki 2003). TGF-β production is also markedly increased in cancerous cells, promoting tumor growth through angiogenesis and the stimulation of VEGF production (Geng et al. 2013). Increased VEGF production results in an increase of nutrient and vasculature supply (essential for tumor progression) to tumorous tissues.

Extracellular matrix deposition activity is also significantly increased, aiding the survival of endothelial cells and maintaining the vasculature to the tumorous cells (Geng et al. 2013). Other side effects involved with increased TGF-β production include suppression of anti-tumor immune response and stimulation of connective tissue formation around the tumour. As a consequence of SMAD4/TGF-β pathway mutations and a result of these effects, tumorous cells have an increased ability to become more invasive and metastasize to distant organs (Geng et al. 2013).

Further experiments have also been conducted investigating the effects of restoring the SMAD4 gene into SMAD4 nulled cell lines *in vitro* and *in vivo* (Schwarte-Waldhoff et al. 2000). Interestingly, an angiogenic switch was observed in both (from pro-angiogenesis to anti-angiogenesis). As expected, the restoration of the SMAD4/TGF-β pathway in vitro resulted in re-established apoptosis activity and growth inhibition. However, this was not observed *in vivo* as sensitivity was not restored to the pathway, VEGF expression was down- regulated and TSP-1 expression was upregulated (Schwarte-Waldhoff et al. 2000). As a result, tumor growth was found to be significantly retarded and vascularization density was markedly impaired in comparison to the original SMAD4 nulled cell lines. Thus, it is theorized that SMAD4, to a certain extent, controls an angiogenic switch with VEGF (pro-angiogenic) and TSP-1 (anti-angiogenic) as the main targets. Although research is yet to confirm how SMAD4 controls VEGF and TSP-1, there are many potential SMAD binding sites on VEGF and future research should be conducted in this area and direction to develop a greater understanding of the SMAD4/TGF-β pathway (Schwarte-Waldhoff et al. 2000).

References

Blobe GC, Schiemann WP, Lodish HF (2000) Role of transforming growth factor beta in human disease. N Engl J Med 342(18):1350–1358. doi:10.1056/nejm200005043421807

Ferrari G, Cook BD, Terushkin V, Pintucci G, Mignatti P (2009) Transforming growth factor-beta 1 (TGF-beta1) induces angiogenesis through vascular endothelial growth factor (VEGF)-mediated apoptosis. J Cell Physiol 219(2):449–458. doi:10.1002/jcp.21706

Geng L, Chaudhuri A, Talmon G, Wisecarver JL, Wang J (2013) TGF-beta suppresses VEGFA-mediated angiogenesis in colon cancer metastasis. PLoS One 8(3):e59918. doi:10.1371/journal.pone.0059918

Miyaki M, Kuroki T (2003) Role of Smad4 (DPC4) inactivation in human cancer. Biochem Biophys Res Commun 306(4):799–804

O'Malley M, LaGuardia L, Kalady MF, Parambil J, Heald B, Eng C, Church J, Burke CA (2012) The prevalence of hereditary hemorrhagic telangiectasia in juvenile polyposis syndrome. Dis Colon Rectum 55(8):886–892. doi:10.1097/DCR.0b013e31825aad32

Schwarte-Waldhoff I, Volpert OV, Bouck NP, Sipos B, Hahn SA, Klein-Scory S, Luttges J, Kloppel G, Graeven U, Eilert-Micus C, Hintelmann A, Schmiegel W (2000) Smad4/DPC4-mediated tumor suppression through suppression of angiogenesis. Proc Natl Acad Sci U S A 97(17):9624–9629

Zhang B, Halder SK, Kashikar ND, Cho YJ, Datta A, Gorden DL, Datta PK (2010) Antimetastatic role of Smad4 signaling in colorectal cancer. Gastroenterology 138(3):969–980.e961–963. doi:10.1053/j.gastro.2009.11.004

Chapter 47
Sprouty-Related, EVH1 Domain-Containing Protein 1 (SPRED-1)

Abstract The *SPRED-1* gene encodes *Sprouty-Related, EVH1 Domain-containing protein 1 (SPRED-1)* in humans. The gene is located on chromosome 15q13.2, consisting of seven coding exons. The activity of *SPRED-1* is primarily regulated by tyrosine phosphorylation, facilitated by haemopoietic factors. Legius syndrome, an autosomal dominant condition, represents with changes in skin pigmentation, attention deficit hyperactivity disorder (ADHD) or attention deficit disorder (ADD) and it is believed to be related to germline mutations in the *SPRED-1* gene. Angiogenesis can be regulated by a variety of genes, one of which is the gene microRNA-126 (miR-126). *SPRED-1* can block the effect of miR-126 in angiogenesis and capillarization in muscle tissue in patients with chronic obstructive pulmonary disease (COPD). *SPRED-1* plays a vital role in the tumorigenesis and metastasis of a stable tumour since it is an inhibitor of Ras-MAPK and RhoA cell signalling pathways. Through further investigation into the mechanism of *SPRED-1* expression, we may be able to better understand and treat the health concerns which its mutation imposes.

Keywords Sprouty-related • EVH1 domain-containing protein 1 • SPRED-1 • Angiogenesis • Normal physiology • Disease • Malignancy

47.1 Sprouty-Related, EVH1 Domain-Containing Protein 1 (SPRED-1)

The *SPRED-1* gene encodes *Sprouty-Related, EVH1 Domain-containing protein 1 (SPRED-1)* in humans. The gene is located on chromosome 15q13.2, consisting of seven coding exons. Mutations in the gene are said to cause Legius syndrome as well as a predisposition to childhood leukaemia. *SPRED-1* was recently discovered to be a negative Ras Mitogen Activated Protein Kinase (Ras-MAPK) pathway regulator associated with a neurofibromatosis 1 (NF-1)-like human syndrome (Phoenix and Temple 2010). It is a tumour suppressor and mainly distributes to the human brain. In addition, it is expressed in the colon, lung and foetal tissues (Brems et al. 2007). The activity of *SPRED-1* is primarily regulated by tyrosine phosphorylation, facilitated by haemopoietic factors (Zhang et al. 2013).

A. Salajegheh, *Angiogenesis in Health, Disease and Malignancy*,
DOI 10.1007/978-3-319-28140-7_47

47.2 SPRED-1 in Disease

Legius syndrome, an autosomal dominant condition, is related to germline mutations in the *SPRED-1* gene resulting in loss of function (Sumner et al. 2011). Legius syndrome is characterized by changes in skin pigmentation. Most individuals who are affected will most likely have several café-au-lait spots, which are flat patches on the skin that are noticeably darker than the surrounding skin. These spots require germline *SPRED-1* mutations as well as a somatic wild-type *SPRED-1* allele, which shows that a full inactivation of *SPRED-1* is involved. Affected individuals may also present with freckles in the groin and armpit regions. Other symptoms associated with Legius syndrome include macrocephaly and unusual facial characteristics. Individuals may also be diagnosed with learning difficulties such as attention deficit hyperactivity disorder (ADHD) or attention deficit disorder (ADD) (Brems et al. 2007). A study showed that *SPRED-1* is also required for hippocampus-dependent learning and synaptic plasticity in mice (Denayer et al. 2008). It must be noted that many of the aforementioned symptoms may also occur in a similar disorder called NF-1. Therefore, it is particularly difficult to distinguish between the two in their earlier stages, making clinical diagnosis between Legius syndrome and NF-1 an emerging challenge. It is important to intricately distinguish between phenotypes of Legius syndrome and NF-1 since the prognoses of each disease are vastly different (Sumner et al. 2011). For example, those with Legius syndrome are not known to have an increased risk of tumours.

Angiogenesis can be regulated by a variety of genes, one of which is the gene MicroRNA-126 (miR-126). This however can be blocked by the expression of an inhibitor of angiogenesis, namely *SPRED-1* (Porlier et al. 2014). In a study, the expression of miR-126 was assessed in relation to capillarization in muscle tissue obtained from a needle biopsy in five patients with mild chronic obstructive pulmonary disease (COPD) as well as five healthy controls of similar ages. It was hypothesized that the expression of *SPRED-1* would be enhanced whereas miR-126 would be downregulated in COPD. Results showed that *SPRED-1* levels were enhanced by 2.1 fold while the number of capillaries tended to be reduced in mild COPD in comparison to the healthy controls. This study shows that there may be a role for *SPRED-1* in muscle capillarization in mild COPD, however further studies more specific to endothelial cells may be required in order to better understand the mechanisms behind muscle angiogenesis (Porlier et al. 2014).

The normal development, response to injury and prevention of tumour progression in healthy tissue requires tight regulation of vascular formation, maintenance and remodelling. It was found that increased *SPRED-1* expression led to inhibition of *VEGF* signalling in embryonic developmental stages in zebrafish, resulting in a loss of vascular integrity and haemorrhage (Fish et al. 2008). Since the inhibition of new vessels is directly related to an inhibition of tumour growth, increased levels of *SPRED-1* was detrimental to the expansion of tumour masses.

The *SPRED-1* gene produces *SPRED-1* protein which acts as a negative regulator of the Ras-MAPK signalling pathway which regulates growth and tumour formation (McCubrey et al. 2007). *SPRED* proteins function downstream of Ras and are membrane-associated suppressors of Ras-MAPK activation (which is normally induced by growth factors). *SPRED-1* inhibits MAPK activation via the suppression of Raf kinase activation. Studies suggest that most *SPRED-1* germline gene mutations result in a "truncated protein" that is incapable of normal direct interaction with the Ras pathway (Brems et al. 2007; Sumner et al. 2011).

47.3 SPRED-1 Role in Malignancy

SPRED-1 plays a vital role in the tumorigenesis and metastasis of a stable tumour since it is an inhibitor of Ras-MAPK and RhoA cell signalling pathways. Recent studies have found that the inactivation of *SPRED-1* may lead to survival time extension, proliferation and induction of angiogenesis of acute myeloid leukaemia (AML) cells. This strongly suggests that *SPRED-1* is related to leukaemogenesis. This is supported by recent studies which have shown that the expression level of *SPRED-1* was decreased in patients who were diagnosed with AML (Zhang et al. 2013).

It has also been suggested that the *SPRED-1* disorder causes a predisposition to leukaemia in children (Pasmant et al. 2009). All conditions are "secondary germline mutations" in genes consisting of encoding components of Ras-MAPK pathway. It is suggested that somatic mutations in genes may be linked to the pathogenesis of juvenile myelomonocytic leukaemia (JMML). Since *SPRED-1* negatively regulates Ras-MAPK signalling by suppressing Raf activation, a mutation in this gene could lead to an uncontrolled MAPK pathway, resulting in the uncontrolled growth of tumours. However, a more recent study has shown that there was an absence of the *SPRED-1* mutation in cases such as early childhood leukaemia including JMML, indicating that the relation between *SPRED-1* mutations and childhood myeloid malignancies requires further clarification (Batz et al. 2010). This opposes the previous claim that a *SPRED-1* disorder may cause predisposition to leukaemia in children.

It can be seen that *SPRED-1* is an important gene with a fundamental role in the regulation of the Ras-MAPK pathway. Mutations in *SPRED-1* have been shown to cause disease as well as cancer through angiogenesis. Through further investigation into the mechanism of *SPRED-1* expression, we may be able to better understand and treat the health concerns which its mutation imposes.

References

Batz C, Hasle H, Bergstrasser E, van den Heuvel-Eibrink MM, Zecca M, Niemeyer CM, Flotho C (2010) Does SPRED1 contribute to leukemogenesis in juvenile myelomonocytic leukemia (JMML)? Blood 115(12):2557–2558. doi:10.1182/blood-2009-12-260901

Brems H, Chmara M, Sahbatou M, Denayer E, Taniguchi K, Kato R, Somers R, Messiaen L, De Schepper S, Fryns JP, Cools J, Marynen P, Thomas G, Yoshimura A, Legius E (2007) Germline loss-of-function mutations in SPRED1 cause a neurofibromatosis 1-like phenotype. Nat Genet 39(9):1120–1126. doi:10.1038/ng2113

Denayer E, Ahmed T, Brems H, Van Woerden G, Borgesius NZ, Callaerts-Vegh Z, Yoshimura A, Hartmann D, Elgersma Y, D'Hooge R, Legius E, Balschun D (2008) Spred1 is required for synaptic plasticity and hippocampus-dependent learning. J Neurosci 28(53):14443–14449. doi:10.1523/jneurosci.4698-08.2008

Fish JE, Santoro MM, Morton SU, Yu S, Yeh RF, Wythe JD, Ivey KN, Bruneau BG, Stainier DY, Srivastava D (2008) miR-126 regulates angiogenic signaling and vascular integrity. Dev Cell 15(2):272–284. doi:10.1016/j.devcel.2008.07.008

McCubrey JA, Steelman LS, Chappell WH, Abrams SL, Wong EW, Chang F, Lehmann B, Terrian DM, Milella M, Tafuri A, Stivala F, Libra M, Basecke J, Evangelisti C, Martelli AM, Franklin RA (2007) Roles of the Raf/MEK/ERK pathway in cell growth, malignant transformation and drug resistance. Biochim Biophys Acta 1773(8):1263–1284. doi:10.1016/j.bbamcr.2006.10.001

Pasmant E, Ballerini P, Lapillonne H, Perot C, Vidaud D, Leverger G, Landman-Parker J (2009) SPRED1 disorder and predisposition to leukemia in children. Blood 114(5):1131. doi:10.1182/blood-2009-04-218503

Phoenix TN, Temple S (2010) Spred1, a negative regulator of Ras-MAPK-ERK, is enriched in CNS germinal zones, dampens NSC proliferation, and maintains ventricular zone structure. Genes Dev 24(1):45–56. doi:10.1101/gad.1839510

Porlier A, Dubé A, Debigaré R, Maltais F (2014) SPRED1 in quadriceps of patients with mild chronic obstructive pulmonary disease (1102.21). FASEB J 28(1 Supplement):1102.21

Sumner K, Crockett DK, Muram T, Mallempati K, Best H, Mao R (2011) The SPRED1 variants repository for Legius syndrome. G3 (Bethesda, MD) 1(6):451–456. doi:10.1534/g3.111.000687

Zhang Y, Li Y, Zhang R (2013) Relationship between Spred1 and acute myeloid leukemia. Zhongguo Shi Yan Xue Ye Xue Za Zhi 21(4):1083–1087. doi:10.7534/j.issn.1009-2137.2013.04.052

Chapter 48
Signal Transducer and Activator of Transcription of 5A and S3 (STAT5 and STATS3)

Abstract STAT genes are named after their protein role as a signal transducer and activator of transcription. The STAT transcription factor family contributes to various phases in cell differentiation, growth and survival, which are activated by JAK (Janus Kinase). The disorderly activated pathway, on the other hand, is found in primary tumours, which frequently leads to the increased lifespan of tumours, increased cases of angiogenesis and immunosuppression. STATS3 gene encodes the signal transducer and activator of transcription 3 protein from the transcription factor family and STAT5A and its protein known for its role as a signal transducer and activator of transcription of 5A, used for coding human proteins and is involved in ischemic heart disease, breast cancer, non-small-cell lung cancer. During embryogenesis process, STAT3 is shown needed for stem cells of embryo to renew itself in mouse during gastrulation and STAT5 and Prolactin (PRL) are working together in a positive autocrine feedback loop, which stimulate angiogenesis, whereby prolactin is induced by the STAT5 signalling cascade. Besides VEGF, other proteins like MMP-9, MMP-2 (matrix metalloproteinase-2), and bFGF (basic fibroblast growth factor) all cause angiogenesis and these are regulated by STATS3. The activation of the STAT5 gene is frequently found in tumours and the angiogenesis process of a malignancy. Better understanding the role of these members of the STAT family specially related to their role in chemotherapeutic sensitivity is always a matter of interest in this field.

Keywords Signal transducer and activator of transcription • STAT5 • STATS3 • Normal physiology • Disease • Malignancy

48.1 Signal Transducer and Activator of Transcription of 5A and S3 (STAT5 and STATS3)

STAT genes are named after their protein role as a signal transducer and activator of transcription. The STAT transcription factor family contributes to various phases in cell differentiation, growth and survival, which are activated by JAK (Janus Kinase) (Bratthauer et al. 2010). The disorderly activated pathway, on the other hand, is

A. Salajegheh, *Angiogenesis in Health, Disease and Malignancy*,
DOI 10.1007/978-3-319-28140-7_48

found in primary tumours, which frequently leads to the increased lifespan of tumours, increased cases of angiogenesis and immunosuppression.

STATS3 gene encodes the signal transducer and activator of transcription 3 protein from the transcription factor family. A transcription factor is a protein, which anchors to DNA and affects the rate of transcription from DNA gene to mRNA molecule. STATS3 gene is located on the chromosome 17, cytogenetic location 17q21.31. STAT5A and its protein known for its role as a signal transducer and activator of transcription of 5A, used for coding human proteins (Grimley et al. 1999). Its mechanisms are currently undergoing studies to determine its effects on ischemic heart disease, breast cancer, non-small cell lung cancer (Sanchez-Ceja et al. 2006; Peck et al. 2012; Yamaura et al. 2003).

As suggested by its name, the STAT5A gene embodies a dual function: signal transduction and the activation of transcription (Grimley et al. 1999). Belonging to the transcription factor STAT family, STAT5A is activated by and mediates cellular responses to cytokines, various cell ligands (IL2, IL3, IL7 GM-CSF), erythropoietin, thrombopoietin, and growth factors (ERBB4 and cytokine KITLG/SCF) (Lin and Leonard 2000). It then becomes phosphorylated by the receptor-associated kinases to form a homodimer or heterodimer structure that translocates into the nucleus- in which they undertake the role to activate transcription (Lin and Leonard 2000). It also binds to the GAS element and activates PRL-induced transcription, while also carrying out the role of standardizing the expression of milk proteins in the process of lactation. Consisting of one SH2 domain, it is able to bind with NR3C1 and interacts with NCOA1, SOCS7 and ERBB4. STAT5A has been discovered to be vital for tumorigenesis, and is independent of cell stimulus when it is activated in lymphoma and myeloma while linked with a TEL/JAK2 gene synthesis (Hodge et al. 2004; Nagy et al. 2006). Through experimentation of STAT5A on mice, it has been discovered that STAT5A advocates the anti-apoptotic functions in the cells (Debierre-Grockiego 2004).

48.2 STATS3 and STAT5A in Normal Physiology

STATS3 protein is found in tissues throughout the body. It responses to cytokines and growth factors and trans-locate into the nucleus of the cell and acts as a transcription activator. It is activated by a phosphorylation process on tyrosine 705. It functions as a key part in division, cell growth, cell movement and apoptosis. For example, it is responsible for the formation of cells specialized in bone modelling and absorption. In body immunity, STATS3 protein transmits signal in inflammation process. During embryogenesis process, STAT3 is shown needed for stem cells of embryo to renew itself in mouse during gastrulation (Takeda et al. 1997).

While the potent angiogenesis inducers and a range of downstream effector molecules that regulate angiogenesis have been found, the autocrine and paracrine signalling pathways are yet to be comprehendible. A 2013 study has established the involvement of STAT5 and Prolactin (PRL) in a positive autocrine feedback loop, which stimulate angiogenesis, whereby prolactin is induced by the STAT5 signalling cascade (Yang et al. 2013). Here, the constitutively active or dominant-negative mutant STAT5A genes in hCMED/D3 human brain endothelial cells were expressed, aiming to outline the signalling pathway that follows STAT5 in humans (Yang et al. 2013). Conditioned medium from CA-STAT5A, which overexpresses endothelial cells, is adequate to stimulate endothelial cell migration and the development of tubes (Yang et al. 2013). However, it was not found to be adequate for proliferation, which denotes that STAT 5A mediates the secretion of autocrine proangiogenic factors. As PRL may be an autocrine factor, CA-STAT5A expression induces PRL formation at protein and RNA levels, where STAT5A binds to PRL promoter areas, which indicates direct transcriptional regulation (Yang et al. 2013).

In the conditioned medium, PRL that has been induced by STAT5A is able to activate STAT5 which suggests there may be a positive feedback loop between STAT5 and PRL which stimulates angiogenesis. In addition, VEGF has been discovered to be a strong proangiogenic factor, which is stimulated by STAT5A activation.

Local secretion of PRL through the pituitary gland has been theorized to undertake roles as an autocrine/paracrine factor in the regulation of angiogenesis. PRL is mediated by the PRL receptor, whereby the STAT5 signalling cascade, along with other signalling pathways, is activated when ligands bond and receptor is dimerised. STAT5 activates the expression of several downstream genes, some of which are active in angiogenesis. STAT5 has been discovered to undertake a role in the STAT5-PRL-VEGF signalling cascade, which mediates growth factor-induced migration, invasion and tube formation of human brain endothelial cells (Yang et al. 2013). STAT5 is activated when PRL is bound to PRLR, which is indicative of an autocrine positive feedback loop. This is believed to accentuate and maintain angiogenic signalling. PRL signalling also brings about VEGF to be secreted, which is vital as a potent proangiogenic factor (Yang et al. 2013).

STAT5 is a vital transcription factor downstream of PRLR which facilitates numerous PRL activities, specifically in mammary gland development and breast cancer. In order to discriminate the development of abnormality, the role of proteases including morphogenic proteins on STAT5 has been explored (Yang et al. 2009). Active STAT5A stimulates PRL to form in endothelial cells; STAT5-induced PRL is bioactive as it stimulates the phosphorylation of PRLR and ERK1/2, and also stimulates the phosphorylation of STAT5. This allows for the establishment of a positive feedback loop that may accentuate or sustain angiogenesis impartial to the original paracrine pro-angiogenic signal. Proteases including morphogenic protein may turn off this loop, where these proteases form full-length PRL to the anti-angiogenic 16 kDa peptide (Yang et al. 2012).

A connection has been drawn between autocrine and paracrine PRL signalling breast carcinoma formation and development (Nitze et al. 2013). Here, PRLR expression has been found in tumour endothelial cells of breast carcinomas (Nitze et al. 2013). PRL has been found to be secreted by malignant glioma cells and present in benign CNS tumours. PRL secreted by EC, as well as PRL from tumour cells may play a role in the complex signalling network that maintains tumour angiogenesis (Yang et al. 2013). The positive feedback loop that binds PRL and STAT5 seems to be an appealing target for therapeutic involvement, where studies have shown that STAT5 blockers or PRLR antagonists can be utilised to hinder the positive feedback loop (Yang and Friedl 2015).

48.3 STATS3 and STAT5A in Disease

Defect in STAT3 gene can cause immunity diseases. STATS3 is required for the differentiation of T-helper cell TH17, which is indicated responsible for autoimmune disease (Yang et al. 2007). Rarely, Mutated STAT3 gene that loss its function causes Job syndrome or Hyperimmunoglobulin E syndrome, characterized by high level of IgE. Condition like this can be inherited too. A mutant gene from one parent is enough to cause this disease since the disease gene is autosomal dominant. Patient with this condition is highly susceptible to recurrent infections like pneumonia and skin eczema. In bone tissue, this can causes abnormalities like hyperextensibility, large range of joint movement, scoliosis, side to side curvature of spine, and osteopenia, weak bone density that can easily fractures. In teeth development, this can cause delay for deciduous teeth to fall out when the secondary dentition erupt (Levy and Loomis 2007). The rate of occurrence of this disease is one case per a million people, making it an extremely rare disease.

Presumptions for the involvement of STAT5A in angiogenesis signalling in ischemic heart disease can be made (Yamaura et al. 2003). The STAT transcription factor family is involved in various phases of differentiation, growth and survival in cells through the activation by JAK (Janus kinase). This rapid activation of JAK and STAT5A respectively occur during myocardial ischemia and reperfusion. In addition, laboratory research on a rat's heart has proven that the rapid activation of the two signalling pathways (which involve the two components, JAK and STAT) result in ischemia (Yamaura et al. 2003).

Research conveyed by the American Heart Association has revealed that STAT5A contributes to ischemic preconditioning (Mascareno et al. 2001). It has been found that in ischemic heart conditions, the selectively activated STAT5A gene from the STAT signalling pathway is coupled with an elevation in the ANG mRNA level, facilitated whereby STATs bind to ANG promotor (Mascareno et al. 2001). STATs activation is blocked by losartan and AG490. After administering losartan, examining the STAT complex formation reveals the reduction of ANG mRNA which may contribute to cardio-protection in ischemic hearts (Mascareno et al. 2001).

In ischemic conditions, STAT5A plays a role in the myocardia through three ways (Mascareno et al. 2001). First, the activation of STATs and their binding to the St domain found in the ANG promoter is related to heart tissue RAS activation through the positive-feedback mechanism. Studies prove there is a clear link between the ANG mRNA level in ischemic heart and JAK2 activation. Second, losartan and AG490 inhibits the function of AT1 receptor, which leads to decreased ANG mRNA level. Furthermore, AG490 selective for JAK2 inhibition leads to the improvement of the heart which implies JAK 2 activation contributes to ischemia (Mascareno et al. 2001). While much research regarding the mechanism of losartan is available, conclusive evidence on the influence of JAK2 activation has not yet been derived. RAS is an important activator which facilitates the cell death of cardiomyocytes. Hence in order to mediate cardiomyocyte apoptotic induction, signalling pathways that affect cell death converges on the STAT/JAK pathway to activate RAS. In addition to the STAT pathway, STAT5A plays a vital role in mediating mechanisms in ischemia. Further studies, however, are required for detailed mechanisms for ischemic heart disease (El-Adawi et al. 2003).

48.4 STATS3 and STAT5A in Malignancy

STATS 3 genes are related to cancer when it is constitutively activated, which means it produces STATS 3 protein continuously. This has proliferative and anti-apoptotic effects on cells (Klampfer 2006). Constitutive STAT3 activation usually indicates poor prognosis of various human cancers (Alvarez et al. 2006). STATS 3 protein promotes oncogenesis through various pathways. On one hand, it causes anti-apoptotic effect by deregulate genes that produce antiapoptotic proteins such as Bcl-xL, Mcl-1, and Bcl-2 (Yin et al. 2006). On the other hand, it causes proliferation of cells by deregulate genes that produce proliferation proteins like Cyclin D1 and Myc (Kusaba et al. 2006).

Upstream Kinases like Src, EGFR (epidermal growth factor receptor), CXCR2 and KSHV-GPCR (Kaposi sarcoma herpes virus G-protein coupled receptor), all can cause constitute activation of STATS3 gene. EGCG from green tea has been shown to inhibit activation of EGFR.

The other way of STATS3 in promoting cancer development is through promoting angiogenesis. This is achieved in 3 levels: by up-regulating the production of VEGF, by inducing endothelial cell proliferation, and by activating stem cell differentiation into endothelial cells. VEGF is a short term for vascular endothelial growth factor. STATS3 acts either as a transcription activator directly on the gene that produces VEGF, or on genes that produce proteins such as VEcadherin which acts on VEGF. Indirectly it also signals a range of cancer cell cytokine/growth factors that induce VEGF production. Studies in cancers such as melanoma, pancreas, gastric, breast, head and neck cancer all show a rise of VEGF directly correlated to STATS3 activity (Chen and Han 2008).

Besides VEGF, other proteins like MMP-9, MMP-2 (matrix metalloproteinase-2), and bFGF (basic fibroblast growth factor) all cause angiogenesis and they are all regulated by STATS3 (Lee et al. 2012). However contradictory, perhaps due to its other role in apoptosis, STATS3 shows a role in tumor suppression in some new studies (Krock et al. 2011). In a journal on brain cancer, or tumor of glioblastoma in man, STATS3 demonstrated to have a dual part either oncogenic or tumor suppressing relying upon the mutation situation of the tumor. The oncogenic pathway is called the LIFRbeta (leukaemia inhibiting factor receptor beta)-STAT3 signalling pathway and the suppressive pathway is called the PTEN –Akt -FOXO axis. A direct link between the two has been portrayed (de la Iglesia et al. 2008). On top of this, two recent studies carried out on APC mutant mice in bowel carcinogenesis demonstrated that STATS3 possess an inhibitory role depended on neoplasm phase (Musteanu et al. 2010).

Angiogenesis in cancer, the activation of the STAT5 gene is frequently found in tumours (Buettner et al. 2002). The relation of STAT5 with angiogenesis is through several mechanisms and factors. Therefore, in order to increase chemotherapeutic sensitivity, STAT5a suppression has been investigated (Buettner et al. 2002).

STAT5A arbitrates the physiological effects of PRL and growth hormones present in mammary glands (Peck et al. 2012). In breast cancer, low levels of nuclear localized and tyrosine phosphorylated STAT4A have been found to be linked with insufficient prognosis and an increased risk of antiestrogen therapy failure. An analysis of unknowing experiments treated with antiestrogen monotherapy showed that low levels of nuclear STAT5A are related to the risk of an undesirable result by at least four times. A study carried out on mice has revealed that the development and differentiation of normal mammary epithelial cells related to pregnancy needs STAT5A (Peck et al. 2012).

The relationship between STAT5A and cancer has been established through laboratory researches and trials. Corresponding to the founding, the reduced STAT5A in a mouse mammary cells delayed the cancer progression. Moreover, the loss of STAT5A allele decreased the tumour incidence resulting in smaller tumours. In reducing STAT5A levels, apoptotic index in adenocarcinomas has increased without affecting cell proliferation or differentiation.

Recent research targeted Genfitinib as an effective inhibitor of STAT5a (Herbst et al. 2004). Also known as epidermal growth factor receptor (EGFR) inhibitor, the mechanism works by causing cell death through the blockage of mechanical signalling within the cancer cells. The purpose of the drug is to treat non-small-cell lung cancer (NSCLC) that contains EGFR mutations on the surface of its cell. In many countries, approval has been made for the treatment targeting patients showing signs of advanced NSCLC as well as other types of solid tumours (Herbst et al. 2004).

References

Alvarez JV, Greulich H, Sellers WR, Meyerson M, Frank DA (2006) Signal transducer and activator of transcription 3 is required for the oncogenic effects of non-small-cell lung cancer-associated mutations of the epidermal growth factor receptor. Cancer Res 66(6):3162–3168. doi:10.1158/0008-5472.can-05-3757

Bratthauer GL, Stamatakos MD, Vinh TN (2010) Cells with minimal expression of the JAK/STAT pathway related proteins STAT5a and the prolactin receptor: evidence of an alternate prolactin receptor isoform in breast disease. Protein Pept Lett 17(1):104–108

Buettner R, Mora LB, Jove R (2002) Activated STAT signaling in human tumors provides novel molecular targets for therapeutic intervention. Clin Cancer Res 8(4):945–954

Chen Z, Han ZC (2008) STAT3: a critical transcription activator in angiogenesis. Med Res Rev 28(2):185–200. doi:10.1002/med.20101

de la Iglesia N, Konopka G, Puram SV, Chan JA, Bachoo RM, You MJ, Levy DE, Depinho RA, Bonni A (2008) Identification of a PTEN-regulated STAT3 brain tumor suppressor pathway. Genes Dev 22(4):449–462. doi:10.1101/gad.1606508

Debierre-Grockiego F (2004) Anti-apoptotic role of STAT5 in haematopoietic cells and in the pathogenesis of malignancies. Apoptosis 9(6):717–728. doi:10.1023/B:APPT.0000045785.65546.a2

El-Adawi H, Deng L, Tramontano A, Smith S, Mascareno E, Ganguly K, Castillo R, El-Sherif N (2003) The functional role of the JAK-STAT pathway in post-infarction remodeling. Cardiovasc Res 57(1):129–138

Grimley PM, Dong F, Rui H (1999) Stat5a and Stat5b: fraternal twins of signal transduction and transcriptional activation. Cytokine Growth Factor Rev 10(2):131–157

Herbst RS, Fukuoka M, Baselga J (2004) Gefitinib--a novel targeted approach to treating cancer. Nat Rev Cancer 4(12):956–965. doi:10.1038/nrc1506

Hodge DR, Xiao W, Wang LH, Li D, Farrar WL (2004) Activating mutations in STAT3 and STAT5 differentially affect cellular proliferation and apoptotic resistance in multiple myeloma cells. Cancer Biol Ther 3(2):188–194

Klampfer L (2006) Signal transducers and activators of transcription (STATs): novel targets of chemopreventive and chemotherapeutic drugs. Curr Cancer Drug Targets 6(2):107–121

Krock BL, Skuli N, Simon MC (2011) Hypoxia-induced angiogenesis: good and evil. Genes Cancer 2(12):1117–1133. doi:10.1177/1947601911423654

Kusaba T, Nakayama T, Yamazumi K, Yakata Y, Yoshizaki A, Inoue K, Nagayasu T, Sekine I (2006) Activation of STAT3 is a marker of poor prognosis in human colorectal cancer. Oncol Rep 15(6):1445–1451

Lee J, Kim JC, Lee SE, Quinley C, Kim H, Herdman S, Corr M, Raz E (2012) Signal transducer and activator of transcription 3 (STAT3) protein suppresses adenoma-to-carcinoma transition in Apcmin/+ mice via regulation of snail-1 (SNAI) protein stability. J Biol Chem 287(22):18182–18189. doi:10.1074/jbc.M111.328831

Levy DE, Loomis CA (2007) STAT3 signaling and the hyper-IgE syndrome. N Engl J Med 357(16):1655–1658. doi:10.1056/NEJMe078197

Lin JX, Leonard WJ (2000) The role of Stat5a and Stat5b in signaling by IL-2 family cytokines. Oncogene 19(21):2566–2576. doi:10.1038/sj.onc.1203523

Mascareno E, El-Shafei M, Maulik N, Sato M, Guo Y, Das DK, Siddiqui MA (2001) JAK/STAT signaling is associated with cardiac dysfunction during ischemia and reperfusion. Circulation 104(3):325–329

Musteanu M, Blaas L, Mair M, Schlederer M, Bilban M, Tauber S, Esterbauer H, Mueller M, Casanova E, Kenner L, Poli V, Eferl R (2010) Stat3 is a negative regulator of intestinal tumor progression in Apc(Min) mice. Gastroenterology 138 (3):1003–1011.e1001–1005. doi:10.1053/j.gastro.2009.11.049

Nagy ZS, Rui H, Stepkowski SM, Karras J, Kirken RA (2006) A preferential role for STAT5, not constitutively active STAT3, in promoting survival of a human lymphoid tumor. J Immunol 177(8):5032–5040

Nitze LM, Galsgaard ED, Din N, Lund VL, Rasmussen BB, Berchtold MW, Christensen L, Panina S (2013) Reevaluation of the proposed autocrine proliferative function of prolactin in breast cancer. Breast Cancer Res Treat 142(1):31–44. doi:10.1007/s10549-013-2731-7

Peck AR, Witkiewicz AK, Liu C, Klimowicz AC, Stringer GA, Pequignot E, Freydin B, Yang N, Ertel A, Tran TH, Girondo MA, Rosenberg AL, Hooke JA, Kovatich AJ, Shriver CD, Rimm DL, Magliocco AM, Hyslop T, Rui H (2012) Low levels of Stat5a protein in breast cancer are associated with tumor progression and unfavorable clinical outcomes. Breast Cancer Res 14(5):R130. doi:10.1186/bcr3328

Sanchez-Ceja SG, Reyes-Maldonado E, Vazquez-Manriquez ME, Lopez-Luna JJ, Belmont A, Gutierrez-Castellanos S (2006) Differential expression of STAT5 and Bcl-xL, and high expression of Neu and STAT3 in non-small-cell lung carcinoma. Lung Cancer (Amsterdam, Netherlands) 54(2):163–168. doi:10.1016/j.lungcan.2006.07.012

Takeda K, Noguchi K, Shi W, Tanaka T, Matsumoto M, Yoshida N, Kishimoto T, Akira S (1997) Targeted disruption of the mouse Stat3 gene leads to early embryonic lethality. Proc Natl Acad Sci U S A 94(8):3801–3804

Yamaura G, Turoczi T, Yamamoto F, Siddqui MA, Maulik N, Das DK (2003) STAT signaling in ischemic heart: a role of STAT5A in ischemic preconditioning. Am J Physiol Heart Circ Physiol 285(2):H476–H482. doi:10.1152/ajpheart.00079.2003

Yang X, Friedl A (2015) A positive feedback loop between prolactin and STAT5 promotes angiogenesis. Adv Exp Med Biol 846:265–280. doi:10.1007/978-3-319-12114-7_12

Yang XO, Panopoulos AD, Nurieva R, Chang SH, Wang D, Watowich SS, Dong C (2007) STAT3 regulates cytokine-mediated generation of inflammatory helper T cells. J Biol Chem 282(13):9358–9363. doi:10.1074/jbc.C600321200

Yang X, Qiao D, Meyer K, Friedl A (2009) STATs mediate fibroblast growth factor induced vascular endothelial morphogenesis. Cancer Res 69(4):1668–1677. doi:10.1158/0008-5472.CAN-07-6385

Yang X, Qiao D, Meyer K, Pier T, Keles S, Friedl A (2012) Angiogenesis induced by signal transducer and activator of transcription 5A (STAT5A) is dependent on autocrine activity of proliferin. J Biol Chem 287(9):6490–6502. doi:10.1074/jbc.M111.254631

Yang X, Meyer K, Friedl A (2013) STAT5 and prolactin participate in a positive autocrine feedback loop that promotes angiogenesis. J Biol Chem 288(29):21184–21196. doi:10.1074/jbc.M113.481119

Yin W, Cheepala S, Roberts JN, Syson-Chan K, DiGiovanni J, Clifford JL (2006) Active Stat3 is required for survival of human squamous cell carcinoma cells in serum-free conditions. Mol Cancer 5:15. doi:10.1186/1476-4598-5-15

Chapter 49
SUFU

Abstract The SUFU gene, also known as the suppressor of fused homolog, is an important part of the sonic hedgehog-signaling pathway. The gene was first located and found to be a suppressor in *Drosophilia melanogaster.* Hedgehog signaling plays a crucial regulating role in angiogenesis and Sonic Hedgehog one of the three genes in this pathway is controlled by several genes, with the SuFu gene being a major down regulator. Mutations with SuFu can cause, among many others, medulloblastoma or cancerous brain tumors, especially in children. It is hypothesized that in cancers, such as breast cancer and prostate cancer, the SuFu gene is affected by genetic lesions which causes activation of the SHH pathway. Therefore, SuFu functions as a potent tumor suppressor gene. SHH increases the growth of blood vessels by increasing the angiogenic factors Ang-1 and Ang-2. Recent studies have noted SuFu to be a crucial player in the complex SHH pathway. Identifying and regulating microRNAs can be the next step towards targeting tumor angiogenesis and cell survival. Perhaps regulating the SHH signaling pathway could be another focal point in cancer research and treatment.

Keywords Suppressor of fused homolog • SUFU • Normal physiology • Disease • Malignancy

49.1 SUFU

The SUFU gene, also known as the suppressor of fused homolog, is an important part of the sonic hedgehog signaling pathway. The gene was first located and found to be a suppressor in *Drosophilia melanogaster* (Cheng and Yue 2008). The SHH pathway is controlled by several genes, with the SuFu gene being a major down regulator. Mutations with this gene can cause, among many others, medulloblastoma or cancerous brain tumors, especially in children. The SuFu gene is comprised of 12 exons and can be found on chromosome 10 at 24.32, also described as 10q24.32 (Collins and Winters 2014; Cheng and Yue 2008).

A. Salajegheh, *Angiogenesis in Health, Disease and Malignancy*,
DOI 10.1007/978-3-319-28140-7_49

49.2 SUFU in Normal Physiology

In order to understand the role and regulatory characteristics of the SuFu gene, the Hedgehog pathway must first be explored. Hedgehog signaling plays a crucial regulating role in angiogenesis. Hedgehog pathway consists of three genes, namely, the Indian, Dessert and the Sonic, of which the Sonic Hedgehog is of the utmost importance (Cheng and Yue 2008). SHH is one of the main signaling pathways in mammals, playing a vital role in the growth and development of mammals. Because it is involved in the cell differentiation of developing limbs, skin and vital organs such as the brain, any form of unusual activity along the signaling pathway can cause tumors (Cheng and Yue 2008). Even in adults, the importance of this pathway remains in the form of regulating the division of stem cells. An increase in activity of the SHH pathway is correlated with development of many common tumors found in humans. When the SHH pathway is activated, it causes destabilization of SuFu. In certain cancers, very little levels of SuFu are expressed due to the high cell turnover rate (Cheng and Yue 2008). It is hypothesized that in cancers, such as breast cancer and prostate cancer, the SuFu gene is affected by genetic lesions which causes activation of the SHH pathway (Cheng and Yue 2008). Hence, SuFu functions as a potent tumor suppressor gene.

49.3 SUFU in Disease

Studies have found that SHH increases the growth of blood vessels by increasing the angiogenic factors Ang-1 and Ang-2 (Lee et al. 2007b). Therefore, it is critical that the SHH pathway is strictly down regulated and negatively controlled at multiple levels, along with genes like SuFu.

In order to express target genes, the regulation of the SuFu gene is closely controlled by the SHH pathway. It was initially thought that SuFu function was opposed via phosphorylation by Fu kinase, however a direct link is yet to be demonstrated. In Drosophilia, over-expression of the hedgehog pathway was observed in instances of phosphorylation of SuFu. One study showed any defect in SuFu alleles compromises the expression of the SuFu protein, which in turn leads to abnormalities in the developmental stage. A study on mice found that homozygous SuFu embryos died within 9.5 days due to developmental abnormalities, whereas heterozygous SuFu embryos survived and developed regularly (Cheng and Yue 2008).

The functional mechanisms of SuFu need to be addressed to get a proper scope of the gene. SuFu genes work in the SHH pathway via distinct mechanisms. SuFu has the ability to bind to GLI1, GLI2 and GLI3 proteins. It is theorized that GLI proteins are bound to SuFu in a head to tail orientation. GL1 is solely found in the nucleus, whereas SuFu can be found expressed in both the nucleus and the cytoplasm. However when they are both expressed together, SuFu brings the GLI1 into the cytoplasm, which forces the nucleus to limit its transcription activity. The cytoplasm anchoring model suggests that GLI1 is anchored to the cytoplasm with the help of SuFu. Likewise, SuFu plays a similar role in Drosophilia by halting nuclear processes. Thus, it was found that SuFu travels back and forth between the cytoplasm and nucleus in both Drosophilia and mammals. A second model of SuFu function in mammals suggests the SuFu gene holds the ability of repressing transcription of GLIs. A study found that SuFu restricts GLI transcription by enlisting the help of the mSin3A HDAC corepressor complex. SuFu was also found to increase the ability of GLI1 to bind to DNA. Therefore, simply repressing the SuFu gene is enough to activate GLI transcription, alluding to its key role in the negative regulation of the SHH signaling pathway (Cheng and Yue 2008).

49.4 SUFU in Malignancy

An abnormality in the angiogenic process leads to a number of problems, including cancer. MicroRNAs are a type of RNA, which inhibit the expression of genes and play a vital role in the development of cancer (Lee et al. 2007a). Of notable interest is miR-378, as it increases tumor angiogenesis and growth and also enhances the survival of these cells, by decreasing the expression of tumor suppression genes like SuFu. A study testing human cell lines found that when SuFu expression levels were high, miR-378 expression levels were low, and when SuFu levels were low, miR-378 levels were high, showcasing that the two were negatively correlated. Because repression of SuFu can lead to rampant proliferation of cells, the inhibition of SuFu is essential in order for miR-378 to promote angiogenesis and increase cell survival (Lee et al. 2007a).

Several experiments have been performed, in order to confirm the repression of SuFu in the presence of miR-378. One experiment found a substantial decrease in SuFu expression levels when cells were transfected with microRNA 378. The results were confirmed via the Western blot technique and immunocytochemical staining. Another experiment was performed by injecting miR-378 on tumors recovered from mice. This study found significant repression of SuFu, showcasing that the physiological role of microRNA 378 in tumour angiogenesis and growth was dependent on SuFu and the SHH pathway. The results were confirmed by H&E staining (Lee et al. 2007a).

It is believed miR-378 functions by targeting the sequence 3′ UTR, from nucleotides 4645 to 4676, in SuFu. In order to confirm this potential target site, an experiment placed the 3′ UTR SuFu sequence into a luciferase reporter vector and found substantial luciferase repression along the microRNA 378 target sequence, when comparing with the control vector. Additionally, cells were transfected with miR-378, grown on a Petri dish and were counted by staining with trypan blue. When the SuFu 3′ UTR sequence was introduced, it was immediately inhibited by the expression of miR-378, further proving SuFu 3′ UTR to be a potential target site of miR-378 (Lee et al. 2007a).

SuFu is an important tumor suppressor gene, which when misbehaves, can cause cancerous brain tumors. Identifying and regulating microRNAs like miR-378 can be the next step towards targeting tumor angiogenesis and cell survival. It also opens the possibility for targeting specific miRNAs for the purpose of gene therapy or perhaps a potent anti-cancer drug that aims to inhibit the survival of tumors. Recent studies have noted SuFu to be a crucial player in the complex SHH pathway as well. Perhaps regulating the SHH signaling pathway could be another focal point in cancer research and treatment.

References

Cheng SY, Yue S (2008) Role and regulation of human tumor suppressor SUFU in Hedgehog signaling. Adv Cancer Res 101:29–43. doi:10.1016/s0065-230x(08)00402-8

Collins CW, Winters JC (2014) AUA/SUFU adult urodynamics guideline: a clinical review. Urol Clin North Am 41(3):353–362, vii. doi:10.1016/j.ucl.2014.04.011

Lee DY, Deng Z, Wang CH, Yang BB (2007a) MicroRNA-378 promotes cell survival, tumor growth, and angiogenesis by targeting SuFu and Fus-1 expression. Proc Natl Acad Sci U S A 104(51):20350–20355. doi:10.1073/pnas.0706901104

Lee SW, Moskowitz MA, Sims JR (2007b) Sonic hedgehog inversely regulates the expression of angiopoietin-1 and angiopoietin-2 in fibroblasts. Int J Mol Med 19(3):445–451

Chapter 50
Tenascins

Abstract A family of four multimeric extracellular matrix (ECM) glycoproteins: tenascin-C, X, R and W, tenascins are diversely expressed during development and have independent mechanism of regulation especially in the adult organisms. Every tenascin member is primarily synthesized by cells in connective tissues. Of all these family members, tenascin-C is the best studied member which is found in the embryo, central nervous system and it is a prominent member of peripheral nerves with an important function in epithelial–mesenchymal interactions and branching morphogenesis. Tenascin-R has been exclusively found in the central nervous system, whereas tenascin-X and –W have the highest levels of expression in the heart, skeletal muscles and kidney. Abnormal expression of these proteins has been associated with various types of disorders including cardiovascular diseases and diabetic retinopathy and cancer development. Tenascin-C is among that protein which has been considered as a potential inducer of tumorigenesis due to its multiple roles in different steps of cancer progression such as angiogenesis.

Keywords Tenascins • Angiogenesis • Normal physiology • Disease • Malignancy

50.1 Tenascins

Another group of regulators of angiogenesis, Tenascins, are a family of oligomeric glycoproteins found in the extracellular matrix (ECM). They are considered to be unique to vertebrates. Important cellular processes tenascins are involved with include cell adhesion, migration and proliferation. There are currently four members of this gene family that have been identified in mammals: Tenascin-C, Tenascin-R, Tenascin-X and Tenascin-W (Hsia and Schwarzbauer 2005).

In the developing embryo, these genes display specialised patterns of expression regulating neural development, skeletogenesis and vasculogenesis (Hsia and Schwarzbauer 2005). However, in the adult they are also involved in normal processes such as wound healing, nerve regeneration, and tissue involution along with pathological conditions like vascular disease, tumorigenesis, and metastasis (Hsia

Dr. Haleh Vosgha contributed equally with Dr. Ali Salajegheh.

A. Salajegheh, *Angiogenesis in Health, Disease and Malignancy*,
DOI 10.1007/978-3-319-28140-7_50

and Schwarzbauer 2005). The following paragraphs underline the role of tenascins with a particular focus on angiogenesis throughout regular physiology, disease and malignancies.

50.2 Tenascins in Normal Physiology

To understand the mechanism by which tenascin proteins relate to angiogenesis it is necessary to first state commonly accepted functions of each protein member (Brellier and Chiquet-Ehrismann 2012). Tenascin-R expression happens primarily during early and adult central nervous system development, while Tenascin-W expression is prominent in the developing and adult metanephric kidney and periosteum around the ribs (Chiquet-Ehrismann et al. 2014). Tenascin-X is the largest known member of the family and in adult expression it is mostly limited to musculoskeletal, cardiac, and dermis tissue. Specifically, it is believed that Tenascin X helps determine how collagen fibrils are deposited in the extracellular matrix (Petersen and Douglas 2013). Deficiency of this particular glycoprotein is strongly linked to an autosomal-recessive form of Ehlers-Danlos syndrome (EDS), where collagen density is reduced resulting in easy bruising, skin hyperflexibility and joint laxity (Bristow et al. 2005).

However, of the four proteins in the Tenascin family, tenascin-C is perhaps the most extensively researched and discussed. It is expressed during neural, skeletal, and vascular morphogenesis yet diminishes in adults with continued basal expression in tendon-associated tissues. However, up-regulation in its expression usually occurs during tissue remodelling processes evident in wound repair or pathological states such as inflammation or tumorigenesis (Bristow et al. 2005). Particular attention should also be given to tenascin-C and its role in angiogenesis, which is essentially the focal point of this chapter.

Tenascin-C (TN-C), the founding member of the Tenascin glycoprotein family, can be induced by inflammatory cytokines, growth factors, mechanical stress and hypoxia. It has been shown to inhibit cell adhesion to fibronectin and thus is often called an adhesion-modulating protein. It is upregulated in normal physiological conditions associated with angiogenesis, including wound healing where it is expressed in the newly formed blood vessels of granulation tissue (Midwood and Orend 2009). Furthermore, studies have also confirmed that TN-C inhibits cardiac endothelial cells from adhering to the ECM. This anti-adhesive property subsequently promotes migration and induces angiogenesis. A 2006 study conducted by Weill Medical College of Cornell University confirmed the importance of TN-C in postnatal cardiac angiogenic function. Findings showed that unlike wild-type mice, mice deficient of TN-C were unable to vascularise cardiac allografts following transplantation (Ballard et al. 2006). Additionally, the results of another study were able to demonstrate the link between TN-C and endothelial progenitor cells (EPC). Their interaction generated angiogenesis in the heart. Bone marrow cells were injected systemically into irradiated young mice and then treated with intramyocardial platelet-derived growth factor (PPDF-AB) 1 month later. This resulted in donor derived cell recruitment with majority of the donor cells integrating at sites of TN-C expression. This confirmed TN-C protein as an essential regulator of endothelial function and promoter of EPC incorporation at sites of angiogenic induction (Ballard et al. 2006).

50.3 Tenascins in Disease

TN-C has been shown to stimulate pro-angiogenic pathways in diseases such as diabetic retinopathy (DR) (Eisma et al. 2015). A study was conducted to investigate the effects of TN-C on normal and diabetic retinal endothelial cells (REC) cultured on reconstituted basement membrane (BM) matrix (Castellon et al. 2002). It was found that it significantly delayed the collapse of REC capillary-like tubes in the BM matrix and improved REC proliferation, secondary sprouting and viability. Adding vascular endothelial growth factor (VEGF) to REC cultures also increased the number of branch points by 300 % suggesting that TN-C has the capability of cooperating with VEGF to promote angiogenesis in vitro. Essentially, resulting data suggests TN-C enhanced the sprouting, migratory, and survival effects of angiogenic growth factors and hence acts as a pro-angiogenic mediator in DR and other pathologic conditions involving neovascularization (Castellon et al. 2002).

Moreover, another study investigated the impaired angiogenic response in the cornea of mice lacking TN-C (Sumioka et al. 2011). Ocular fibroblasts and macrophages of wild-type and TN-C null mice were compared. The effects were then evaluated after cauterisation of the centre of the cornea in mice. Findings show that absence of TN-C attenuated the expression of VEGF and TGF and suppressed the neovascularisation of injury-induced corneal stroma, hence confirming TN-C as a modulator of angiogenic pathways (Sumioka et al. 2011).

50.4 Tenascins in Malignancy

Finally, the role of Tenascin proteins in malignant cases needs to be addressed. Research has proven the supportive role that tenascin proteins play in tumour growth, metastases and angiogenesis (Chiquet-Ehrismann 2004). It is believed that the presence of TN-C in the vascular bed may induce the maturing of tumour vessels and could apply scaffolding and remodelling functions during primitive stages of tumour vessel formation. In fact, carcinomas of the breast, colon, glioma and lung directly correlate a high TN-C level with a low survival prognosis (Midwood et al. 2011). A study focusing on TN expression and distribution patterns in astrocytomas, a type of brain cancer, was conducted using 59 astrocytomas and 11 samples of normal brain by western blot analysis. Proliferation indices were recorded using computer-based image technology (Midwood et al. 2011). Enhanced intercellular expression of TN-C was consistently observed in glioblastomas multiforme (GM) and was associated with a higher tumour grade with higher proliferation indices. The strong association of TN and vascular hyperplasia suggests that TN may play a crucial role in angiogenesis. Moreover, it has been demonstrated that tenascin-C can regulate expression of pro-angiogenic factors pertinent to cancer outcome, including the prototypic angiogenic cytokine, VEGF (Midwood and Orend 2009). One relevant study conducted in 2003 by Tanaka et al. from Jikei University School of Medicine focused on the role of TN-C in melanoma stem cells. Significantly lower tumour formation was recorded under skin of nude mice lacking TN-C upon transplanting melanoma cells, in comparison to nude mice showing tenascin-C expression. This has been explained by the notion of reduced VEGF expression and thus reduced tumour angiogenesis (Tanaka et al. 2004).

References

Ballard VL, Sharma A, Duignan I, Holm JM, Chin A, Choi R, Hajjar KA, Wong SC, Edelberg JM (2006) Vascular tenascin-C regulates cardiac endothelial phenotype and neovascularization. FASEB J 20(6):717–719. doi:10.1096/fj.05-5131fje

Brellier F, Chiquet-Ehrismann R (2012) How do tenascins influence the birth and life of a malignant cell? J Cell Mol Med 16(1):32–40. doi:10.1111/j.1582-4934.2011.01360.x

Bristow J, Carey W, Egging D, Schalkwijk J (2005) Tenascin-X, collagen, elastin, and the Ehlers-Danlos syndrome. Am J Med Genet C Semin Med Genet 139c(1):24–30. doi:10.1002/ajmg.c.30071

Castellon R, Caballero S, Hamdi HK, Atilano SR, Aoki AM, Tarnuzzer RW, Kenney MC, Grant MB, Ljubimov AV (2002) Effects of tenascin-C on normal and diabetic retinal endothelial cells in culture. Invest Ophthalmol Vis Sci 43(8):2758–2766

Chiquet-Ehrismann R (2004) Tenascins. Int J Biochem Cell Biol 36(6):986–990. doi:http://dx.doi.org/10.1016/j.biocel.2003.12.002

Chiquet-Ehrismann R, Orend G, Chiquet M, Tucker RP, Midwood KS (2014) Tenascins in stem cell niches. Matrix Biol 37:112–123. doi:10.1016/j.matbio.2014.01.007

Eisma JH, Dulle JE, Fort PE (2015) Current knowledge on diabetic retinopathy from human donor tissues. World J Diabetes 6(2):312–320. doi:10.4239/wjd.v6.i2.312

Hsia HC, Schwarzbauer JE (2005) Meet the tenascins: multifunctional and mysterious. J Biol Chem 280(29):26641–26644. doi:10.1074/jbc.R500005200

Midwood KS, Orend G (2009) The role of tenascin-C in tissue injury and tumorigenesis. J Cell Commun Signal 3(3–4):287–310. doi:10.1007/s12079-009-0075-1

Midwood KS, Hussenet T, Langlois B, Orend G (2011) Advances in tenascin-C biology. Cell Mol Life Sci 68(19):3175–3199. doi:10.1007/s00018-011-0783-6

Petersen JW, Douglas JY (2013) Tenascin-X, collagen, and Ehlers-Danlos syndrome: Tenascin-X gene defects can protect against adverse cardiovascular events. Med Hypotheses 81(3):443–447. doi:10.1016/j.mehy.2013.06.005

Sumioka T, Fujita N, Kitano A, Okada Y, Saika S (2011) Impaired angiogenic response in the cornea of mice lacking tenascin C. Invest Ophthalmol Vis Sci 52(5):2462–2467. doi:10.1167/iovs.10-5750

Tanaka K, Hiraiwa N, Hashimoto H, Yamazaki Y, Kusakabe M (2004) Tenascin-C regulates angiogenesis in tumor through the regulation of vascular endothelial growth factor expression. Int J Cancer 108(1):31–40. doi:10.1002/ijc.11509

Chapter 51
Tissue Inhibitor of Metalloproteinases 1 and 2 (TIMP-1 and TIMP-2)

Abstract As an ancient eukaryotic protein, tissue inhibitors of metalloproteinases (TIMPs) have been detected in a wide range of species such as drosophila, zebra fish, *C. elegans* and humans. Mammalian TIMP family consists of four paralogous genes encoding TIMP-1 to TIMP-4 which share significant homology and structural identity at the protein level. These proteins are well-known as major regulators and inhibitors of matrix metalloproteinases (MMPs). TIMP-1 main function is to inhibit the enzymatic activity of the MMPs by non-covalently binding to MMPs binding site and interacting with the Zn+2. TIMP-2 is the only member of the TIMP family which can not only inhibit MMPs but it is required for the cellular mechanism and involved in cancer progression and metastasis. Overexpression of TIMP-1 is seen in many cancers, it is also linked to inaccurate diagnosis of mammary cancer. Recent studies have demonstrated that TIMP-2 can hinder the mitogenic response of human microvascular endothelial cells to growth factors. Evidence showed that cancer patients with great level of TIMP-1had a decreased survival rate and increased levels of TIMP-1 in cancer cells were linked to decreased recurrence. It has been found that there is an endogenous inhibitory function of TIMP-2 for both angiogenesis mechanism and tumour growth, resulting in suppression the activity of receptor tyrosine kinases VEGFR2 and FGFR1.

Keywords Tissue inhibitor of metalloproteinases • TIMP-1 • TIMP-2 • Angiogenesis • Normal physiology • Disease • Malignancy

51.1 Tissue Inhibitor of Metalloproteinases 1 and 2 (TIMP-1 and TIMP-2)

Tissue inhibitor of metalloproteinases-1 (TIMP-1) belongs to TIMP gene family and has wide spectrum of functions in different tissues (Olson et al. 1997). The main function of this gene is to inhibit the enzymatic activity of the MMPs by non-covalently binding to MMPs binding site and interacting with the Zn+2 (Olson et al. 1997). MMPs are a family of zinc-dependent enzymes involved in degradation of

Dr. Haleh Vosgha contributed equally with Dr. Ali Salajegheh.

A. Salajegheh, *Angiogenesis in Health, Disease and Malignancy*,
DOI 10.1007/978-3-319-28140-7_51

extracellular matrix at physiological PH (Birkedal-Hansen et al. 1993). Matrix metalloproteinase (MMP) activity is essential for extracellular matrix turnover and is linked with both physiologic and pathologic tissue restoring (Stetler-Stevenson 2008). Tissue inhibitors of metalloproteinase (TIMPs) hinder MMP activity. TIMP-2 annuls angiogenic factor-induced endothelial cell proliferation *in vitro* and angiogenesis *in vivo* (Stetler-Stevenson and Seo 2005). This happens irrespective of MMP inhibition. These processes need α3β1 integrin-mediated binding of TIMP-2 to endothelial cells. TIMP-2 inhibits total protein tyrosine phosphatase (PTP) activity linked with β1 integrin subunits as well as dissociation of the phosphatase SHP-1 from β1. TIMP-2 however stimulates PTP activity linked with tyrosine kinase receptors FGFR-1 and KDR. Also, an unpredicted MMP-independent pathway for TIMP-2 inhibition of endothelial cell proliferation *in vitro* shows an important component of the antiangiogenic effect of TIMP2 *in vivo* (Stetler-Stevenson 2008).

51.2 Tissue Inhibitor of Metalloproteinases 1 and 2 (TIMP-1 and TIMP-2) in Normal Physiology

TIMP-1 has been shown to exert growth factor activities. In the absence of serum and exogenous growth factors this gene is capable of stimulating cellular proliferation (Gasson et al. 1985). A growing body of experimental evidence suggests that this gene inhibits apoptosis by binding to its receptor CD63 on the cell membrane and activating Akt signaling pathway (Lee et al. 2014). Akt stimulates the phosphorylation of numerous downstream substrates, all resulting in anti-apoptotic effects. TIMP-1 also plays an important role in B cell survival and growth (Lee et al. 2014). It has been reported TIMP-1 that is expressed by stromal cells in germinal centre are capable of inhibiting the apoptosis of B cells, as the reduction in expression of TIMP-1 correlates with increasing the apoptosis of B cells (Lee et al. 2014). This gene is shown to stimulate the differentiation of germinal centre B cell by up-regulating CD40 and CD23 and down-regulating CD77 (Guedez et al. 2001). Moreover, TIMP-1 induces the expression of interleukin-10, suppresses the expression of germinal centre markers CD10, Bcl-6, PAX-5, up-regulates plasma cell-associated antigens CD138, and X-box protein 1 (Guedez et al. 2001). Another study has shown that TIMP-1 induces the cell cycling pathway that is involved in alteration of gene expression and inhibition of cell growth (Taube et al. 2006).

An extracellular matrix receptor, α3β1, and a soluble extracellular matrix component, TIMP-2, work together with a RTK, either FGFR-1 or KDR, to inhibit the mitogenic response of hMVECs to angiogenic stimuli (FGF-2 or VEGF-A), in a PTP-dependent pathway (Seo et al. 2003). This reaction is TIMP-2 dependant and does not involve MMP-inhibitory activity. It is possible to predict a cell surface receptor mechanism for TIMP-2 that would bring about these selective effects on cell proliferation (Seo et al. 2003).

The role of α3β1 is not clear in angiogenesis. This receptor is expressed on both quiescent and activated endothelial cells (da Silva et al. 2010). This laminin binding integrin can have a direct or indirect role in cell migration on collagen, laminin, and fibronectin (da Silva et al. 2010). The angiogenesis inhibitor thrombospondin use it as a major receptor. Direct binding of TIMP-2 to α3β1 produces a signalling pathway resulting in PTP inactivation of RTK stimulation. This reduces the link between HSP60 and SHP-1 with α3β1. Therefore, binding of TIMP-2 changes the interaction of the α3β1 with these proteins. Moreover, integrin activation coincides with dissociation of HSP60 from α3β1. This is shown by the fact that TIMP-2 interaction with α3β1 leads to a decreased SHP-1 association. TIMP-2 combined with hMVEC had no effect on the connection of these cells to ECM components nor does it cause apoptosis *in vitro*. When β1 null GD25 murine fibroblasts are treated with TIMP-2, PDGF-stimulated proliferation of these cells could not be prevented, and reformation of β1 expression in GD25-1A cells re-established TIMP-2 responsiveness. Consequently, this indicates that TIMP-2 is not only an antagonist of α3β1; it also binds to this receptor to change the signal transduction those results in suppression of hMVEC proliferation (Nakagami et al. 2002).

In addition, TIMP-2 suppresses FGFR-1 and KDR phosphorylation when combined with cognate ligands, FGF-2 and VEGF-A, respectively (Kim et al. 2012). Decreasing the RTK phosphorylation makes it sensitive to the PTP inhibitor orthovanadate for both FGF-2 and VEGF-A- stimulated hMVECs. TIMP-2 binding to α3β1 strengthens the link between the PTP SHP-1 with the RTKs, FGFR-1 and KDR, compared to angiogenic factor treatment solely (Kim et al. 2012).

51.3 Tissue Inhibitor of Metalloproteinases 1 and 2 (TIMP-1 and TIMP-2) in Disease

Studies have shown that TIMP1 gene is involved in numerous diseases (Papazafiropoulou and Tentolouris 2009). Elevated TIMP1 level have been observed in stable coronary, carotid, peripheral artery and acute coronary syndrome (Papazafiropoulou and Tentolouris 2009). A study by Sundstra have suggested that there is an association between plasma TIMP-1 levels with major cardiovascular risk factors, left ventricular hypertrophy indices and systolic dysfunction (Jackson et al. 1998; Walsh et al. 1999). Through degradation of extracellular matrix, elevated level of TIMP1 leads to cardiovascular destruction and remodelling (Jackson et al. 1998; Walsh et al. 1999). TIMP1 is also involved in inflammatory diseases such as rheumatoid arthritis and inflammatory bowel disease (IBD) (Myers et al. 2004). Myars conducted a study to determine the effect of this gene on rheumatoid arthritis (Myers et al. 2004). The obtained results showed that the level of TIMP1was significantly higher in serum that was isolated from patient with rheumatoid arthritis as compared to normal subjects (Myers et al. 2004). In addition, elevated expression of this gene in IBD patient was associated with increased amount of inflammatory markers such as C reactive protein (CRP) and serum amyloid A (SAA) (Brew et al. 2000). These mediators are involved in degrading and modifying bowel wall structures and are capable of inducing the development of chronic inflammatory lesions of digestive tract (Brew et al. 2000). Kapsoritakis have reported that the RNA expression of TIMP1 is elevated in diseased fibrotic liver specimen (Kapsoritakis et al. 2008). Another study have shown that the serum level of TIMP1 was increased in liver fibrosis in alcoholics and severity of chronic liver disease increased with elevated level of TIMP1 (Walsh et al., 1999).

The TIMP-2 stimulated association of SHP-1 with these RTKs is essential (Seo et al. 2006). TIMP-2 increased total PTP activity linked with either FGFR-1 of KDR more than that which occurs during stimulation with either angiogenic factor alone. In situations where PTP activity in the FGFR-1 or KDR complexes is enhanced, TIMP -2 decreases total PTP activity associated with the $\alpha3\beta1$. Reduced SHP-1 association with $\alpha3\beta1$ immunoprecipitates after treating with TIMP-2 indicates that TIMP-2 has the potential to stimulate a transfer of SHP-1 activity from $\alpha3\beta1$ to either FGFR-1 or KDR. SHP-1 can be expressed in several cell types such as hematopoietic cells, endothelial cells and epithelial cells (Seo et al. 2006). Overexpression of catalytically inactive (dominant-negative) SHP-1 leads to continuous mitogenic signalling. In endothelial cells SHP-1 activation by TNF- prevents the proliferation of VEGF- and EGF-. dnSHP-1 expression reduced the inhibitory effect of TNF-on endothelial cell proliferation. In the same way, expression of dnSHP-1 in hMVEC increases basal proliferation and annuls the suppressive effects of TIMP-2 cell proliferation. These show that timp-2 treatment leads to a rise in SHP-1 and an increase in total PTP activity associated with the FGFR-1 and KDR complexes. Together, these indicate that SHP-1 has an essential role in improving the suppressive effects of TIMP-2 on hMVEC mitogenic responses (Seo et al. 2006).

51.4 Tissue Inhibitor of Metalloproteinases 1 and 2 (TIMP-1 and TIMP-2) in Malignancy

Timp-1 is not only an intracellular matrix metalloproteinases inhibitor but it is a prospective biological indicator of many kinds of human cancers (Ma et al. 2014). Studies linking expression of TIMP-1 and characteristics of laryngeal squamous cell carcinoma seen in a clinical setting are scarcely investigated (Ma et al. 2014). Laryngeal squamous cell carcinoma [LSCC] frequently occurs in many cases of cancer of the head and neck (Ma et al. 2014). LSCC comes second in frequencies of cancers associated with the respiratory tract (Ma et al. 2014). Recent studies have linked TIMP-1 as a cell surface marker that aids in the diagnosis of many human cancers (Ma et al. 2014). These cancers include mammary cancer, colon cancer, cancer of the stomach, hepatocellular cancer, cancer of the rectus, glioblastoma, myeloma and lymphoid associated cancers (Ma et al. 2014). Many researches have shown high TIMP-1 expression in LSCC associated with the head and neck and in the growth and infiltration of tumours (Ma et al. 2014). More exploration is needed to confirm if TIMP-1 can be useful in the identification of decreased survival for molecules aimed at the treatment of LSCC (Ma et al. 2014).

Evidence from a study involving qPCR showed that mRNA expression quantities of TIMP-1 in LSCC groups of cells were greater than in unaffected cells neighbouring tumour tissue. This correlates with former research that showed TIMP-1 is expressed in malignant tumour tissue (Ma et al. 2014). Surprisingly, other literature has shown that TIMP-1 may not be depended upon as a diagnostic indicator of cancer. This controversy can be as a result of the presence of different tumour categories used and the methods of experiments (Ma et al. 2014).

As overexpression of TIMP-1 is seen in many cancers, it is also linked to inaccurate diagnosis of mammary cancer (Bigelow et al. 2009). Diagnostic research of the entire amount of TIMP-1 protein gave support to the linking of increased tumorigenic tissue TIMP-1 amounts and a decrease in accuracy in diagnosing breast cancer (Wurtz et al. 2005). The manner in which TIMP-1 encourages tumour formation is uncertain (Bigelow et al. 2009).

Some regions where TIMP-1 can bind are found on breast cancer cells. Evidence has shown that TIMP-1 causes the proliferation of cells by interacting with currently unknown receptors on cells (Wurtz et al. 2005). This interaction is not associated with MMP activity (Wurtz et al. 2005). To prove this mechanism, the proliferative action of timp-1 on breast cancer cells, BC-3A and BC 61 given varying tumorigenic potential was investigated (Wurtz et al. 2005). Results showed that only the more malignant cells [BC-61] were affected by TIMP-1 and had greater proliferation according to the dose of TIMP-1 used (Wurtz et al. 2005). Therefore the activation of TIMP-1 receptors can be linked to the extent of malignancies (Wurtz et al. 2005).

Much research has shown that TIMP-1 expression is greater in breast cancer tissue than benign or healthy breast tissue (Wurtz et al. 2005). Yoshiji et al showed that TIMP-1 expressed on mRNA was greater in malignant mammary tissue than in non-malignant mammary tissue (Wurtz et al. 2005). In many malignant breast cancers increased levels of TIMP-1 on mRNA coincided greatly with lymph node metastases and decreased survival rate, however this did not occur with other diagnostic factors that include the status of menopause, the size of the cancerous tissue and the level of malignancy (Wurtz et al. 2005). In studies conducted to compare the amounts of TIMP-1 by ELISA in samples of breast cancers, fibroadenomas and nodal malignancies, it was shown that the amount of TIMP-1 is much greater in breast cancer and nodal metastases than benign fibroadenomas (Wurtz et al. 2005). Moreover, evidence showed that cancer patients with great level of TIMP-1had a decreased survival rate that was disease-free as well a general decreased survival rate. It was also showed that increased levels of TIMP-1 in cancer cells were linked to decreased recurrence free survival (Wurtz et al. 2005). Studies in the future should be directed towards identifying the diagnostic levels of TIMP-1 in blood samples because the use of tissue samples can be difficult due to collection and storage of samples and heterogeneity of sample material (Wurtz et al. 2005).

Many findings have proved that TIMPs can suppress angiogenic responses *in vitro* (Jiang et al. 2002; Nakagami et al. 2002; Murphy et al. 1993). Tumour growth and vascularity can be reduced *in vivo* by TIMP-2 (Imren et al. 1996). The mechanism involves suppressing MMP activity needed for endothelial cell migration and invasion, that is, an indirect inhibition of angiogenesis. TIMP-2 prevents angiogenic activities *in vivo* for FGF-2 and VEGF-A using an MMP independent orthovanadate-sensitive pathway (Seo et al. 2006). This shows that the pathway for TIMP-2 suppression of angiogenesis *in vivo* is essentially the same as that involved in suppressing the proliferation of hMVEC *in vitro*. Moreover, administration of orthovanadate did not arrest MMP activity nor did it interrupt TIMP-2 inhibition of MMP activity showing that it does not compromise the MMP inhibitory activity of TIMP-2 (Seo et al. 2006).

It has been shown by studies that TIMP-2 inhibition of angiogenesis *in vivo* and the *in vitro* mechanism of endothelial cell growth suppression have two similar properties (Seo et al. 2008). These shared features are MMP inhibition is not involved (TIMP-2 is antiangiogenic) and both being easily affected by the PTP inhibitor orthovanadate (Seo et al. 2008). As a result, both MMP inhibitor functions and MMP – independent activities (mainly growth suppression) can be carried out by TIMP-2 i*n vivo. In vitro* experiments have proved in the past that TIMP-2 associated with α3β1 can bring about a signalling cascade that leads to improved PTP activity linked with angiogenic factor receptors (Oh et al. 2006).

TIMP-2 inhibition of endothelial cell proliferation is a crucial factor of the TIMP-2 anti-angiogenesis (Seo et al. 2003). Suppression of angiogenesis also requires downstream effects. The effects of TIMP-2 on additional endothelial cell functions required for angiogenesis, such as effects on cell survival, invasion, differentiation, remain to be detergiogenic therapies. FGF-2 or VEGF-A stimulation *in vitro* by TIMP-2 leads to suppression of endothelial cell proliferation (Seo et al. 2003, 2008). *In vivo*, TIMP-2 hinder angiogenic responses that have been brought about by either FGF-2 or VEGF-A. The PTP inhibitor orthovanadate annuls both the *in vitro* and *in vivo* effects, thereby suggesting that TIMP-2 induced SHP-1 mediated inactivation of RTKs, instead of anti MMP activity, is responsible for these responses (Seo et al. 2006). Finally, the principal antiangiogenic effects of TIMP-2 are not totally due to anti-MMP activity but also due to TIMP-2 induced SHP-1 mediated inactivation of RTKs. The suppressive effect of TIMP-2 on angiogenesis involves PTP activity and this indicates that partial manipulation of SHP-1 activity in endothelial cells may be a target for therapies in several pathologic conditions associated with angiogenic responses. At last, synthetic metalloproteinase inhibitors have been developed for clinical trials in cancer therapy and this was due to TIMP-2 inhibiting tumour growth.

References

Bigelow RL, Williams BJ, Carroll JL, Daves LK, Cardelli JA (2009) TIMP-1 overexpression promotes tumorigenesis of MDA-MB-231 breast cancer cells and alters expression of a subset of cancer promoting genes in vivo distinct from those observed in vitro. Breast Cancer Res Treat 117(1):31–44. doi:10.1007/s10549-008-0170-7

Birkedal-Hansen H, Moore WG, Bodden MK, Windsor LJ, Birkedal-Hansen B, DeCarlo A, Engler JA (1993) Matrix metalloproteinases: a review. Crit Rev Oral Biol Med 4(2):197–250

Brew K, Dinakarpandian D, Nagase H (2000) Tissue inhibitors of metalloproteinases: evolution, structure and function. Biochim Biophys Acta 1477(1–2):267–283

da Silva RG, Tavora B, Robinson SD, Reynolds LE, Szekeres C, Lamar J, Batista S, Kostourou V, Germain MA, Reynolds AR, Jones DT, Watson AR, Jones JL, Harris A, Hart IR, Iruela-Arispe ML, DiPersio CM, Kreidberg JA, Hodivala-Dilke KM (2010) Endothelial α3β1-integrin represses pathological angiogenesis and sustains endothelial-VEGF. Am J Pathol 177(3):1534–1548. doi:10.2353/ajpath.2010.100043

Gasson JC, Golde DW, Kaufman SE, Westbrook CA, Hewick RM, Kaufman RJ, Wong GG, Temple PA, Leary AC, Brown EL et al (1985) Molecular characterization and expression of the gene encoding human erythroid-potentiating activity. Nature 315(6022):768–771

Guedez L, Mansoor A, Birkedal-Hansen B, Lim MS, Fukushima P, Venzon D, Stetler-Stevenson WG, Stetler-Stevenson M (2001) Tissue inhibitor of metalloproteinases 1 regulation of interleukin-10 in B-cell differentiation and lymphomagenesis. Blood 97(6):1796–1802

Imren S, Kohn DB, Shimada H, Blavier L, DeClerck YA (1996) Overexpression of tissue inhibitor of metalloproteinases-2 retroviral-mediated gene transfer in vivo inhibits tumor growth and invasion. Cancer Res 56(13):2891–2895

Jackson CJ, Arkell J, Nguyen M (1998) Rheumatoid synovial endothelial cells secrete decreased levels of tissue inhibitor of MMP (TIMP1). Ann Rheum Dis 57(3):158–161

Jiang Y, Goldberg ID, Shi YE (2002) Complex roles of tissue inhibitors of metalloproteinases in cancer. Oncogene 21(14):2245–2252

Kapsoritakis AN, Kapsoritaki AI, Davidi IP, Lotis VD, Manolakis AC, Mylonis PI, Theodoridou AT, Germenis AE, Potamianos SP (2008) Imbalance of tissue inhibitors of metalloproteinases (TIMP) - 1 and - 4 serum levels, in patients with inflammatory bowel disease. BMC Gastroenterol 8:55. doi:10.1186/1471-230x-8-55

Kim SH, Cho YR, Kim HJ, Oh JS, Ahn EK, Ko HJ, Hwang BJ, Lee SJ, Cho Y, Kim YK, Stetler-Stevenson WG, Seo DW (2012) Antagonism of VEGF-A-induced increase in vascular permeability by an integrin alpha3beta1-Shp-1-cAMP/PKA pathway. Blood 120(24):4892–4902. doi:10.1182/blood-2012-05-428243

Lee SY, Kim JM, Cho SY, Kim HS, Shin HS, Jeon JY, Kausar R, Jeong SY, Lee YS, Lee MA (2014) TIMP-1 modulates chemotaxis of human neural stem cells through CD63 and integrin signalling. Biochem J 459(3):565–576. doi:10.1042/bj20131119

Ma J, Wang J, Fan W, Pu X, Zhang D, Fan C, Xiong L, Zhu H, Xu N, Chen R, Liu S (2014) Upregulated TIMP-1 correlates with poor prognosis of laryngeal squamous cell carcinoma. Int J Clin Exp Pathol 7(1):246–254

Murphy AN, Unsworth EJ, Stetler-Stevenson WG (1993) Tissue inhibitor of metalloproteinases-2 inhibits bFGF-induced human microvascular endothelial cell proliferation. J Cell Physiol 157(2):351–358. doi:10.1002/jcp.1041570219

Myers A, Lakey R, Cawston TE, Kay LJ, Walker DJ (2004) Serum MMP-1 and TIMP-1 levels are increased in patients with psoriatic arthritis and their siblings. Rheumatology (Oxford) 43(3):272–276. doi:10.1093/rheumatology/keh032

Nakagami H, Cui T-X, Iwai M, Shiuchi T, Takeda-Matsubara Y, Wu L, Horiuchi M (2002) Tumor necrosis factor-α inhibits growth factor–mediated cell proliferation through SHP-1 activation in endothelial cells. Arterioscler Thromb Vasc Biol 22(2):238–242

Oh J, Diaz T, Wei B, Chang H, Noda M, Stetler-Stevenson WG (2006) TIMP-2 upregulates RECK expression via dephosphorylation of paxillin tyrosine residues 31 and 118. Oncogene 25(30):4230–4234. doi:10.1038/sj.onc.1209444

Olson MW, Gervasi DC, Mobashery S, Fridman R (1997) Kinetic analysis of the binding of human matrix metalloproteinase-2 and -9 to tissue inhibitor of metalloproteinase (TIMP)-1 and TIMP-2. J Biol Chem 272(47):29975–29983

Papazafiropoulou A, Tentolouris N (2009) Matrix metalloproteinases and cardiovascular diseases. Hippokratia 13(2):76–82

Seo DW, Li H, Guedez L, Wingfield PT, Diaz T, Salloum R, Wei BY, Stetler-Stevenson WG (2003) TIMP-2 mediated inhibition of angiogenesis: an MMP-independent mechanism. Cell 114(2):171–180

Seo DW, Li H, Qu CK, Oh J, Kim YS, Diaz T, Wei B, Han JW, Stetler-Stevenson WG (2006) Shp-1 mediates the antiproliferative activity of tissue inhibitor of metalloproteinase-2 in human microvascular endothelial cells. J Biol Chem 281(6):3711–3721. doi:10.1074/jbc.M509932200

Seo D-W, Kim SH, Eom S-H, Yoon HJ, Cho Y-R, Kim P-H, Kim YK, Han J-W, Diaz T, B-y W, Stetler-Stevenson WG (2008) TIMP-2 disrupts FGF-2-induced downstream signaling pathways. Microvasc Res 76(3):145–151. doi:10.1016/j.mvr.2008.07.003

Stetler-Stevenson WG (2008) The tumor microenvironment: regulation by MMP-independent effects of tissue inhibitor of metalloproteinases-2. Cancer Metastasis Rev 27(1):57–66. doi:10.1007/s10555-007-9105-8

Stetler-Stevenson WG, Seo DW (2005) TIMP-2: an endogenous inhibitor of angiogenesis. Trends Mol Med 11(3):97–103. doi:10.1016/j.molmed.2005.01.007

Taube ME, Liu XW, Fridman R, Kim HR (2006) TIMP-1 regulation of cell cycle in human breast epithelial cells via stabilization of p27(KIP1) protein. Oncogene 25(21):3041–3048. doi:10.1038/sj.onc.1209336

Walsh KM, Timms P, Campbell S, MacSween RN, Morris AJ (1999) Plasma levels of matrix metalloproteinase-2 (MMP-2) and tissue inhibitors of metalloproteinases -1 and -2 (TIMP-1 and TIMP-2) as noninvasive markers of liver disease in chronic hepatitis C: comparison using ROC analysis. Dig Dis Sci 44(3):624–630

Wurtz SO, Schrohl AS, Sorensen NM, Lademann U, Christensen IJ, Mouridsen H, Brunner N (2005) Tissue inhibitor of metalloproteinases-1 in breast cancer. Endocr Relat Cancer 12(2):215–227. doi:10.1677/erc.1.00719

Chapter 52
Tissue Factor Gene

Abstract Primary role of tissue factor gene is to preserve the vascular integrity in response to injury by being the principal initiator of the coagulation cascade. It is an integral membrane protein produced by both healthy and tumor cells. Its expression can be induced in endothelial cells, macrophages or vascular smooth muscle cells by a variety of stimuli including biogenic amines, growth factors and cytokines. Endothelial cells do not physiologically express active TF and only when damage of the vascular wall by chemical or mechanical means occurs, TF is expressed to blood flow to be circulated by binding to plasma factor VIIa when eventually in downstream, activation of haemostatic protease complexes, leading to both the activation of platelets and the generation of a fibrin clot. Involvement of tissue factor gene in regulation of tumor angiogenesis and metastasis has been described previously although this is independent of its function as the initiator of the coagulation cascade. Angiogenesis can also be promoted by TF either indirectly by altering the growth regulatory molecules acting on vascular endothelial cells and hence regulating the angiogenic properties of tumour cells, or directly by "clotting-independent mechanisms". Due to these finding TF is now considered as a potential target in the treatment of several malignancies.

Keywords Tissue factor gene • TGF • Angiogenesis • Normal physiology • Disease • Malignancy

52.1 Tissue Factor Gene

Tissue factor (TF) gene's primary role is to preserve the vascular integrity in response to injury by being the principal initiator of the coagulation cascade. It is an integral membrane protein produced by both healthy and tumor cells (Butenas 2012). The expression of TF can be induced in endothelial cells, macrophages or vascular smooth muscle cells by a variety of stimuli including biogenic amines, growth factors and cytokines (Steffel et al. 2006). It is either present in the bloodstream as circulating TF or on a cellular level as encrypted, surface intracellular protein (Steffel et al. 2006). TF serves as a procoagulant by binding "factor VIIa resulting in activation of factor IX and factor X, ultimately leading to fibrin

A. Salajegheh, *Angiogenesis in Health, Disease and Malignancy*,
DOI 10.1007/978-3-319-28140-7_52

formation" (Steffel et al. 2006). In addition, it contributes to processes that lead to malignant disease progression. This includes "tumour haemostasis, angiogenesis, metastasis and malignant cell survival" (Bluff et al. 2008). Hence, in order to gain a greater understanding of certain disease and malignancies associated with TF, the specific role as well as the regulation and inhibition of this gene must be further explored.

52.2 Tissue Factor Gene and Normal Physiology

In normal physiology, "cells in contact with blood do not express physiologically active TF" (Butenas 2012). Only when damage of the vascular wall by chemical or mechanical means occurs, TF is expressed to blood flow to be circulated by binding to plasma factor VIIa (Butenas 2012). Following this, thrombin is generated by the downstream activation of hemostatic protease complexes, leading to both the activation of platelets and the generation of a fibrin clot (Bluff et al. 2008). TF avoids the inhibition by serine proteases inhibitors due to its "poor enzymatic qualities" (Butenas 2012). In addition to TF, calcium is required for the formation of the fibrin clot and subsequent vessel stabilization after injury. Under physiologic conditions, TF is found in astrocytes of brain tissue, epithelial cells of the lung and cardiomyocytes of the heart and endothelial cells of placenta in high levels (Butenas 2012). The regulation of TF is complex, being regulated by various transcription factors at distinct regions of the gene. These include the induction of TF in monocytes and endothelial cells exposed to bacteria is mediated by a distal enhancer and the control of epithelial cells by a proximal enhancer (Mackman 1995).

Alternatively, it is suggested that TF actually protects cells from apoptosis and promotes tumor growth and angiogenesis by inhibiting the "negative regulatory control of PAR-2-mediated signaling" (Bluff et al. 2008). The expression of TF is also be associated with various diseases, including cancer, spetic shock and atherosclerosis (Mackman 1995).

52.3 Tissue Factor Gene and Disease

The tissue factor genes expression, carried by circulating microparticles, is related to numerous diseases including venous thrombosis, diabetes, atherosclerosis and varying other cardiovascular diseases. Venous thromboembolism (VTE) is "a leading cause of morbidity and mortality worldwide" (Manly et al. 2011). It is elicited by change in blood flow, alteration in blood composition and activation of endothelium (Manly et al. 2011). In the absence of vessel injury, fibrin-rich clots, deemed venous thrombi are deposited on the surface of the epithelium. Nonetheless, either functional or physical agitation of the endothelium "promote thrombosis due to reduced expression of anticoagulants and the induction of TF expression" (Manly et al. 2011). Individuals with acute coronary syndromes or cardiovascular risk factors including smoking, diabetes, hypertension and dyslipidemia typically display elevated levels of TF (Steffel et al. 2006).

By the promotion of thrombus formation, TF can be considered to be involved in the pathogenesis of atherosclerosis as it may "induce migration and proliferation of vascular smooth muscle cells" (Steffel et al. 2006). On a cellular level, monocytes transform into both foam cells and macrophages, as in the inflammatory indication of atherosclerosis. With this inflammatory environment, interleukins and cytokines are released and thereby prompt the expression of TF (Zhang et al. 2011). In addition, the expression of TF is associated with "apoptosis of macrophages in lipid-rich plaques" which, together with inflammation, determine plaque thrombogenicity (Steffel et al. 2006). Lastly, tumours associated with "high bloodborne TF levels are associated with a greater risk of thromboembolic disease and with decreased overall survival" (Cole and Bromberg 2013).

Patients with unstable angina or myocardial infarction have increased levels of TF compared to those with stable angina (Steffel et al. 2006). In such acute coronary syndromes, inflammatory cells are increased at the site of occlusion, thereby inducing TF (Steffel et al. 2006).

In healthy humans, TF expression in monocytes can be regulated by glucose intake (Zhang et al. 2011). Advanced glycation end-products are formed by hyperglycaemia, which subsequently induce endothelial cells to express TF.

Thus, tissue factor gene, through the expression on circulating microparticles, leads to the progression of diseases. With an understanding of this genes mechanism and regulation, targeted treatment can be determined for the specific associated diseases.

52.4 Tissue Factor Gene and Malignancy

Tumor angiogenesis and metastasis have been found to be regulated by the tissue factor gene. This is independent of its function as the initiator of the coagulation cascade (Zhang et al. 2011). Due to this, TF is now considered a "potential target in the treatment of several malignancies" (Cole and Bromberg 2013). The elevated levels of TF in the presence of a tumor are the result of tumor-induced upregulation of TF expression in endothelial cells and monocytes, the shedding of TF-carrying microparticles via permeable tumor blood vessels as well as the "upregulation of endothelial cell TF expression by chemotherapeutic agents" (Cole and Bromberg 2013). Frequently, complications associated with cancer include haemorrhage and thromboembolism (Bluff et al. 2008). These complications are thought to involve the two opposing mechanisms of control of haemostasis: tissue factor pathway inhibitor (TFPI) and TF (Bluff et al. 2008). It has been suggested that a low-dose chemotherapy regime may reduce "the incidence of adverse clotting events associated with antiangiogenic/chemotherapy combination therapies" (Bluff et al. 2008). In contrast, a high-dose chemotherapy regime may encourage the vascular endothelial growth factor receptor inhibitor to trigger a procaoagulant conversion, resulting in dramatic increases in their surface TFPI expression (Bluff et al. 2008).

Angiogenesis can also be promoted by TF either indirectly by altering the growth regulatory molecules acting on vascular endothelial cells and hence regulating the angiogenic properties of tumour cells, or directly by "clotting-independent mechanisms" (Bluff et al. 2008). TF has been implicated in neoplastic angiogenesis due to its presence on vasculature of breast cancer tumor, but not on "vasculature of benign breast masses" (Cole and Bromberg 2013). In terms of metastasis, the capacity to spread and infiltrate the bloodstream and lymphatic vessels, is deemed highly reliant on components of the coagulation cascade (Bluff et al. 2008). Therefore, it is clear that TF "supports metastasis through a fibrinogen-dependent and platelet-dependent restriction in natural killer cell-mediated clearance of micrometastases" (Bluff et al. 2008). The specific malignancies that have been associated with TF expression include breast cancer as well as thromboembolism development in both ovarian and pancreatic cancer.

Though the regulating mechanism of tissue factor is unclear, it has been found that the gene is "selectively expressed in highly invasive breast cancer cells" (Zhang et al. 2011). One hypothesis is that microRNA-19 regulates TF specific expression in invasive breast cancer cells. In such highly invasive breast cancer lines, tissue factor gene promoter was activated, forcing expression of "tissue factor cDNA", implying that "the 3′-UTR of the tissue factor transcript is responsible for the suppression of tissue factor expression" (Zhang et al. 2011). Tissue factor also assists in the migration of breast cancer cells through the formation of the tissue factor VIIa complex via cellular signalling (Bluff et al. 2008). Factor VIIa is also shown to "exert an antiapoptotic effect and play a key role in tumor growth and metastasis" (Cole and Bromberg 2013).

With a greater understanding of TF's overall interaction with the immune response, as well as its role in tumor growth and metastases, may lead to the development of targeted TF treatment for breast cancer and other malignancies, which may provide a promising alternative considering the current lack of effective treatment options (Cole and Bromberg 2013). As well as breast cancer, tissue factor expression on microparticles has been detected in up to two thirds of pancreatic carcinoma patients, especially those with venous thromboembolism (Zwicker et al. 2009). It has been concluded that tissue factor particles derived from tumors in cancer patients may be "central to the pathogenesis of cancer-associated thrombosis" (Zwicker et al. 2009).

In summary, tissue factor gene plays an important role in the coagulation cascade in normal physiology, thereby maintaining vascular integrity. However, it also has been linked to the progression of various cardiovascular diseases as well as diabetes. Lastly, it is believe to play an integral role in tumor growth and metastases in a certain malignancies, especially breast cancer. With further research, targeted TF gene treatment for breast cancer which is currently lacking from current treatment options.

References

Bluff JE, Brown NJ, Reed MW, Staton CA (2008) Tissue factor, angiogenesis and tumour progression. Breast Cancer Res 10(2):204. doi:10.1186/bcr1871

Butenas S (2012) Tissue factor structure and function. Scientifica 2012:964862. doi:10.6064/2012/964862

Cole M, Bromberg M (2013) Tissue factor as a novel target for treatment of breast cancer. Oncologist 18(1):14–18. doi:10.1634/theoncologist.2012-0322

Mackman N (1995) Regulation of the tissue factor gene. FASEB J 9(10):883–889

Manly DA, Boles J, Mackman N (2011) Role of tissue factor in venous thrombosis. Annu Rev Physiol 73:515–525. doi:10.1146/annurev-physiol-042210-121137

Steffel J, Luscher TF, Tanner FC (2006) Tissue factor in cardiovascular diseases: molecular mechanisms and clinical implications. Circulation 113(5):722–731. doi:10.1161/circulationaha.105.567297

Zhang X, Yu H, Lou JR, Zheng J, Zhu H, Popescu NI, Lupu F, Lind SE, Ding WQ (2011) MicroRNA-19 (miR-19) regulates tissue factor expression in breast cancer cells. J Biol Chem 286(2):1429–1435. doi:10.1074/jbc.M110.146530

Zwicker JI, Liebman HA, Neuberg D, Lacroix R, Bauer KA, Furie BC, Furie B (2009) Tumor-derived tissue factor-bearing microparticles are associated with venous thromboembolic events in malignancy. Clin Cancer Res 15(22):6830–6840. doi:10.1158/1078-0432.ccr-09-0371

Chapter 53
Transforming Growth Factor α and β (TGF-α and TGF-β)

Abstract Transforming Growth Factor alpha (TGF-α) is a protein structurally similar to Epidermal Growth Factor (EGF) and a ligand for EGF-receptors, competing with Epidermal Growth Factors. It has been found in a variety of non-neoplastic and neoplastic disorders where angiogenesis is involved or contributes to part of the pathology. TGF-β signaling pathway is involved in many cellular processes in the developing embryo. TGF-α is involved in various corneal diseases and is responsible for the angiogenesis and epidermal hyperplasia in psoriasis and has influence in protection and healing of gastro-intestinal lesions. TGF-β is a potent inducer or inhibitor of angiogenesis by increasing pro-angiogenic growth factors such as VEGFs or by enhancing anti-angiogenic factor productions. It has been found in focal ischemia in stroke and in Alzheimer's disease with attenuation of beta-amyloid (Aβ), TGF-β has shown potential neuroprotective role in reducing accumulation of Aβ deposition indirectly via angiogenesis. TGF- α overexpression has been detected in oesophageal squamous cell carcinoma and is linked to invasive breast carcinoma. TGF-β promotes the formation of tumor microenvironment by interacting with other growth factors (PDGFs, FGFs and VEGFs) and extracellular matrix (ECM) remodeling and its regulatory role in hepatocyte carcinogenesis has been reported recently. The mediatory role of transforming growth factors in angiogenic signaling pathway in tumor progression or anti-tumor angiogenesis is an emerging therapeutic target.

Keywords Transforming growth factor • TGF-α • TGF-β • Angiogenesis • Normal physiology • Disease • Malignancy

53.1 Transforming Growth Factor α and β (TGF-α and TGF-β)

Transforming Growth Factor alpha (TGF-α) is a protein structurally similar to Epidermal Growth Factor (EGF) and a ligand for EGF-receptors, competing with Epidermal Growth Factors. It has been found in a variety of non-neoplastic and neoplastic disorders where angiogenesis is involved or contributes to part of the pathology. TGF-β signaling pathway is involved in many cellular processes in the

A. Salajegheh, *Angiogenesis in Health, Disease and Malignancy*,
DOI 10.1007/978-3-319-28140-7_53

developing embryo and the adult organism (de Groot and Kruijer 1990). The processes include cell growth, differentiation, homeostasis, apoptosis and other cellular functions (Attisano et al. 1993).

53.2 TGF-α and TGF-β in Normal Physiology

Many cells synthesize TGF-βs and have specific receptors for TGF-β (Doyle et al. 2010). It regulates the actions of many other peptide growth factors and influence a positive or negative direction of their effects (Sporn et al. 1986). TGF-β signaling pathway regulates TGF-β superfamily ligands bind to a type II receptor which recruits and phosphorylates a type I receptor (Wrana et al. 1992). The type I receptor phosphorylates receptor-regulated SMADs (R-SMADs) and bind the coSMAD SMAD4. R-SMAD/coSMAD complexes accumulated in the nucleus act as transcription factors and participate in the regulation of target gene expression (Otten et al. 2010).

Similar to FGF-2, TGF-β plays a role in regulating tooth development and acts as a synergic growth factor. TGF-β is known as an inducer for odontoblast differentiation within dentine-pulp complex and dentine formation (Nakashima et al. 1998; Smith et al. 1998). Reduction in the TGF-β expression can be observed from the odontoblast layer after the completion of tooth development.

53.3 TGF-α and TGF-β in Disease

TGF-α has been found to be related to various corneal diseases through the use of indirect immunohistochemistry, where it was seen in cornea specimens with neovascularization resulting from scarring after keratitis, graft rejection, acute necrotising keratitis or scarring after mechanical/chemical injury (Yang et al. 2011; Li et al. 2000). This suggests uniform distribution and aetiological contribution among all the corneal disorders/injuries.

Additionally, it is also evident that TGF- α is responsible for the angiogenesis and epidermal hyperplasia in psoriasis (Elder et al. 1989). The correlation was discovered by analysing the amount of TGF- α and its messenger RNA which reflect the level of TGF- α gene expression. The amount of TGF- α and its messenger RNA are much more abundant in specimens of psoriatic epidermidis than those of normal skin of psoriatic patients (Elder et al. 1989).

TGF- α has been shown to influence protection and healing of gastro-intestinal lesions. It helps maintain integrity of the gastro-intestinal mucosa and repair acute and chronic mucosal lesions. TGF- α is released locally in the gastric mucosa and particularly when topical irritants have been exposed to the mucosa (Konturek et al. 1995). TGF- α stimulates proliferation of mucosal cells and protects the gastro intestine. It also vasodilates and permits gastric adaptation to noxious substances, and heals acute and chronic lesions as well as inhibits the secretion of gastric acids (Konturek et al. 1995; Filipe et al. 1995). EGF receptors increase in number in the ulcer region and therefore migration of cells and TGF- α (which binds to EGF receptors) to the ulcer margin leads to formation of granulation tissue and micro vessels (i.e. angiogenesis) during the healing process (Konturek et al. 1995). There was also increased expression of TGF- α in gastric carcinogenesis (Filipe et al. 1995).

Many studies have identified dual functions of TGF-β as a potent inducer or inhibitor of angiogenesis by increasing pro-angiogenic growth factors such as VEGFs or by enhancing anti-angiogenic factor productions respectively (Piek et al. 2001; Nakagawa et al. 2004; Otten et al. 2010; Li et al. 2002). The TGF-β signaling pathways involve ALK1 (TGF-β receptor I) activated Smads (Nakagawa et al. 2004). R-Smads 2 and R-Smads3 are important for TGF-β signaling that potentiates production of pro-angiognenic growth factors like VEGFs. Meanwhile, the inhibitory angiogenic pathway is mediated by R-Smads6 and R-Smads7 activation (Li et al. 2002). A study by Ferrari et al. (2009) found that defects in ALK1 receptor in mice resulted in death due to absence of angiogenesis. Similarly, this absence of vessel growth in certain areas of capillary bed was also shown in humans, where mutations to this ALK1 receptor led to hereditary haemorrhagic telangiectasia (Fierlbeck et al. 2003).

Increase in the expression levels of TGF-β has been reported following focal ischemia in animal models of stroke (Issa et al. 2005; Navaratna et al. 2009; Chen et al. 2014; Zhou et al. 2001; Doyle et al. 2010; Dobolyi et al. 2012). Mice subjected to MCAO demonstrated up-regulation of TGF-β in the penumbral regions (Dobolyi et al. 2012; Navaratna et al. 2009). This increase in the TGF-β expression levels was correlated with reduction in brain infarct size (Dobolyi et al. 2012). Furthermore, human clinical data showed significant increase in the TGF-β mRNA expression levels in hippocampal CA1 regions in patients with transient global ischemia (Chen et al. 2014; Christoforidis et al. 2005). Subsequently, 6 h after the onset, TGF-β expression levels were detected throughout the brain (Doyle et al. 2010; Dobolyi et al. 2012; Zhou et al. 2001). In response to such pathological events, brain cells such as astrocytes and microglia have been reported to express TGF-β predominantly (Doyle et al. 2010). In Alzheimer's Disease (AD), attenuation of beta-amyloid (Aβ) plaques has been linked to the levels of TGF-β expression levels in both *in vitro* and *in vivo* models of AD (Caraci et al. 2011). The potential neuroprotective mechanism by which TGF-β reduce accumulation of Aβ deposition is suggested to be the TGF-β Smad2/3 signaling which may promote blood to brain trafficking of microglia indirectly via angiogenesis (Caraci et al. 2012; Tesseur et al. 2006). Nevertheless, identifying this mechanism is still under investigation. Nonetheless, the ability of TGF-β in providing neuroprotection indicates potential role of TGF-β-induced angiogenic signaling as means to delay progression of neurodegenration (Caraci et al. 2011, 2012).

Previous studies suggested that TGF-β induced wound healing angiogenesis post dental injuries by its release from demineralized dentine in to pulp tissue. During dental tissue repair, TGF-β expression targets the receptor in odontoblast to induce the production of reparative dentine in mature teeth (Melin et al. 2000; Magloire et al. 2001). Further studies by Piattelli et al. (2004) was done to investigate the expression of TGF-β1 with irreversible pulpitis, inflammation of dental pulp due to pathological or physiological dental injury. Their study found higher expression of TGF-β1 in the odontoblastic and subodontoblastic layer within the irreversible pulpitis than normal healthy pulp tissue. This suggested that TGF-β1 induced pulp tissue healing as a result of pulpal inflammation. In addition, Prime et al. (2004) showed TGF-β not only regulates the wound healing responses due to inflammation, but also is responsible for bacterial infection by regulating the immune cells such as leukocytes.

53.4 TGF-α and TGF-β in Malignancy

TGF- α overexpression is also frequently detected in oesophageal squamous cell carcinoma by immunohistochemical staining in the tissue specimens (Cabrijan et al. 2013). The survival analysis of patients suggested that TGF- α expression was a significant prognostic factor in oesophageal squamous cell carcinoma whereas multivariate analysis failed to detect a significant correlation. Therefore TGF- α expression may be linked to angiogenesis and progression and metastases (Ciardiello et al. 1991) of oesophageal squamous cell carcinoma (Soares et al. 2000).

TGF- α overexpression has also been linked to invasive breast carcinomas (Ciardiello et al. 1991) whereby its presence was found in the epithelial and adjacent stromal cells. A strong correlation was found between TGF- α expression and tumour angiogenesis, however, there is not yet any evidence suggesting correlation to other clinicopathological features (Soares et al. 2000). Due to its similarity with epidermal growth factor, TGF- α has not been identified as a distinct drug target as drugs targeting EGFR similarly affect it.

While the TGF-β signaling through Smads2/3 provide neuroprotection via pro-angiogenic activity in neurodenerative disorders, Geng et al. (2013) has found that TGF-β also mediates suppression of VEGF-activated angiogenesis in colon cancer. This was demonstrated by abrogation of TGF-β, which enabled increased expression of VEGF levels in colon cancer specimens of humans. Furthermore, the expression levels of phosphorylated Smads2 were significantly reduced in colon cancer in comparison to the normal (Geng et al. 2013). Therefore, the inhibitory effect of TGF-β on VEGF-mediated angiogenesis provides novel approaches in the development of future therapeutics regarding metastasis and tumor progression.

TGF-β promotes the formation of tumor microenvironment by interacting with other growth factors (PDGFs, FGFs and VEGFs) and extracellular matrix (ECM) remodeling (Seystahl et al. 2015). Kumar et al. (Seystahl et al. 2015) reported that TRACON105 (TRC105), a humanized monoclonal antibody can enhance the endoglin shedding and membrane-type 1 matrix metalloproteinase (MMP)-14 at the cell surface to release soluble endoglin (sEng), an antiangiogenic factor, which is considered a good therapeutic target to impair tumor angiogenesis (Seystahl et al. 2015).

In recent experimental studies, it was reported that Apigenin, an identified chemopreventive bioflavonoid could not only inhibit the hypoxia-induced surge of VEGF gene expression, but also block the prostate carcinogenesis via the blockage of the TGF-β pathway (Mirzoeva et al. 2008, 2014). A research group recently demonstrated the Arsenic trioxide (As_2O_3) induced the MicroRNA (miR)-491 inhibition of the TGF-β/SMAD3/nuclear factor kappa-light-chain-enhancer of activated B cells (NF-κB) pathway (VEGF-related) and resulted in the anti-angiogenesis in Hepatocellular carcinoma (HCC) model (Jiang et al. 2014b). The same group also performed another study investigating the effects of Glabridin (GLA) on HCCs found negative regulation of the miR-148a in TGF-β/SMAD2 signaling pathway (Jiang et al. 2014a). Besides these two studies, others have indicated that microRNA is also very important target to regulate the tumor angiogenesis. For example, miR 17–92 cluster via Myc signaling in Colon cancer (Dews et al. 2006), miR-92a inhibits the angiogenesis by targeting the mRNAs of several proangiogenic factors (integrin subunit alpha5) (Bonauer et al. 2009) and miR-181b, which is mediated by TGF-β, can promote carcinogenesis in hepatocyte and this process can be negatively regulated by tissue inhibitor of metalloprotease 3 (TIMP3) that also inhibits angiogenesis (Wang et al. 2010). To sum up these findings, using the microRNA techniques to regulate the TGF-β-mediated angiogenic signaling pathway in tumor progression or anti-tumor angiogenesis is also another emerging therapeutic targets.

References

Attisano L, Carcamo J, Ventura F, Weis FMB, Massague J, Wrana JL (1993) Identification of human activin and TGFP Type I receptors that form heteromeric kinase complexes with Type II receptors. Cell 75:9

Bonauer A, Carmona G, Iwasaki M, Mione M, Koyanagi M, Fischer A, Burchfield J, Fox H, Doebele C, Ohtani K, Chavakis E, Potente M, Tjwa M, Urbich C, Zeiher AM, Dimmeler S (2009) MicroRNA-92a controls angiogenesis and functional recovery of ischemic tissues in mice. Science 324(5935):1710–1713. doi:10.1126/science.1174381

Cabrijan L, Lipozencic J, Batinac T, Lenkovic M, Zgombic ZS (2013) Differences between keratoacanthoma and squamous cell carcinoma using TGF-alpha. Coll Antropol 37(1):147–150

Caraci F, Battaglia G, Bruno V, Bosco P, Carbonaro V, Giuffrida ML, Drago F, Sortino MA, Nicoletti F, Copani A (2011) TGF-beta1 pathway as a new target for neuroprotection in Alzheimer's disease. CNS Neurosci Ther 17:237–249

Caraci F, Spampinato S, Sortino MA, Bosco P, Battaglia G, Bruno V, Drago F, Nicoletti F, Copani A (2012) Dysfunction of TGF-beta1 signaling in Alzheimer's disease: perspectives of neuroprotection. Cell Tissue Res 347:291–301

Chen J, Venkat P, Zacharek A, Chopp M (2014) Neurorestorative therapy for stroke. Front Hum Neurosci 8:382. doi:10.3389/fnhum.2014.00382

Christoforidis GA, Mohammad Y, Kehagias D, Avutu B, Slivka AP (2005) Angiographic assessment of pial collaterals as a prognostic indicator following intra-arterial thrombolysis for acute ischemic stroke. AJNR Am J Neuroradiol 26:1789–1797

Ciardiello F, Kim N, McGeady ML, Liscia DS, Saeki T, Bianco C, Salomon DS (1991) Expression of transforming growth factor alpha (TGF alpha) in breast cancer. Ann Oncol 2(3):169–182

de Groot RP, Kruijer W (1990) Transcriptional activation by TGF beta 1 mediated by the dyad symmetry element (DSE) and the TPA responsive element (TRE). Biochem Biophys Res Commun 168(3):1074–1081

Dews M, Homayouni A, Yu D, Murphy D, Sevignani C, Wentzel E, Furth EE, Lee WM, Enders GH, Mendell JT, Thomas-Tikhonenko A (2006) Augmentation of tumor angiogenesis by a Myc-activated microRNA cluster. Nat Genet 38(9):1060–1065. doi:10.1038/ng1855

Dobolyi A, Vincze C, Pal G, Lovas G (2012) The neuroprotective functions of transforming growth factor Beta proteins. Int J Mol Sci 13(7):8219–8258. doi:10.3390/ijms13078219

Doyle KP, Cekanaviciute E, Mamer LE, Buckwalter MS (2010) TGFbeta signaling in the brain increases with aging and signals to astrocytes and innate immune cells in the weeks after stroke. J Neuroinflammation 7:62. doi:10.1186/1742-2094-7-62

Elder JT, Fisher GJ, Lindquist PB, Bennett GL, Pittelkow MR, Coffey RJ Jr, Ellingsworth L, Derynck R, Voorhees JJ (1989) Overexpression of transforming growth factor alpha in psoriatic epidermis. Science (New York) 243(4892):811–814

Ferrari G, Cook BD, Terushkin V, Pintucci G, Mignatti P (2009) Transforming growth factor-beta 1 (TGF-beta1) induces angiogenesis through vascular endothelial growth factor (VEGF)-mediated apoptosis. J Cell Physiol 219(2):449–458. doi:10.1002/jcp.21706

Fierlbeck W, Liu A, Coyle R, Ballermann BJ (2003) Endothelial cell apoptosis during glomerular capillary lumen formation in vivo. J Am Soc Nephrol 14(5):1349–1354

Filipe MI, Osborn M, Linehan J, Sanidas E, Brito MJ, Jankowski J (1995) Expression of transforming growth factor alpha, epidermal growth factor receptor and epidermal growth factor in precursor lesions to gastric carcinoma. Br J Cancer 71(1):30–36

Geng L, Chaudhuri A, Talmon G, Wisecarver JL, Wang J (2013) TGF-Beta suppresses VEGFA-mediated angiogenesis in colon cancer metastasis. PLoS One 8(3):e59918. doi:10.1371/journal.pone.0059918

Issa R, AlQteishat A, Mitsios N, Saka M, Krupinski J, Tarkowski E, Gaffney J, Slevin M, Kumar S, Kumar P (2005) Expression of basic fibroblast growth factor mRNA and protein in the human brain following ischaemic stroke. Angiogenesis 8(1):53–62. doi:10.1007/s10456-005-5613-8

Jiang F, Mu J, Wang X, Ye X, Si L, Ning S, Li Z, Li Y (2014a) The repressive effect of miR-148a on TGF beta-SMADs signal pathway is involved in the glabridin-induced inhibition of the cancer stem cells-like properties in hepatocellular carcinoma cells. PLoS One 9(5):e96698. doi:10.1371/journal.pone.0096698

Jiang F, Wang X, Liu Q, Shen J, Li Z, Li Y, Zhang J (2014b) Inhibition of TGF-beta/SMAD3/NF-kappaB signaling by microRNA-491 is involved in arsenic trioxide-induced anti-angiogenesis in hepatocellular carcinoma cells. Toxicol Lett 231:55–61. doi:10.1016/j.toxlet.2014.08.024

Konturek PC, Konturek SJ, Brzozowski T, Ernst H (1995) Epidermal growth factor and transforming growth factor-alpha: role in protection and healing of gastric mucosal lesions. Eur J Gastroenterol Hepatol 7(10):933–937

Li Z, Shimada Y, Uchida S, Maeda M, Kawabe A, Mori A, Itami A, Kano M, Watanabe G, Imamura M (2000) TGF-alpha as well as VEGF, PD-ECGF and bFGF contribute to angiogenesis of esophageal squamous cell carcinoma. Int J Oncol 17(3):453–460

Li JH, Zhu HJ, Huang XR, Lai KN, Johnson RJ, Lan HY (2002) Smad7 inhibits fibrotic effect of TGF-Beta on renal tubular epithelial cells by blocking Smad2 activation. J Am Soc Nephrol 13(6):1464–1472

Magloire H, Romeas A, Melin M, Couble ML, Bleicher F, Farges JC (2001) Molecular regulation of odontoblast activity under dentin injury. Adv Dent Res 15:46–50

Melin M, Joffre-Romeas A, Farges JC, Couble ML, Magloire H, Bleicher F (2000) Effects of TGFbeta1 on dental pulp cells in cultured human tooth slices. J Dent Res 79(9):1689–1696

Mirzoeva S, Kim ND, Chiu K, Franzen CA, Bergan RC, Pelling JC (2008) Inhibition of HIF-1 alpha and VEGF expression by the chemopreventive bioflavonoid apigenin is accompanied by Akt inhibition in human prostate carcinoma PC3-M cells. Mol Carcinog 47(9):686–700. doi:10.1002/mc.20421

Mirzoeva S, Franzen CA, Pelling JC (2014) Apigenin inhibits TGF-beta-induced VEGF expression in human prostate carcinoma cells via a Smad2/3- and Src-dependent mechanism. Mol Carcinog 53(8):598–609. doi:10.1002/mc.22005

Nakagawa T, Li JH, Garcia G, Mu W, Piek E, Bottinger EP, Chen Y, Zhu HJ, Kang DH, Schreiner GF, Lan HY, Johnson RJ (2004) TGF-beta induces proangiogenic and antiangiogenic factors via parallel but distinct Smad pathways. Kidney Int 66(2):605–613. doi:10.1111/j.1523-1755.2004.00780.x

Nakashima M, Toyono T, Murakami T, Akamine A (1998) Transforming growth factor-beta superfamily members expressed in rat incisor pulp. Arch Oral Biol 43(9):745–751

Navaratna D, Guo S, Arai K, Lo EH (2009) Mechanisms and targets for angiogenic therapy after stroke. Cell Adh Migr 3(2):216–223

Otten J, Bokemeyer C, Fiedler W (2010) Tgf-Beta superfamily receptors-targets for antiangiogenic therapy? J Oncol 2010:317068. doi:10.1155/2010/317068

Piattelli A, Rubini C, Fioroni M, Tripodi D, Strocchi R (2004) Transforming growth factor-beta 1 (TGF-beta 1) expression in normal healthy pulps and in those with irreversible pulpitis. Int Endod J 37(2):114–119

Piek E, Ju WJ, Heyer J, Escalante-Alcalde D, Stewart CL, Weinstein M, Deng C, Kucherlapati R, Bottinger EP, Roberts AB (2001) Functional characterization of transforming growth factor beta signaling in Smad2- and Smad3-deficient fibroblasts. J Biol Chem 276(23):19945–19953. doi:10.1074/jbc.M102382200

Prime SS, Pring M, Davies M, Paterson IC (2004) TGF-beta signal transduction in oro-facial health and non-malignant disease (part I). Crit Rev Oral Biol Med 15(6):324–336

Seystahl K, Tritschler I, Szabo E, Tabatabai G, Weller M (2015) Differential regulation of TGF-beta-induced, ALK-5-mediated VEGF release by SMAD2/3 versus SMAD1/5/8 signaling in glioblastoma. Neuro Oncol 17:254–265. doi:10.1093/neuonc/nou218

Smith AJ, Matthews JB, Hall RC (1998) Transforming growth factor-beta1 (TGF-beta1) in dentine matrix. Ligand activation and receptor expression. Eur J Oral Sci 106(Suppl 1):179–184

Soares R, Pereira MB, Silva C, Amendoeira I, Wagner R, Ferro J, Schmitt FC (2000) Expression of TGF-alpha and EGFR in breast cancer and its relation to angiogenesis. Breast J 6(3):171–177

Sporn MB, Roberts AB, Wakefield LM, Assoian RK (1986) Transforming growth factor-B biological function and chemical structure. Science 233:2

Tesseur I, Zou K, Esposito L, Bard F, Berber E, Can JV, Lin AH, Crews L, Tremblay P, Mathews P (2006) Deficiency in neuronal TGF-beta signaling promotes neurodegenration and Alzheimer's pathology. J Clin Invest 116:3060–3069

Wang B, Hsu SH, Majumder S, Kutay H, Huang W, Jacob ST, Ghoshal K (2010) TGFbeta-mediated upregulation of hepatic miR-181b promotes hepatocarcinogenesis by targeting TIMP3. Oncogene 29(12):1787–1797. doi:10.1038/onc.2009.468

Wrana JL, Attisano L, Carcamo J, Zentella A, Doody J, Laiho M, Wang X-F, Massague J (1992) TGFβ signals through a heteromeric protein kinase receptor complex. Cell 71:11

Yang SY, Miah A, Pabari A, Winslet M (2011) Growth factors and their receptors in cancer metastases. Front Biosci (Landmark Ed) 16:531–538

Zhou F, Xiang Z, Peiling L, Junjie J, Zhen LX (2001) The expression and changes of heat shock protein 70, MDA and haemorheology in rat cortex after diffuse axonal injury with secondary insults. J Clin Neurosci 8(3):250–252. doi:10.1054/jocn.2000.0760

Chapter 54
Transforming Growth Factor Beta-Receptor Type II (TGFβR2)

Abstract Transforming growth factor beta-receptor type II (TGFβR2) is a human tumour-suppressor gene consisting of seven exons and with a locus at chromosome 3p22. The TGFβR2 gene encodes transforming growth factor beta-receptor Type II (TβR-II), a transmembrane protein that contains serine-threonine protein kinases in its intracellular domain. TβR-II is an important component of the TGF-β signalling pathway, which regulates several cellular processes, including cell growth and differentiation, arrest of cell cycle, apoptosis and formation of extra cellular matrix. The pleiotropic effects of TGF-β signaling and its importance for general development means that there are various phenotypic consequences if this signaling is disturbed, including neoplastic transformations, deregulation of immune cells, or defects in distinct organs. Mutations in this gene can lead to several systemic disorders that can have life-threatening consequences, such as FTAAD, MFS and LDS, as well as various cancers. Future investigations regarding the molecular mechanisms of the TGF-β pathway and how it leads to various phenotypes could lead to better understanding of the functions of this pathway and the development of better treatments for various conditions.

Keywords Transforming growth factor beta-receptor type II • TGFβR2 • Angiogenesis • Normal physiology • Disease • Malignancy

54.1 Transforming Growth Factor Beta-Receptor Type II (TGFβR2)

Transforming growth factor beta-receptor type II (TGFβR2) is a human tumour-suppressor gene consisting of seven exons and with a locus at chromosome 3p22 (Numata et al. 2008). The TGFβR2 gene encodes transforming growth factor beta-receptor Type II (TβR-II), a transmembrane protein that contains serine-threonine protein kinases in its intracellular domain (Elliott and Blobe 2005).

Dr. Moein Amin contributed equally with Dr. Ali Salajegheh.

A. Salajegheh, *Angiogenesis in Health, Disease and Malignancy*,
DOI 10.1007/978-3-319-28140-7_54

54.2 TGFβR2 in Normal Physiology

TβR-II is an important component of the TGF-β signalling pathway, which regulates several cellular processes, including cell growth and differentiation, arrest of cell cycle, apoptosis and formation of extra cellular matrix (Numata et al. 2008). There are three high-affinity cell surface receptors to which the TGF-β ligand binds to, which are TGF-β receptor Type I (TβR-I), TβR-II and TGF-β receptor Type III (TβR-III). Together, these receptors allow for the transmission of signals from the cell surface to the cell's interior by signal transduction.

The high-affinity direct binding of TGF-β ligand onto TβR-II activates the receptor, which then recruits and phosphorylates TβR-I. An alternative pathway is for TGF-β to bind to TβR-III, which then presents TGF-β to TβR-II. Once phosphorylated, TβR-I propagates the signal to activate downstream targets, called SMAD proteins (Frederic et al. 2008). Specifically, SMAD2 and SMAD3 are phosphorylated and activated by TβR-I so they can then translocate to the nucleus to form transcriptional complexes to control target genes. TGF-β is also capable of activating non-SMAD intracellular pathways (Hellbach et al. 2014).

TGF-β signaling in endothelial cells (EC) is essential for angiogenesis in early embryonic development. This was exhibited by an investigation by Allinson et al. (2012) that showed loss of TGFβR2 in EC in embryos causes cerebral hemorrhage in the forebrain, as well as embryonic death (Allinson et al. 2012). In postnatal life, the roles of TGF-β signaling in EC are more diverse and complex. TGF-β affects EC in a concentration-dependent manner, where low concentrations promote angiogenesis and higher concentrations inhibit it. TGF-β also affects other signalling pathways by promoting other factors that influence angiogenesis. One example of this is how inhibition of TGF-β signalling causes the simultaneous activation of VEGF signalling, which leads to increased angiogenesis (Hellbach et al. 2014). Currently, the molecular mechanisms of the TGF-β pathway that lead to various phenotypes are not yet fully understood, and further investigations are especially required to understand the process of intra-neural angiogenesis. Improved understanding of this process could lead to the development of better treatments for premature human infants who develop spontaneous cerebral micro-hemorrhage (Allinson et al. 2012).

54.3 TGFβR2 in Disease

The pleiotropic effects of TGF-β signaling and its importance for general development means that there are various phenotypic consequences if this signaling is disturbed, including neoplastic transformations, deregulation of immune cells, or defects in distinct organs (Hellbach et al. 2014). Some of the more well-known health conditions related to changes in TGFβR2 gene include Loeys-Dietz syndrome, Marfan syndrome and Familial thoracic aortic aneurysms and dissection, which are described in the following section.

Loeys-Dietz syndrome (LDS) is a hereditary disorder of connective tissue that can have symptoms occurring throughout the body. LDS is most commonly characterised by vascular effects, such as arterial aneurysms and/or dissections, and skeletal manifestations. This disorder can be caused by heterozygous genetic mutations in TGFβR1 or TGFβR2, either of which leads to the disruption of growth and development of the body's connective tissue and body systems. Around 70 % of LDS cases are attributed to sequence variant mutations in the TGFβR2 gene, with over 80 mutations of the TGFβR2 gene being found to cause LDS. Most of these are missense alterations of highly conserved amino acids either adjacent to or within the serine/threonine kinase domains of TβR-II (Loeys and Dietz 1993). Three quarters of LDS cases are the result of a de nevo (non-inherited) gene mutations, while the rest of LDS patients have an affected parent. These mutations result in TβR-II receptors having significantly reduced function (Santibanez et al. 2011). It was initially thought that these mutations would decrease TGF-β activity because of disruption of the kinase domain (Judge et al. 2011), but experimental evidence suggests that the manifestations of LDS are largely associated with excess activation of and signaling by TGF-β (Loeys and Dietz 1993). For example, nuclear accumulation of phosphorylated SMAD2 and levels of connective tissue growth factor are increased, both of which are normally induced by TGF-β (Numata et al. 2008). Researchers speculate that the TGF-β proteins increase their activity to compensate for the reduction in TβR-II receptor activity.

TGFβR2 has also been loosely associated with Marfan syndrome (MFS), which is a systemic disorder affecting connective tissue. MFS shares some common features with LDS, but there are important differences in regards to surgical management of the symptoms. Six TGFβR2 mutations have been implicated in MFS, including point mutations, deletions or insertions, and splice mutations (Frederic et al. 2008). The majority of MFS symptoms are the result of heterozygous mutations in the FBN1 gene, which causes the up-regulation of TGF-β signalling in affected tissues. A study conducted by Judge et al. hypothesized that increased TGF-β signalling was a cause of thickening and prolongation of the mitral valve leaflets of the heart, which is a manifestation of MFS that is not readily attributable to mutations in FBN1 (Judge et al. 2011). The investigation showed that treatment of mutant mice with a TGF-β-neutralizing antibody was successful in bringing the thickness and length of the mitral valve leaflets back to normal. This supported that increased TGF-β signalling could be a cause of these mitral valve defects and suggested the possibility of using therapies that reduce TGF-β levels to treat mitral valve diseases. However, this is not conclusive, as mice studies cannot be directly translated to human disease.

Both MFS and LDS can be underlying conditions that make patients more susceptible to developing thoracic aortic aneurysms and dissection (TAAD). This condition is characterised by the enlargement of the aorta in the thoracic cavity, which can lead to a sudden tear of the inner wall of the aorta. It has been estimated that 19–20 % of people with TAAD have a genetic predisposition to it, with the disease showing autosomal dominant inheritance (Spin 2011). When TAAD is inherited, it is called Familial TAAD (FTAAD). TGFβR2 mutations that lead to aneurysms and dissections have been found to occur predominantly in the functionally important intracellular serine-threonine kinase domain, with few mutations being described in the extracellular domain (Santibanez et al. 2011). At least nine variations in the TGFβR2 gene have been identified in FTAAD that disturb signal transduction, which can impair cell growth and development.

54.4 TGFβR2 in Malignancy

Human cancer cells are cells that have acquired the ability to evade the body's immune system and disturb the cellular homeostasis that is normally regulated through signal transduction pathways, such as the TGF-β signaling pathway. The formation of tumour results from genetic, epigenetic, or somatic alterations that disrupt these pathways (Elliott and Blobe 2005).

TGFβR2 gene is important in suppressing the development of tumours in its early stages due to its tight control of the cell cycle and its potent growth inhibitory effect on epithelial and lymphoid tissues, from which most human cancers arise. However, it has been shown that TGF-β signaling may be able to induce tumour progression in its later stages due to its ability to promote epithelial-to-mesenchymal transition, inhibit apoptosis caused by growth factor deprivation, and its immunosuppressant function. For example, TGF-β stimulates the generation of myofibroblasts from mesenchymal precursors in the tumour stroma, which facilitates tumour cell production and invasion, and promotes neoangiogenesis in tumour tissue (Santibanez et al. 2011).

Mutations in the TGFβR2 gene are implicated in several malignancies. Somatic gene mutations, which acquired during one's lifetime as opposed to being inherited, are only present in certain cells and can disrupt the expression of various genes that are components of the TGF-β signalling pathway. Thus, the growth inhibition and tumour suppression function is lost and cell growth becomes unregulated (Frederic et al. 2008). This can potentially lead to tumour formation and tends to increase a person's risk of developing various cancers, especially when TGFβR2 gene mutations are present in the colon, rectum or oesophagus. In colon cancer, frequent alterations in TGFβR2 gene and less common mutations in SMAD2 gene have been seen (Elliott and Blobe 2005). Both of these are related to the TGF-β signaling pathway. Approximately 30 % of malignant colon tumour cells contain a TGFβR2 gene mutation. TGFβR2 gene mutations also occur in colorectal, breast, pancreatic and gastric cancers, as well as head and neck tumors (Santibanez et al. 2011). These mutations can result in absent TGβ-II receptor expression on the cell surface, leading to defective TGF-β signalling pathways and the inactivation of the tumour suppressive action of TGF-β (Balasubramanian et al. 2002).

TGFβR2 gene plays an important role in regulating the normal cell cycle and suppressing the formation of tumours. Mutations in this gene can lead to several systemic disorders that can have life-threatening consequences, such as FTAAD, MFS and LDS, as well as various cancers. Future investigations regarding the molecular mechanisms of the TGF-β pathway and how it leads to various phenotypes could lead to better understanding of the functions of this pathway and the development of better treatments for various conditions.

References

Allinson KR, Lee HS, Fruttiger M, McCarty JH, Arthur HM (2012) Endothelial expression of TGFbeta type II receptor is required to maintain vascular integrity during postnatal development of the central nervous system. PLoS One 7(6):e39336. doi:10.1371/journal.pone.0039336

Balasubramanian SP, Brown NJ, Reed MW (2002) Role of genetic polymorphisms in tumour angiogenesis. Br J Cancer 87(10):1057–1065. doi:10.1038/sj.bjc.6600625

Elliott RL, Blobe GC (2005) Role of transforming growth factor Beta in human cancer. J Clin Oncol 23(9):2078–2093. doi:10.1200/jco.2005.02.047

Frederic MY, Hamroun D, Faivre L, Boileau C, Jondeau G, Claustres M, Beroud C, Collod-Beroud G (2008) A new locus-specific database (LSDB) for mutations in the TGFBR2 gene: UMD-TGFBR2. Hum Mutat 29(1):33–38. doi:10.1002/humu.20602

Hellbach N, Weise SC, Vezzali R, Wahane SD, Heidrich S, Roidl D, Pruszak J, Esser JS, Vogel T (2014) Neural deletion of Tgfbr2 impairs angiogenesis through an altered secretome. Hum Mol Genet 23(23):6177–6190. doi:10.1093/hmg/ddu338

Judge DP, Rouf R, Habashi J, Dietz HC (2011) Mitral valve disease in Marfan syndrome and related disorders. J Cardiovasc Transl Res 4(6):741–747. doi:10.1007/s12265-011-9314-y

Loeys BL, Dietz HC (1993) Loeys-Dietz syndrome. In: Pagon RA, Adam MP, Ardinger HH et al (eds) GeneReviews(R). University of Washington, Seattle

Numata S, Ueno S, Iga J, Yamauchi K, Hongwei S, Hashimoto R, Takeda M, Kunugi H, Itakura M, Ohmori T (2008) TGFBR2 gene expression and genetic association with schizophrenia. J Psychiatr Res 42(6):425–432. doi:10.1016/j.jpsychires.2007.04.002

Santibanez JF, Quintanilla M, Bernabeu C (2011) TGF-beta/TGF-beta receptor system and its role in physiological and pathological conditions. Clin Sci (London: 1979) 121(6):233–251. doi:10.1042/cs20110086

Spin JM (2011) Gene mutations and familial thoracic aortic aneurysms: a walk on the mild side. Circ Cardiovasc Genet 4(1):4–6. doi:10.1161/circgenetics.110.959171

Chapter 55
Tsp-1

Abstract A family of thrombospondins (TSPs) contains five matricellular proteins which are involved in regulation of cellular phenotype and tissue remodelling through interaction with a broad range of matrix proteins and cell-surface receptors. These proteins have been attributed to a myriad of physiological and pathological processes such as embryonic development, inflammation, wound healing, synaptogenesis, angiogenesis and neoplasia. Thrombospondin-1 (Tsp-1) is the most studied member of Tsp family which is initially purified from thrombin-activated platelet releasates. As it was later found as a secretion product from various epithelial and mesenchymal cells, it was primarily introduced as a regulator of cell-matrix and cell-cell interactions which affects platelet function. Tsp-1 was then recognized as the first natural inhibitor of angiogenesis. Due to its different cell receptors in addition to its multifunctional nature, Tsp-1 has been shown to both stimulate and inhibit angiogenesis, resulting in to have roles in both tumour progression and inhibition. Therefore, evidence has implicated Tsp-1 in having a key role in tumour biology and vascular diseases.

Keywords Thrombospondin-1 • Tsp-1 • Angiogenesis • Normal physiology • Disease • Malignancy

55.1 Thrombospondin-1 (Tsp-1)

Thrombospondin-1 (Tsp-1) is an angiogenic inhibitor and it belongs to the extra cellular matrix proteins (Eichhorn et al. 2007). Its first discovery was in activated platelets and it is the most investigated protein of the Tsp family (Lawler et al. 1978). It has been suggested that the Tsp family consist of five extracellular calcium binding proteins and those are Tsp-1, Tsp-2, Tsp-3, Tsp-4 and Tsp-5 (Lopez-Dee et al. 2011). The structures of Tsp-1 and Tsp-2 are similar and they both become expressed on the cell surface during physiological events (Lopez-Dee et al. 2011). Moreover, Tsp-1 has two globular domains, an N-terminal globular domain and a C-terminal globular domain (Lopez-Dee et al. 2011). Tsp-1 N-terminal domain binds to integrins and different types of cellular proteins that have important roles in

Dr. Haleh Vosgha contributed equally with Dr. Ali Salajegheh.

A. Salajegheh, *Angiogenesis in Health, Disease and Malignancy*,
DOI 10.1007/978-3-319-28140-7_55

angiogenesis, cell motility and adhesion (Calzada et al. 2003). One of the important cellular receptors of N-terminal of Tsp-1 is CD36 which is present on the microvascular endothelium and is important for the anti-angiogenic effect (Tonini et al. 2003). In addition to its role in fatty acid and glucose metabolism, it has been suggested that CD36 has an important role in signal transduction upon binding to Tsp-1 (Gruarin et al. 2000). The anti-angiogenic effect of Tsp-1 has been mainly linked to its ability to bind to CD36. In addition to Tsp-1 CD36 mediated antiangiogenic effect, Tsp-1 can also induce cell cycle arrest (Eichhorn et al. 2007). The structural domain that contains the inhibitory activity of Tsp-1 is known as "Tsp type 1 repeat" (Eichhorn et al. 2007). The C-terminal domain of Tsp-1 binds to CD47 which has a great impact on angiogenesis (Isenberg et al. 2006). This binding has been linked to the modulation of nitric oxide which has important roles in both normal and pathological processes (Carmeliet 2005). Therefore, the C-terminal domain of Tsp-1 has crucial role in chemotaxis and vasodilation.

55.2 Tsp-1 in Normal Physiology

Tsp-1 is secreted by different types of cells such as endothelial cells, fibroblasts, smooth muscle cells, adipocytes, macrophages and malignant glioma cells (Naganuma et al. 2004). After the expression of Tsp-1 on the cell surface, it binds with fibronectin which is protein component of the extracellular matrix (Lopez-Dee et al. 2011). Therefore, Tsp-1 becomes stored in the extracellular matrix where it undergoes protein folding and changing in its confirmation. A number of intracellular events become activated when Tsp- 1 interacts with its receptor leading to endothelial cell apoptosis (Sargiannidou et al. 2001). It has been shown that Tsp-1 is involved in inflammation, cancer and angiogenesis (Lopez-Dee et al. 2011).

55.3 Tsp-1 in Disease

Several studies have shown that pharmacological suppression or genetic depletion of TSP1 increases chronic kidney disease progression in animal models. This protein is capable of inhibiting angiogenesis, activating MMP-dependent ECM turnover and facilitating fibroblast activation and migration. These factors are all identified as important contributors to the progression of this disease (Zeisberg et al. 2014). Gaussem and co-workers have shown that the plasma concentrations of TSP1 in patients with peripheral arterial disease are significantly higher compared to the control group (Smadja et al. 2011). Smadja have reported that the transgenic mice (TSP1-/-) were clinically and histologically protected from tissue necrosis caused by limb ischemia (Smadja et al. 2011). Other studies have shown that TSP1 can act as inflammatory or anti-inflammatory protein. Cadena conducted a study to compare the plasma level of TSP1 in rheumatoid arthritis patients and control groups (Rico et al. 2008). The result suggested that plasma levels of

TSP1 were statistically higher in rheumatoid arthritis patients as compared to the control groups (Rico et al. 2008). Human studies indicate that TSP-1 expression is higher in diabetic vessels than non-diabetic vessels. The author states that the increased expression of TSP-1 in the blood vessel is correlated to the augmented levels of glucose in diabetic patients (Rico et al. 2008). Also the mRNA expression of TSP-1 by epithelial progenitor cell is shown to be significantly higher in diabetic mice as compared to wild type mice (Choi et al. 2012). Studies in Local production or inhibition of TSP1 might be considered as a future treatment for many diseases.

55.4 Tsp-1 in Malignancy

There are conflicting results regarding the role of this gene in tumour progression, while most studies in literature suggest that it inhibits tumour growth (Miyanaga et al. 2002), some other studies suggest that TSP-1 stromal levels are associated with tumour progression (Qian et al. 2001). Obviously, the role of TSP-1 in malignancies depends on the presence of certain factors plus TSP-1 expression levels within the tumour tissues. Regarding angiogenesis many factors will determine the role of TSP-1 in angiogenesis such as TSP-1 levels, availability of angiogenic stimulators and their levels for example basic fibroblast growth factor (bFGF). The location of TSP-1 is crucial as in a rat aorta model, matrix TSP-1 stimulates angiogenesis while in rat cornea it inhibits bFGF-induced angiogenesis.it was proved that the proteolytic state of TSP-1 will determine its inhibitory or stimulatory effect (Sargiannidou et al. 2001; Esemuede et al. 2004).

There are two ways through which TSP1 inhibits tumour growth. By directly acting on endothelial cell migration and survival, as it was shown that TSP-1 secreted from stromal fibroblasts, endothelial cells and immune cells, curtails tumour progression. However, by indirectly affecting the growth factors mobilization that are responsive to TGF-Beta (Transforming Growth Factor-Beta), TSP1 supresses their growth through activation of TGF-Beta (Lawler 2002). The different mechanisms through which TSP-1 affects tumour growth opens the doors for new anti-angiogenic therapeutics, that can be used to treat many cancers along with other diseases where disturbance in angiogenesis happens such as diabetic retinopathy (Lawler 2002).

In colorectal cancer; it was proved that there is a strong association between decreased TSP-1 expression and malignant progression along with significant decrease in the survival rates (Rojas et al. 2008; Miyanaga et al. 2002). Different independent studies revealed that reduced TSP1 expression was associated with increased microvascular densities, increased invasion and poor prognosis (Miyanaga et al. 2002). The reduction in TSP-1 expression was associated with its methylation, suggesting what there is no "methylation-induced expression silencing". It is believed that this methylation leads to a decrease in concentration of secreted active TGF-beta and attenuates its signalling growth as well (Li et al. 1999).

Moving to Non-small cell lung cancer, when comparing NSCLC tissues with normal ones significantly lower levels of TSP-1 and higher microvessel densities were observed. The same thing was observed when comparing late-stage NSCLC with early-stage NSCLC and in those with lymph node metastasis compared with those without metastasis. However this significantly inverse correlation between TSP-1 expression and microvessels density was observed in squamous cell carcinoma but not in adenocarcinoma. These results suggest an intimate relationship between micro vessel density and NSCLC progression, it also suggest that high expression of TSP-1 may play a crucial role in inhibiting tumour occurrence and development (Rojas et al. 2008). Furthermore, TXR1/TSP1 expression can be used to predict taxanes' resistance in NSCLC treatment, where a study showed a correlation between TSP1 expression levels and response to docetaxel-based first–line chemotherapy treatment in advanced/metastatic non-small-cell lung cancer patients. Patients with high TSP1 had longer progression free-survival, overall survival and response rate (Chen et al. 2009).

Regarding the role of TSP-1 in oral squamous cell carcinoma, there are controversial results in literature, where one of the studies suggested that MVD (microvascular density) counts were significantly lower in tumours with high levels of TSP-1 expression. It was also proved that expression levels were inversely correlated with the pattern of tumour invasion and with lymph nodal status. But there was no correlation between its expression and T category and the histologic grade. In addition, Kaplan –Meier analysis reveal that the 5-year survival rate of TSP-1 high expression groups was significantly higher than TSP-1 low expression groups (Papadaki et al. 2011). On the other hand, studies suggested that TSP-1 expression levels were high in stroma of oral SCC and it's produced by mesenchymal cells, but not by the epithelial cells. This increase in stromal TSP-1 levels was associated with an increase in the MMP-9 and thus enhancing the proteolytic activity in a paracrine manner along with an increase in motility (Hayashido et al. 2003). To conclude with oral SCC, TSP-1 expression is considered as a valuable tool to assess the aggressiveness and prognosis of oral SCCs because of its inhibitory effect on tumour vascularity (Yao et al. 2000). However, the stromal TSP-1 levels gave a totally different indication (Hayashido et al. 2003).

Results regarding the role of TSP-1 in pancreatic adenocarcinoma support the results seen in some oral SCC studies (Qian et al. 2001). High TSP- 1 stromal levels were observed in those carcinomas, where it was mostly produced by stromal cells as well as carcinoma cells (Mao et al. 2013; Kazerounian et al. 2008). Similarly, with Hepatocellular carcinoma, it was found that TSP-1 was synthesized in endothelial cells, Hepatocellular carcinoma cells and in fibroblasts that surround the tumour. This suggested that TSP-1 is synthesized basically by non-parenchymal cells that surround the tumour, like fibroblasts and endothelial cells (Hayashi et al. 1997).

Interestingly, TSP-1 expression in cervical cancer was frequently reduced, suggesting its intimate correlation with the malignant progression of cervical cancers. It was shown that the abnormal reduction of TSP-1 mRNA expression was caused by transcriptional down regulation of the gene such as epigenetic gene silencing, while a deletion or mutational changes of TSP-1 were rarely seen. As with other types of cancer the inverse correlation between TSP-1 and microvascular density is seen here as well, it was thought that it is associated with an angiogenic phenotype in cervical cancer (Kodama et al. 2001).

Finally, its association with ovarian carcinoma was suggested as well. Data on this particular type of cancer revealed that a reduction in TSP-1 expression was associated with advanced epithelial ovarian carcinoma therefore the inhibitory effect of TSP1 was seen here as well (Alvarez et al. 2001). There is also a strong correlation between the absence of TSP-1 expression with FIGO (International Federation of Gynaecology and Obstetrics) staging and histological grading of ovarian cancer (Wei et al. 2012).

References

Alvarez AA, Axelrod JR, Whitaker RS, Isner PD, Bentley RC, Dodge RK, Rodriguez GC (2001) Thrombospondin-1 expression in epithelial ovarian carcinoma: association with p53 status, tumor angiogenesis, and survival in platinum-treated patients. Gynecol Oncol 82(2):273–278. doi:10.1006/gyno.2001.6287

Calzada MJ, Sipes JM, Krutzsch HC, Yurchenco PD, Annis DS, Mosher DF, Roberts DD (2003) Recognition of the N-terminal modules of thrombospondin-1 and thrombospondin-2 by alpha-6beta1 integrin. J Biol Chem 278(42):40679–40687. doi:10.1074/jbc.M302014200

Carmeliet P (2005) Angiogenesis in life, disease and medicine. Nature 438(7070):932–936. doi:10.1038/nature04478

Chen ZJ, Le HB, Zhang YK, Qian LY, Li WD (2009) Microvessel density and expression of thrombospondin-1 in non-small cell lung cancer and their correlation with clinicopathological features. J Int Med Res 37(2):551–556

Choi K-Y, Kim D-B, Kim M-J, Kwon B-J, Chang S-Y, Jang S-W, Cho E-J, Rho T-H, Kim J-H (2012) Higher plasma thrombospondin-1 levels in patients with coronary artery disease and diabetes mellitus. Korean Circ J 42(2):100–106. doi:10.4070/kcj.2012.42.2.100

Eichhorn ME, Kleespies A, Angele MK, Jauch KW, Bruns CJ (2007) Angiogenesis in cancer: molecular mechanisms, clinical impact. Langenbecks Arch Surg/Deutsche Gesellschaft fur Chirurgie 392(3):371–379. doi:10.1007/s00423-007-0150-0

Esemuede N, Lee T, Pierre-Paul D, Sumpio BE, Gahtan V (2004) The role of thrombospondin-1 in human disease1. J Surg Res 122(1):135–142. doi:http://dx.doi.org/10.1016/j.jss.2004.05.015

Gruarin P, Thorne RF, Dorahy DJ, Burns GF, Sitia R, Alessio M (2000) CD36 is a ditopic glycoprotein with the N-terminal domain implicated in intracellular transport. Biochem Biophys Res Commun 275(2):446–454. doi:10.1006/bbrc.2000.3333

Hayashi K, Kurohiji T, Shirouzu K (1997) Localization of thrombospondin in hepatocellular carcinoma. Hepatology (Baltimore) 25(3):569–574. doi:10.1002/hep.510250313

Hayashido Y, Nakashima M, Urabe K, Yoshioka H, Yoshioka Y, Hamana T, Kitano H, Koizumi K, Okamoto T (2003) Role of stromal thrombospondin-1 in motility and proteolytic activity of oral squamous cell carcinoma cells. Int J Mol Med 12(4):447–452

Isenberg JS, Ridnour LA, Dimitry J, Frazier WA, Wink DA, Roberts DD (2006) CD47 is necessary for inhibition of nitric oxide-stimulated vascular cell responses by thrombospondin-1. J Biol Chem 281(36):26069–26080. doi:10.1074/jbc.M605040200

Kazerounian S, Yee KO, Lawler J (2008) Thrombospondins in cancer. Cell Mol Life Sci 65(5):700–712. doi:10.1007/s00018-007-7486-z

Kodama J, Hashimoto I, Seki N, Hongo A, Yoshinouchi M, Okuda H, Kudo T (2001) Thrombospondin-1 and -2 messenger RNA expression in invasive cervical cancer: correlation with angiogenesis and prognosis. Clin Cancer Res 7(9):2826–2831

Lawler J (2002) Thrombospondin-1 as an endogenous inhibitor of angiogenesis and tumor growth. J Cell Mol Med 6(1):1–12

Lawler JW, Slayter HS, Coligan JE (1978) Isolation and characterization of a high molecular weight glycoprotein from human blood platelets. J Biol Chem 253(23):8609–8616

Li Q, Ahuja N, Burger PC, Issa JP (1999) Methylation and silencing of the Thrombospondin-1 promoter in human cancer. Oncogene 18(21):3284–3289. doi:10.1038/sj.onc.1202663

Lopez-Dee Z, Pidcock K, Gutierrez LS (2011) Thrombospondin-1: multiple paths to inflammation. Mediators Inflamm 2011:10. doi:10.1155/2011/296069

Mao Y, Keller ET, Garfield DH, Shen K, Wang J (2013) Stroma Cells in tumor microenvironment and breast cancer. Cancer Metastasis Rev 32:303–315. doi:10.1007/s10555-012-9415-3

Miyanaga K, Kato Y, Nakamura T, Matsumura M, Amaya H, Horiuchi T, Chiba Y, Tanaka K (2002) Expression and role of thrombospondin-1 in colorectal cancer. Anticancer Res 22(6c):3941–3948

Naganuma H, Satoh E, Asahara T, Amagasaki K, Watanabe A, Satoh H, Kuroda K, Zhang L, Nukui H (2004) Quantification of thrombospondin-1 secretion and expression of alphavbeta3 and alpha3beta1 integrins and syndecan-1 as cell-surface receptors for thrombospondin-1 in malignant glioma cells. J Neurooncol 70(3):309–317

Papadaki C, Tsaroucha E, Kaklamanis L, Lagoudaki E, Trypaki M, Tryfonidis K, Mavroudis D, Stathopoulos E, Georgoulias V, Souglakos J (2011) Correlation of BRCA1, TXR1 and TSP1 mRNA expression with treatment outcome to docetaxel-based first-line chemotherapy in patients with advanced/metastatic non-small-cell lung cancer. Br J Cancer 104(2):316–323. doi:10.1038/sj.bjc.6606027

Qian X, Rothman VL, Nicosia RF, Tuszynski GP (2001) Expression of thrombospondin-1 in human pancreatic adenocarcinomas: role in matrix metalloproteinase-9 production. Pathol Oncol Res 7(4):251–259

Rico MC, Manns JM, Driban JB, Uknis AB, Kunapuli SP, Dela Cadena RA (2008) Thrombospondin-1 and transforming growth factor beta are pro-inflammatory molecules in rheumatoid arthritis. Transl Res 152(2):95–98. doi:10.1016/j.trsl.2008.06.002

Rojas A, Meherem S, Kim YH, Washington MK, Willis JE, Markowitz SD, Grady WM (2008) The aberrant methylation of TSP1 suppresses TGF-beta1 activation in colorectal cancer. Int J Cancer 123(1):14–21. doi:10.1002/ijc.23608

Sargiannidou I, Zhou J, Tuszynski GP (2001) The role of thrombospondin-1 in tumor progression. Exp Biol Med (Maywood) 226(8):726–733

Smadja DM, d'Audigier C, Bieche I, Evrard S, Mauge L, Dias JV, Labreuche J, Laurendeau I, Marsac B, Dizier B, Wagner-Ballon O, Boisson-Vidal C, Morandi V, Duong-Van-Huyen JP, Bruneval P, Dignat-George F, Emmerich J, Gaussem P (2011) Thrombospondin-1 is a plasmatic marker of peripheral arterial disease that modulates endothelial progenitor cell angiogenic properties. Arterioscler Thromb Vasc Biol 31(3):551–559. doi:10.1161/atvbaha.110.220624

Tonini T, Rossi F, Claudio PP (2003) Molecular basis of angiogenesis and cancer. Oncogene 22(42):6549–6556. doi:10.1038/sj.onc.1206816

Wei W, Kong B, Qu X (2012) Alteration of HGF and TSP-1 expression in ovarian carcinoma associated with clinical features. J Obstet Gynaecol Res 38(1):57–64. doi:10.1111/j.1447-0756.2011.01695.x

Yao L, Zhao YL, Itoh S, Wada S, Yue L, Furuta I (2000) Thrombospondin-1 expression in oral squamous cell carcinomas: correlations with tumor vascularity, clinicopathological features and survival. Oral Oncol 36(6):539–544

Zeisberg M, Tampe B, LeBleu V, Tampe D, Zeisberg EM, Kalluri R (2014) Thrombospondin-1 deficiency causes a shift from fibroproliferative to inflammatory kidney disease and delays onset of renal failure. Am J Pathol 184(10):2687–2698. doi:10.1016/j.ajpath.2014.06.014

Chapter 56
Tumour Necrosis Factor-α (TNF-α)

Abstract Tumour necrosis factor-α (TNF) has been discovered as a major regulator of inflammation. This protein was also identified as an essential role player in the cytokine network. As a pluripotent mediator, TNF- α affects number of cellular/molecular pathways including adhesion, migration, angiogenesis and apoptosis. It is a member of TNF family which is largely produced by T-cells and macrophages. Its main function is the regulation of immune system by activating T-cells, B-cells and macrophages in order to induce other cytokine's expression and cell adhesion molecules. Furthermore, the prominent expression of TNF- α in blood vessels endothelial cells reflects its role during wound healing. It has been reported that TNF- α has an indirect angiogenic role, depending on the synthesis of secondary mediators such as vascular endothelial growth factor (VEGF). However, its dual angiogenic (pro/anti) roles in different conditions have been made it difficult to understand. Overall, TNF- α dysregulation has been connected to various human inflammatory diseases in addition to several malignancies.

Keywords Tumour necrosis factor-α • TNF-α • Angiogenesis • Normal physiology • Disease • Malignancy

56.1 Tumour Necrosis Factor-α (TNF-α)

Tumour necrosis factor (TNF-α) is a type of chemical mediators known as adipokine involved in inflammation and acute phase reaction. It is secreted primarily by activated macrophages, however, it can also be produced by other immune cells such as natural killer cell (NK), lymphocytes and neutrophils (Mocellin et al. 2005). The primary role of TNF-α is the regulation of immune cells. Its clinical importance lies with its association with wound healing and various human diseases such as inflammatory bowel disease (IBD), prostate cancer.

Dr. Haleh Vosgha contributed equally with Dr. Ali Salajegheh.

A. Salajegheh, *Angiogenesis in Health, Disease and Malignancy*,
DOI 10.1007/978-3-319-28140-7_56

56.2 TNF- α in Normal Physiology

TNF-α is an essential to stimulates local containment of infection, and increases the rate of wound healing cascade (Guo and DiPietro 2010). It also stimulate endothelial cells at the injury site to express proteins that initiates the coagulation cascade in the local capillaries to reduce blood flow and prevent further loss of blood. It is crucial in preventing the pathogens from entering the blood stream (bacteraemia), which would enable them to spread systemically causing septic shocks. On the other hand, the fluid that has entered into local tissue carries the pathogen enclosed phagocytic cells, especially dendritic cells through the lymphatic system to regional lymph nodes, hence, initiating the adaptive immune response. Also, innate immune response is triggered by the local tissue destruction, which stimulates the recruitment of monocytes to the injury site (Guo and DiPietro 2010). They secrete TNF-α, which speeds up the pathogen exclusion and wound healing process. Once most of the pathogens have been excluded, the release of TNF-α is reduced, and fibroblasts begins to proliferate to facilitate the tissue regeneration/wound healing cascade (Sinno and Prakash 2013).

Sepsis is the condition whereby there is presence of pathogens in the bloodstream (Thalheimer et al. 2005; Schulte et al. 2013). It can result in the systemic release of TNF-α causing vasodilation and reduction in plasma volume due to increased vascular permeability, which can trigger a septic shock (Thalheimer et al. 2005; Schulte et al. 2013). In these situations, disseminated intravascular coagulation and blood clotting are also triggered by TNF-α, resulting in the depletion of clotting factors. It compromise in the normal coagulation cascade mechanism, so is often result in the failure of vital organs, for example, heart, liver, etc. and associated with death (Schulte et al. 2013).

56.3 TNF-α in Disease

TNF-α is believed to plays a significant role in inflammatory bowel disease (IBD) as it is a pro-inflammatory factor. One theory is that the over-imbalance of pro-inflammatory cytokines contributes to an over-active immune system and chronic inflammation of the gastrointestinal tract (GIT) (Komatsu et al. 2001). Another theory of IBD is that the patient's immune system is constantly being stimulated due to an unknown reason, hence is always in an activated state (Rubin et al. 2000). The binding of TNF-α with surface TNF-α receptor 1 and TNF- α receptor 2 stimulates inflammation of the GIT. Hence, it is contributing to the development of ulcers and persistent inflammation, which are characteristic of IBD (Papadakis and Targan 2000; Komatsu et al. 2001).

Therefore, controlling the inflammation becomes an integral to the treatment of inflammatory bowel disease (IBD). One way of controlling or modifying the inflammation is through medications. Medications that modifies the normal response of immune system are called, immunomodulators, and those that specifically targets TNF-α are known as anti-TNF agents. Modifying or supressing the imbalance in the immune system can restore the immune system and therefore, prevents ongoing ulcer development and inflammation. Anti-TNF medications works by binding to free floating TNF-α in the extracellular space and transmembrane TNF-α found on the cell surface, which prevents their subsequent binding to the target receptors, hence, reduces the effects of inflammation. Examples of anti-TNF medication used in IBD include Infliximab (monoclonal antibody), Certolizumab Pegol (PEGylated Fab'), etc. Survey has shown that ulcers can be healed within a few weeks in patients receiving anti-TNF therapy. Common side effects of anti-TNF therapy may include developing tolerance of the medication, which occurs in 50 % of the patients after undergoing anti-TNF therapy for 12 month. Therefore, it is very important for the patients to be under the advice of their physicians when undertaking this type of therapy (Braegger et al. 1992).

56.4 TNF-α in Malignancy

Tumour necrosis factor-alpha (TNF-α) exhibits dual effects in the development of tumours. It is cytotoxic to tumour cells in higher dose; however, it is also involved in tumour formation and angiogenesis in chronic release of low dosage in the tumour environment, such as in prostate cancer (Nakashima et al. 1998; Bertazza and Mocellin 2010; Fajardo et al. 1992; Ricote et al. 2003).

TNF-α has a dose-dependent opposing effects on angiogenesis. Animal studies indicates that TNF-α induces in vitro chemotaxis of adrenal capillary endothelial cells and initiates angiogenesis in lower doses (0.05–0.5 U/mL). However, at higher TNF doses (0.5 mg/L) prohibits angiogenesis and tumour formation (Nakashima et al. 1995). These characteristics of TNF-α make it desirable in potential anti-cancer treatments as the delivery of high dose of TNF-α could slow down or inhibit the progression of carcinogenic growth by reducing its vasculature, and like its name suggests, causing necrosis of tumours. In the contrary, these research also indicates that chronic release of low dose of TNF-α can increase the ability of cancerous tissue to metastasise due to its angiogenic effects (Nakashima et al. 1995). Therefore, two types of treatments have been developed based on current understanding of TNF-α: One is to reduce TNF-α level in patients to prevent further growth and metastasis of the tumour, and the other is to delivery high dose of TNF-α to the patient to induce tumour necrosis effects.

Emerging evidence in recent studies indicate that TNF-α also has a significant role in castration-induced regression of the normal prostate (Debes and Tindall 2004). Prostate cancer development is regulated by androgen dependent gene pathways. Consequently, androgen removal was believed to be an effective method for the treatment of prostate. The treatment of prostate cancer has changed from castration (removal of testis) to interventions that are directly associate with the hypothalamic-pituitary-gonadal axis, which secretes testosterone (Szlosarek and

Balkwill 2003). Initially, these therapies are very effective and results in improvement lasting 2–3 years. However, all patients eventually develop castrate resistant prostate cancer (CRPC), which cannot be treated at this stage (Wang and Lin 2008; van Horssen et al. 2006). TNF-α is involved the progression of prostate cancer to a castrate resistant state, at least in some patients, as evidence suggests that long term exposure to low amount of TNF-α contributes to hypersensitivity to androgen-sensitive human prostate adenocarcinoma cells, thus induces the resistance (Smyth et al. 2004).

References

Bertazza L, Mocellin S (2010) The dual role of tumor necrosis factor (TNF) in cancer biology. Curr Med Chem 17(29):3337–3352

Braegger CP, Nicholls S, Murch SH, Stephens S, MacDonald TT (1992) Tumour necrosis factor alpha in stool as a marker of intestinal inflammation. Lancet 339(8785):89–91

Debes JD, Tindall DJ (2004) Mechanisms of androgen-refractory prostate cancer. N Engl J Med 351(15):1488–1490. doi:10.1056/NEJMp048178

Fajardo LF, Kwan HH, Kowalski J, Prionas SD, Allison AC (1992) Dual role of tumor necrosis factor-alpha in angiogenesis. Am J Pathol 140(3):539–544

Guo S, DiPietro LA (2010) Factors affecting wound healing. J Dent Res 89(3):219–229. doi:10.1177/0022034509359125

Komatsu M, Kobayashi D, Saito K, Furuya D, Yagihashi A, Araake H, Tsuji N, Sakamaki S, Niitsu Y, Watanabe N (2001) Tumor necrosis factor-alpha in serum of patients with inflammatory bowel disease as measured by a highly sensitive immuno-PCR. Clin Chem 47(7):1297–1301

Mocellin S, Rossi CR, Pilati P, Nitti D (2005) Tumor necrosis factor, cancer and anticancer therapy. Cytokine Growth Factor Rev 16(1):35–53. doi:10.1016/j.cytogfr.2004.11.001

Nakashima J, Tachibana M, Ueno M, Baba S, Tazaki H (1995) Tumor necrosis factor and coagulopathy in patients with prostate cancer. Cancer Res 55(21):4881–4885

Nakashima J, Tachibana M, Ueno M, Miyajima A, Baba S, Murai M (1998) Association between tumor necrosis factor in serum and cachexia in patients with prostate cancer. Clin Cancer Res 4(7):1743–1748

Papadakis KA, Targan SR (2000) Tumor necrosis factor: biology and therapeutic inhibitors. Gastroenterology 119(4):1148–1157

Ricote M, Royuela M, Garcia-Tunon I, Bethencourt FR, Paniagua R, Fraile B (2003) Pro-apoptotic tumor necrosis factor-alpha transduction pathway in normal prostate, benign prostatic hyperplasia and prostatic carcinoma. J Urol 170(3):787–790. doi:10.1097/01.ju.0000082712.41945.17

Rubin GP, Hungin AP, Kelly PJ, Ling J (2000) Inflammatory bowel disease: epidemiology and management in an English general practice population. Aliment Pharmacol Ther 14(12):1553–1559

Schulte W, Bernhagen J, Bucala R (2013) Cytokines in sepsis: potent immunoregulators and potential therapeutic targets – an updated view. Mediators Inflamm 2013:165974. doi:10.1155/2013/165974

Sinno H, Prakash S (2013) Complements and the wound healing cascade: an updated review. Plast Surg Int 2013:7. doi:10.1155/2013/146764

Smyth MJ, Cretney E, Kershaw MH, Hayakawa Y (2004) Cytokines in cancer immunity and immunotherapy. Immunol Rev 202:275–293. doi:10.1111/j.0105-2896.2004.00199.x

Szlosarek PW, Balkwill FR (2003) Tumour necrosis factor alpha: a potential target for the therapy of solid tumours. Lancet Oncol 4(9):565–573

Thalheimer U, Triantos CK, Samonakis DN, Patch D, Burroughs AK (2005) Infection, coagulation, and variceal bleeding in cirrhosis. Gut 54(4):556–563. doi:10.1136/gut.2004.048181

van Horssen R, Ten Hagen TL, Eggermont AM (2006) TNF-alpha in cancer treatment: molecular insights, antitumor effects, and clinical utility. Oncologist 11(4):397–408. doi:10.1634/theoncologist.11-4-397

Wang X, Lin Y (2008) Tumor necrosis factor and cancer, buddies or foes? Acta Pharmacol Sin 29(11):1275–1288. doi:10.1111/j.1745-7254.2008.00889.x

Chapter 57
Urokinase Plasminogen Activator

Abstract The urokinase plasminogen activator system (uPAS) is an essential mechanism in a wide variety of biological processes such as fibrinoloysis, inflammation, atherosclerotic plaque formation, matrix remodeling during wound healing, tumor invasion, angiogenesis, and metastasis. The urokinase plasminogen activator (uPA) is a member of the urokinase plasminogen activator system (uPAS) which can bind to its receptor (uPAR) and initiates a proteolytic cascade resulting in the conversion of plasminogen to plasmin. Plasmin with its proteolytic function leads to degradation a number of extracellular basement membrane components as well as activation of metalloproteinases. This enzymatic function of uPA is involved in a range of physiopathological processes requiring basement membrane (BM) and/or extracellular matrix (ECM) remodeling, including tumor progression and metastasis. It has been frequently reported that the overexpression of uPA and its receptors is associated with tumour invasion and metastasis. Moreover, an increased expression of uPA, uPAR, and PAI-1is correlated with poor prognosis in several cancers including breast cancer. Therefore, due to the involvement of uPA in cancer progression, uPA system can be a potent candidate for manipulation and cancer treatment as a targeted therapy.

Keywords Urokinase plasminogen activator • uPA • Angiogenesis • Normal physiology • Disease • Malignancy

57.1 Urokinase Plasminogen Activator (uPA)

Urokinase plasminogen activator (uPA) is a serine protease, which has been shown to play key roles in numerous physiological processes, including the stimulation of angiogenesis (Ulisse et al. 2009; Huang et al. 2015; Choong and Nadesapillai 2003). uPA forms part of the urokinase plasminogen activator system (uPAS), along with the receptor uPAR and its two inhibitors PAI-1 and PAI-2 (Duffy 2004). This chapter is focused on the protease uPA, which is responsible for regulating the migration, proliferation and adhesion of cells in the process of tissue remodeling. uPA is a key

Dr. Haleh Vosgha contributed equally with Dr. Ali Salajegheh.

A. Salajegheh, *Angiogenesis in Health, Disease and Malignancy*,
DOI 10.1007/978-3-319-28140-7_57

regulator of angiogenesis, but affects many other cellular processes including protein expression, receptor shedding, cytokine modulation, tissue remodeling and phenotypic modulation (Parfenova et al. 2009; Lund et al. 2011).

57.2 uPA in Normal Physiology

The role of uPA in angiogenesis is well established (Hildenbrand et al. 1995; Kaneko et al. 2003; Min et al. 1996; Heymans et al. 1999; Parfyonova et al. 2002; Traktuev et al. 2007). For angiogenesis to occur endothelial cells must proliferate, the extracellular matrix must be degraded, endothelial cells must migrate and a new matrix must be synthesized. uPA has been shown to play an important role in extracellular matrix remodeling and the release of growth factors for angiogenesis (Mekkawy et al. 2014). uPA binds to uPAR at the cell surface and causes a protein cascade including plasmin formation. Plasmin formation results in the activation or release of matrix metalloproteases (MMPs), such as MMP-3 and MMP-12 (Mekkawy et al. 2014; Duffy 2004). MMPs cause extracellular proteolysis and degradation of basement membrane proteins, including fibronectin. This process where the vessel is destabilized is a necessary step so that endothelial cells can migrate to allow the generation of new capillaries (Parfenova et al. 2009). Plasmin activation through uPA also results in the release of growth factors for angiogenesis, including hepatocyte growth factor (HGF), vascular endothelial growth factor (VEGF), FGF2 and TGFß. These growth factors stimulate angiogenesis. For example, VEGF promotes endothelial cell proliferation following the degradation of the extracellular matrix by uPA (Duffy 2004). A study by Tsokolaeva et al. examined mice with infarcted hearts and revealed that those deficient in uPA had impaired angiogenesis and that this process was not stimulated by VEGF. This indicated that VEGF angiogenic activity is dependent upon uPA (Traktuev et al. 2007). This reveals the importance of uPA for myocardial angiogenesis.

Through these mechanisms uPA plays an essential role in regulating angiogenesis in normal physiology. The expression of uPA is correlated with angiogenesis in the female reproductive tract (Parfyonova et al. 2002). The fundamental role uPA plays in normal physiology is also seen in its facilitation of wound healing through regulation of angiogenesis. A study by Piao et al., revealed that the lipid sphingomyelin metabolite, sphingosylphosphorylcholine (SPC), is able to improve wound healing through upregulation of uPA (Piao et al. 2005). SPC increased the transcriptional and translational levels of uPA. These increased uPA levels stimulated angiogenesis and this was shown to significantly accelerate the closure of a wound in an *in vitro* model. This study revealed the importance of uPA in angiogenesis for wound healing. uPA stimulates extracellular matrix degradation and cell migration, facilitating angiogenesis. This provides a route for nutrients and immune cells to reach the injury site, improving wound repair (Piao et al. 2005).

57.3 uPA in Disease

uPA plays a complex and important role in angiogenesis in inflammation, tissue ischemia and malignant tumours. Hence, intervening with the role of uPA holds therapeutic promise. Excessive expression of uPA has been shown to promote tumour growth and metastasis, thus suggesting that locally suppressing the activity of uPA may be a target for anti-cancer therapies (Parfyonova et al. 2002). On the other hand, uPA overexpression has been suggested as a method of stimulating angiogenesis in ischemic tissues (Traktuev et al. 2007). This complex role in disease and malignancy is discussed in the following paragraphs.

The role of uPA in stimulating angiogenesis suggests its therapeutic significance as a thrombolytic drug in infarction. Numerous studies have shown that uPA provides tissue protection and improved blood flow to ischemic tissue. Tsokolaeva *et al.* studied mice with experimentally produced hindlimb ischemia (Traktuev et al. 2007). It was concluded that wild type mice were better able to restore blood flow than mice with uPA deficiency. Additionally, uPA did not cause oedema or hypervascularisation of nonischemic regions (Traktuev et al. 2007). Such studies indicate the relevance for uPA in treating tissue ischemia. Heymans et al. (1999) showed that uPA is critical for angiogenesis following myocardial infarction. This study concluded that mice with uPA deficiency had impaired myocardial angiogenesis following heart infarction. They suffered depressed cardiac contractility, ischemia and arrhythmias, resulting in death (Heymans et al. 1999).

57.4 uPA in Malignancy

The involvement of uPA in tumor malignancy has been a focus of much research (Choong and Nadesapillai 2003; Ulisse et al. 2009; Duffy et al. 1988; Mekkawy et al. 2014; Duffy 2004). Current data indicates that through stimulating angiogenesis uPA plays a fundamental role in tumor growth and progression (Duffy 2004). Mekkawy et al. found that high uPA levels in tumour tissue correlated with tumor aggressiveness and survival (Mekkawy et al. 2014). uPA is involved at numerous steps in cancer progression by facilitating the migration of endothelial, smooth muscle, inflammatory and cancerous cells – therefore facilitating invasion, angiogenesis and cancer metastasis (Mekkawy et al. 2014). Proteolysis of the extracellular matrix and basement membrane, which is caused by uPA, is necessary to facilitate angiogenesis, invasion and metastasis of a tumor (Parfenova et al. 2009).

Duffy et al., described uPA to be an extremely significant prognostic factor for breast cancer (Duffy et al. 1988). A study of 8377 patients with breast cancer revealed that uPA levels measured is a stronger predictor of breast cancer outcome than tumor grade or tumor size. The concentration of uPA in breast tissue extracts of benign tumours was found to be 19 times lower than in cancer tissue (Duffy et al. 1988). This is due to uPA involvement in angiogenesis but also in other steps of tumor development (Mekkawy et al. 2014). Jelisavac-Cosic et al. (2011) completed a seven and a half year follow up study on patients with invasive breast cancer, which showed a correlation between elevated uPA levels and aggressiveness of the tumour. Additionally, recent investigations have associated uPA with metastasis and poor outcome for melanoma patients. The levels of uPA found in melanomas are significantly elevated compared to normal levels (Gershtein et al. 2001).

The results of such studies have indicated that uPA could be an attractive target in tackling cancer. Research has primarily focused on inhibiting uPA via two main mechanisms – either by preventing uPA binding to uPAR or by inhibiting the catalytic activity of uPA (Mekkawy et al. 2014). A study by Min et al. showed how uPAR antagonists can be successful in inhibiting tumour angiogenesis and progression (Min et al. 1996). In an *in vitro* model, the growth factor domain of uPA was bound to the Fc portion of IgG. This molecule antagonized the action of uPA binding its receptor. Angiogenesis was inhibited and B16 melanoma growth was suppressed in syngeneic mice (Min et al. 1996).

While the link between uPA levels and tumour aggressiveness is well established for some cancers, some studies on the relevance of uPA is certain cancers have yielded conflicting results (Mekkawy et al. 2014). Numerous studies have shown that uPA expression is higher in gastric tumour than in normal tissue and that high uPA levels are linked with shorter survival (Mekkawy et al. 2014). Kaneko et al. showed that increased levels of uPA correlated with tumor angiogenesis and also with a decreased survival rate for patients with gastric cancer (Kaneko et al. 2003). This study on tumors from 101 patients concluded that survival of patients with uPA protein expression was lower than patients with negative expression. uPA was shown to be correlated with tumor microvessel density (MVD), suggesting an important role of uPA in angiogenesis in gastric cancer (Kaneko et al. 2003). However, the results of other studies have suggested that prognosis of gastric cancer is not correlated with uPA levels (Mekkawy et al. 2014). Therefore, further research is required in this area. In 2015, a study for the first time has found the correlation between elevated expression of uPA and tumour invasiveness of endometrial endometrioid adenocarcinoma (EEC) (Huang et al. 2015). However, unlike the uPA angiogenic involvement in other malignancies such as breast cancer, the mechanism of this gene in tumorigenesis and metastasis of EEC is not completely clear.

References

Choong PF, Nadesapillai AP (2003) Urokinase plasminogen activator system: a multifunctional role in tumor progression and metastasis. Clin Orthop Relat Res (415 Suppl):S46–S58. doi:10.1097/01.blo.0000093845.72468.bd

Duffy MJ (2004) The urokinase plasminogen activator system: role in malignancy. Curr Pharm Des 10(1):39–49

Duffy MJ, O'Grady P, Devaney D, O'Siorain L, Fennelly JJ, Lijnen HJ (1988) Urokinase-plasminogen activator, a marker for aggressive breast carcinomas. Preliminary report. Cancer 62(3):531–533

Gershtein ES, Medvedeva SV, Babkina IV, Kushlinskii NE, Trapeznikov NN (2001) Tissue- and urokinase-type plasminogen activators and type 1 plasminogen activator inhibitor in melanomas and benign skin pigment neoplasms. Bull Exp Biol Med 132(1):670–674

Heymans S, Luttun A, Nuyens D, Theilmeier G, Creemers E, Moons L, Dyspersin GD, Cleutjens JP, Shipley M, Angellilo A, Levi M, Nube O, Baker A, Keshet E, Lupu F, Herbert JM, Smits JF, Shapiro SD, Baes M, Borgers M, Collen D, Daemen MJ, Carmeliet P (1999) Inhibition of plasminogen activators or matrix metalloproteinases prevents cardiac rupture but impairs therapeutic angiogenesis and causes cardiac failure. Nat Med 5(10):1135–1142. doi:10.1038/13459

Hildenbrand R, Dilger I, Horlin A, Stutte HJ (1995) Urokinase plasminogen activator induces angiogenesis and tumor vessel invasion in breast cancer. Pathol Res Pract 191(5):403–409. doi:10.1016/s0344-0338(11)80726-0

Huang CY, Chang MC, Huang WY, Huang CT, Tang YC, Huang HD, Kuo KT, Chen CA, Cheng WF (2015) Urokinase-type plasminogen activator resulting from endometrial carcinogenesis enhances tumor invasion and correlates with poor outcome of endometrial carcinoma patients. Sci Rep 5:10680. doi:10.1038/srep10680

Jelisavac-Cosic S, Sirotkovic-Skerlev M, Kulic A, Jakic-Razumovic J, Kovac Z, Vrbanec D (2011) Prognostic significance of urokinase-type plasminogen activator (uPA) and plasminogen activator inhibitor (PAI-1) in patients with primary invasive ductal breast carcinoma – a 7.5-year follow-up study. Tumori 97(4):532–539. doi:10.1700/950.10409

Kaneko T, Konno H, Baba M, Tanaka T, Nakamura S (2003) Urokinase-type plasminogen activator expression correlates with tumor angiogenesis and poor outcome in gastric cancer. Cancer Sci 94(1):43–49

Lund IK, Nielsen BS, Almholt K, Rono B, Hald A, Illemann M, Green KA, Christensen IJ, Romer J, Lund LR (2011) Concomitant lack of MMP9 and uPA disturbs physiological tissue remodeling. Dev Biol 358(1):56–67. doi:10.1016/j.ydbio.2011.07.021

Mekkawy AH, Pourgholami MH, Morris DL (2014) Involvement of urokinase-type plasminogen activator system in cancer: an overview. Med Res Rev 34(5):918–956. doi:10.1002/med.21308

Min HY, Doyle LV, Vitt CR, Zandonella CL, Stratton-Thomas JR, Shuman MA, Rosenberg S (1996) Urokinase receptor antagonists inhibit angiogenesis and primary tumor growth in syngeneic mice. Cancer Res 56(10):2428–2433

Parfenova EV, Plekhanova OS, Men'shikov M, Stepanova VV, Tkachuk VA (2009) Regulation of growth and remodeling of blood vessels: the unique role of urokinase. Rossiiskii fiziologicheskii zhurnal imeni IM Sechenova/Rossiiskaia akademiia nauk 95(5):442–464

Parfyonova YV, Plekhanova OS, Tkachuk VA (2002) Plasminogen activators in vascular remodeling and angiogenesis. Biochem Biokhimiia 67(1):119–134

Piao YJ, Lee CH, Zhu MJ, Kye KC, Kim JM, Seo YJ, Suhr KB, Park JK, Kim CD, Lee JH (2005) Involvement of urokinase-type plasminogen activator in sphingosylphosphorylcholine-induced angiogenesis. Exp Dermatol 14(5):356–362. doi:10.1111/j.0906-6705.2005.00272.x

Traktuev DO, Tsokolaeva ZI, Shevelev AA, Talitskiy KA, Stepanova VV, Johnstone BH, Rahmat-Zade TM, Kapustin AN, Tkachuk VA, March KL, Parfyonova YV (2007) Urokinase gene transfer augments angiogenesis in ischemic skeletal and myocardial muscle. Mol Ther 15(11):1939–1946. doi:10.1038/sj.mt.6300262

Ulisse S, Baldini E, Sorrenti S, D'Armiento M (2009) The urokinase plasminogen activator system: a target for anti-cancer therapy. Curr Cancer Drug Targets 9(1):32–71

Chapter 58
Vascular Endothelial Growth Factor (VEGF)

Abstract One of the best-validated signalling pathways in angiogenesis is vascular endothelial growth factor (VEGF) and its receptors. VEGF family encompasses 7 VEGF glycoproteins members including, VEGF-A, VEGF-B, VEGF-C, VEGF-D, and VEGF-E and placental growth factors 1 and 2 (PLGF). Of these, the very well-characterized member is VEGF-A. Vascular endothelial growth factor, also known as vascular permeability factor (VPF), is an endothelial cell-specific mitogen and motogen which promotes angiogenesis and is a potent mediator of vascular permeability as well as being a key mediator of tumour-associated neoangiogenesis and progression. VEGF has been shown as an autocrine stimulator of proliferation, migration and survival in hematological, pancreatic and prostatic tumour cells. In addition to this, its expression is upregulated in the majority of human tumours and many other disorders. Therefore, inhibition of VEGF in xenografts brought a promising horizon for antineoplastic and antiangiogenic methods of cancer treatment and revealed the regulating role of VEGF in angiogenesis and tumour growth.

Keywords Vascular endothelial growth factor • VEGF • Angiogenesis • Normal physiology • Disease • Malignancy

58.1 Vascular Endothelial Growth Factor (VEGF)

Vascular endothelial growth factor (VEGF) is a molecule that activates the process of angiogenesis (Folkman and Shing 1992; Folkman 1995). Under normal physiological conditions, VEGF protein functions to form new blood vessels during embryonic development, new blood vessels after injuries, formation of new vessels to bypass blocked vessels (collateral circulation) and form muscle following exercise (Hoeben et al. 2004). VEGF is also responsible for stimulating vascular endothelial cell growth, survival and proliferation (Boehm et al. 1997).

VEGF belongs to a family of proteins that are all structurally related (Hoeben et al. 2004). They all regulate growth of different components of the vascular system, in particular blood and lymph vessels. VEGF-B, VEGF-C and VEGF-D,

---

Dr. Haleh Vosgha contributed equally with Dr. Ali Salajegheh.

A. Salajegheh, *Angiogenesis in Health, Disease and Malignancy*,
DOI 10.1007/978-3-319-28140-7_58

VEGF-E (viral factor) and Placental Growth Factors (PLGF) are all members of this family (Ferrara and Davis-Smyth 1997). However, of these, VEGF-A (simply known as VEGF) is the most angiogenic protein inducing migration of endothelial cells, mitosis of endothelial cells, increasing metalloproteinase activity and creation of blood vessel lumen (Stuttfeld and Ballmer-Hofer 2009). In addition to VEGF-A's mitogenic (endothelial cells) biological properties, it also is a vasodilator, thus inducing increased microvascular permeability (Stuttfeld and Ballmer-Hofer 2009).

58.2 VEGF in Normal Physiology

There are a wide variety of human and animal tissues, expressing low levels of VEGF-A, while some tissues such as fetal tissues, the placenta, corpus luteum and in many human tumours, requiring angiogenesis produce high levels of VEGF-A (Roskoski 2007; Maharaj et al. 2006). VEGF-A plays an important role in postnatal angiogenic processes; wound healing, ovulation, menstruation, maintenance of blood pressure and pregnancy, to name a few. This protein enhances the permeability of endothelial cells through increasing the activity of vesicular-vacuolar organelles which are clustered vesicles to promote the transport of metabolites between luminal and abluminal plasma membrane. VEGF-A can also increase the permeability of endothelial cells by affecting the arrangement of cadherin/catenin complexes in order to weaken the adherence of endothelial cell junctions (Esser et al. 1998; Dvorak et al. 1995; Kevil et al. 1998).

All Vascular Endothelial Growth Factor ligands elicit their angiogenic response by binding to VEGF specific receptors, which are all tyrosine kinase receptors (VEGFRs). The binding of the VEGF ligand to VEGFR causes dimerization and activation via transphosphorylation and signal transduction (Cébe-Suarez et al. 2006). VEGFR-1, VEGFR-2 and VEGFR-3 are known as the primary receptors for VEGF. VEGFR-1 and VEGFR-2 are more closely related to angiogenesis whereas VEGFR-3 is associated with Lymphangiogenesis (Hicklin 2004). VEGF receptors vary in their endothelial expression. VEGFR-2 is expressed on almost all endothelial cells, however VEGFR-1 and VEGFR-3 are expressed in distinct vascular beds (Witmer et al. 2002). Extracellular co-receptors Neuropilin-1 and Neuropilin-2 are believed to somewhat increase ligand binding affinity to primary receptors (VEGFR-1,2,3) but their specific role in angiogenesis is still not clearly understood (Kumar et al. 2009). VEGFR-1 is an important receptor for embryonic angiogenesis (vessel formation) but does not however associate closely with any pathologic conditions. VEGFR-2 on the other hand is responsible for the majority of angiogenic effects given off by VEGF including microvascular permeability, endothelial cell proliferation, migration, invasion and survival of blood vessels (Wiesmann et al. 1997).

VEGF-A is one of the main growth factors released by activated platelet during inflammation. The gene attracts monocytes and neutrophils. VEGFR-1 is expressed on inflammatory cells (Dvorak 2002). Specific role or mechanism that VEGF-A plays in the process is not very clear but it is known that the neutrophils and monocytes that are recruited from the circulation produce a number of pro-inflammatory cytokines such as IL-1B and TNF-a (Hoeben et al. 2004). These cytokines also up-regulate gene expression of VEGF-A in keratinocytes at wound margins. The formation of new blood vessels plays an integral part in wound healing.

58.3 VEGF in Disease

VEGF has been shown to be involved in several diseases including eye diseases, proliferative diabetic retinopathy (PDR), diabetic macular edema (DME), heart failure, and hydrocephalus (Taimeh et al. 2013; Osaadon et al. 2014; Gupta et al. 2013; Penn et al. 2008; Testa et al. 2008; Shim et al. 2014).

Development of both proliferative diabetic retinopathy (PDR) and diabetic macular edema (DME) is associated with the central role of VEGF (Gupta et al. 2013; Osaadon et al. 2014). VEGF can lead to altering retinal capillary permeability through increasing the phosphorylation of some proteins involved in tight-junctions such as zonula occludens (Antonetti et al. 1999). High levels of VEGF result in activation of mitogen-activated protein (MAP) and endothelial cell proliferation. VEGFR-2 induction has been also reported to activate the phophatidylinositol 3-kinase (PI3)/Akt pathway. These cascades can promote endothelial cells to release matrix metalloproteinases (MMPs) and urokinase-type plasminogen activator which subsequently lead to basement membranes degradation and cell migration. Basement membranes for the newly formed capillaries have been synthesized following by proliferation and migration of endothelial cells (Witmer et al. 2003). Pericytes and smooth muscle cells that are regulated by platelet-derived growth factor (PDGF) are key factor to control the stability of these capillaries (Osaadon et al. 2014; Gupta et al. 2013). There are several available drugs against VEGF with the purpose of reducing neovascularization and vessel leakage by blocking its effects. Bevacizumab, Ranibizumab, Pegaptanib and Aflibercept are these commercial drugs in this class which are effective in inhibiting neovascularization associated with various retinal proliferative vascular diseases (Osaadon et al. 2014; Gupta et al. 2013).

Heart failure has several underlying causes among which dysfunctional blood vessel formation is a main problem (Taimeh et al. 2013). According to some researches, initial upregulated of VEGF has been observed in the initial phases of heart failure, however decreased expression of VEGF-A is connected to disproportionate microvascular growth and decreased capillary density (Murohara et al. 1998). Valuable information has been gathered about the potential function of targeting VEGF as a treatment option for heart failure. However, it has not, to date, published the clinical benefits of VEGF gene therapy in patients with coronary artery disease or peripheral artery disease (Testa et al. 2008).

Hydrocephalus is a medical condition characterized by abnormal accumulation of cerebrospinal fluid (CSF) in the brain ventricles (Shim et al. 2014). It has been noticed that high expression of VEGF is linked to CSF of neonatal and pediatric patients with hydrocephalus (Naureen et al. 2014). Furthermore, elevated VEGF has been reported in the cerebral tissues adjacent to the neuroepithelium of premature infants with ventriculomegaly in the presence of hemorrhage and in the periventricular white matter following subarachnoid hemorrhage (Chu et al. 2011). However, hydrocephalus is believed a secondary event next to infection and traumatic brain injury. Therefore, elevated VEGF in the CSF could be a consequence of hydrocephalus and a number of injury mechanisms could increase VEGF and can play as a repair response to insufficient angiogenesis, cerebral injury or neural degeneration (Shim et al. 2014).

58.4 VEGF in Malignancy

VEGF is a principal regulator of tumour angiogenesis. VEGF can establish new vasculature in early stages of tumour development (Karamysheva 2008). By recruiting bone-marrow-derived progenitor cells, VEGF is able to stimulate tumour growth at primary and metastatic sites (Hoeben et al. 2004). After the establishment of the blood vessels, VEGF can continue to help the vessels grow via endothelial cell proliferation, migration and invasion. Furthermore, VEGF also help the supporting blood vessels to survive by inhibiting endothelial cell apoptosis and fuelling the tumour with adequate nutrients (Felmeden et al. 2003).

Under hypoxic conditions the human body up regulates Erythropoietin (EPO) gene transcription, the hormone that increases number of erythrocytes to meet the body's metabolic needs (Haase 2010, 2013). It is quite common to observe localised hypoxia within a specific region or organ due to insufficient perfusion. VEGF-A plays a key role angiogenesis and neo-vascularisation, increasing oxygen and energy substrate perfusion (Carmeliet and Jain 2000; Hillen and Griffioen 2007). VEGF mRNA transcription is up regulated when cells are under hypoxic or hypoglycaemic pressure. Usually hypoxia decreases the global rate of translation, however VEGF mRNA's have an internal ribosome entry site (IRES) that drives protein synthesis during time of low oxygen (Semenza 2001). Hypoxic inducible protein complex (HIF-1) produced by hypoxic cells binds to enhancer sequences of VEGF-A gene, EPO gene and other important genes such as one for crucial glucose transporters and glycolytic enzymes (Haase 2010, 2013). VEGF is released as a result. The HIF-1 gene is a heterodimer made up of HIF-1a and HIF-1B subunits. Under normal levels of oxygen HIF-1a is degraded via hydroxylation then ubiquitination (Hoeben et al. 2004). HIF-1b is always expressed and is very stable. Growth factors such as insulin like growth factors 1 and 2 and bFGF have also been studied and shown to increase expression of HIF-1a (Carmeliet et al. 2001). VEGFR-1 up regulation expression is also hypoxia induced as well as VEGFR- Placental growth factor (PIGF) a gene that binds to VEGFR1 apparently also is hypoxia induced in several body tissues and along with VEGF has synergistic effects on angiogenesis (Enholm et al. 1997).

Tumour Necrosis Factor-Alpha (TNF-a) is an inflammatory cytokine that triggers the release of a number of angiogenic molecules (bFGF, PAF, VEGF-A and

VEGF-C) as well as up regulating other proteolytic systems (Hoeben et al. 2004). More importantly, TNF-a has actually been proven to increase VEGFR-2 transcription within endothelial cells (Sainson et al. 2008; Giraudo et al. 1998). Other growth factors that have been shown to increase VEGF-A mRNA expression include Tissue Growth factor (TGF), Platelet Derived Growth factor BB (PDGF-BB) and Epidermal Growth factor (EGF). According to a study published in 1998, VEGF-A mRNA is triggered *in vivo* in wounds by PDGF and also by EGF and TGF but to a lesser effect (Carmeliet and Collen 1998). Another study also proposed that IL-1 in human synovial fibroblasts and IL-6 in tumour cell lines also induce VEGF-A expression (Hyder 2002).

Of the many hormone studied to be involved with VEGF, estrogens show the most clear angiogenic biological properties (Barnabas et al. 2013). Estrogens have been shown to trigger gene transcription of VEGF-A whilst also stabilising VEGF-A mRNA which extends the half- life of the transcripts (Ruohola et al. 1999). The 5′ regulatory regions within the VEGF-A gene are not associated with the Estrogen response; rather the regulatory regions with AP-1 and Sp1 sites are the ones that are associated with Estrogen mechanism (Hoeben et al. 2004). The effect of testosterone on VEGF-A gene transcription has been studied in androgen-dependent S115 mouse breast cancer cell line as well as human prostatic tissue. According to the study, it was found that testosterone increased VEGF-A expression whilst VEGF-B and VEGF-C are not affected by the hormonal treatment to the same effect of VEGF-A (Scaldaferri et al. 2009).

As explained in the introductory section, tumour growth and metastatic dissemination are closely linked with angiogenesis – a vital supply of blood, nutrients and oxygen (Carmeliet and Jain 2000). VEGF pathway is a key regulator of this process and has been well documented in scientific literature (Senger et al. 1993; Roskoski 2007). The activation of the VEGF/VEGF-receptor pathway starts a number of signalling processes that initiate endothelial cell growth, migration, mitogenesis, vessel permeability, migration and differentiation and pre-existing vasculature survival (Sakurai and Kudo 2011). Overexpression of VEGF has been shown to have a link with tumour development and growth in colorectal carcinomas, thyroid, prostate cancer, pancreatic carcinomas gastric carcinomas, breast cancer, cervical cancer and melanoma (Salajegheh et al. 2013, 2014, 2015; Tomao et al. 2014; Dana et al. 2015; Hirakawa et al. 2005; Nishida et al. 2006).

VEGF is an extremely powerful inducer of vascular permeability – in fact 50,000 fold more potent than histamine. This biological property of VEGF is of upmost importance with regards to hyper permeability of vessel of tumour that can be attributed to tumour cell expression of VEGF (Matsumoto and Ema 2014). The permeability is said to induce leakage of many plasma proteins into the interstitial fluid. These include fibrinogen and clotting proteins (Gerber et al. 1998). The deposit of fibrin within extravascular spaces can alter normally antiangiogenic stroma in typical tissues into a pro-angiogenic environment.

VEGF can induce a several different effects on vascular endothelial cells including morhophological changes, alterations in cytoskeleton and stimulation of cell growth and migration (Matsumoto and Ema 2014). VEGF is known to upregulate the expression of many endothelial cells genes such as procoagulant tissue factor, fibronylytic pathway proteins, GLUT-1 glucose transporters, nitric oxide synthase (Gerber et al. 1998). VEGF also shows vasodilation properties through the release of nitric oxide (NO) and prostaglandins. Increased heart rate, hypotension comes as a result (Yang et al. 2002).

VEGF has been clinically proven to be an apoptosis inhibitor by increasing gene expression of anti-apoptic proteins including bcl-2 and activating a pathway by the name of P13K-Akt in vitro (Hicklin 2004). Endothelial cells of the newly formed vessels of tumour are known to show particular dependence of VEGF but not those that are well established.

VEGF acting as a mitogen for endothelial cells induces proliferation mainly involving VEGFR-2 as they are able to activate a few extracellular kinases. Protein Kinase C pathways also seems another pathway by which proliferation is mediated (Garcia-Barros 2003).

The bone marrow is responsible for production of hematopoietic stem cells (HSC's), endothelial cells, osteoblasts and osteoclasts all which are responsive to VEGF (Yoder 2012). Endothelial progenitor cells (EPC)'s originate from HSC's and when stimulated can enter the circulation (Yoder 2012). Once in the circulation EPC's are referred to as circulating EPCs (CEPs). Scientific evidence suggests that EPCs that are bone marrow made aid tumour vascularisation as they are taken to neovascularisation areas and complement angiogenesis (Dias et al. 2000).

VEGFR's on the endothelium of tumours are stimulated by VEGF (Olsson et al. 2006). Numerous studies have also investigated the role of VEGF on stimulation of VEGFRs on tumour cells themselves on non-small lung cell carcinomas, prostate carcinomas, breast carcinomas and melanomas (Garcia-Barros 2003; Adams et al. 2000; Rajabi et al. 2012; Achen and Stacker 2006; Alevizakos et al. 2013; Botelho et al. 2010; Das and Wakelee 2012). VEGFR expression by tumour cells in addition to already known capacity for malignancies to express VEGF introduces a potential role for a VEGF/VEGFR autocrine loop within these tumours. Treatment with Bevacizumab shows that these autocrine loops lie within human leukaemia and support cell survival and migration in vivo cell leukaemia (Zahiragic et al. 2007). Clinical blocking of the VEGFR-2 gene via a synthetic antibody administered to mice bearing human leukemic xenografts resulted in reduced tumour growth (Sullivan and Brekken 2010).

The VEGF pathways close association with tumour related angiogenesis poses an obvious therapeutical exploitation area in oncology studies (Ferrara and Kerbel 2005). By inhibiting the VEGF pathway, anti- angiogenesis treatment can be achieved to arrest or slow down tumour growth. One of the most well-known of the angiogenesis inhibitors is a drug called Bevacizumab (Avastin) as a monoclonal antibody – a laboratory synthesised molecule that is made in such a way that they mimic the antibodies that the human body naturally produces as part of the immune system (Shih and Lindley 2006). The antibody binds to VEGF and blocks signalling of the molecule, leading to supressed formation of new blood vessel growth (Wild et al. 2000). Reduced nutrient supply to the tumour can slow or stop its growth. The U.S. Food and Drug Administration (FDA) approves of Bevacizumab to be used alone for glioblastoma (tumours that originate from connective tissue in brain) and when in combination with other drugs for treatment of non-small cell lung cancers, metastatic colorectal cancer as well as metastatic renal cell cancer (Summers et al. 2010; Cohen et al. 2007a, b; Gil-Gil et al. 2013). Anti-VEGFR-2 antibodies have also been used and shown to decrease signalling, angiogenesis and primary metastatic growth (Niu and Chen 2010). Another option is to block VEGFR tyrosine kinase activity via tyrosine kinase inhibitors (Dias et al. 2000). Antiangiogenic conditions can also be achieved via inhibition of other regulators of VEGF gene regulation such as HIF-1a and COX-2 (Messmer-Blust et al. 2009; Wang et al. 2004). Tumour specific toxins can also be administered which induce vascular damage within tumour cells without producing toxic effects to normal tissues (Hoeben 2004). An example is VEGFA165-DT385 conjugate, which has been shown to inhibit tumour growth *in vivo* (Hoeben 2004). Apoptosis induced drug delivery (AIDD), a relatively new mechanism undergoing research aims to efficiently deliver drugs from carrier cells to tumour cells (Ma and Gallo 1998). Drug delivery is optimised in this way as tumour cells engulf apoptotic bodies filled with drugs via phagocytosis and since the apoptotic membrane is more permeable drug transport is improved. Rapamycin, a drug that impairs tumour cell VEGF-A secretion can also be used in certain cases as well as DNA vaccinations (Vanneman and Dranoff 2012). VEGF/VEGFR targeted therapies alone have only been known to show very limited effectivity in human patients. Further cytotoxic therapy that is able to act on the rapidly proliferating neoplastic should be introduced for optimal regressions.

To conclude, angiogenesis plays a critical role in the development of the capillary network in all human beings: essentially, it builds on the existing network that has been formed by vasculogenesis (Papetti and Herman 2002). After all, every tissue in the body does require a blood supply. This blood supply is formed through either sprouting angiogenesis, or the more recently discovered intussusceptive angiogenesis (Karamysheva 2008).

Yet, although fundamental throughout life, angiogenesis does not always have positive effects for the human body. Evidently, there are instances when angiogenesis should be promoted, but there are also cases where its inhibition is beneficial, but more so therapeutic. Inhibition of angiogenesis is highly beneficial in many diseases and cancer (Ferrara and Kerbel 2005). The majority of research into angiogenesis has centred on controlling the process with regards to these pathologies. This is obviously due to the plethora of research papers conducted in regards to them with the hope of finding effective curative procedures involving drugs and genetic therapies for the diseases and cancers they are involved in. Therefore, the future research into angiogenesis lies in establishing more drugs and specific inhibitors of the angiogenesis process in both disease and cancer. It appears what is most urgent to not only medicine but also humanity is finding a cure for metastatic tumour growth in the body through drugs that can effectively inhibit angiogenesis in these tumours while causing minimal side effects in the body.

References

Achen MG, Stacker SA (2006) Tumor lymphangiogenesis and metastatic spread-new players begin to emerge. Int J Cancer 119(8):1755–1760. doi:10.1002/ijc.21899

Adams J, Carder PJ, Downey S, Forbes MA, MacLennan K, Allgar V, Kaufman S, Hallam S, Bicknell R, Walker JJ, Cairnduff F, Selby PJ, Perren TJ, Lansdown M, Banks RE (2000) Vascular endothelial growth factor (VEGF) in breast cancer: comparison of plasma, serum, and tissue VEGF and microvessel density and effects of tamoxifen. Cancer Res 60(11):2898–2905

Alevizakos M, Kaltsas S, Syrigos K (2013) The VEGF pathway in lung cancer. Cancer Chemother Pharmacol 72(6):1169–1181. doi:10.1007/s00280-013-2298-3

Antonetti DA, Barber AJ, Hollinger LA, Wolpert EB, Gardner TW (1999) Vascular endothelial growth factor induces rapid phosphorylation of tight junction proteins occludin and zonula occluden 1. A potential mechanism for vascular permeability in diabetic retinopathy and tumors. J Biol Chem 274(33):23463–23467

Barnabas O, Wang H, Gao X-M (2013) Role of estrogen in angiogenesis in cardiovascular diseases. Journal of Geriatric Cardiology 10(4):377–382. doi:10.3969/j.issn.1671-5411.2013.04.008

Boehm T, Folkman J, Browder T, O'Reilly MS (1997) Antiangiogenic therapy of experimental cancer does not induce acquired drug resistance. Nature 390(6658):404–407. doi:10.1038/37126

Botelho F, Pina F, Lunet N (2010) VEGF and prostatic cancer: a systematic review. Eur J Cancer Prev 19(5):385–392. doi:10.1097/CEJ.0b013e32833b48e1

Carmeliet P, Collen D (1998) Vascular development and disorders: molecular analysis and pathogenic insights. Kidney Int 53(6):1519–1549

Carmeliet P, Jain RK (2000) Angiogenesis in cancer and other diseases. Nature 407(6801):249–257. doi:10.1038/35025220

Carmeliet P, Moons L, Luttun A, Vincenti V, Compernolle V, De Mol M, Wu Y, Bono F, Devy L, Beck H, Scholz D, Acker T, DiPalma T, Dewerchin M, Noel A, Stalmans I, Barra A, Blacher S, Vandendriessche T, Ponten A, Eriksson U, Plate KH, Foidart J-M, Schaper W, Charnock-Jones DS, Hicklin DJ, Herbert J-M, Collen D, Persico MG (2001) Synergism between vascular endothelial growth factor and placental growth factor contributes to angiogenesis and plasma extravasation in pathological conditions. Nat Med 7(5):575–583. doi:10.1038/87904

Cébe-Suarez S, Zehnder-Fjällman A, Ballmer-Hofer K (2006) The role of VEGF receptors in angiogenesis; complex partnerships. Cell Mol Life Sci 63(5):601–615. doi:10.1007/s00018-005-5426-3

Chu SH, Feng DF, Ma YB, Zhang H, Zhu ZA, Li ZQ, Zhang ZH (2011) Expression of HGF and VEGF in the cerebral tissue of adult rats with chronic hydrocephalus after subarachnoid hemorrhage. Mol Med Rep 4(5):785–791. doi:10.3892/mmr.2011.500

Cohen MH, Gootenberg J, Keegan P, Pazdur R (2007a) FDA drug approval summary: bevacizumab (Avastin) plus Carboplatin and Paclitaxel as first-line treatment of advanced/metastatic recurrent nonsquamous non-small cell lung cancer. Oncologist 12(6):713–718. doi:10.1634/theoncologist.12-6-713

Cohen MH, Gootenberg J, Keegan P, Pazdur R (2007b) FDA drug approval summary: bevacizumab plus FOLFOX4 as second-line treatment of colorectal cancer. Oncologist 12(3):356–361. doi:10.1634/theoncologist.12-3-356

Dana N, Javanmard Sh H, Rafiee L (2015) Antiangiogenic and antiproliferative effects of black pomegranate peel extract on melanoma cell line. Res Pharm Sci 10(2):117–124

Das M, Wakelee H (2012) Targeting VEGF in lung cancer. Expert Opin Ther Targets 16(4):395–406. doi:10.1517/14728222.2012.669752

Dias S, Hattori K, Zhu Z, Heissig B, Choy M, Lane W, Wu Y, Chadburn A, Hyjek E, Gill M, Hicklin DJ, Witte L, Moore MAS, Rafii S (2000) Autocrine stimulation of VEGFR-2 activates human leukemic cell growth and migration. J Clin Investig 106(4):511–521. doi:10.1172/jci8978

Dvorak HF (2002) Vascular permeability factor/vascular endothelial growth factor: a critical cytokine in tumor angiogenesis and a potential target for diagnosis and therapy. J Clin Oncol 20(21):4368–4380. doi:10.1200/jco.2002.10.088

Dvorak HF, Brown LF, Detmar M, Dvorak AM (1995) Vascular permeability factor/vascular endothelial growth factor, microvascular hyperpermeability, and angiogenesis. Am J Pathol 146(5):1029–1039

Enholm B, Paavonen K, Ristimäki A, Kumar V, Gunji Y, Klefstrom J, Kivinen L, Laiho M, Olofsson B, Joukov V, Eriksson U, Alitalo K (1997) Comparison of VEGF, VEGF-B, VEGF-C and Ang-1 mRNA regulation by serum, growth factors, oncoproteins and hypoxia. Oncogene 14(20):2475–2483. doi:10.1038/sj.onc.1201090

Esser S, Lampugnani MG, Corada M, Dejana E, Risau W (1998) Vascular endothelial growth factor induces VE-cadherin tyrosine phosphorylation in endothelial cells. J Cell Sci 111(Pt 13):1853–1865

Felmeden DC, Blann AD, Lip GY (2003) Angiogenesis: basic pathophysiology and implications for disease. Eur Heart J 24(7):586–603

Ferrara N, Davis-Smyth T (1997) The biology of vascular endothelial growth factor. Endocr Rev 18(1):4–25. doi:10.1210/edrv.18.1.0287

Ferrara N, Kerbel RS (2005) Angiogenesis as a therapeutic target. Nature 438(7070):967–974. doi:10.1038/nature04483

Folkman J (1995) Angiogenesis in cancer, vascular, rheumatoid and other disease. Nat Med 1(1):27–31

Folkman J, Shing Y (1992) Angiogenesis. J Biol Chem 267(16):10931–10934

Garcia-Barros M (2003) Tumor response to radiotherapy regulated by endothelial cell apoptosis. Science 300(5622):1155–1159. doi:10.1126/science.1082504

Gerber HP, Dixit V, Ferrara N (1998) Vascular endothelial growth factor induces expression of the antiapoptotic proteins Bcl-2 and A1 in vascular endothelial cells. J Biol Chem 273(21):13313–13316. doi:10.1074/jbc.273.21.13313

Gil-Gil MJ, Mesia C, Rey M, Bruna J (2013) Bevacizumab for the treatment of glioblastoma. Clin Med Insight Oncol 7:123–135. doi:10.4137/CMO.S8503

Giraudo E, Primo L, Audero E, Gerber HP, Koolwijk P, Soker S, Klagsbrun M, Ferrara N, Bussolino F (1998) Tumor necrosis factor-alpha regulates expression of vascular endothelial growth factor receptor-2 and of its co-receptor neuropilin-1 in human vascular endothelial cells. J Biol Chem 273(34):22128–22135

Gupta N, Mansoor S, Sharma A, Sapkal A, Sheth J, Falatoonzadeh P, Kuppermann BD, Kenney MC (2013) Diabetic retinopathy and VEGF. Open Ophthalmol J 7:4–10. doi:10.2174/1874364101307010004

Haase VH (2010) Hypoxic regulation of erythropoiesis and iron metabolism. Am J Physiol Renal Physiol 299(1):F1–F13. doi:10.1152/ajprenal.00174.2010

Haase VH (2013) Regulation of erythropoiesis by hypoxia-inducible factors. Blood Rev 27(1):41–53. doi:10.1016/j.blre.2012.12.003

Hicklin DJ (2004) Role of the vascular endothelial growth factor pathway in tumor growth and angiogenesis. J Clin Oncol 23(5):1011–1027. doi:10.1200/jco.2005.06.081

Hillen F, Griffioen AW (2007) Tumour vascularization: sprouting angiogenesis and beyond. Cancer Metastasis Rev 26(3–4):489–502. doi:10.1007/s10555-007-9094-7

Hirakawa S, Kodama S, Kunstfeld R, Kajiya K, Brown LF, Detmar M (2005) VEGF-A induces tumor and sentinel lymph node lymphangiogenesis and promotes lymphatic metastasis. J Exp Med 201(7):1089–1099

Hoeben A (2004) Vascular endothelial growth factor and angiogenesis. Pharmacological Reviews 56(4):549–580. doi:10.1124/pr.56.4.3

Hoeben A, Landuyt B, Highley MS, Wildiers H, Van Oosterom AT, De Bruijn EA (2004) Vascular endothelial growth factor and angiogenesis. Pharmacol Rev 56(4):549–580. doi:10.1124/pr.56.4.3

Hyder SM (2002) The role of steroid hormones on the regulation of vascular endothelial growth factorm. Am J Pathol 161(1):345–346. doi:10.1016/s0002-9440(10)64186-7

Karamysheva AF (2008) Mechanisms of angiogenesis. Biochem Biokhimiia 73(7):751–762

Kevil CG, Payne DK, Mire E, Alexander JS (1998) Vascular permeability factor/vascular endothelial cell growth factor-mediated permeability occurs through disorganization of endothelial junctional proteins. J Biol Chem 273(24):15099–15103

Kumar I, Staton CA, Cross SS, Reed MWR, Brown NJ (2009) Angiogenesis, vascular endothelial growth factor and its receptors in human surgical wounds. Br J Surg 96(12):1484–1491. doi:10.1002/bjs.6778

Ma J, Gallo JM (1998) Delivery of cytotoxic drugs from carrier cells to tumour cells by apoptosis. Apoptosis 3(3):195–202

Maharaj AS, Saint-Geniez M, Maldonado AE, D'Amore PA (2006) Vascular endothelial growth factor localization in the adult. Am J Pathol 168(2):639–648. doi:10.2353/ajpath.2006.050834

Matsumoto K, Ema M (2014) Roles of VEGF-A signalling in development, regeneration, and tumours. J Biochem 156(1):1–10. doi:10.1093/jb/mvu031

Messmer-Blust A, An X, Li J (2009) Hypoxia-regulated angiogenic inhibitors. Trends Cardiovasc Med 19(8):252–256. doi:10.1016/j.tcm.2010.02.006

Murohara T, Horowitz JR, Silver M, Tsurumi Y, Chen D, Sullivan A, Isner JM (1998) Vascular endothelial growth factor/vascular permeability factor enhances vascular permeability via nitric oxide and prostacyclin. Circulation 97(1):99–107

Naureen I, Waheed Kh A, Rathore AW, Victor S, Mallucci C, Goodden JR, Chohan SN, Miyan JA (2014) Fingerprint changes in CSF composition associated with different aetiologies in human neonatal hydrocephalus: inflammatory cytokines. Childs Nerv Syst 30(7):1155–1164. doi:10.1007/s00381-014-2415-6

Nishida N, Yano H, Nishida T, Kamura T, Kojiro M (2006) Angiogenesis in cancer. Vasc Health Risk Manag 2(3):213–219

Niu G, Chen X (2010) Vascular endothelial growth factor as an anti-angiogenic target for cancer therapy. Curr Drug Targets 11(8):1000–1017

Olsson AK, Dimberg A, Kreuger J, Claesson-Welsh L (2006) VEGF receptor signalling – in control of vascular function. Nat Rev Mol Cell Biol 7(5):359–371. doi:10.1038/nrm1911

Osaadon P, Fagan XJ, Lifshitz T, Levy J (2014) A review of anti-VEGF agents for proliferative diabetic retinopathy. Eye (Lond) 28(5):510–520. doi:10.1038/eye.2014.13

Papetti M, Herman IM (2002) Mechanisms of normal and tumor-derived angiogenesis. Am J Physiol Cell Physiol 282(5):C947–C970. doi:10.1152/ajpcell.00389.2001

Penn JS, Madan A, Caldwell RB, Bartoli M, Caldwell RW, Hartnett ME (2008) Vascular endothelial growth factor in eye disease. Prog Retin Eye Res 27(4):331–371. doi:10.1016/j.preteyeres.2008.05.001

Rajabi P, Neshat A, Mokhtari M, Rajabi MA, Eftekhari M, Tavakoli P (2012) The role of VEGF in melanoma progression. J Res Med Sci Off J Isfahan Univ Med Sci 17(6):534–539

Roskoski R Jr (2007) Vascular endothelial growth factor (VEGF) signaling in tumor progression. Crit Rev Oncol Hematol 62(3):179–213. doi:10.1016/j.critrevonc.2007.01.006

Ruohola JK, Valve EM, Karkkainen MJ, Joukov V, Alitalo K, Härkönen PL (1999) Vascular endothelial growth factors are differentially regulated by steroid hormones and antiestrogens in breast cancer cells. Mol Cell Endocrinol 149(1–2):29–40. doi:10.1016/s0303-7207(99)00003-9

Sainson RC, Johnston DA, Chu HC, Holderfield MT, Nakatsu MN, Crampton SP, Davis J, Conn E, Hughes CC (2008) TNF primes endothelial cells for angiogenic sprouting by inducing a tip cell phenotype. Blood 111(10):4997–5007. doi:10.1182/blood-2007-08-108597

Sakurai T, Kudo M (2011) Signaling pathways governing tumor angiogenesis. Oncology 81(Suppl 1):24–29. doi:10.1159/000333256

Salajegheh A, Pakneshan S, Rahman A, Dolan-Evans E, Zhang S, Kwong E, Gopalan V, Lo CY, Smith RA, Lam AK (2013) Co-regulatory potential of vascular endothelial growth factor-A and vascular endothelial growth factor-C in thyroid carcinoma. Hum Pathol 44(10):2204–2212. doi:10.1016/j.humpath.2013.04.014

Salajegheh A, Vosgha H, Rahman MA, Gopalan V, Smith RA, Lam AK-Y (2014) The modulatory role of miRNA 126 in thyroid cancer angiogenesis [Abstract]. In: Proceedings of the 105th annual meeting of the American Association for Cancer Research. Cancer Res 74(19 Suppl):Abstract nr 528. doi:10.1158/1538-7445.AM2014-528

Salajegheh A, Vosgha H, Md Rahman A, Amin M, Smith RA, Lam AK (2015) Modulatory role of miR-205 in angiogenesis and progression of thyroid cancer. J Mol Endocrinol 55(3):183–196. doi:10.1530/JME-15-0182

Scaldaferri F, Vetrano S, Sans M, Arena V, Straface G, Stigliano E, Repici A, Sturm A, Malesci A, Panes J, Yla–Herttuala S, Fiocchi C, Danese S (2009) VEGF-A links angiogenesis and inflammation in inflammatory bowel disease pathogenesis. Gastroenterology 136(2):585–595.e585. doi:10.1053/j.gastro.2008.09.064

Semenza GL (2001) Regulation of hypoxia-induced angiogenesis: a chaperone escorts VEGF to the dance. J Clin Investig 108(1):39–40. doi:10.1172/jci200113374

Senger DR, Van de Water L, Brown LF, Nagy JA, Yeo KT, Yeo TK, Berse B, Jackman RW, Dvorak AM, Dvorak HF (1993) Vascular permeability factor (VPF, VEGF) in tumor biology. Cancer Metastasis Rev 12(3–4):303–324

Shih T, Lindley C (2006) Bevacizumab: an angiogenesis inhibitor for the treatment of solid malignancies. Clin Ther 28(11):1779–1802. doi:10.1016/j.clinthera.2006.11.015

Shim JW, Sandlund J, Madsen JR (2014) VEGF: a potential target for hydrocephalus. Cell Tissue Res 358(3):667–683. doi:10.1007/s00441-014-1978-6

Stuttfeld E, Ballmer-Hofer K (2009) Structure and function of VEGF receptors. IUBMB Life 61(9):915–922. doi:10.1002/iub.234

Sullivan LA, Brekken RA (2010) The VEGF family in cancer and antibody-based strategies for their inhibition. mAbs 2(2):165–175

Summers J, Cohen MH, Keegan P, Pazdur R (2010) FDA drug approval summary: Bevacizumab plus interferon for advanced renal cell carcinoma. Oncologist 15(1):104–111. doi:10.1634/theoncologist.2009-0250

Taimeh Z, Loughran J, Birks EJ, Bolli R (2013) Vascular endothelial growth factor in heart failure. Nat Rev Cardiol 10(9):519–530. doi:10.1038/nrcardio.2013.94

Testa U, Pannitteri G, Condorelli GL (2008) Vascular endothelial growth factors in cardiovascular medicine. J Cardiovasc Med (Hagerstown) 9(12):1190–1221. doi:10.2459/JCM.0b013e3283117d37

Tomao F, Papa A, Rossi L, Zaccarelli E, Caruso D, Zoratto F, Benedetti Panici P, Tomao S (2014) Angiogenesis and antiangiogenic agents in cervical cancer. OncoTargets Ther 7:2237–2248. doi:10.2147/ott.s68286

Vanneman M, Dranoff G (2012) Combining immunotherapy and targeted therapies in cancer treatment. Nat Rev Cancer 12(4):237–251. doi:10.1038/nrc3237

Wang YQ, Luk JM, Ikeda K, Man K, Chu AC, Kaneda K, Fan ST (2004) Regulatory role of vHL/HIF-1alpha in hypoxia-induced VEGF production in hepatic stellate cells. Biochem Biophys Res Commun 317(2):358–362. doi:10.1016/j.bbrc.2004.03.050

Wiesmann C, Fuh G, Christinger HW, Eigenbrot C, Wells JA, de Vos AM (1997) Crystal structure at 1.7 A resolution of VEGF in complex with domain 2 of the Flt-1 receptor. Cell 91(5):695–704

Wild R, Dhanabal M, Olson TA, Ramakrishnan S (2000) Inhibition of angiogenesis and tumour growth by VEGF121-toxin conjugate: differential effect on proliferating endothelial cells. Br J Cancer 83(8):1077–1083. doi:10.1054/bjoc.2000.1439

Witmer AN, Dai J, Weich HA, Vrensen GFJM, Schlingemann RO (2002) Expression of vascular endothelial growth factor receptors 1, 2, and 3 in quiescent endothelia. J Histochem Cytochem 50(6):767–777. doi:10.1177/002215540205000603

Witmer AN, Vrensen GF, Van Noorden CJ, Schlingemann RO (2003) Vascular endothelial growth factors and angiogenesis in eye disease. Prog Retin Eye Res 22(1):1–29

Yang R, Ogasawara AK, Zioncheck TF, Ren Z, He GW, DeGuzman GG, Pelletier N, Shen BQ, Bunting S, Jin H (2002) Exaggerated hypotensive effect of vascular endothelial growth factor in spontaneously hypertensive rats. Hypertension 39(3):815–820

Yoder MC (2012) Human endothelial progenitor cells. Cold Spring Harb Perspect Med 2(7):a006692. doi:10.1101/cshperspect.a006692

Zahiragic L, Schliemann C, Bieker R, Thoennissen NH, Burow K, Kramer C, Zuhlsdorf M, Berdel WE, Mesters RM (2007) Bevacizumab reduces VEGF expression in patients with relapsed and refractory acute myeloid leukemia without clinical antileukemic activity. Leukemia 21(6):1310–1312. doi:10.1038/sj.leu.2404632

Chapter 59
Vascular Cell Adhesion Molecule-1 (VCAM-1)

Abstract Expressed on activated endothelial cells, vascular cell adhesion molecule-1 (*VCAM-1*) is an immunoglobulin-like adhesion molecule. It is not expressed under baseline conditions; however it is quickly induced in response to inflammatory stimuli, such as the cytokines tumour necrosis factor and in proatherosclerotic conditions including in early lesions. *VCAM-1* displays a variety of roles that are related to angiogenesis with its main role in cell-to-cell interactions, promoting migration and adhesion of leukocytes to the blood vessels. This protein is also involved in rolling-type adhesion and firm adhesion mainly through binding to α_4 integrin which is expressed by most circulating leucocytes. Because of its wide distribution in human tissues and organs, the role of *VCAM-1* is indicated in heart disease, inflammation in osteoarthritis and in the metastasis of cancers, especially breast and colon cancers. By understanding the functions of this gene in these various common pathologies, treatments can be developed for these diseases.

Keywords Vascular cell adhesion molecule-1 • VCAM-1 • Angiogenesis • Normal physiology • Disease • Malignancy

59.1 Vascular Cell Adhesion Molecule-1 (VCAM-1)

The vascular cell adhesion molecule-1 (*VCAM-1*) is a human gene that plays a role in various cell-to-cell adhesion interactions. It is located on chromosome 1 (p32-p31) and has three known isoforms determined by the splicing of introns along the DNA sequence (Tatusova et al. 2014). *VCAM-1* is speculated to be involved in inflammatory responses causing conditions such as heart diseases and osteoarthritis (Conde et al. 2012; Lu et al. 2010). Studies have also shown that *VCAM-1* is commonly expressed on breast, renal and gastric tumours, giving the tissue metastatic ability (Ding et al. 2003; Schlesinger and Bendas 2014; Hynes 2011) which can turn a benign tumour malignant or re-activate dormant cancers.

Dr. Haleh Vosgha contributed equally with Dr. Ali Salajegheh.

A. Salajegheh, *Angiogenesis in Health, Disease and Malignancy*,
DOI 10.1007/978-3-319-28140-7_59

59.2 VCAM-1 in Normal Physiology

VCAM-1 is expressed at low levels on endothelial cells of healthy tissue (Schlesinger and Bendas 2014; Li and Chen 2012). It demonstrates high affinity for the α4 integrin, also known as very late activation antigen-4 (*VLA-4*), through which most of the cell-to-cell interactions are mediated (Hynes 2011). When stimulated, a variety of leukocytes are activated and brought to the site of inflammation via adhesion and migration through the endothelial wall which is an important step for angiogenesis (Schlesinger and Bendas 2014). Furthermore, *VCAM-1* may become upregulated on vascular endothelium when stimulated by the pro-angiogenic growth factor, *VEGF* (Schlesinger and Bendas 2014).

59.3 VCAM-1 in Diseases

Atherosclerosis is a disease in which plaque accumulates in the endothelial wall of arteries which can occlude them and lead to increased risk of heart diseases, such as angina pectoris, if the plaque builds up in coronary blood vessels (Lusis 2000). Cellular adhesion molecules such as *VCAM-1* may play a role in the development of this disease. Inflammation of the endothelium of blood vessels increases *VCAM-1* expression to attract circulating leukocytes and platelets to the site (Schlesinger and Bendas 2014; Li and Chen 2012). A study by on 128 patients demonstrated increased levels of *VCAM-1* in the serum with incidence of coronary heart disease (Lu et al. 2010). Researchers further looked at how results compared with varying degrees of severity of atherosclerosis, including healthy control, stable angina pectoris, unstable angina pectoris and acute myocardial infarction patients. Results showed positive correlation between increases in severity of coronary heart disease symptoms with expression of *VCAM-1* in the serum (Lu et al. 2010).

According to World Health Organisation, ischaemic heart disease was the leading cause of death worldwide in 2012, making knowledge of potential causes very important to development of treatments. Ischemia occurs when there is a lack of blood supply to a particular organ and is especially significant if it occurs in the heart (Oldenburg et al. 2004). Not only does *VCAM-1* have a role in development of certain diseases, but it may also play a part in therapeutic treatment of this ischaemic heart disease (Grieve et al. 2013; Stanimirovic et al. 1997). An investigation involved injection of multi-potent adult progenitor cells (MAPC) after 3 days and then after 21 days into ischaemic muscle of some mice and revealed increased expression of *VCAM-1* on both instances when compared with the control (Davidson et al. 2013). In comparison to those mice that did not receive a MAPC injection, the ischaemic tissue demonstrated greater adhesion of leukocytes to the post-capillary venules. *In vivo* studies have shown that MAPCs derived from non-endothelial basement membrane stem cells play a role in angiogenesis (Reyes et al. 2002). The relation between injection of MAPC and the subsequent expression of *VCAM-1* indicates the potential therapeutic role of this gene in preparing for an angiogenic therapy for ischaemia using stem cells.

Healthy joint function is important for mobility and general health. Unfortunately, diseases such as osteoarthritis can affect this as the cartilage in the joints gradually degenerates resulting in pain and stiffness of joints. A study by analysed the expression of *VCAM-1* on chondrocytes of cartilage in response to adipokines released by adipocytes as obesity is a predisposing factor for osteoarthritis (Conde et al. 2012). It was shown that adipokines (leptin, adiponectin) and various other soluble kinases caused an increase in *VCAM-1* expression. As *VCAM-1* is involved in cell adhesion of leukocytes, greater expression of *VCAM-1* on the chondrocytes of joints can thus be related to the inflammation causing the breakdown of cartilage in osteoarthritis (Conde et al. 2012). High levels of adipose tissue can be observed in obesity which has a role in secretion of leptin and adiponectin (Kershaw and Flier 2004). Since these secretions increase *VCAM-1* expression, obesity could be a factor affecting the rate of progression of osteoarthritis.

A further investigation into hand osteoarthritis in particular demonstrates the correlation between *VCAM-1*, angiogenesis and progression of osteoarthritis (Pulsatelli et al. 2013). Angiogenesis can be seen not only in affected joints but also in healthy joints elsewhere in the body. Joints of hand osteoarthritis patients were compared with healthy individuals and it was found that soluble *VCAM-1* was present at higher levels in the serum of affected patients. Greater levels of *VCAM-1* in patients presenting with erosive hand osteoarthritis amongst those affected was also observed, suggesting that *VCAM-1* expression may also be related to an increased severity of osteoarthritis (Pulsatelli et al. 2013).

59.4 VCAM-1 in Malignancy

Cancer is a major concern as cancerous tissue has the ability to grow almost infinitely which is facilitated by formation of blood vessels via angiogenesis, providing it with its nutrient and metabolic needs (Ding et al. 2003). Colon cancer and its relation to *VCAM-1* have been well investigated. A study observed tissue samples from patients with gastric cancer, gastric ulcers and healthy patients for *VCAM-1* expression and compared it with microvessel density (MVD) (Ding et al. 2003). The tissue samples were analysed to find increased *VCAM-1* expression with increased MVD. Tissues that had increased MVD were those taken from patients with gastric cancer. Nonetheless, it is noted that whilst MVD measurements may be good prognostic indicators for tumours and its progression, metastasis and recurrence, it holds its limitations as it is not a direct measure of the angiogenic activity of the tumour (Hlatky et al. 2002). High MVD however, is related to a poor prognosis for patients with breast cancers (Morabito et al. 2004). Another study demonstrated that even after undergoing minimally invasive colorectal resection (MICR), colorectal cancer patients displayed sustained higher levels of plasma *VCAM-1* (Shantha Kumara et al. 2012). As *VCAM-1* is associated with high MVD, there is an increased chance for these patients that underwent MICR to develop a secondary cancer later on. Thus, it is suggested that *VCAM-1* expression can be correlated to the development and progression of cancer and possibly angiogenesis.

A tumour growth becomes harmful when it metastasises to form secondary tumours elsewhere in the body and the key component for metastasis to occur is angiogenesis. Reviews indicate how metastasis of breast cancer cells often form secondary tumours in the bone and lungs and is related to a high level of *VCAM-1* expression and high numbers of leukocytes being attracted to the site (Hynes 2011; Schlesinger and Bendas 2015). This was confirmed in a more recent review which outlined the mechanism by which *VCAM-1* promotes this metastasis. The breast cancer cells expressed *VCAM-1* to promote interaction with *VLA-4* to recruit tumour-associated macrophages and monocytes to the lung tissue (Schlesinger and Bendas 2014). The tumour associated macrophages appear to act as a protection for the tumour from the body's immune system. In the bone, soluble *VCAM-1* will bring pre-osteoclasts to the site while *VLA-4* positive pre-osteoclasts bind to *VCAM-1* receptors on the tumour cells. This paracrine signalling activates the pre-osteoclasts to become mutated osteoclasts that destroy the bone matrix for its growth (Schlesinger and Bendas 2014; Hynes 2011).

References

Conde J, Scotece M, Lopez V, Gomez R, Lago F, Pino J, Gomez-Reino JJ, Gualillo O (2012) Adiponectin and leptin induce VCAM-1 expression in human and murine chondrocytes. PLoS One 7(12):e52533. doi:10.1371/journal.pone.0052533

Davidson BP, Ryu JC, Xie A, Belcik JT, Zhao Y, Yue Q, Woda J, Ting A, Lindner JR (2013) Contrast ultrasound molecular imaging of the pro-angiogenic paracrine effects of stem cell therapy in Limb Ischemia. Circulation 127(6):710–719

Ding YB, Chen GY, Xia JG, Zang XW, Yang HY, Yang L (2003) Association of VCAM-1 over-expression with oncogenesis, tumor angiogenesis and metastasis of gastric carcinoma. W J Gastroenterol 9(7):1409–1414

Grieve SM, Lonborg J, Mazhar J, Tan TC, Ho E, Liu CC, Lay W, Gill AJ, Kuchel P, Bhindi R, Figtree GA (2013) Cardiac magnetic resonance imaging of rapid VCAM-1 up-regulation in myocardial ischemia-reperfusion injury. Eur Biophys J 42(1):61–70. doi:10.1007/s00249-012-0857-x

Hlatky L, Hahnfeldt P, Folkman J (2002) Clinical application of antiangiogenic therapy: microvessel density, what it does and doesn't tell us. J Natl Cancer Inst 94(12):883–893

Hynes RO (2011) Metastatic cells will take any help they can get. Cancer Cell 20(6):689–690. doi:10.1016/j.ccr.2011.11.022

Kershaw EE, Flier JS (2004) Adipose tissue as an endocrine organ. J Clin Endocrinol Metab 89(6):2548–2556. doi:10.1210/jc.2004-0395

Li J, Chen B (2012) Oct4 was a novel target of Wnt signaling pathway. Mol Cell Biochem 362(1–2):233–240. doi:10.1007/s11010-011-1148-z

Lu HH, Sheng ZQ, Wang Y, Zhang L (2010) Levels of soluble adhesion molecules in patients with various clinical presentations of coronary atherosclerosis. Chin Med J (Engl) 123(21):3123–3126

Lusis AJ (2000) Atherosclerosis. Nature 407(6801):233–241

Morabito A, Sarmiento R, Bonginelli P, Gasparini G (2004) Antiangiogenic strategies, compounds, and early clinical results in breast cancer. Crit Rev Oncol Hematol 49(2):91–107. doi:10.1016/s1040-8428(03)00168-9

Oldenburg WA, Lau LL, Rodenberg TJ, Edmonds HJ, Burger CD (2004) Acute mesenteric ischemia: a clinical review. Arch Intern Med 164(10):1054–1062. doi:10.1001/archinte.164.10.1054

Pulsatelli L, Assirelli E, Dolzani P, Addimanda O, Mancarella L, Pavloska B, Facchini A, Meliconi R (2013) Increased serum vascular cell adhesion molecule (VCAM)-1 levels in patients with erosive hand osteoarthritis. Rheumatology 52(2):400–402. doi:10.1093/rheumatology/kes309

Reyes M, Dudek A, Jahagirdar B, Koodie L, Marker PH, Verfaillie CM (2002) Origin of endothelial progenitors in human postnatal bone marrow. J Clin Invest 109(3):337–346. doi:10.1172/jci14327

Schlesinger M, Bendas G (2014) Vascular cell adhesion molecule-1 (VCAM-1)-an increasing insight into its role in tumorigenicity and metastasis. Int J Cancer 136(11):2504–2514. doi:10.1002/ijc.28927

Schlesinger M, Bendas G (2015) Vascular cell adhesion molecule-1 (VCAM-1)–an increasing insight into its role in tumorigenicity and metastasis. Int J Cancer 136(11):2504–2514. doi:10.1002/ijc.28927

Shantha Kumara HM, Tohme ST, Herath SA, Yan X, Senagore AJ, Nasar A, Kalady MF, Baxter R, Whelan RL (2012) Plasma soluble vascular adhesion molecule-1 levels are persistently elevated during the first month after colorectal cancer resection. Surg Endosc 26(6):1759–1764. doi:10.1007/s00464-011-2112-4

Stanimirovic DB, Wong J, Shapiro A, Durkin JP (1997) Increase in surface expression of ICAM-1, VCAM-1 and E-selectin in human cerebromicrovascular endothelial cells subjected to ischemia-like insults. Acta Neurochir Suppl 70:12–16

Tatusova T, Ciufo S, Fedorov B, O'Neill K, Tolstoy I (2014) RefSeq microbial genomes database: new representation and annotation strategy. Nucleic Acids Res 42(Database issue):D553–D559. doi:10.1093/nar/gkt1274

Chapter 60
Vascular Endothelial (VE)-Cadherin

Abstract The vascular endothelial (VE)-cadherin which is originally called cadherin-5 is a class of transmembrane calcium-dependent adhesive proteins or cadherin family. It is expressed by endothelial cells and located at endothelial cell's junction in order to play an integral role in endothelium integrity as well as the control of vascular permeability essentially during embryogenic angiogenesis. Apart from its dominant responsibility in cell-cell contacts, VE-cadherin can mediate various cellular and molecular processes including cell proliferation and apoptosis and modulates vascular endothelial growth factor receptor functions. Using the cytoplasmic domain to attach to several protein partners, including p120, β-catenin and plakoglobin, VE-cadherin partly regulates the endothelial permeability. VE-cadherin is also associated with tumour progression, epithelial to mesenchymal transition (EMT) and angiogenesis. It is of pivotal importance in angiogenesis mechanism mainly through altering the role of some essential angiogenic factors and modulation of vascular endothelial growth factor receptors signalling, resulting in increased cell survival and vascular permeability in a wide range of human diseases and cancers.

Keywords Vascular endothelial (VE)-cadherin • VE- cadherin • Angiogenesis • Normal physiology • Disease • Malignancy

60.1 Vascular Endothelial (VE)-Cadherin (VE-Cadherin)

Vascular endothelial (VE)-Cadherin, Cadherin 5, type 2 or CD 144, like Monobutyrin, is a significant component of angiogenesis. Its biological activity remains important for normal development as well as in a number of pathologies such as diabetic retinopathy, atherosclerosis, rheumatoid arthritis and the formation of malignancies.

VE-Cadherin is a type of Cadherin belongs to a class of transmembrane calcium-dependent adhesive proteins known as cadherins. Cadherins are cell-cell adhesion proteins involved in mechanical adhesion between cells as well as tissue morphogenesis. During development, cadherins assist in cell recognition and sorting

Dr. Haleh Vosgha contributed equally with Dr. Ali Salajegheh.

A. Salajegheh, *Angiogenesis in Health, Disease and Malignancy*,
DOI 10.1007/978-3-319-28140-7_60

whereby cells are properly positioned. After development, cadherins are involved in maintaining and forming boundaries, organising cell movements, and processes that maintain tissue integrity. These processes include initiating and maintaining the structure and function of the cell as well as tissue polarity. Additionally, Cadherins also form and maintain diverse tissues and organs.

Cadherins are long proteins, which extend from the surface of the cell. They are typically characteristic of extracellular cadherin repeats, that is, a single membrane-spanning segment and a cytoplasmic region (Shapiro and Weis 2009). The extracellular domain contains 3–5 internal repeats of approximately 110 amino acids, which are responsible for the homophilic and heterophilic interactions during adhesion and cell sorting (Halbleib and Nelson 2006). The extracellular domains are aligned one after another and connected with three calcium ions bound between each domain, which rigidifies the structure of the cadherin. Calcium is important for the cadherin function whereby the removal of calcium renders cadherin vulnerable to proteases (protein-cutting enzymes) (Halbleib and Nelson 2006). Cadherins can be categorized into classical, desmosomal, protocadherins and unconventional classes. As a classical cadherin, VE-Cadherin is associated with actin via interaction with the cytoplasmic tail. VE-cadherin is typically tightly bound to cytoplasmic catenins: beta-catenin, plakoglobin (Bravi et al. 2014) and which form the cytoplasmic tail. Composed of these three proteins of the armadillo family, the cytoplasmic tail is responsible for interaction with the cytoskeleton and for junctional strength. The end of the cytoplasmic tail contains a tryptophan amino acid that binds to cadherins on adjacent cells, therefore mechanically adhering neighbouring cells together. The strength of cell adhesion is influenced by the tyrosine phosphorylation status of cadherin, b-catenin and p120 (Blaschuk and Rowlands 2000).

60.2 VE-Cadherin in Normal Physiology

As an endothelial cell adhesion protein, VE-Cadherin has a critical role in maintaining the integrity of the endothelium and the regulation of vascular permeability when found in stable vessels. VE-Cadherins are able to control the movement of endogenous mediators of inflammation such as leukocytes, histamines, thrombin, TNF-α and platelet-activating factors. It participates in the regulation of cellular process including cell proliferation and apoptosis, cell signaling and the modulation of vascular endothelial growth factor receptor functions. VE-cadherin behaves like mesenchymal cells during angiogenesis and is fundamental during embryonic angiogenesis.

Angiogenesis involves the subsequent growth and remodelling of blood vessel from a network of pre-existing vessels. Within blood vessels, the endothelium of the vessels act as a selective barrier between the bloodstream and the tissues underneath. Here the endothelium acts as the site for new vessel formation. The endothelium consists of endothelial cells joined together by cell-cell junctions though transmembrane adhesion proteins. Endothelial cells must adhere firmly to one another in order to form intact and fully functional blood vessels (Blaschuk and

Rowlands 2000). VE-cadherin acts as this transmembrane component for adhesion of these endothelial cells and thus VE-cadherin is a key modulator of the angiogenesis and in the maintaining the structural integrity of blood vessels. Disruption of VE-cadherin-mediated endothelial cell adhesion inhibits angiogenesis.

Angiogenesis is regulated through the interaction of proangiogenic factors and angiogenesis inhibitors. VE-cadherin drives endothelial cell rearrangement during vessel formation and its activity directly affects angiogenesis through the modulation of vascular permeability and transcription factors. VE-cadherin increases the effects of angiogenic factors on endothelial cells. They regulate angiogenesis through their ability to alter the action of the vascular fibroblast growth factor receptor and endothelial growth factor (VEGF). VEGF enhances endothelial cell layer permeability and increase endocytosis of VE-cadherin. Vascular permeability is also amplified by inflammatory cytokines, including tumor necrosis factor α (TNFα), histamine, and platelet activating factor (PAF).

VE-cadherin is shown to control vascular development and angiogenesis through the genetic activation and inactivation of transcription factors. Genetic inactivation of the VE-cadherin gene leads to embryonic death. This is the result from vascular faults as well as migration of antibodies to VE-cadherin. VE-cadherin expression is regulated by Ets transcription factors.

60.3 VE-Cadherin in Disease

Angiogenesis is critical in a number of pathological processes and subsequent dissociation of VE-cadherin from their junctions can lead to diseases including diabetic retinopathy and rheumatoid arthritis. These pathologies may be a consequence of inflammatory stimuli, mutations in regulators of VE-cadherin and the down regulation of transcription and expression level of VE-cadherin.

In rheumatoid arthritis, the soluble VE-Cadherin in rheumatoid arthritis patients correlates with disease activity (Sidibé et al. 2011). Rheumatoid arthritis (RA) involves the persistent inflammation of the synovial joint resulting in cartilage destruction and bone erosion. This autoimmune, chronic, systemic inflammatory disorder causes the synovial membrane to become inflamed, increasing its mass as the lining of cells undergoes hyperplasia. Excess synovial fluid and the development of synovial fibrovascular tissue (pannus) occurs in the synovial membrane. The RA synovial membrane is characteristic of many leukocytes adjoining to blood vessels and involves the movement of leukocytes from the bloodstream into the synovial membrane. The proliferation of new blood vessels is critical in the development of the synovial pannus, which allows access for leukocytes causing inflammation. These leukocytes secrete mediators such as various growth factors, pro-inflammatory cytokines and chemokines, which affect EC adhesion, affecting endothelium permeability and angiogenesis for blood vessel formation. The major mediator for RA is tumor necrosis factor α (TNF-α). TNF-α targets VE-cadherin, causing VE-cadherin to undergo proteolysis. Normally, the endothelium is maintained through tight junctions that limit the diffusion of lipids, proteins and solutes. However, alterations in these junctions influence endothelial cell motility, vascular morphogenesis and permeability. Since VE-cadherin is critical in the adhesion of endothelial cell adhesion for the endothelium, cleavage of VE-cadherin increases the dissociation of endothelial cells and increases permeability, allowing leukocyte ingress and subsequent inflammation characteristic of RA. The mechanism for VE-cadherin occurs though VE-cadherin tyrosine phosphorylation from tyrosine kinases such as the Src family kinases. VE-cadherin proteolysis with the stimulation of TNF-α suggests the possible clinical use of VE-cadherin in managing rheumatoid arthritis.

Increased angiogenesis has been associated with inflammatory diseases such as RA. As established previously, VE-cadherin also contributes to the process of angiogenesis.

The role of VE-cadherin and angiogenesis in pathological conditions can be further observed in diabetic retinopathy. Diabetic retinopathy is a complication of diabetes which results in damage to the blood vessels in the retina. Damage to these blood vessels can result from poor circulation due to damages to the endothelial cells or from the abnormal growth of new vessels.

Damage to the endothelial cells results from the degeneration of the blood retinal barrier (BRB) causing leakage of vessels and macular oedema to develop. The breakdown of the BRB causes distortion and the loss of vision. These effects arise from the up regulation of proteinases that cleave VE-cadherin from the retinal endothelial cell surface, leading to increased vascular permeability (Zachary 2001).

Normally the tight junctions of the endothelial cells maintain the integrity of the BRB allowing the normal function of the retina. However, increased permeability of the BRB associated with a decreased in VE-cadherin (from protein cleavage) weakens these junctions, resulting in the vision loss experienced in diabetic retinopathy.

Through the process of ocular angiogenesis in response to tissue ischemia, changes in retinal vasculature to form new blood vessels is commonly seem in diabetic retinopathy causing vision loss from the formation of scar tissue. As previously established, VE-cadherin has a pivotal function in angiogenesis acting as the adhesive junction between endothelial cells in blood vessels.

The activity of VE-cadherin in diabetic retinopathy has instigated the use of VE-cadherin antagonists to reduce retinal angiogenesis. As a reduction in VE-cadherin suppresses vessel development in endothelial cells, antagonist treatment also decreases the migration of cells.

60.4 VE-Cadherin in Malignancy

Cancer growth is dependent on a source of blood supply for the tumor to survive, develop and metastasize. Tumor angiogenesis is required for the formation of this network and involves the initiation of new blood vessel growth to form a growing tumor mass. VE-cadherin is considered to play a significant role in tumor angiogenesis by facilitating the formation of vascular tubes from the adhesion of endothelial cells. VE-cadherin has been detected in breast cancer, melanoma cells and osteosarcoma cells.

The growth of a tumor is triggered by intratumoral hypoxia causing the release of a number of angiogeneic factors and subsequent angiogenesis. Epithelial-mesenchymal transition (EMT) is a significant process during tumor development and leads to increase tumor cell malignancy. EMT is the process by which epithelial cells lose their cell polarity and cell-cell adhesion, and gain migratory and invasive properties to become mesenchymal stem cells (Kalluri and Weinberg 2009). The EMT process is characteristic of the loss or reorganization of tight junction proteins resulting in the highly mobile and invasive properties of tumor cells. This leads to an increase in metastatic potential and the progression of the cancer. During EMT, VE-cadherin is stimulated, promoting tumor progression. The role of VE-cadherin is particularly true in mammary tumor cells where VE-cadherin is induced during EMT and is abnormally expressed in breast cancer. Using a mouse mammary carcinoma model to demonstrate tumor progression and EMT, investigations that VE-cadherin enhances the proliferation of tumor cell through the formation of cord-like invasive structure and adhesion to endothelial cells (Labelle et al. 2008). These events are important contributors to their increase in malignancy and metastatic potential. The mechanism of VE-cadherin occurs through its interaction with transforming growth factor-β (TGF- β). Therefore, it is clear that VE-cadherin is critical in tumor angiogenesis through the up regulation in vasculature of cancers such as in breast cancer.

The significance of VE-cadherin expression and function in cancer progression is similarly observed in aggressive human melanoma cells and osteosarcoma cells. Through experimentation with osteosarcoma cells, the significance of VE-cadherin in facilitating tumor malignant progress is further exemplified. By inhibiting VE-cadherin gene expression in osteosarcoma cells with siRNA, osteosarcoma cells were no longer able to induce angiogenic processes that form vascular networks to attain oxygen and nutrients.

The structure of normal blood vessels differs from that of tumor blood vessels. Normally, mature blood vessels consist of pericytes lodged in the basement membrane and are required to regulate blood flow and maintain the vasculature. In tumor blood vessels, these pericytes are absent and lack a well-formed basement membrane. Tumor vasculature is also characteristic of open interendothelial junctions. This decrease in endothelial cell adhesion may be caused by the tumor secreted vascular endothelial growth factor (VEGF) (Blaschuk and Rowlands 2000). VEGF stimulates VE-cadherin as well as b-catenin and p120 in endothelial cells by activating tyrosine phosphorylation, causing endothelial intercellular junctions to open through the weakening of the adhesive strength between cells. By altering VE-cadherin and angiogenic growth of blood vessels, the abnormal structure of tumor blood vessels and decreased adhesion between tumor endothelial cell increases tumor cell malignancy and its migratory and invasive properties.

The establishment of the idea of angiogenesis-driven tumor growth via VE-cadherin has instigated research for the potential use of VE-cadherin as a therapy for tumor growth. VE-cadherin is essential in angiogenesis for blood vessel formation and integrity. Therefore blocking its function with antibodies remains a potential approach against tumor angiogenesis and growth. Such drugs could be used to selectively disrupt the abnormal vasculature of tumours without affecting normal blood vessels (Blaschuk and Rowlands 2000).

As observed in aggressive melanoma cells, the down regulation of VE-cadherin revoked the formation of vascular networks, providing support for the use of anti-VE cadherin approaches in therapy.

References

Blaschuk OW, Rowlands TM (2000) Cadherins as modulators of angiogenesis and the structural integrity of blood vessels. Cancer Metastasis Rev 19(1–2):1–5

Bravi L, Dejana E, Lampugnani MG (2014) VE-cadherin at a glance. Cell Tissue Res 355(3):515–522. doi:10.1007/s00441-014-1843-7

Halbleib JM, Nelson WJ (2006) Cadherins in development: cell adhesion, sorting, and tissue morphogenesis. Genes Dev 20(23):3199–3214. doi:10.1101/gad.1486806

Kalluri R, Weinberg RA (2009) The basics of epithelial-mesenchymal transition. J Clin Investig 119(6):1420–1428. doi:10.1172/jci39104

Labelle M, Schnittler HJ, Aust DE, Friedrich K, Baretton G, Vestweber D, Breier G (2008) Vascular endothelial cadherin promotes breast cancer progression via transforming growth factor signaling. Cancer Res 68(5):1388–1397. doi:10.1158/0008-5472.can-07-2706

Shapiro L, Weis WI (2009) Structure and biochemistry of cadherins and catenins. Cold Spring Harb Perspect Biol 1(3):a003053

Sidibé A, Mannic T, Arboleas M, Subileau M, Gulino-Debrac D, Bouillet L, Jan M, Vandhuick T, Le Loët X, Vittecoq O, Vilgrain I (2011) Soluble VE-cadherin in rheumatoid arthritis patients correlates with disease activity: evidence for tumor necrosis factor α-induced VE-cadherin cleavage. Arthritis Rheum 64(1):77–87. doi:10.1002/art.33336

Zachary I (2001) Signaling transduction mechanisms mediating biological actions of the vascular endothelial growth factor family. Cardiovasc Res 49(3):568–581. doi:10.1016/s0008-6363(00)00268-6

Chapter 61
Zinc Finger E-Box Binding Homeobox 1 (ZEB1)

Abstract Zinc Finger E-Box Binding Homeobox 1 (ZEB1) is a vital member of the ZEB family of transcription factors which is involved in a regulation of key factors during embryonic development and cell differentiation. ZEB1 has been identified as a central inducer of epithelial-to-mesenchymal transition (EMT) process which can be linked with increased aggressiveness, metastasis and a wide variety of human carcinomas. It can control the epithelial basement membrane components- in particular E-cadherin- whose disruption is an essential stage in tumour invasiveness. In addition, it has been found that ZEB1 has an ability to regulate replicative senescence, maintenance of cancer cell stemness and tumour angiogenesis. The correlation between ZEB1 and several well-known microRNAs has been also described in the regulation of angiogenesis mechanism. Although the potential angiogenic regulatory function of ZEB1 has not well recognized, the crosstalk between ZEB1 and vascular endothelial growth factor (VEGF) expression and direct inducing angiogenesis has been recently reported.

Keywords Zinc finger E-box binding homeobox 1 • ZEB1 • Angiogenesis • Normal physiology • Disease • Malignancy

61.1 Zinc Finger E-Box Binding Homeobox 1 (ZEB1)

ZEB1 is a protein coding gene that has its genomic location at 10p 11.2 (Vincent et al. 2009). This gene encodes a zinc finger transcription factor that represses the expression of a number of different genes. For instance, the ZEB family proteins serve as a transcriptional repressor of interleukin 2 (IL-2) and E-cadherin, and Smad target genes (Liu et al. 2008a). ZEB1 is especially well known as the EMT-inducing transcription factor that contributes to tumour cell dissemination and tumour-initiating capacity (Liu et al. 2008a). EMT stands for epithelial-mesenchymal transition, and is associated in "wound healing, organ fibrosis and initiation of metastasis

Dr. Haleh Vosgha contributed equally with Dr. Ali Salajegheh.

A. Salajegheh, *Angiogenesis in Health, Disease and Malignancy*,
DOI 10.1007/978-3-319-28140-7_61

for cancer progression." (Wellner et al. 2009) As ZEB1 suppress expression of E-cadherin, which is a cell-cell recognition molecule, the cells become more receptive to initiation of EMT. This results in the epithelial cells becoming invasive, metastatic cancers (Wellner et al. 2009). Also, ZEB1 transcriptionally represses the Coxsackie virus and Adenovirus receptor (CAR) expression in pancreatic and breast human cancer cells (Lacher et al. 2011). Mutations in this gene are found to have potential to cause neurological disorders, dysmorphic features, megacolon and other malformations, and also Hirschsprung's disease (Cacheux et al. 2001).

61.2 ZEB1 in Normal Physiology

Zeb1 (Zinc-finger E-box-binding homeobox 1) is a transcriptional factor, containing two Kruppel-type zinc finger domains by which it can bind to target DNAs (Soini et al. 2011).

ZEB1 is mainly expressed in endothelial and mesenchymal cells of the villous structures during the last two trimesters. This suggested that ZEB1 has a crucial role in the development of placental blood vessels (Pirinen and Soini 2014). ZEB1 has been connected to morphogenesis mechanism in several processes such as embryonal development and tissue scar formation by inducing the development of neural tissues, chondrocytes, skeletal muscle cells and hematolymphoid tissues (Vandewalle et al. 2009; Pirinen and Soini 2014). The role of ZEB1 in palatal and tooth developments has been also reported by some studies (Shin et al. 2012; Arima et al. 2012). ZEB1 expression can be induced by estrogen and progesterone suggesting that it plays an important role in female physiology (Saykally et al. 2009). It has the highest expression in normal adult bladder and uterus, whereas during embryonic development its highest expression is found in lung, thymus and heart (Hurt et al. 2008).

61.3 ZEB1 in Disease

ZEB1 gene mutations and its abnormal expression can affect complex human health conditions mainly through the induction of epithelial mesenchymal transition (Kurihara et al. 2015). Fuchs endothelial corneal dystrophy and posterior polymorphous corneal dystrophy are among those human diseases in which ZEB1 functions as a strong inducer of EMT (Okumura et al. 2015; Liu et al. 2008b). The mechanisms of these disorders are related to reduction of extracellular matrix proteins and expression of epithelial genes which cause corneal defects such as corneal thickening and endothelial and keratocyte proliferation due to abnormal expression of ZEB1 (Okumura et al. 2015; Liu et al. 2008b).

61.4 ZEB1 in Malignancy

ZEB1 proteins play a substantial role in promoting angiogenesis and tumourigenicity. One of its means to elevate angiogenesis is the transcriptional repression of the expression of microRNA-200 (miRNA-200) (Wellner et al. 2009). miRNAs are non-coding single-stranded RNA molecules, that play an important role in physiological and pathological processes in the body (Dong et al. 2013). miRNA-200 family members (miR-200a, miR-200b, miR-200c, miR-141 and miR-429) inhibit angiogenesis through direct and indirect mechanisms and have effects on decreasing metastasis formation in a number of cancers (Pecot et al. 2013). The miRNA-200 family targets and regulates interleukin-8 (CXCL8) and CXCL1 which are the key pro-angiogenic cytokines secreted by tumour endothelial and cancer cells (Pecot et al. 2013). These two cytokines have their pro-angiogenic effects through endothelial CXCR1 and CXCR2 (interleukin-8 receptors) in an autocrine fashion (Pecot et al. 2013). IL-8 is not only secreted by malignant cancer cells but by leukocytes, fibroblasts, and endothelial cells too (Xie 2001). By binding to CXCR1 and CXCR2 G-protein-coupled receptors, which are expressed on a range of cells including cancer and endothelial cells, it activates the receptors and is followed by phosphorylation, desensitisation, and internalisation of the receptors (Xie 2001). Both CXCR1 and CXCR2 receptors mediate the angiogenic activity by chemokines, while the effect is mainly through CXCR2, the ligands of which are both IL-8 and CXCL1 (Heidemann et al. 2003). It has been found that the inhibition of CXCR2 leads to reduced vascularization in tumour (Heidemann et al. 2003).

IL-8 and other CXC chemokines also induce angiogenic functions in endothelial cells directly through migration, proliferation, tube formation, and release of VEGF (Heidemann et al. 2003). Neovascularization is indirectly promoted by IL-8 indirectly, for instance by inducing VEGF production in tumour cells (Heidemann et al. 2003). The elevated levels of IL-8 is found to have association with worsened clinical outcomes in ovarian, renal and lung adenocarcinomas, and the high expression of IL-8 and CXCL1 have a strong correlation with worse overall survival in lung and renal cancers (Pecot et al. 2013). ZEB1 and miR-200 repress each other ("reciprocal feedback loop"); that is, ZEB1 represses miR-200 family but is also targeted by miR-200. This suggests that miR-200-based therapies could be a treatment method for tumours and cancers by the down-regulation of ZEB1 (Pecot et al. 2013).

The ectopic expression of ZEB1 is found to up-regulate angiogenesis in human breast cancer cells as it recruits Sp1 transcription factor to its binding sites on VEGF promoter, inducing its promoter activity. This elevates VEGF expression at both mRNA and the protein levels in MDA-MB-231 cells, inducing tumour angiogenesis in breast cancer.(54) Hence, a potential new strategy for treating tumour angiogenesis could be the down-regulation of ZEB1 expression to inhibit VEGF expression, which in turn will reduce angiogenesis activity in tumour tissues (Clarhaut et al. 2009).

References

Arima Y, Hayashi H, Sasaki M, Hosonaga M, Goto TM, Chiyoda T, Kuninaka S, Shibata T, Ohata H, Nakagama H, Taya Y, Saya H (2012) Induction of ZEB proteins by inactivation of RB protein is key determinant of mesenchymal phenotype of breast cancer. J Biol Chem 287(11):7896–7906. doi:10.1074/jbc.M111.313759

Cacheux V, Dastot-Le Moal F, Kaariainen H, Bondurand N, Rintala R, Boissier B, Wilson M, Mowat D, Goossens M (2001) Loss-of-function mutations in SIP1 Smad interacting protein 1 result in a syndromic Hirschsprung disease. Hum Mol Genet 10(14):1503–1510

Clarhaut J, Gemmill RM, Potiron VA, Ait-Si-Ali S, Imbert J, Drabkin HA, Roche J (2009) ZEB-1, a repressor of the semaphorin 3F tumor suppressor gene in lung cancer cells. Neoplasia (New York) 11(2):157–166

Dong H, Lei J, Ding L, Wen Y, Ju H, Zhang X (2013) MicroRNA: function, detection, and bioanalysis. Chem Rev 113(8):6207–6233. doi:10.1021/cr300362f

Heidemann J, Ogawa H, Dwinell MB, Rafiee P, Maaser C, Gockel HR, Otterson MF, Ota DM, Lugering N, Domschke W, Binion DG (2003) Angiogenic effects of interleukin 8 (CXCL8) in human intestinal microvascular endothelial cells are mediated by CXCR2. J Biol Chem 278(10):8508–8515. doi:10.1074/jbc.M208231200

Hurt EM, Saykally JN, Anose BM, Kalli KR, Sanders MM (2008) Expression of the ZEB1 (deltaEF1) transcription factor in human: additional insights. Mol Cell Biochem 318(1–2):89–99. doi:10.1007/s11010-008-9860-z

Kurihara K, Isobe T, Yamamoto G, Tanaka Y, Katakura A, Tachikawa T (2015) Expression of BMI1 and ZEB1 in epithelial-mesenchymal transition of tongue squamous cell carcinoma. Oncol Rep 34(2):771–778. doi:10.3892/or.2015.4032

Lacher MD, Shiina M, Chang P, Keller D, Tiirikainen MI, Korn WM (2011) ZEB1 limits adenoviral infectability by transcriptionally repressing the coxsackie virus and adenovirus receptor. Mol Cancer 10:91. doi:10.1186/1476-4598-10-91

Liu Y, El-Naggar S, Darling DS, Higashi Y, Dean DC (2008a) ZEB1 links epithelial-mesenchymal transition and cellular senescence. Development 135(3):579–588. doi:10.1242/dev.007047

Liu Y, Peng X, Tan J, Darling DS, Kaplan HJ, Dean DC (2008b) Zeb1 mutant mice as a model of posterior corneal dystrophy. Invest Ophthalmol Vis Sci 49(5):1843–1849. doi:10.1167/iovs.07-0789

Okumura N, Minamiyama R, Ho LT, Kay EP, Kawasaki S, Tourtas T, Schlotzer-Schrehardt U, Kruse FE, Young RD, Quantock AJ, Kinoshita S, Koizumi N (2015) Involvement of ZEB1 and Snail1 in excessive production of extracellular matrix in Fuchs endothelial corneal dystrophy. Lab Invest 95(11):1291–1304. doi:10.1038/labinvest.2015.111

Pecot CV, Rupaimoole R, Yang D, Akbani R, Ivan C, Lu C, Wu S, Han HD, Shah MY, Rodriguez-Aguayo C, Bottsford-Miller J, Liu Y, Kim SB, Unruh A, Gonzalez-Villasana V, Huang L, Zand B, Moreno-Smith M, Mangala LS, Taylor M, Dalton HJ, Sehgal V, Wen Y, Kang Y, Baggerly KA, Lee JS, Ram PT, Ravoori MK, Kundra V, Zhang X, Ali-Fehmi R, Gonzalez-Angulo AM, Massion PP, Calin GA, Lopez-Berestein G, Zhang W, Sood AK (2013) Tumour angiogenesis regulation by the miR-200 family. Nat Commun 4:2427. doi:10.1038/ncomms3427

Pirinen E, Soini Y (2014) A survey of zeb1, twist and claudin 1 and 4 expression during placental development and disease. APMIS 122(6):530–538. doi:10.1111/apm.12187

Saykally JN, Dogan S, Cleary MP, Sanders MM (2009) The ZEB1 transcription factor is a novel repressor of adiposity in female mice. PLoS One 4(12):e8460. doi:10.1371/journal.pone.0008460

Shin JO, Nakagawa E, Kim EJ, Cho KW, Lee JM, Cho SW, Jung HS (2012) miR-200b regulates cell migration via Zeb family during mouse palate development. Histochem Cell Biol 137(4):459–470. doi:10.1007/s00418-012-0915-6

Soini Y, Tuhkanen H, Sironen R, Virtanen I, Kataja V, Auvinen P, Mannermaa A, Kosma V-M (2011) Transcription factors zeb1, twist and snail in breast carcinoma. BMC Cancer 11:73. doi:10.1186/1471-2407-11-73

Vandewalle C, Van Roy F, Berx G (2009) The role of the ZEB family of transcription factors in development and disease. Cell Mol Life Sci 66(5):773–787. doi:10.1007/s00018-008-8465-8

Vincent AL, Niederer RL, Richards A, Karolyi B, Patel DV, McGhee CN (2009) Phenotypic characterisation and ZEB1 mutational analysis in posterior polymorphous corneal dystrophy in a New Zealand population. Mol Vis 15:2544–2553

Wellner U, Schubert J, Burk UC, Schmalhofer O, Zhu F, Sonntag A, Waldvogel B, Vannier C, Darling D, zur Hausen A, Brunton VG, Morton J, Sansom O, Schuler J, Stemmler MP, Herzberger C, Hopt U, Keck T, Brabletz S, Brabletz T (2009) The EMT-activator ZEB1 promotes tumorigenicity by repressing stemness-inhibiting microRNAs. Nat Cell Biol 11(12):1487–1495. doi:10.1038/ncb1998

Xie K (2001) Interleukin-8 and human cancer biology. Cytokine Growth Factor Rev 12(4):375–391

Chapter 62
Miscellaneous Genes Involved in Angiogenesis in Normal Physiology, Disease and Malignancy

Abstract In addition to the major series of genes involved in normal angiogenesis, there are series of miscellaneous genes and their products that interact and are involved in the process of angiogenesis. These are somehow effective and important element of the angiogenesis process, however they are not essentially the direct influential genes or solely are the major role players in this process. Here we described a series of genes as the body of evidence suggests their roles in angiogenesis in normal physiological setting as well as in diseases and different malignancies. In this chapter the role of number of these genes including CDC25, CRK, Factor V, FUS 1, HGS, HSP20, Lactate, MIP-1β, Nicotinamide, P120RasGAP, PAK1, Sema6A, Sprouty1 were investigated and summarised and their involvement and interaction in the process of angiogenesis in normal physiology, disease and malignancy were explained. Further understanding of the role of these vital regulators of angiogenesis will enlighten the future of medical research and the therapeutic approach towards anti- angiogenic treatment.

Keywords CDC25 • CRK • Factor V • FUS 1 • HGS • HSP20 • Lactate • MIP-1β • Nicotinamide • P120RasGAP • PAK1 • P21-activated kinase 1 • Sema6A • Sprouty1

62.1 Introduction

In addition to the major series of genes involved in normal angiogenesis, there are series of miscellaneous genes and their products that interact and are involved in the process of angiogenesis. These are somehow effective and important element of the angiogenesis process, however they are not essentially the direct influential genes or solely are the major role players in this process. Here we described a series of genes as the body of evidence suggests their roles in angiogenesis in normal physiological setting as well as in diseases and different malignancies. **CDC25** plays a major role to allow the appropriate progression of cells through the different phases of the cell cycle. The **CRK** signaling pathways is involved with are related to the recruitment of cytoplasmic proteins near tyrosine kinase. Coagulation Factor V commonly known as the **Factor V** (*F5*) gene, is an essential protein found in the coagulation system of the body. Factor V is

A. Salajegheh, *Angiogenesis in Health, Disease and Malignancy*,
DOI 10.1007/978-3-319-28140-7_62

relatively involved in the process of angiogenesis, via the production of fibrin. Fibrin plays an active role in the facilitation of angiogenesis, especially in that of repair angiogenesis, wound healing in physiological angiogenesis and pathological and tumour angiogenesis. **FUS1** gene is a tumour suppressor gene and its transcription is stimulated during nitrogen starvation of the tissue. **HGS** codes an enzyme known as Hepatocyte growth factor-regulated tyrosine kinase substrate and is primarily involved in regulating the trafficking of growth factor receptors within a cell. HRS protein is involved in degradation of growth factor receptors such as platelet derived growth factor receptor (PDGFR) and vascular endothelial growth factor receptor (VEGFR), and hence anti-angiogenesis. Heat Shock Protein, **HSP20** is an indirect role player in angiogenesis through VEGF. HSP20 has two primary roles of angiogenic and anti-apoptotic roles in the development of malignant tissues. **Lactate** has been recognised as merely a metabolic end product of glycolysis, but further research shows that, this gene also plays a role in controlling and stimulating tissue healing, vascular genesis or angiogenesis. **MIP1-β** is important in the inflammatory response, especially in the migration of leukocytes to the areas of infection and can cause a rise in platelets and vascular smooth muscle cells. The protein product of the RASA1 gene is known as **P120RasGAP**. Ras is considered a proto-oncogene and therefore, P120RasGap protein is also vital in normal cells as it acts as a regulator for a signaling transaction pathway known as the RAS pathway. When this pathway becomes overstimulated by a deficiency in the activation mechanism of P120RasGap, cells begin to grow uncontrollably, dividing and causing metastasis, as well as increasing angiogenesis to support the neoplastic cells. **Nicotinamide** is a water-soluble molecule, which is the amide form of Niacin and Nicotinamide phosphoribosyltransferase expression takes place under hypoxia. This process can occur under the influence of VEGF induction and formation of capillary-like tubes in certain tissues. **PAK1** gene (P21-activated kinase 1) codes for Serine/threonine-protein kinase enzyme and serves as a target for the small GTP binding proteins Rac (including Rac1) PAK1 controls cell-cell adhesion as well as cell motility and morphology. PAK1 and Rac1 genes can be used in therapeutic treatment of tumour cells and reduction in the spread of unwanted angiogenesis. **Sema6A** also known as Semaphorin 6A is transmembrane bound member of the semaphorins family which are implicated in the development of the neural tissue, regulation of the immune responses and tumorigenesis. **Sprouty1** also known as protein sprouty homolog 1 is a membrane-bound member of Sprouty (Spry) family of proteins. Spry1 was initially identified in Drosophila as an antagonist of the fibroblast growth factor (FGF) and epidermal growth factor (EGF) receptor signalling pathways. Spry1 activation halted endothelial cell proliferation and differentiation by inhibition of the ERK/MAPK pathway induced by angiogenic stimulators basic fibroblast growth factor (bFGF) and VEGF.

62.2 CDC25 and Angiogenesis in Normal Physiology, Disease and Malignancy

The human gene cell division cycle 25 (CDC25) is a multigene family which consists of CDC25A, CDC25B, and CDC25C. The genes encodes for the CDC25 phosphatase family that regulates the progression of cell division cycle through dephosphorylation of cyclin-dependent kinase (CDK) (Donzelli and Draetta 2003). CDC25 phosphatase contains an N terminal region which acts as the site for phosphorylation and ubiquitination. The C terminal region of CDC25 is responsible for the catalytic function of CDC25. CDC25 function to activate CDK through the removal of the 2 phosphate groups. CDK binds with cyclin to form CDK complexes which are key mediators of the cell cycle progression. They are normally regulated by the phosphorylation of target protein. For example, cyclin-CDK complexes in earlier cell cycle phase assist in activating the cyclin-CDK complexes later in the cell cycle phase (Lammer et al. 1998).

CDC25 plays a major role to allow the appropriate progression of cells through the different phases of the cell cycle. CDC25A is vital to ensure the cell progress from the G1 to S in the Synthesis phase of cell cycle (Boutros et al. 2006). CDC25B and CDC25C are required for the cell to enter into mitosis (G2 and M phase) (Lammer et al. 1998).

The CDC25 also has an essential role in identifying replicated or damaged DNA. The phosphatase family is inactivated during DNA damage and this prevents the cell from progressing through the cell cycle until the damaged DNA is repaired (Donzelli and Draetta 2003; Boutros et al. 2006; Rudolph 2007; Nagata et al. 1991). Being a key regulator of cell cycle progression, CDC25 is has a strong oncogenic role. The transition between different phases of the cell cycle is very tightly controlled and failure to do so will contribute to genomic instability. Over expression of CDC25 has shown to cause cells to be pushed rapidly from S or G2 phase into mitosis and the cells end up with incompletely replicated DNA (Karlsson et al. 1999). Increased expression of CDC25A and CDC25B (but not CDC25C) has been reported in many different types of cancer such as breast, lung, ovarian, prostate, colorectal, oesophageal, thyroid, laryngeal, pancreatic, glioma and neuroblastoma (Boutros et al. 2007). Furthermore, overexpression of CDC25A and CDC25B is often correlated with poor prognosis and more aggressive disease (Nagata et al. 1991; Galaktionov et al. 1995). The overexpression of CDC25 is isoform specific. For example, overexpression of CDC25A is more related to breast and oesophageal cancer. CDC25B on the other hand is more related to lung, thyroid carcinoma and gliomas (Boutros et al. 2007). Many cancers and their relationship with CDC25 expression levels were studied in great details; one example is the non-small cell lung carcinoma (NSCLC). The expression level of CDC 25B in NSCLC was studied to see if there is any correlation between CDC25B's expression level and the prognosis and progression of the disease. The study showed that increase expression of CDC25B predicted poor survival. Furthermore, a significant correlation was found between the upregulation of CDC25B and high number of microvessels within the tumour. This suggests that CDC25B may have a key role in angiogenesis and CDC25B's expression level can be used to determine the prognosis for NSCLC patients (Boldrini et al. 2007). The exact role of CDC25B in angiogenesis is still unclear and more research needs to be done. However, previous studies have shown that the level and phosphorylation of CDC25A is related to the action of ET-1 a major angiogenic factor in NSCLC (Stannard et al. 2003). CDC25 phosphatases, being an activator of the cell cycle regulator CDK is an attractive opportunity for new antiproliferative cancer drugs. It is suggested that CDC25 Inhibitory compounds when combined with other chemotherapeutic compounds may have portent therapeutic potential (Cazales et al. 2007; Eckstein 2000).

62.3 CRK and Angiogenesis in Normal Physiology, Disease and Malignancy

CRK is a human gene located on chromosome 17, which is a suspect in relation to cancer. It encodes Adapter Molecule Crk, or proto-oncogene c-Crk (Al-Mahmood et al. 2009). This adapter molecule Crk is part of the family of adapter proteins. Adapter proteins are involved in various signalling pathways and are accessories to main proteins. The signalling pathways Crk is involved with are related to the recruitment of cytoplasmic proteins near tyrosine kinase.

The tyrosine kinase enzyme is an on-off switch that can phosphorylate a protein inside a cell. It does this by transferring a phosphate group to the amino acid Tyrosine on the protein from a molecule of ATP. Protein phosphorylation by a kinase is a mechanism important in signalling and communication within a cell to regulate cellular activity, namely cell division. These protein kinase enzymes may become mutated, and in doing so can get fixed in the on position; causing cell growth to be unregulated, a prerequisite to cancer.

The Adapter Molecule Crk was originally isolated in a virus called Avian Sarcoma Virus as the v-Crk onco-protein. In this form a viral protein gets irreversibly bound to the regulatory domains of the v-Crk. This renders it unable to be regulated, yet still able to stimulate the Tyrosine phosphorylation via Tyrosine kinase (Avraamides et al. 2008). The name itself stands for "CT10 Regulator of Kinase", where the CT10 stands for the avian virus mentioned. Since its discovery in chickens, this Crk Adaptor Protein has been found to exist in many eukaryotic organisms in all tissue types. From then, this protein has been found to take part in many different and complex physiological processes. Important processes such as being involved in cell motility by regulating the actin cytoskeleton, the cell cycle and apoptosis are among its repertoire (Hos et al. 2011). All of these functions of this gene can be related to malignancies.

There are several specific types of cancer that the Crk gene has been related to. These include lung cancer, breast cancer, sarcomas and ovarian cancer. The involvement of Crk in lung and breast cancer will be explored in this review. In regards to lung cancer, a study done by Miller et al. profiled the expression of Crk mRNA in the adenocarcinomas of 86 patients. This revealed that a significant rise in expression of Crk mRNA was present in stage three cancers compared to stage one. Also, a substantial increase in the more invasive tumours was observed. This study also found that there was an elevation of Crk expression in the high risk tumours compared with the low risk ones. Moreover, the poorly differentiated tumours displayed greater expression compared with the moderate to well defined tumours. The cancerous lung tissue also presented with levels of Crk protein much higher than that of healthy lung tissue (Cursiefen et al. 2004). These attributes of cancer risk and tumour differentiation correlate highly with poor survival rates. Another investigation done by Kim et al. studied the genomic profiling of 128 malignant tumours in the lung. This showed that an amplification of the adaptor protein CrkL, a homologue of Crk, was frequently revealed to be amplified in the lung cancer patients. The scientists of this investigation demonstrated that as well as this, they could knock out several of the aggressive characteristics of the tumour such as cell motility, cell survival and progression of the cell cycle. This was done by blocking the expression CrkL by a process called "siRNA-mediated knockdown". Conversely, over expression of CrkL in malignant bronchial epithelial cells greatly boosted cell growth without the need of EGF (Epidermal growth factor); leaving the cells to proliferate unchecked (Matsuda et al. 1991).

Breast cancer is another common and deadly form of malignancy that has been related to the expression of Crk. In a study by Rodrigues et al., analogous to lung cancer, it found that there is a strong correlation between the expression of Crk and the progressing stages of breast cancer. In this study, it revealed that the levels of Crk protein were raised in 60 % of the tumours compared with the normal breast tissue in the same patient. It was also found, as it was in lung cancer, that "siRNA-mediated knockdown" of Crk expression considerably reduced the migration and invasion capacity of the breast cancer cells (Nakamotc et al. 1996). In a similar report studying the overexpression of Crk in breast cancer cell lines, it was shown that this overexpression would result in the differential expression of numerous other genes that are involved in similar processes of tumour progression and their biology. These include cell proliferation, migration, adhesion and cell cycle regulation. Furthermore, this same overexpression of Crk promotes the dispersal of malignant cells and their attainment of a more "mesenchymal phenotype"; meaning the cell would lose their adheren junctions, giving them increased motility and metastatic properties (Birge et al. 2009). As these studies suggest that Crk expression is intimately involved with tumourigenicity, other reports explore the mechanisms of supressing the breast cancer tumours by targeting Crk. It has been found that the modification of post-translational Crk via phosphorylation of one of its tyrosine receptors seems to distinctly highlight a mechanism of tumour suppression (Mayer and Hanafusa 1990). This method suppresses cellular migration and induces apoptosis of the malignant cells by EphrinB2. EphrinB2 is a ligand for the EphB4 receptor on the Crk protein, required for phosphorylation of Crk by another protein called Alb. It is believed that this binding of the ligand EphrinB2 disassembles the Crk mediated complexes, inhibiting its function in the cancer cells (Kim et al. 2010).

Evidently, there seems to be a correlation between the expression of the Crk gene and several types of malignant tumours. The involvement of Crk in other cancerous tissues will no doubt be the subject of many further studies. This will eventually lead to, as was already explored in the studies cited above, efforts of designing therapeutic agents to specifically target Crk and its related signalling proteins. As proof of the effectiveness of such drugs already exists, Crk looks to be a promising avenue to pursue more research into the prevention of its related cancers, potentially leading to breakthroughs in the field of oncology.

62.4 Factor V and Angiogenesis in Normal Physiology, Disease and Malignancy

Coagulation Factor V (proaccelerin, labile factor), commonly known as the Factor V (*F5*) gene, is an essential protein found in the coagulation system of the body. *F5* is synthesised in the liver (Suehiro et al. 2005) and can be found in two variations; $F5_1$ and $F5_2$, these factors perform synergistically in both procoagulant and anticoagulant pathways (Hoekema et al. 1997; Varadi et al. 1996) in order to maintain haemostatic balance (Duga et al. 2004).

In normal physiology the *F5* protein circulates in the bloodstream, remaining dormant up until the coagulation system is triggered by sites of vascular injury, at which point levels of *F5* become highly concentrated (Duga et al. 2004; Nicolaes and Dahlback 2002). When *F5* is activated it interacts with another coagulation factor, known as the Coagulation Factor X (Stuart-Prower) gene (*F10*), to produce the active forms; Factor Va and Factor Xa. Together active forms Factor Va and Factor Xa form a complex which converts coagulation protein prothrombin to its active form; thrombin. Thrombin is responsible for the conversion of fibrinogen into fibrin, the material from which blood clots are formed (Pittman et al. 1994).

Furthermore *F5* has an additional role in the regulation of the coagulation system via its interaction with Activation Protein C (APC). APC is responsible for regulating coagulation by cleaving *F5* at specific sites in order to inactivate the coagulation process (Esmon 2003). The termination of the function of APC, disallows the overgrowth of blood clots due over stimulation of the coagulation system by *F5* (Pittman et al. 1994; Esmon 2003; Segers et al. 2007).

As a factor which functions in both the intrinsic and extrinsic pathways of coagulation, maintenance of *F5* and the balance of its cofactor activity is of key importance to the coagulation system of the body. An imbalance in these factors can lead to perturbed *F5* expression and function, resulting in the potential of both, thrombotic and haemophilic clotting disorders (Segers et al. 2007; Vos 2006; Asselta et al. 2006).

Factor V Leiden (FVL) Thrombophilia, also known as APC Resistance, Leiden Type or Hereditary Resistance to APC, is an inherent mutation in the *F5* gene. FVL involves having an abnormal version of the *F5* gene, this abnormality results in the inability of *F5* to interact with APC, therefore resisting its action (Segers et al. 2007; Pittman et al. 1994; van Hinsbergh et al. 2001). As afore mentioned, APC regulates the function of *F5* to prevent over clotting during coagulation, however as APC is unable to properly interact with *F5* in patients with FVL, this results in the overproduction of fibrin and increased risk of venous thromboembolism (Ornstein and Cushman 2003). As fibrin supports the process of angiogenesis and the facilitation of existing blood vessel networks, over production of fibrin further enhances this process, facilitating the formation of neovascularisation (van Hinsbergh et al. 2001; Cheresh et al. 1989). FVL can also lead to a number of conditions including, increased risk of deep vein thrombosis, colorectal cancer, recurrent pregnancy loss and thromboembolic events.

Recent studies have demonstrated the involvement of *F5* in Peritoneal Fibrosis via the expression of *F5* and tissue factor, alongside *F10* deposition. A study found that following inflammatory stimulation, profibrotic and procoagulant mechanisms in the peritoneum that involved factors *F5, F10* and tissue factor, resulted in the progression of angiogenesis and the accumulation of extracellular matrix components. The study was carried out on an inducible rat model and therefore the same outcomes cannot be concluded in human Peritoneal Fibrosis, however due to the shared genetic similarity between rats and humans, it is postulated that similar findings will be demonstrated in humans (Saito et al. 2009).

The role of *F5*, like that in pathological angiogenesis, plays a similar role in malignant or tumour angiogenesis. Similarly *F5* is relatively involved in the process, via the production of fibrin. As previously mentioned, fibrin plays an active role in the facilitation of angiogenesis, especially in that of repair angiogenesis (wound healing in physiological angiogenesis and pathological and tumour angiogenesis). Fibrin possesses the ability to protect cancer cells during circulation against mechanical stress as well as in the immune system. Fibrin further promotes tumour angiogenesis via the formation of a matrix for tumours to develop within (Klerk et al. 2007; van Hinsbergh et al. 2001).

A number of coagulation factors and markers have been found to be present in the hypercoagulable state in cancer patients (Iversen et al. 1996; Iversen and Thorlacius-Ussing 2002). These increased levels of coagulation factors and markers have an increased adverse effect in patients. This is due to the ability of tissue factor and thrombin to not only enhance tumour angiogenesis, but to further increase the adhesiveness of malignant cells to the microvascular endothelium (Rickles et al. 2003; Klepfish et al. 1993). In 2004, a report noted an increased risk of 3.3 fold, in the development of digestive cancers in male patients who presented recurrent activation of the coagulation system (Miller et al. 2004).

Little is currently understood about the mechanisms by which coagulation genes contribute to the risk of colorectal cancer, however, a number of studies have found FVL to be a strong prothrombotic mutation in cancer (Paspatis et al. 2002; Loktionov et al. 2003; Hubner and Houlston 2007; Bertina et al. 1994). Incidence of colorectal cancer in patients with FVL were found to be 45 % higher (12 in 44) than incidences in healthy patients, 36 in 600 (Mozsik et al. 2005). Increased risk of colorectal cancer appears to be associated with increased levels of fibrinogen (Grande et al. 2008;

Kockar et al. 2005) known to be associated with the facilitation of metastasis during tumour angiogenesis (Palumbo et al. 2000).

62.5 FUS1 and Angiogenesis in Normal Physiology, Disease and Malignancy

The FUS1 gene is a tumour suppressor gene found on human chromosome 3p21.3 (Ji and Roth 2008). It is highly associated with the prevention of lung cancers and has been consistently found to have an inhibiting function in breast cancers also (Ji and Roth 2008). The cancer reducing potential of FUS1 products are investigated primarily with angiogenesis being an accepted effect of developing tumours. How angiogenesis specifically varies with the various causes of FUS1 inhibition is described below (Lee et al. 2007).

The FUS1 gene encodes a 1372-amino acid protein that was found to be produced only during the transcription and translation of FUS1 (Petersen et al. 1995). The protein product is a myristoylated protein that is associated with cell membrane interaction and cell fusion. It inhibits tumour cell-induced cloning preventing replication of cancerous cells and has an anti-metastases function limiting the potential for malignancy. Presence of the protein was shown to stimulate apoptosis in the cancerous cells additionally and to reduce the angiogenesis of the location (Ji and Roth 2008).

Transcription of FUS1 is stimulated during nitrogen starvation of the tissue and is initiated by a human pheromone signal. Loss of this pheromone communication, such as a human with a mutant pheromone pathway, is found to prevent transcription and is therefore another mechanism by which the FUS1 gene's tumour suppressor function is lost (Ji and Roth 2008).

Myristoylation is a modification process where, in FUS1 specifically, a 14-carbon myristoyl group is added to the NH2-terminal of the protein product. Without this modification, all tumour-suppressing functions were found to be reduced significantly.

Some micro RNA molecules were shown to inhibit the function of tumour suppressor gene SuFu as well as FUS1 and their expression of tumour suppression products. Micro RNA molecules are commonly associated with the inhibition of mRNA molecules and therefore translation of the product. One experiment investigated the effects of miR-378 (a micro RNA molecule that inhibits the expression of FUS1) on cancerous tissue activity (Lee et al. 2007). Transfection with miR-378 greatly improved cancerous cell survival and angiogenesis with comparison to a control tissue. To further investigate this, caspase-3 enzyme presence and activity assay was analysed, as it is an integral enzyme in apoptosis. The results demonstrated that the mi-378 transfected cells had greatly reduced levels of caspase-3 indicating that the apoptotic function related to the gene FUS1 was greatly reduced (Lee et al. 2007).

The function of the FUS1 in prevention and treatment of lung tumours is highly accepted. The method by which the gene can be inserted is currently being investigated. Viral vectors have limited viability in the delivery of the gene to the host cells due to the host's immune response to the delivery virus and transgene. It is therefore currently only applicable to localised tumours (Gopalan et al. 2004).

Angiogenesis is a generally accepted occurrence with developing tumours. There are, however, specific relationships between FUS1 expression and the development and modification of blood vessels supplying the cancerous tissue. With the use of mi-378 in another experiment, the survival and angiogenesis of tissue tumours with and without FUS1 were compared. A group of mice with the FUS1 gene activated was compared with a group of mice with miR-378 administered (inhibiting translation to the FUS1 protein product). After 2 weeks, difference in tumour was already significant (Lee et al. 2007). After 4 weeks, the blood vessels supplying the tissues were analyzed and the results were as follows. The blood vessels supplying the control increased in number and size corresponding to the tumour growth. The blood vessels supplying the FUS1 inactivated tumour increased in number as expected but showed a dramatically disproportionate increase in size to the area of the tumour (Lee et al. 2007). This suggests that FUS1 plays a significant role in the prevention of enlargement of blood vessels serving tumorous tissues.

The FUS1 gene is an important tumour-suppressing gene with relation mostly to lung and breast cancers (Li et al. 2011; Kondo et al. 2001). Its function is to reduce tumour cell metastases by inhibiting tumour cell replication and to stimulate apoptosis in cancerous cells. The disabling of the function of the FUS1 protein is an indication in the development of lung cancer and can occur in many ways (Ito et al. 2004; Kondo et al. 2001). These include mutation in the FUS1 gene, a mutant pheromone-signalling pathway, loss of myristoylation ability of the cell and the presence of some miRNAs. As a treatment and prevention strategy, introduction of the FUS1 gene is highly promising. The specific effect of FUS1 on the angiogenesis of a tumour has not been explored as significantly as its tumour suppressing capabilities. Current research indicates that the FUS1 gene plays an important role in limiting blood flow to tumour cells by reducing the expansion in size of blood vessels, but not in preventing the formation of new ones.

62.6 HGS: Hepatocyte Growth Factor-Regulated Tyrosine Kinase Substrate and Angiogenesis in Normal Physiology, Disease and Malignancy

HGS is a gene which codes for an enzyme known as Hepatocyte growth factor-regulated tyrosine kinase substrate (HRS protein) (Nagata et al. 2014). The HRS protein is primarily involved in regulating the trafficking of growth factor receptors within a cell (Rayala et al. 2006). It is believed that HRS protein interacts with a protein known as PELP1, and this interaction induces the activation of Src proteins (including the c-Src protein that was mentioned above) as well as the MAPK pathway (Rayala et al. 2006). The MAPK pathway is a signal transduction pathway that involves activation of kinase proteins which then activate transcription of certain genes that are required for cell growth and proliferation (Rayala et al. 2006). Clearly, any up regulation of this pathway could cause normal cells to become neoplastic, and eventually malignant. Hence, genes involved in the MAPK pathway are considered to be oncogenes. HRS has been linked to the MAPK pathway via the PELP1 protein. When researchers undertook the task of examining this correlation, they did so by seeing how PELP1 protein activation would be affected in cells that had depleted HRS. What they found was that depleting HRS via HRS specific small interfering RNA, was that MAPK proteins were reduced, and hence they concluded that HRS influenced MAPK activity (Rayala et al. 2006). What was not known was whether this was due to a confounding factor other than the PELP1 protein, so researchers decided to examine whether an increase in HRS in cells that heavily expressed PELP1 would result in up regulation of the MAPK pathway. They concluded that PELP1 protein is indeed the cause of the higher activation of the MAPK pathway since cells that expressed PELP1 in large amounts were the cells that were heavily affected (Rayala et al. 2006). Studies have demonstrated that once HRS protein activates the MAPK pathway, a specific translation factor named Elk-1 is activated (Rayala et al. 2006). The activation of Elk-1 results in transcription activation, resulting in the manufacture of proteins involved in breast cancer (Chai et al. 2001). This pathway is involved in the induction of neoplastic changes; however, it is still only poorly understood. A more comprehensive understanding of ERS is in the field of angiogenesis. Up regulation of growth factor receptors is of vital importance in normal physiology. It allows for hypertrophy of normal cells in times when it is required (such as in strength training) as well as hyperplasia in times of wound repair. This however is contrasted in pathologic up regulation of growth factor receptors. In pathologic up regulation of the growth factor receptors, cells become more responsive to external stimuli and proliferate more easily despite an increase in the molecules that signal their increase. This is of vital importance in pathologic angiogenesis required in neoplasms. Endothelial cells are required for angiogenesis within neoplasms, and research suggests that HRS protein is heavily involved. Research involving a cancer named Glioma, which is a result of malignant glial cells in the brain, has shown that the HRS protein is involved in degradation of growth factor receptors, and hence anti-angiogenesis. Platelet derived growth factor receptor (PDGFR) as well as vascular endothelial growth factor receptor (VEGFR) are amongst the two most heavily expressed receptor proteins found in endothelial cells that are involved in angiogenesis (Wurdinger et al. 2008). The HRS protein is involved in this mechanism due to its protein sorting function. One of the functions of HRS protein is that it directs growth factor receptors to lysosomes within the cell. This allows for these receptors to become degraded, and helps control angiogenesis from becoming under regulated (Wurdinger et al. 2008). If HRS protein is depleted by reduced translation or perhaps by an enzyme that degrades it, the growth factor receptors remain on the cells and angiogenesis occurs. This is was evident when angiogenesis involved in glioma was investigated (Wurdinger et al. 2008).

Researchers hypothesized that a type of catalytic RNA was involved in degrading the HRS protein, which resulted in vascularization of the tumour. To test their hypothesis, they used molecules which inhibited a catalytic RNA named miR-296 (Wurdinger et al. 2008). miR-296 is a micro RNA that has a specific binding site to the HRS protein. The inhibition of miR-296 was accomplished by the use of a molecule named Ambion (Wurdinger et al. 2008). Ambion binds to miR-296 and renders it inactive. The researchers observed that the reduced miR-296 resulted in a reduced degradation of HRS, which consequently resulted in an increase in growth factor receptors. It is believed that miR-296 is impaired in most cancers (including the glioma cancer that was investigated), and hence its interaction with HRS is vital to understanding angiogenesis in pathologic malignancies.

62.7 HSP20 and Angiogenesis in Normal Physiology, Disease and Malignancy

Heat Shock Proteins (HSPs) are a class of proteins that play important roles both inside and outside the cells. As their name suggests, HSPs are primarily employed in response to cellular stress that threatens to denature proteins, particularly thermal stress. Their action within the cell is that of a molecular chaperone; they both refold denatured proteins and wrap around them to prevent denaturation (Schmitt et al. 2007). Extracellular HSPs are thought to possess limited immune functionality as they aid in the formation of peptide antigens (Carmeliet and Jain 2000). In addition to their role in cellular protection, HSPs perform in other aspects of protein lifecycle including folding and distribution. HSPs are divided into many subclasses based on molecular weight (Wang et al. 2009). HSP20, also known as HSPB6, is a smaller polypeptide that is active in response to both physiological and pathological cellular stresses. HSP20 can be found in many tissues but is most prevalent in muscle cells of airways, blood vessels, skeletal muscle, the colon and especially the heart. Direct effects of HSP20 revolve around supporting and protecting cardiomyocytes and other muscle cells (Goukassian et al. 2001). Research also indicates it plays a role in cardiac angiogenesis as noted during insertion of MSC featuring recombinant HSP20 DNA into areas of cardiac infarction.

At this stage the majority of research involving HSP20's role in angiogenesis is focused on molecular pathways with relatively little on HSP20's role in angiogenic malignancies and pathologies. Wang et al. (2009) were able to create recombinant DNA in the mesenchymal stem cells (MSC) of adult male rats. Using adenoviral vectors they were able to express code for both HSP20 and Green Fluorescent Protein (GFP, used as a control). In vitro cultures of both MSC strains then underwent oxidative stress for using hydrogen peroxide at varying concentrations and for varying time periods. Results pertaining to angiogenesis included the increased secretion of pro-angiogenic factors by MSC. Cells expressing HSP20 secreted 80 %, 55 % and 39 % more VEGF, FGF-2 and IGF-1 respectively when compared to the GFP control. This was further expanded by inserting the aforementioned MSC into the left ventricle of female rats adjacent to regions of myocardial infarction. HSP20 MSCs resulted in enhanced angiogenesis and a 43 % increase in vascularisation compared to the GFP control. Whilst these results were promising, it was acknowledged that MSCs were responsible for the secretion of many growth factors and chemokines and that more research was required for HSP20 in understanding of both growth factor secretion as well as myocardial infarction response (Wang et al. 2009).

Further clarity was gained through the use of extracellular HSP20 on Human Umbilical Vein Endothelial Cells (HUVEC). In this instance Bovine Serum Albumin was used as a control. Results indicated that increased concentrations of HSP20 as well as increased dosage time of HSP20 helped with the initial stages of angiogenesis. Following from these results of analytical techniques such as competitive binding assays and the use of immunofluorescent staining indicated that HSP20 also promotes angiogenesis by binding to VEGF-receptor 2 (VEGF-R2). To further confirm this hypothesis HUVEC were first treated with CBO-P11, a VEGF-R inhibitor, before HSP20 dosage was applied. Angiogenic cell migration and other stimulatory effects of HSP20 previously experienced decreased in these HUVEC, in agreement with the notion that at least some of HSP20's angiogenic function is based on interactions with VEGF-R2. Finally it was observed that when comparing hearts of wild type mice, those with transgenic HSP20 overexpression benefitted from increased vascularisation. Limitations acknowledged by the authors and possible future areas for research include unknown stability of HSP20 outside of the cell, how HSP20 may be delivered to the heart for therapeutic purposes if at all possible and if there is in fact an advantage in using HSP20 rather than simply VEGF (Zhang et al. 2012).

There is much progress that can be made in our understanding of HSP20's role in angiogenesis, especially in relation to pathological conditions. The majority of publications involving angiogenesis and HSP20 emphasise the molecular steps involved in activating HSP20. One study found that cardiomyocytes of rats with type II diabetes released exosomes containing microRNA 320 (miR-320) to communicate with endothelial tissues. Simply put, microRNAs are very small, non-coding RNA nucleotide sequences. miR-320 was found to have an anti-angiogenic action through down regulation of many growth factors as well as HSP20. The aforementioned exosomes appear to have exciting therapeutic potential with respect angiogenic regulation; in addition to miR-320 they may also contain the pro-angiogenic miR-126 (Wang et al. 2014). The same interaction between miR-320 containing vesicles and HSP20 was also noticed in ischaemic heart disease (Ren et al. 2009). Substantially less is known about HSP20 in pathologic conditions not pertaining to the heart. Researchers have observed that HSPs are involved with diabetic retinopathy, however HSP20 is not listed amongst the key factors. Furthermore for the HSPs that play a role in diabetic retinopathy, it is through their non-angiogenic functionality (Heise and Fort 2011). Similarly, although HSPs are active in the pathology of arthritic inflammation (Lambrecht et al. 2014), our current understanding does not include a link to HSP facilitated angiogenesis nor HSP20.

HSP20 has two primary roles in the development of malignant tissues; firstly the angiogenic mechanisms outlined earlier and secondly as an anti-apoptotic factor. It is obvious that the effects of HSP play a vital role in improving the resilience of the cell with respect to stress. However this would be an overly simplistic analysis of the role of HSPs in human health; cell preservation for the most part is beneficial, however, the function of HSPs also serves to extend the life of cancerous cells (Chen et al. 2014). Both functionalities were demonstrated by Chen et al. (2014) through the study of progress and spread of tumours in mice. All mice were administered with first Lewis Lung Carcinoma Cells before doses of either HSP20 or a saline control solution were injected into the peritoneal cavity. After a time period tumours were weighed in a quantitative assessment of growth. Equally, cervical lymph nodes were used to assess metastasis. The authors observed that the HSP20 mice secreted more of the pro-angiogenic VEGF and Basic FGF when compared to the control mice. Conclusions of the report state that HSP20 had anti-apoptotic and angiogenic functionality, both of which would aid in tumour development (Chen et al. 2014). Additionally, HSP20 can be regulated by other factors and complex pathways. Tumour tissues are found to create and release Prostaglandin E2 (PGE2) that binds to E-Type Prostaglandin 4 (EP4), a G-protein receptor. This in turn lowers the activity of intracellular cAMP Dependent Kinase (PKA). It was found that that PKA is responsible for the activation of HSP20 and many other angiogenic factors. These findings lead us to believe that tumours are able to influence their own blood supply through the use of PGE2 and EP4 and indirectly trigger the release of HSP20 and correspondingly the release of VEGF. As such future therapeutic efforts may place emphasis on the mediation of PGE2 and EP4 (Zhang and Daaka 2011).

In summary, investigation of HSP20's role in angiogenesis detailed interactions with VEGF, VEGF-R2 and FGF. Studies of HSP20's role in pathological angiogenesis has revealed that HSP20 is in turn mediated by numerous endogenic factors such as miR-320, miR-126, PKA and EP4. Identification of the pathways regulating pathologic and malignant angiogenesis as well as the key chemical mediators involved provides many feature targets for therapeutic strategies. This small review of HSP20's role in angiogenesis has brought to light a delicate balance and a vast array of complex interactions. This clearly illustrates the importance of both past and future studies to improve our understanding and ultimately to future therapeutic developments.

62.8 Lactate and Angiogenesis in Normal Physiology, Disease and Malignancy

For several years, lactate has been recognised as merely a metabolic end product of glycolysis. However, enlightened by Thomas K. Hunt and his team, they connected a relationship between metabolism and angiogenesis; where they concluded that lactate has been formally proven to be an important factor that contributes to tissue healing (Trabold et al. 2003). Angiogenesis has three main roles in the human body and occurs in: normal physiology, which is general wound healing and vascular genesis; diseases such as several ischemia and malignant tumours. However, before discussing lactate in angiogenesis, a little more background on this chemical compound should be understood.

Lactate or lactic acid is a chemical compound constantly produced from the end product, pyruvate, of glycolysis in animals. In animals, during vigorous exercise, the main source of energy is generated through glycolysis. There are two types of glycolysis: aerobic and anaerobic. Most of the energy is produced via aerobic glycolysis. This causes the animal to begin to breathe faster as the requirement for oxygen in the working muscle increases. However, as the energy required exceeds the amount of energy oxygen can generate, the body starts to undergo anaerobic glucose metabolism to meet the demands of the muscles (Ziemann et al. 2011). Usually in the human body under physiological conditions, lactate concentration levels are usually controlled around 1.8 millimoles (mM). However, in tissues that are being regenerated, undergoing high levels of aerobic or anaerobic glycolysis and in malignant tissues and tumours, the levels of lactate may rise up to 10–15 mM (Gottfried et al. 2006).

Thomas K. Hunt's research explains how lactate plays a major role in wound healing, vascular regeneration or angiogenesis in tissues (Hunt et al. 2007, 2008; Ghani et al. 2004; Beckert et al. 2006). During immune activation and cell proliferation, lactate accumulates due to metabolism. Healing through stimulating regenerative cells and also endothelial cells and the tube formation in vitro, circulating vascular progenitor cells and vascular morphogenesis in vivo are all stimulated by lactate. One of the processes initiated by lactate, is angiogenesis (Porporato et al. 2012). Angiogenesis occurs when there is tissue trauma due to chemical, physical pathological disturbances, to regenerate via growth factors. When there is a wound, lactate accumulates due to glycolysis after circulation disruption, immune system activation and increase cell proliferation and differentiation; and with this, scientists had begun to research more on ways to exploit lactate therapeutically to increase the rate of wound healing. Although, recent tests have shown that lactate was able to communicate with endothelial cells directly and furthermore stimulated the phosphoinositide 3 kinase (PI3K)/Akt pathway in human endothelial cells via ligand mediated activation of three receptor tyrosine kinases (Polet and Feron 2013). The tyrosine receptors initiate, nerve growth factors, epidermal growth factors, fibroblast growth factors and other growth factors related to angiogenesis. Although however, over expression of gene or mutation in gene expression on tyrosine receptors may lead to diseases or malignant growth (Polet and Feron 2013).

An abnormally high accumulation of lactate, may lead to lactic acidosis and septic shock, but more significantly, a sign of malignancy. In areas of malignancy, there tends to be higher levels of lactate, as lactate is the end product of glucose metabolism. The accumulated lactate then act to stabilize hypoxia-inducible factor 1α to up-regulate the expressions on pro-angiogenic genes (Lolmede et al. 2003). One example of these pro-angiongenic genes is vascular endothelial cell growth factor which generally restores blood supply to areas where circulation is blocked.

All cells in the body require energy to do their respective jobs in the body including proliferation, and therefore resulting in lactate being produced through glycolysis (Burroughs et al. 2013). However, cancer proliferate and grow a lot faster than normal cells do and primarily use aerobic glycolysis or the Warburg effect to produce the energy they need to grow and establish new vascular supplies (Lunt and Vander Heiden 2011). Although tumour cells have high survivability, even under hypoxic and low oxygen environments, cancer cells sometimes become too large and start to necrotize from the inside due to little produced energy from anaerobic metabolism. To prevent this, cancer cells tend to generate their own blood vessels for oxygen. They do this by using lactate to stimulate the vascular endothelial cell growth factors which give tumours a sustainable environment for growth (Porporato et al. 2012).

The lactate gene is a by-product in glycolysis, but further research shows that, this gene also plays a role in controlling and stimulating tissue healing, vascular genesis or angiogenesis. However, on the downside, lactate also plays a role in promoting cancer growth. The fact, lactate stimulates angiogenesis and uncontrolled cancer growth when mutated, had been recently discovered means there is more to find out. Furthermore, if the lactate gene is able to initiate angiogenesis, then there may be a chance to slow down or completely inhibit cancer growth.

62.9 MIP-1β and Angiogenesis in Normal Physiology, Disease and Malignancy

Chemokines, or chemotactic cytokines are factors found as part of the inflammatory response, occurring during disease or infection. Leukocytes transgress the vascular endothelium and move into the peripheral tissues. Moreover, chemokines play a crucial role in activities such as angiogenesis and haematopoiesis, and in diseases such as HIV. The most important characteristic of the chemokines is conservation of four cysteines. There are two main groups, namely CXC and CC chemokines. The subtype depends on whether or not an amino acid divides the first two NH2-terminal cysteines (Menten et al. 2002).

Macrophage inflammatory protein-1β, or MIP-1β, is also known as Chemokine C-C motif ligand 4 (CCL4). It is a human protein which is encoded by CCL4. In susceptible cells CCL4 is capable of inhibiting HIV-1 infection. Type MIP-1β is of the four chemokine subfamily; others include CCL3 (MIP-1α), CCL9/10 (MIP-1δ) and CCL15 (MIP-1γ). They consist of a variety of cells, such as dendritic cells, lymphocytes and macrophages (Maurer and von Stebut 2004).

Macrophage inflammatory protein 1 was discovered in 1988 (Odorisio et al. 2006) and gets its name according to its inflammatory properties in vitro, as well as in vivo. It is essential in the immune response and as part of the inflammation process. MIP triggers human cells, such as basophils, eosinophils, and neutrophils, which in turn lead to neutrophilic inflammation.

MIP-1 is 8–10 kDa in size (Maurer and von Stebut 2004), and is closely linked to the LD78α gene located on chromosome 17 (Maurer and von Stebut 2004; Menten et al. 2002). Their structural organization is head to head and they are alienated by 14 kb. The chromosome consists of three exons and two introns (Maurer and von Stebut 2004). There is a single copy per gaploid genome of the LD78α gene. Many people present with an extra non-allelic human MIP1-β, however, it is not well researched and explained. 3′ untranslated regions of the MIP1-β cDNAs consist of a polyadenylation site (AATAAA) and a small amount of AT-rich sequences (Menten et al. 2002). MIP1-β is synthesized as 92 amino acid precursors (Odorisio et al. 2006; Pralhad et al. 2003) the mature protein of MIP1-β is structured of 69 residues and has a M_r of 7826.9 Da (Menten et al. 2002).

Cells like hematopoietic cells (dendritic cells, NK cells, mast cells) play a crucial role in the inflammatory response. During stimulation cells such as astrocytes, microglia, and epithelial cells generate a lower amount of CCL4. A large amount of MIP1-β is formed when monocytes are stimulated with LPS or IL-7. T-cells and B-cells are activated by the binding of the antigen, which secretes MIP1-β. NK cells use physiological activation processes such as lysis of the target cells to generate MIP1-β. After stimulation with LPS, TNF-α, IFN-γ, or IL-1β brain microvessel endothelial cells produce the MIP1-β gene. Moreover dendritic cells are active in releasing the gene in response to LPS, TNF-α, or CD40 ligand (Menten et al. 2002).

High affinity for proteoglycans is a similar trait of the chemokines, such as heparin and heparin sulfate. The increase of chemokines' bioactivity is a result of binding with proteoglycans at the surface area of endothelial cells or into the extracellular matrix. Interactions between them are known as electrostatic forces and involve amino acids. MIP1-β has 3 vital basic amino acids for heparin binding: Arg18, Arg46 and Arg48 (Odorisio et al. 2006). The fourth Lys45, however, is non vital. The amino acid Arg46 is proved to be essential for binding MIP1-β and heparin (Menten et al. 2002).

MIP1-β is important in the inflammatory response, especially in the migration of leukocytes to the areas of infection. Local inflammatory reaction is a consequence of the chemokines targeting the leukocytes. Migration of T lymphocytes towards MIP-1β depends on the integrin-mediated adhesion of lymphocytes to the endothelium. In this particular case MIP1-β is a potent chemokine which is responsible for the activation and augmenting adhesion of T lymphocytes to vascular cell molecule VCAM-1 (Menten et al. 2002).

MIP-1β causes a rise in platelets and vascular smooth muscle cells, but has no effect on neutrophils and basophils. Stimulation of the NK with MIP1-β in vitro led to chemokine stimulated leukocytes with the highest activity concentration of 100–1000 ng/ml (Maurer and von Stebut 2004), releasing granzyme A and *N*-acetyl-β-D-glucosaminidase.

For immature dendritic cells MIP1-β is a very powerful chemoattractant, however with maturation, chemotactic response is weakened and the susceptibility of the cells to MIP-1β is lost. Such changes correspond with the modifications in the chemokine receptor expression on the dendritic cells.

The human immunodeficiency virus type 1 (HIV-1) involves the availability of specific chemokine receptors as well as CD4 to affect the target cell. The receptor CCR5 is known to act with strains of HIV-1. It has the maximum effect during the asymptomatic stages of the disease, replacing CCR5 receptor with the N-terminus of CCR2 reacting with MIP-1β (Farzan et al. 1997). Nonetheless, it does not sustain viral entry. Consequently, chemokine signaling and HIV-1 entry are dismountable processes. Only the *N*-terminal domain of CCR5 is needed for the viral entry (Farzan et al. 1997).

Rheumatoid arthritis is a systemic inflammatory disorder, which affects organs and tissues, especially synovial joints. MIP-1β takes a part in the inflammatory reaction observed in rheumatoid arthritis. In particular, certain leukocyte subsets, such as monocyte and T-cells are targeted. Raised levels of chemokine MIP-1β are present in the synovial fluid and tissue (Menten et al. 2002).

Bacterial meningitis is a disease caused by infection with bacteria; the most distinct characteristic is the inflammation of the cerebrospinal fluid (CSF). Various types of chemokines, including MIP-1β, take part in the accumulation of leukocytes in the CSF (Odorisio et al. 2006). In patients with Alzheimer's disease MIP-1β is upregulated in reactive astrocytes (Menten et al. 2002). One of the most important haematological diseases is sepsis, which indicates the existence of microorganisms in the blood. The level of MIP-1β is high in the case of sepsis (Menten et al. 2002), however it is not relevant to the prognosis of the patient.

62.10 Nicotinamide and Angiogenesis in Normal Physiology, Disease and Malignancy

Vascular growth often takes place in normal physiology during wound repair and perinatal development. Hypoxia can also lead to the formation of new vascular structures in the human body. During hypoxia, Nicotinamides become accumulated in the ischemic tissues to induce vascular growth (Morris 1989).

Nicotinamide is a water-soluble molecule, which is the amide form of Niacin. Nicotinamides are available in the diet (vitamin B3) and in commercial products; in the form of vitamin supplements and cosmetics (Surjana et al. 2010). The focus of this section of the paper is to study the physiological and pathological aspects of the Nicotinamide derivatives in regard to angiogenesis.

Nicotinamides are the primary precursor of Nicotinamide adenine dinucleotide (NAD^+), Nicotinamide N-methyltransferase (NNMT), Nicotinamide phosphoribosyltransferase (NAMPT) and NAD(P)H oxidase (Garten et al. 2009). NAD^+ is an essential coenzyme in Adenosine Triphosphate (ATP) production and cellular respiration. It is also the main substrate for the Poly [ADP-ribose] polymerase 1 (PARP-1), which is encoded by the PARP-1 gene. This gene has been recognised as an active transcription region of chromatin. It is believed that together, PARP-1 and

NAD^+, play an important role in cancer formation via manipulating the cells' reaction toward the genotoxicity (Surjana et al. 2010).

NAD(P)H oxidase has been proven to be an important factor during cell proliferation and migration of endothelial cells (EC). Recent studies link Nicotinamide activity and Vascular Endothelial Growth Factor (VEGF) signalling to confirm the role of the reactive oxygen species in angiogenesis. The reactive oxygen species (ROS) are involved in the signalling pathways, which lead to angiogenesis. ROS includes molecules such as hydrogen peroxide and superoxide anion, that are produced during hypoxia or reoxygenation from a Nicotinamide derivative (NADPH oxidase) in the ECs (Ushio-Fukai et al. 2002). In order to activate ROS, NAD(P)H oxidase has to first be activated by growth factors such as VEGF and angiopoietin-1. VEGF plays the most critical role in angiogenesis during tumour formation. It has been discovered that by targeting NAD(P)H oxidase, the process of angiogenesis can be disrupted, thus disruption of tumour formation. The effectiveness of this strategy is demonstrated by the dietary administration of antioxidants (anti-NADPH oxidase) (Ushio-Fukai and Nakamura 2008).

NAMPT expression takes place under hypoxia. This process can occur under the influence of VEGF induction and formation of capillary-like tubes in certain tissues, which have the potential to operate as an important mechanism for tumour formation. It has been found that NAMPTs plays a role in pathologic conditions. NAMPTs are found in the special fat tissue (visfatin) of obese patients as well as in patients with type II diabetes. Interestingly, this phenomenon could protect the body during the myocardial infarction by angiogenesis (Waltenberger and Pardali 2012). In addition to pathologic conditions; NAMPTS plays a crucial role in cancer formation by enhancing angiogenesis. An inhibitory molecule called FK-866 has been identified as inducing apoptosis in certain cells by inhibiting NAMPTs. It is known that NAMPTs act as a main convertor of Nicotinamide to NAD mediators and as previously mentioned, NAD is required for angiogenesis. As a result, inhibition of angiogenesis and tumour formation can be explained by the inhibition of NAMPTs. It is concluded that FK-866 is tightly linked with cell apoptosis without change in the cell metabolism by interfering with Nicotinamide activity towards angiogenesis. More clinical trials are needed to appreciate the antineoplastic role of FK-866 for cancer treatment (Garten et al. 2009).

Another Nicotinamide derivative involved in cancer stimulation by angiogenesis is known as NNMT. NNMT is involved in methylation of Nicotinamides in order to recycle NAD and as previously discussed, NAD is involved with angiogenesis. Overproduction of NNMT has been recorded in the mesenchymal cells of tumour growing tissues in several malignancies (Thirant et al. 2012). As a result, this enzyme has been used as an indicator of the invasiveness of a tumour and a prognostic tool of some cancers such as renal or thyroid cancer (Aras et al. 2012). Results from studies surrounding the control of lung cancer have indicated that NNMT could be an "interesting" target for this type of cancer.

Furthermore, studies have found, Short Hairpin RNA (shRNA)-mediated gene is used to down-regulate the concentration of NNMT, resulting in a significant reduction of tumour production (Sartini et al. 2015). Another study has successfully revealed other roles of NNMT in the involvement of the formation of the neoplasm. These include roles such as: inhibition of apoptosis, increasing ATP levels and promoting ROS, which are considered important factors of angiogenesis. As previously mentioned, NNMT is likely to become a target for some cancers such as colorectal malignancy, however, involvement of NNMT in other common cancers, such as breast cancer, is yet to be distinguished and therefore requires further investigation (Xie et al. 2014).

There is a different view in regards to the function of a Nicotinamide derivative, which may act as a potent anti-angiogenesis agent. This Nicotinamide derivative is known as 2-[1-[1-(6-chloro-5-fluoropyrimidin-4-yl)ethyl]piperidin-4-ylamino]-N-(3-chlorophenyl) pyridine-3-carboxamide (BRN-250). BRN-250 and BRN-103 are believed to inhibit VEGF in ECs. The main physiological contribution of BRN-250 is to inhibit the proliferation of the Human Umbilical Vascular Endothelial Cells (HUVECs). It is suggested, this anti-angiogenesis activity could be used in the cancer treatment in some certain malignancies (Choi et al. 2013).

62.11 P120RasGAP and Angiogenesis in Normal Physiology, Disease and Malignancy

The RASA1 gene belongs to a family of genes known as the Ras genes (Chan and Chen 2012). Ras is an abbreviation for "rat sarcoma", which alludes to the fact that these genes are heavily involved in neoplasms and are considered proto-oncogenes (Chan and Chen 2012). The RASA1 gene is located on chromosome number 5. The protein that this gene codes for is known as P120RasGAP (Lapinski et al. 2007). Technically speaking, the P120RasGAP protein is part of a family of proteins known as small GTPase. Small GTPases are a class of enzymes that hydrolyze guanosine triphosphate. This hydrolysis acts as a molecular switch, either activating certain proteins, or deactivating them (Lapinski et al. 2007). P120RasGap protein is vital in normal cells as it acts as a regulator for a signaling transaction pathway known as the RAS pathway (Cooper and Kashishian 1993; Pakneshan et al. 2013; Rahman et al. 2015). The RAS signal transduction is normally deactivated by P120RasGap, but when it is activated, it leads to an activation of several effector molecules and proteins. These effectors include Ral-GEF, GAFs, PI 3-K, and MEKK (Reuter et al. 2000). All of these effectors are enzymes, and are responsible for activating genes that contribute to cellular growth, proliferation, and differentiation respectively (Reuter et al. 2000). When the RAS pathway is tightly regulated by the inactivation of P120RasGap protein, cells proliferate and differentiate in a controlled manner. This leads to normal angiogenesis, and normal replacement of cells. However, when this pathway becomes overstimulated by a deficiency in the activation mechanism of P120RasGap, cells begin to grow uncontrollably, dividing and causing metastasis, as well as increasing angiogenesis to support the neoplastic cells. A second pathway that P120RasGap helps regulate is through the protein inactivation of c-Src protein. The protein abbreviated as c-Src stands for "cellular Src" or "cellular sarcoma" and it is a non-receptor tyrosine kinase protein (Chan and Chen 2012). c-Src is involved in the phosphorylation of cellular proteins, which are implicated in a variety of cellular functions similar to the protein functions that are observed in the activation of the Ras pathway. The c-Src protein is correlated with hyperactivity in human neoplasias, however, genetic mutations for the gene responsible for c-Src are only rarely observed (Chan and Chen 2012). Since the c-Src protein is only rarely mutated but is highly correlated with human neoplasia, it is hypothesized that the protein which regulates it may be involved, hence the P120RasGap protein. This was first hypothesized as tests were conducted on cultivated neoplastic human cell lines that contained mutations in the P120RasGap protein as a result of mutation in the RASA1 gene (Chan and Chen 2012). The hypothesis was tested by inhibiting the P120RasGap protein with a molecule known to inhibit the enzyme farnesyltransferase (Chan and Chen 2012). Farnesyltransferase is an enzyme that targets proteins of the Ras family, and as a result, is used frequently in carcinogenesis research. It modifies proteins of the Ras family by adding a lipid group. This allows proteins such as the P120RasGap to associate to specific membranes within a cell such as the golgi apparatus membrane and endoplasmic reticulum membrane (Chan and Chen 2012). In one study in which Farnesyltransferase was inhibited, proteins of the Ras family such as P120RasGap were unable to activate the c-Src protein, and it is believed that inhibition of P120RasGap is the reason as to why this occurs. The inability of P120RasGap protein to bind to membranes appears to be integral to activating c-Src. It is important to note that Farnesyltransferase inhibition does not inactivate the P120RasGap protein, but simply does not change it into the confirmation required to bind to membranes of cellular organelles (Chan and Chen 2012). Thus from this study, it is evident that P120RasGap protein, and proteins from the Ras family, when bound to cellular membranes activate c-Src and cause angiogenesis and carcinogenesis within neoplastic cells. However, the question as to how Ras family proteins cause the activation of c-Src protein is still not very clear. It was first hypothesised that Ras family proteins create free radicals within a cell known as reactive oxygen species (ROS), which then induce changes of the c-Src protein, and hence cause angiogenesis along with carcinogenesis (Chan and Chen 2012). This hypothesis was proven to be incorrect as a molecule which reacts with and captures ROS species was introduced, yet c-Src was still activated despite there being any ROS (Chan and Chen 2012). A second hypothesis was that c-Src was actuated through autocrine molecules. Autocrine molecules are molecules released from a cell to the extracellular environment, with the same cell that released the molecule acting as the target cell (Weigand et al. 2005). This hypothesis was also disproven as cells were put in a conditioned medium which destroyed autocrine factors, yet the c-Src protein was still able to become activated (Chan and Chen 2012). The last hypothesis on the molecular activation of c-Src via Ras family proteins was that perhaps the proteins of the Ras family such as P120RasGap protein induced c-Src protein activation by a mechanism involving cell adhesion. This hypothesis seemed plausible because one of the effects of P120RasGap protein is the induction of cell adhesion which results in angiogenesis via endothelial cells.

Once again, this hypothesis was proven to also be incorrect, as cells that were placed in a fluid suspension were still able to activate the c-Src protein despite any cellular adhesion or contact. In conclusion, evidence supports the fact that proteins of the Ras family such as P120RasGap induce neoplastic changes within cells, however, the exact biochemical mechanism of how this happens is still unresolved, and requires further research.

62.12 PAK1 and Angiogenesis in Normal Physiology, Disease and Malignancy

The PAK1 gene codes for Serine/threonine-protein kinase enzyme (also known as P21-activated kinase 1) in humans (Rodrigues et al. 2005; Fathers et al. 2010). The PAK proteins are the quintessential effectors that join the Rho GTPases to cytoskeleton reorganisation and nuclear signalling (Klemke et al. 1998). The Rho GTPases are a group of signalling G Proteins and their constituents have been shown to regulate multiple aspects of intracellular actin Dynamics, and are therefore present in all eukaryotic organisms. All of the G proteins act as molecular switches and are involved in cytoskeletal development, cell movement and organelle development. PAK1 effects the Rac1 component of the GTPases, and Rac1 is involved in the control of cell growth and the reorganisation of the cytoskeleton, as well as having a key role in the pleiotropic regulation of the cell cycle and cell-cell adhesion. It also produces Lamellipodia a cytoskeletal Actin projection on the mobile edge of the cell (Noren et al. 2006).

Due to Rac proteins having such an important role in the regulation of cell motility and slight mutation or change in their gene expression can result in the facilitation of tumour growth and the onset of cancer (Brown et al. 1996; Bekri et al. 1997; Boureux et al. 2007). For cancer cells to grow and spread by invading new tissues deregulation of the cells motility is essential. If there were a mutation or a disturbance in the expression of the Rac1 gene then the resulting abnormal cell motility could result in epithelial mesenchymal transition, a predisposing factor for tumour metastasis as well as drug resistant tumour relapse (Small et al. 2002). As a result a few recent studies have tried to initiate tumour therapy involving the pharmacological inhibition of the Rac1 proteins activity.

Epithelial mesenchymal transition is the process by which epithelial cells lose their cell polarity and their cell-cell adhesion, this gives the cells invasive and migratory properties and as a result is linked heavily in the initiation of metastasis in tumours and the progression of cancer (Boureux et al. 2007; Small et al. 2002).

As PAK1 serves as a target for the small GTP binding proteins Rac (including Rac1) it controls the expression of Rac1 gene and can be said to have control over cell-cell adhesion as well as its own control over cell motility and morphology. Therefore if there was a mutation in the PAK1 gene or in the small GTP binding proteins (Rac) it could result in uncontrolled cell motility and loss of cell-cell adhesion. This would then facilitate the growth and spread of tumour cells through epithelial-mesenchymal transition (EMT), resulting in the migration of tumour cells through the basement membrane and entering the blood stream. Once in circulation they would become circulating tumour cells until they find a new site to colonise and undergo clonal outgrowth at these metastatic sites. In theory due to the fact that the tumour cells have now been mobilised by a fault in the PAK1 gene or in the small GTP binding proteins and settled in new metastatic sites, they would need a new blood supply in order to grow and gain the nutrients they require. To get around this problem they would initiate angiogenesis to form tumour blood vessels. These are not as efficient as normal blood vessels and often present with problems such as peri-vascular detachment, vessel dilation and an irregular shape. The vessels however are not believed to supply sufficient oxygen to all the tissues around them (Small et al. 2002).

Due to the mutation in PAK1 and Rac1 and the loss of cell-to-cell adhesion between the constituent cells of the tumour, the migration of the tumour into blood vessels between the deep lying tumour cells will happen. These will then go on to provide them with nutrients allowing them to grow and divide, increasing the tumour size and its ability to spread to other areas in the body. The tumour blood vessels are formed through sprouting angiogenesis (Boureux et al. 2007; Small et al. 2002).

PAK1 and Rac1 gene can be used in therapeutic treatment of tumour cells and reduction in the spread of unwanted angiogenesis. Evidence suggests that the normal endothelial cell compartments (which establish the new blood vessels in angiogenesis) can be targeted using radiation therapy (Jordan et al. 1999; Hanahan and Weinberg 2011; Stallings-Mann et al. 2012). This works by targeting the genomically stable endothelial cells, and due to the fact that new blood vessel formation is a fragile process vulnerable to disruption on many levels the therapy acts as a selection agent to kill the cell compartment. Tumour cells have the ability to evolve resistance at a rapid rate, due to a short generation time (in this case days) and a high frequency of genetic instability causing variation between the constituent tumour cells. This can result in cells with more resistance to traditional therapies and an increase in the occurrence of PAK1 and Rac1 gene mutations, resulting in unwanted angiogenesis leading to tumour growth and the risk of metastasis. Whereas if normal endothelial cells were targeted which have a longer generation time (months) and more genomic stability making them more susceptible to treatment such as radiation therapy (Hanahan and Weinberg 2011; Stallings-Mann et al. 2012).

The result of targeting the endothelial cells leads to their destruction/loss of function, preventing effective angiogenesis to the tumour cells reducing there nutrients and increasing self-pollution due to the loss of the vessels as their waste pathways. So as a way of treating the tumour cells that may present mutations in both the PAK1 and Rac1 genes target the healthy and genetically stable endothelial cells involved in angiogenesis. This stops tumour blood vessel formation and deprives the tumour cells (with PAK1 and Rac1 mutations) of nutrients resulting in their eventually death and preventing their replication and division.

62.13 Sema6A and Angiogenesis in Normal Physiology, Disease and Malignancy

Sema6A also known as Semaphorin 6A is transmembrane bound member of the semaphorins family which are implicated in the development of the neural tissue, regulation of the immune responses and tumorigenesis. Semaphorins are composed of Sema functional domain, a 500 amino-acid residue N-terminal domain, an α integrin-like extracellular domain and a receptor domain that determines binding specificity (Koppel et al. 1997). In the family of semaphorins there are 8 classes, with class 1–2 found in invertebrates, class 3–7 in vertebrates and viral variant encoded by some viruses. Signal transduction of semaphorins are performed through plexin receptor present on the cell surface. The plexin receptor plays an important role cell-cell interactions and can initiate multiple signalling pathways in responses to semaphorin activation. The class 6 semaphorin are composed of sub-class of 4 transmembrane proteins Sema6A-Sema6D which interact with plexin-2A, plexin 4A and plexin A1 (Sema6C and Sema6D) respectively. The class 6 variants of semaphorins have been reported to be expressed in endothelial cells and play a role in both physiology and pathology forms of angiogenic expression. The Sema6B and Sema6D were reported to stimulate endothelial growth by interaction of pro-angiogenic growth factors VEGF2 and FGF2 respectively. The Sema6A and plexin A2/plexin A4 was reported to inhibit endothelial cell proliferation and halt vascular development in tumour cells (Urbich et al. 2012; Kigel et al. 2011). The mechanisms of Sema6A regulation of angiogenic expression were explored in the study by Urbich et al. (2012) which attribute Sema6A inhibitory capacity to its ability to block the VEGFR2. Their data also suggest that inhibition of sprouting angiogenesis was observed with the up regulation of Sema6A expression (Urbich et al. 2012).

Based on the data, we can conclude that Sema6A has critical role in modulation of vessel development inhibiting endogenous signalling proteins preventing expression of angiogenesis. Sema6A may be a potential therapeutic target for angiogenic-associated disease such as tumour angiogenesis.

62.14 Sprouty1 (Spry1) and Angiogenesis in Normal Physiology, Disease and Malignancy

Sprouty1 also known as protein sprouty homolog 1 is a membrane-bound member of Sprouty (Spry) family of proteins. There are four human homologs Spry proteins (Spry1–4) which have been described and are ubiquitously expressed in embryonic and adult tissues. Spry is composed of highly conserved cysteine-rich Spry functional domain at C-terminus typical expressed in all members of Spry proteins and a variable N terminal (Impagnatiello et al. 2001; Edwin et al. 2009). Spry proteins specifically inhibit ligand-inducible receptor tyrosine kinase (RTK)-dependant Ras/

MAPK signalling pathways. Spry1 was initially identified in Drosophila as an antagonist of the fibroblast growth factor (FGF) and epidermal growth factor (EGF) receptor signalling pathways (Minowada et al. 1999). Spry1 plays a key role in a multitude of cellular event including embryonic development, tissue/organ formation and postnatal growth and maintenance. However, in down regulation of the expression of SPRY1 gene can be observed in several different tumours such as breast, prostate and liver. *In vitro* studies have shown that overexpression of Spry1 can inhibit cellular proliferation, migration and limit tumour growth *in vivo*. Therefore, due to the inhibitory capacity on the signalling cascade Ras/MAPK pathway Spry can act as a tumour suppressor (Minowada et al. 1999).

The mechanism by which the membrane-bound Spry1 proteins inhibits the RTK signalling pathways begins through binding to Grb-2 a secondary messenger which initiate Ras/MAPK signalling cascade (Cabrita and Christofori 2008). The active Spry1 protein prevents dimerization of Grb-2 and Sos-1 consequently interrupting the interaction with the Ras kinase and the downstream secondary mediators Raf-1, Mek1, Erk1/2 suspending gene expression. Activation of Spry protein requires phosphorylation which initiates Spry relocation at the Grb-2 and Sos-1 interaction site (Hanafusa et al. 2002). The inhibition of growth factors signalling appears specific to the Ras pathways neither affecting other pathways such as PI3 kinase mediated or growth factor induced RTK. The mechanism of which Spry inhibits Ras activation may be due to the reversible modification of nucleotide sequence Sos protein (Gross et al. 2001). Spry expression is dependent on ligand-induced activation of Ras signalling cascade to activate SPRY gene expression providing a regulatory feedback mechanism to the Ras pathway activation. Along with the inhibition of Ras pathway Spry protein reacts with other endogenous secondary messengers which mediate inhibition of cellular migration by the activation of the PTP1B protein (Gross et al. 2001; Impagnatiello et al. 2001).

Studies of the role of Spry1 in angiogenesis have not been fully characterized but several factors indicate that Spry may be good candidate as a potential inhibitor for therapeutic use (Lo et al. 2004). The earliest performed study of Spry1 protein described the regulatory role in the branching of trachea development in Drosophila which carries similar morphological similarities in mechanisms of utilized in angiogenesis. Studies performed on structural homologs Spry2 and Spry4 of Sprouty family have been reported to play a role in angiogenesis. In addition, as a potent and specific inhibitor of the Ras/MAPK of ligand-induced RTK Spry1 may play a regulatory role in the expression angiogenesis (Lo et al. 2004). In the study by Impagnatillo et al. (2001) Spry1 activation halted endothelial cell proliferation and differentiation by inhibition of the ERK/MAPK pathway induced by angiogenic stimulators basic fibroblast growth factor (bFGF) and VEGF (Impagnatiello et al. 2001). This was further supported by study by Sabatel et al. (2010) which showed that silencing gene expression Spry1 proteins resulted in activation of endothelial allowing these cells to proliferate, migrate and form a network of capillaries *in vivo* which synonymous to the process of angiogenesis.. Based on these properties Spry1 may be a potential endogenous inhibitor to angiogenic expression (Sabatel et al. 2010).

References

Al-Mahmood S, Colin S, Farhat N, Thorin E, Steverlynck C, Chemtob S (2009) Potent in vivo antiangiogenic effects of GS-101 (5′-TATCCGGAGGGCTCGCCATGCTGCT-3′), an antisense oligonucleotide preventing the expression of insulin receptor substrate-1. J Pharmacol Exp Ther 329(2):496–504. doi:10.1124/jpet.108.147496

Aras U, Gandhi YA, Masso-Welch PA, Morris ME (2012) Chemopreventive and anti-angiogenic effects of dietary phenethyl isothiocyanate in an N -methyl nitrosourea-induced breast cancer animal model. Biopharm Drug Dispos 34(2):98–106. doi:10.1002/bdd.1826

Asselta R, Tenchini ML, Duga S (2006) Inherited defects of coagulation factor V: the hemorrhagic side. J Thromb Haemost 4(1):26–34. doi:10.1111/j.1538-7836.2005.01590.x

Avraamides CJ, Garmy-Susini B, Varner JA (2008) Integrins in angiogenesis and lymphangiogenesis. Nat Rev Cancer 8(8):604–617. doi:10.1038/nrc2353

Beckert S, Farrahi F, Aslam RS, Scheuenstuhl H, Konigsrainer A, Hussain MZ, Hunt TK (2006) Lactate stimulates endothelial cell migration. Wound Repair Regen 14(3):321–324. doi:10.1111/j.1743-6109.2006.00127.x

Bekri S, Adelaide J, Merscher S, Grosgeorge J, Caroli-Bosc F, Perucca-Lostanlen D, Kelley PM, Pebusque MJ, Theillet C, Birnbaum D, Gaudray P (1997) Detailed map of a region commonly amplified at 11q13.q14 in human breast carcinoma. Cytogenet Cell Genet 79(1–2):125–131

Bertina RM, Koeleman BP, Koster T, Rosendaal FR, Dirven RJ, de Ronde H, van der Velden PA, Reitsma PH (1994) Mutation in blood coagulation factor V associated with resistance to activated protein C. Nature 369(6475):64–67. doi:10.1038/369064a0

Birge RB, Kalodimos C, Inagaki F, Tanaka S (2009) Crk and CrkL adaptor proteins: networks for physiological and pathological signaling. Cell Commun Signal 7:13. doi:10.1186/1478-811x-7-13

Boldrini L, Gisfredi S, Ursino S, Lucchi M, Mussi A, Fontanini G (2007) CDC25B: relationship with angiogenesis and prognosis in non-small cell lung carcinoma. Hum Pathol 38(10):1563–1568. doi:10.1016/j.humpath.2007.03.006

Boureux A, Vignal E, Faure S, Fort P (2007) Evolution of the Rho family of ras-like GTPases in eukaryotes. Mol Biol Evol 24(1):203–216. doi:10.1093/molbev/msl145

Boutros R, Dozier C, Ducommun B (2006) The when and wheres of CDC25 phosphatases. Curr Opin Cell Biol 18(2):185–191. doi:10.1016/j.ceb.2006.02.003

Boutros R, Lobjois V, Ducommun B (2007) CDC25 phosphatases in cancer cells: key players? Good targets? Nat Rev Cancer 7(7):495–507. doi:10.1038/nrc2169

Brown JL, Stowers L, Baer M, Trejo J, Coughlin S, Chant J (1996) Human Ste20 homologue hPAK1 links GTPases to the JNK MAP kinase pathway. Curr Biol 6(5):598–605

Burroughs SK, Kaluz S, Wang D, Wang K, Van Meir EG, Wang B (2013) Hypoxia inducible factor pathway inhibitors as anticancer therapeutics. Fut Med Chem 5(5):10.4155/fmc.4113.4117. doi:10.4155/fmc.13.17

Cabrita MA, Christofori G (2008) Sprouty proteins, masterminds of receptor tyrosine kinase signaling. Angiogenesis 11(1):53–62. doi:10.1007/s10456-008-9089-1

Carmeliet P, Jain RK (2000) Angiogenesis in cancer and other diseases. Nature 407(6801):249–257. doi:10.1038/35025220

Cazales M, Boutros R, Brezak MC, Chaumeron S, Prevost G, Ducommun B (2007) Pharmacologic inhibition of CDC25 phosphatases impairs interphase microtubule dynamics and mitotic spindle assembly. Mol Cancer Ther 6(1):318–325. doi:10.1158/1535-7163.mct-06-0299

Chai Y, Chipitsyna G, Cui J, Liao B, Liu S, Aysola K, Yezdani M, Reddy ES, Rao VN (2001) c-Fos oncogene regulator Elk-1 interacts with BRCA1 splice variants BRCA1a/1b and enhances BRCA1a/1b-mediated growth suppression in breast cancer cells. Oncogene 20(11):1357–1367. doi:10.1038/sj.onc.1204256

Chan PC, Chen HC (2012) p120RasGAP-mediated activation of c-Src is critical for oncogenic Ras to induce tumor invasion. Cancer Res 72(9):2405–2415. doi:10.1158/0008-5472.can-11-3078

Chen S, Huang H, Yao J, Pan L, Ma H (2014) Heat shock protein B6 potently increases non-small cell lung cancer growth. Mol Med Rep 10(2):677–682. doi:10.3892/mmr.2014.2240

Cheresh DA, Berliner SA, Vicente V, Ruggeri ZM (1989) Recognition of distinct adhesive sites on fibrinogen by related integrins on platelets and endothelial cells. Cell 58(5):945–953

Choi H-E, Choi J-H, Lee JY, Kim JH, Kim JH, Lee JK, Kim GI, Park Y, Chi YH, Paik SH, Lee JH, Lee K-T (2013) Synthesis and evaluation of nicotinamide derivative as anti-angiogenic agents. Bioorg Med Chem Lett 23(7):2083–2088. doi:10.1016/j.bmcl.2013.01.125

Cooper JA, Kashishian A (1993) In vivo binding properties of SH2 domains from GTPase-activating protein and phosphatidylinositol 3-kinase. Mol Cell Biol 13(3):1737–1745

Cursiefen C, Chen L, Borges LP, Jackson D, Cao J, Radziejewski C, D'Amore PA, Dana MR, Wiegand SJ, Streilein JW (2004) VEGF-A stimulates lymphangiogenesis and hemangiogenesis in inflammatory neovascularization via macrophage recruitment. J Clin Invest 113(7):1040–1050. doi:10.1172/jci20465

Donzelli M, Draetta GF (2003) Regulating mammalian checkpoints through Cdc25 inactivation. EMBO Rep 4(7):671–677. doi:10.1038/sj.embor.embor887

Duga S, Asselta R, Tenchini ML (2004) Coagulation factor V. Int J Biochem Cell Biol 36(8):1393–1399. doi:10.1016/j.biocel.2003.08.002

Eckstein JW (2000) Cdc25 as a potential target of anticancer agents. Invest New Drugs 18(2):149–156

Edwin F, Anderson K, Ying C, Patel TB (2009) Intermolecular interactions of sprouty proteins and their implications in development and disease. Mol Pharmacol 76(4):679–691. doi:10.1124/mol.109.055848

Esmon CT (2003) The protein C pathway. Chest 124(3 Suppl):26s–32s

Farzan M, Choe H, Martin KA, Sun Y, Sidelko M, Mackay CR, Gerard NP, Sodroski J, Gerard C (1997) HIV-1 entry and macrophage inflammatory protein-1beta-mediated signaling are independent functions of the chemokine receptor CCR5. J Biol Chem 272(11):6854–6857

Fathers KE, Rodrigues S, Zuo D, Murthy IV, Hallett M, Cardiff R, Park M (2010) CrkII transgene induces atypical mammary gland development and tumorigenesis. Am J Pathol 176(1):446–460. doi:10.2353/ajpath.2010.090383

Galaktionov K, Lee AK, Eckstein J, Draetta G, Meckler J, Loda M, Beach D (1995) CDC25 phosphatases as potential human oncogenes. Science (New York, NY) 269(5230):1575–1577

Garten A, Petzold S, Körner A, S-i I, Kiess W (2009) Nampt: linking NAD biology, metabolism and cancer. Trends Endocrinol Metab 20(3):130–138. doi:10.1016/j.tem.2008.10.004

Ghani QP, Wagner S, Becker HD, Hunt TK, Hussain MZ (2004) Regulatory role of lactate in wound repair. Methods Enzymol 381:565–575. doi:10.1016/s0076-6879(04)81036-x

Gopalan B, Ito I, Branch CD, Stephens C, Roth JA, Ramesh R (2004) Nanoparticle based systemic gene therapy for lung cancer: molecular mechanisms and strategies to suppress nanoparticle-mediated inflammatory response. Technol Cancer Res Treat 3(6):647–657

Gottfried E, Kunz-Schughart LA, Ebner S, Mueller-Klieser W, Hoves S, Andreesen R, Mackensen A, Kreutz M (2006) Tumor-derived lactic acid modulates dendritic cell activation and antigen expression. Blood 107(5):2013–2021. doi:10.1182/blood-2005-05-1795

Goukassian D, Diez-Juan A, Asahara T, Schratzberger P, Silver M, Murayama T, Isner JM, Andres V (2001) Overexpression of p27(Kip1) by doxycycline-regulated adenoviral vectors inhibits endothelial cell proliferation and migration and impairs angiogenesis. FASEB J 15(11):1877–1885. doi:10.1096/fj.01-0065com

Grande M, Milito G, Attina GM, Cadeddu F, Muzi MG, Nigro C, Rulli F, Farinon AM (2008) Evaluation of clinical, laboratory and morphologic prognostic factors in colon cancer. World J Surg Oncol 6:98. doi:10.1186/1477-7819-6-98

Gross I, Bassit B, Benezra M, Licht JD (2001) Mammalian sprouty proteins inhibit cell growth and differentiation by preventing ras activation. J Biol Chem 276(49):46460–46468. doi:10.1074/jbc.M108234200

Hanafusa H, Torii S, Yasunaga T, Nishida E (2002) Sprouty1 and Sprouty2 provide a control mechanism for the Ras/MAPK signalling pathway. Nat Cell Biol 4(11):850–858. doi:10.1038/ncb867

Hanahan D, Weinberg RA (2011) Hallmarks of cancer: the next generation. Cell 144(5):646–674. doi:10.1016/j.cell.2011.02.013

Heise EA, Fort PE (2011) Impact of diabetes on alpha-crystallins and other heat shock proteins in the eye. J Ocular Biol Dis Info 4(1–2):62–69. doi:10.1007/s12177-011-9073-7

Hoekema L, Nicolaes GA, Hemker HC, Tans G, Rosing J (1997) Human factor Va1 and factor Va2: properties in the procoagulant and anticoagulant pathways. Biochemistry 36(11):3331–3335. doi:10.1021/bi9623284

Hos D, Regenfuss B, Bock F, Onderka J, Cursiefen C (2011) Blockade of insulin receptor substrate-1 inhibits corneal lymphangiogenesis. Invest Ophthalmol Vis Sci 52(8):5778–5785. doi:10.1167/iovs.10-6816

Hubner RA, Houlston RS (2007) MTHFR C677T and colorectal cancer risk: a meta-analysis of 25 populations. Int J Cancer 120(5):1027–1035. doi:10.1002/ijc.22440

Hunt TK, Aslam RS, Beckert S, Wagner S, Ghani QP, Hussain MZ, Roy S, Sen CK (2007) Aerobically derived lactate stimulates revascularization and tissue repair via redox mechanisms. Antioxid Redox Signal 9(8):1115–1124. doi:10.1089/ars.2007.1674

Hunt TK, Aslam R, Hussain Z, Beckert S (2008) Lactate, with oxygen, incites angiogenesis. Adv Exp Med Biol 614:73–80. doi:10.1007/978-0-387-74911-2_9

Impagnatiello MA, Weitzer S, Gannon G, Compagni A, Cotten M, Christofori G (2001) Mammalian sprouty-1 and -2 are membrane-anchored phosphoprotein inhibitors of growth factor signaling in endothelial cells. J Cell Biol 152(5):1087–1098

Ito I, Ji L, Tanaka F, Saito Y, Gopalan B, Branch CD, Xu K, Atkinson EN, Bekele BN, Stephens LC, Minna JD, Roth JA, Ramesh R (2004) Liposomal vector mediated delivery of the 3p FUS1 gene demonstrates potent antitumor activity against human lung cancer in vivo. Cancer Gene Ther 11(11):733–739. doi:10.1038/sj.cgt.7700756

Iversen LH, Thorlacius-Ussing O (2002) Relationship of coagulation test abnormalities to tumour burden and postoperative DVT in resected colorectal cancer. Thromb Haemost 87(3):402–408

Iversen LH, Okholm M, Thorlacius-Ussing O (1996) Pre- and postoperative state of coagulation and fibrinolysis in plasma of patients with benign and malignant colorectal disease--a preliminary study. Thromb Haemost 76(4):523–528

Ji L, Roth JA (2008) Tumor suppressor FUS1 signaling pathway. J Thorac Oncol 3(4):327–330. doi:10.1097/JTO.0b013e31816bce65

Jordan P, Brazao R, Boavida MG, Gespach C, Chastre E (1999) Cloning of a novel human Rac1b splice variant with increased expression in colorectal tumors. Oncogene 18(48):6835–6839. doi:10.1038/sj.onc.1203233

Karlsson C, Katich S, Hagting A, Hoffmann I, Pines J (1999) Cdc25B and Cdc25C differ markedly in their properties as initiators of mitosis. J Cell Biol 146(3):573–584

Kigel B, Rabinowicz N, Varshavsky A, Kessler O, Neufeld G (2011) Plexin-A4 promotes tumor progression and tumor angiogenesis by enhancement of VEGF and bFGF signaling. Blood 118(15):4285–4296. doi:10.1182/blood-2011-03-341388

Kim YH, Kwei KA, Girard L, Salari K, Kao J, Pacyna-Gengelbach M, Wang P, Hernandez-Boussard T, Gazdar AF, Petersen I, Minna JD, Pollack JR (2010) Genomic and functional analysis identifies CRKL as an oncogene amplified in lung cancer. Oncogene 29(10):1421–1430. doi:10.1038/onc.2009.437

Klemke RL, Leng J, Molander R, Brooks PC, Vuori K, Cheresh DA (1998) CAS/Crk coupling serves as a "molecular switch" for induction of cell migration. J Cell Biol 140(4):961–972

Klepfish A, Greco MA, Karpatkin S (1993) Thrombin stimulates melanoma tumor-cell binding to endothelial cells and subendothelial matrix. Int J Cancer 53(6):978–982

Klerk CP, Smorenburg SM, Spek CA, Van Noorden CJ (2007) Colon cancer metastasis in mouse liver is not affected by hypercoagulability due to Factor V Leiden mutation. J Cell Mol Med 11(3):561–568. doi:10.1111/j.1582-4934.2007.00046.x

Kockar C, Kockar O, Ozturk M, Dagli M, Bavbek N, Kosar A (2005) Global fibrinolytic capacity increased exponentially in metastatic colorectal cancer. Clin Applied Thromb Hemost 11(2):227–230

Kondo M, Ji L, Kamibayashi C, Tomizawa Y, Randle D, Sekido Y, Yokota J, Kashuba V, Zabarovsky E, Kuzmin I, Lerman M, Roth J, Minna JD (2001) Overexpression of candidate tumor suppressor gene FUS1 isolated from the 3p21.3 homozygous deletion region leads to G1 arrest and growth inhibition of lung cancer cells. Oncogene 20(43):6258–6262. doi:10.1038/sj.onc.1204832

Koppel AM, Feiner L, Kobayashi H, Raper JA (1997) A 70 amino acid region within the semaphorin domain activates specific cellular response of semaphorin family members. Neuron 19(3):531–537

Lambrecht S, Juchtmans N, Elewaut D (2014) Heat-shock proteins in stromal joint tissues: innocent bystanders or disease-initiating proteins? Rheumatology (Oxford) 53(2):223–232. doi:10.1093/rheumatology/ket277

Lammer C, Wagerer S, Saffrich R, Mertens D, Ansorge W, Hoffmann I (1998) The cdc25B phosphatase is essential for the G2/M phase transition in human cells. J Cell Sci 111(Pt 16):2445–2453

Lapinski PE, Bauler TJ, Brown EJ, Hughes ED, Saunders TL, King PD (2007) Generation of mice with a conditional allele of the p120 Ras GTPase-activating protein. Genesis (New York, NY: 2000) 45(12):762–767. doi:10.1002/dvg.20354

Lee DY, Deng Z, Wang CH, Yang BB (2007) MicroRNA-378 promotes cell survival, tumor growth, and angiogenesis by targeting SuFu and Fus-1 expression. Proc Natl Acad Sci U S A 104(51):20350–20355. doi:10.1073/pnas.0706901104

Li G, Kawashima H, Ji L, Ogose A, Ariizumi T, Umezu H, Xu Y, Hotta T, Endo N (2011) Frequent absence of tumor suppressor FUS1 protein expression in human bone and soft tissue sarcomas. Anticancer Res 31(1):11–21

Lo TL, Yusoff P, Fong CW, Guo K, McCaw BJ, Phillips WA, Yang H, Wong ES, Leong HF, Zeng Q, Putti TC, Guy GR (2004) The ras/mitogen-activated protein kinase pathway inhibitor and likely tumor suppressor proteins, sprouty 1 and sprouty 2 are deregulated in breast cancer. Cancer Res 64(17):6127–6136. doi:10.1158/0008-5472.can-04-1207

Loktionov A, Watson MA, Stebbings WS, Speakman CT, Bingham SA (2003) Plasminogen activator inhibitor-1 gene polymorphism and colorectal cancer risk and prognosis. Cancer Lett 189(2):189–196

Lolmede K, Dunand de Saint Front V, Galitzhy J, Lafontan M, Bouloumie A (2003) Effects of hypoxia on the expression of proangiogenic factors in differentiated 3T3-F442A adipocytes. Int J Obes Relat Metab Disord 27(10):1187–1195

Lunt SY, Vander Heiden MG (2011) Aerobic glycolysis: meeting the metabolic requirements of cell proliferation. Annu Rev Cell Dev Biol 27:441–464. doi:10.1146/annurev-cellbio-092910-154237

Matsuda M, Mayer BJ, Hanafusa H (1991) Identification of domains of the v-crk oncogene product sufficient for association with phosphotyrosine-containing proteins. Mol Cell Biol 11(3):1607–1613

Maurer M, von Stebut E (2004) Macrophage inflammatory protein-1. Int J Biochem Cell Biol 36(10):1882–1886. doi:10.1016/j.biocel.2003.10.019

Mayer BJ, Hanafusa H (1990) Association of the v-crk oncogene product with phosphotyrosine-containing proteins and protein kinase activity. Proc Natl Acad Sci U S A 87(7):2638–2642

Menten P, Wuyts A, Van Damme J (2002) Macrophage inflammatory protein-1. Cytokine Growth Factor Rev 13(6):455–481

Miller GJ, Bauer KA, Howarth DJ, Cooper JA, Humphries SE, Rosenberg RD (2004) Increased incidence of neoplasia of the digestive tract in men with persistent activation of the coagulant pathway. J Thromb Haemost 2(12):2107–2114. doi:10.1111/j.1538-7836.2004.01011.x

Minowada G, Jarvis LA, Chi CL, Neubuser A, Sun X, Hacohen N, Krasnow MA, Martin GR (1999) Vertebrate Sprouty genes are induced by FGF signaling and can cause chondrodysplasia when overexpressed. Development 126(20):4465–4475

Morris P (1989) Angiogenic potency of nucleotide metabolites: potential role in ischemia-induced vascular growth. J Mol Cell Cardiol 21(4):351–358. doi:10.1016/0022-2828(89)90645-7

Mozsik G, Rumi G, Domotor A, Figler M, Gasztonyi B, Papp E, Par A, Par G, Belagyi J, Matus Z, Melegh B (2005) Involvement of serum retinoids and Leiden mutation in patients with esophageal, gastric, liver, pancreatic, and colorectal cancers in Hungary. World J Gastroenterol 11(48):7646–7650

Nagata A, Igarashi M, Jinno S, Suto K, Okayama H (1991) An additional homolog of the fission yeast cdc25+ gene occurs in humans and is highly expressed in some cancer cells. New Biol 3(10):959–968

Nagata T, Murata K, Murata R, Sun SL, Saito Y, Yamaga S, Tanaka N, Tamai K, Moriya K, Kasai N, Sugamura K, Ishii N (2014) Hepatocyte growth factor regulated tyrosine kinase substrate in the peripheral development and function of B-cells. Biochem Biophys Res Commun 443(2):351–356. doi:10.1016/j.bbrc.2013.11.029

Nakamotc T, Sakaii R, Ozawa K, Yazaki Y, Hirai H (1996) Direct binding of C-terminal region of p130Cas to SH2 and SH3 domains of Src kinase. J Biol Chem 271(15):8959–8965

Nicolaes GA, Dahlback B (2002) Factor V and thrombotic disease: description of a janus-faced protein. Arterioscler Thromb Vasc Biol 22(4):530–538

Noren NK, Foos G, Hauser CA, Pasquale EB (2006) The EphB4 receptor suppresses breast cancer cell tumorigenicity through an Abl-Crk pathway. Nat Cell Biol 8(8):815–825. doi:10.1038/ncb1438

Odorisio T, Cianfarani F, Failla CM, Zambruno G (2006) The placenta growth factor in skin angiogenesis. J Dermatol Sci 41(1):11–19. doi:10.1016/j.jdermsci.2005.08.008

Ornstein DL, Cushman M (2003) Cardiology patient page. Factor V Leiden. Circulation 107(15):e94–e97. doi:10.1161/01.cir.0000068167.08920.f1

Pakneshan S, Salajegheh A, Smith RA, Lam AK (2013) Clinicopathological relevance of BRAF mutations in human cancer. Pathology 45(4):346–356. doi:10.1097/PAT.0b013e328360b61d

Palumbo JS, Kombrinck KW, Drew AF, Grimes TS, Kiser JH, Degen JL, Bugge TH (2000) Fibrinogen is an important determinant of the metastatic potential of circulating tumor cells. Blood 96(10):3302–3309

Paspatis GA, Sfyridaki A, Papanikolaou N, Triantafyllou K, Livadiotaki A, Kapsoritakis A, Lydataki N (2002) Resistance to activated protein C, factor V leiden and the prothrombin G20210A variant in patients with colorectal cancer. Pathophysiol Haemost Thromb 32(1):2–7. doi:57282

Petersen J, Weilguny D, Egel R, Nielsen O (1995) Characterization of fus1 of Schizosaccharomyces pombe: a developmentally controlled function needed for conjugation. Mol Cell Biol 15(7):3697–3707

Pittman DD, Tomkinson KN, Michnick D, Selighsohn U, Kaufman RJ (1994) Posttranslational sulfation of factor V is required for efficient thrombin cleavage and activation and for full procoagulant activity. Biochemistry 33(22):6952–6959

Polet F, Feron O (2013) Endothelial cell metabolism and tumour angiogenesis: glucose and glutamine as essential fuels and lactate as the driving force. J Intern Med 273(2):156–165. doi:10.1111/joim.12016

Porporato PE, Payen VL, De Saedeleer CJ, Preat V, Thissen JP, Feron O, Sonveaux P (2012) Lactate stimulates angiogenesis and accelerates the healing of superficial and ischemic wounds in mice. Angiogenesis 15(4):581–592. doi:10.1007/s10456-012-9282-0

Pralhad T, Madhusudan S, Rajendrakumar K (2003) Concept, mechanisms and therapeutics of angiogenesis in cancer and other diseases. J Pharm Pharmacol 55(8):1045–1053. doi:10.1211/0022357021819

Rahman MA, Salajegheh A, Smith RA, Lam AK (2015) Multiple proliferation-survival signalling pathways are simultaneously active in BRAF V600E mutated thyroid carcinomas. Exp Mol Pathol 99(3):492–497. doi:10.1016/j.yexmp.2015.09.006

Rayala SK, Hollander P, Balasenthil S, Molli PR, Bean AJ, Vadlamudi RK, Wang RA, Kumar R (2006) Hepatocyte growth factor-regulated tyrosine kinase substrate (HRS) interacts with PELP1 and activates MAPK. J Biol Chem 281(7):4395–4403. doi:10.1074/jbc.M510368200

Ren XP, Wu J, Wang X, Sartor MA, Qian J, Jones K, Nicolaou P, Pritchard TJ, Fan GC (2009) MicroRNA-320 is involved in the regulation of cardiac ischemia/reperfusion injury by targeting heat-shock protein 20. Circulation 119(17):2357–2366. doi:10.1161/circulationaha.108.814145

Reuter CW, Morgan MA, Bergmann L (2000) Targeting the Ras signaling pathway: a rational, mechanism-based treatment for hematologic malignancies? Blood 96(5):1655–1669

Rickles FR, Patierno S, Fernandez PM (2003) Tissue factor, thrombin, and cancer. Chest 124(3 Suppl):58s–68s

Rodrigues SP, Fathers KE, Chan G, Zuo D, Halwani F, Meterissian S, Park M (2005) CrkI and CrkII function as key signaling integrators for migration and invasion of cancer cells. Mol Cancer Res 3(4):183–194. doi:10.1158/1541-7786.mcr-04-0211

Rudolph J (2007) Cdc25 phosphatases: structure, specificity, and mechanism. Biochemistry 46(12):3595–3604. doi:10.1021/bi700026j

Sabatel C, Cornet AM, Tabruyn SP, Malvaux L, Castermans K, Martial JA, Struman I (2010) Sprouty1, a new target of the angiostatic agent 16K prolactin, negatively regulates angiogenesis. Mol Cancer 9:231. doi:10.1186/1476-4598-9-231

Saito H, Kitamoto M, Kato K, Liu N, Kitamura H, Uemura K, Nogaki F, Takeda T, Mori N, Ono T (2009) Tissue factor and factor v involvement in rat peritoneal fibrosis. Perit Dial Int 29(3):340–351

Sartini D, Seta R, Pozzi V, Morganti S, Rubini C, Zizzi A, Tomasetti M, Santarelli L, Emanuelli M (2015) Role of nicotinamide N-methyltransferase in non-small cell lung cancer: in vitro effect of shRNA-mediated gene silencing on tumourigenicity. Biol Chem 396(3):225–234. doi:10.1515/hsz-2014-0231

Schmitt E, Gehrmann M, Brunet M, Multhoff G, Garrido C (2007) Intracellular and extracellular functions of heat shock proteins: repercussions in cancer therapy. J Leukoc Biol 81(1):15–27. doi:10.1189/jlb.0306167

Segers K, Dahlback B, Nicolaes GA (2007) Coagulation factor V and thrombophilia: background and mechanisms. Thromb Haemost 98(3):530–542

Small JV, Stradal T, Vignal E, Rottner K (2002) The lamellipodium: where motility begins. Trends Cell Biol 12(3):112–120

Stallings-Mann ML, Waldmann J, Zhang Y, Miller E, Gauthier ML, Visscher DW, Downey GP, Radisky ES, Fields AP, Radisky DC (2012) Matrix metalloproteinase induction of Rac1b, a key effector of lung cancer progression. Sci Transl Med 4(142):142ra195. doi:10.1126/scitranslmed.3004062

Stannard C, Soskic V, Godovac-Zimmermann J (2003) Rapid changes in the phosphoproteome show diverse cellular responses following stimulation of human lung fibroblasts with endothelin-1. Biochemistry 42(47):13919–13928. doi:10.1021/bi035414u

Suehiro Y, Veljkovic DK, Fuller N, Motomura Y, Masse JM, Cramer EM, Hayward CP (2005) Endocytosis and storage of plasma factor V by human megakaryocytes. Thromb Haemost 94(3):585–592

Surjana D, Halliday GM, Damian DL (2010) Role of nicotinamide in DNA damage, mutagenesis, and DNA repair. J Nucleic Acid 2010:1–13. doi:10.4061/2010/157591

Thirant C, Galan-Moya E-M, Dubois LG, Pinte S, Chafey P, Broussard C, Varlet P, Devaux B, Soncin F, Gavard J, Junier M-P, Chneiweiss H (2012) Differential proteomic analysis of human glioblastoma and neural stem cells reveals HDGF as a novel angiogenic secreted factor. Stem Cells 30(5):845–853. doi:10.1002/stem.1062

Trabold O, Wagner S, Wicke C, Scheuenstuhl H, Hussain MZ, Rosen N, Seremetiev A, Becker HD, Hunt TK (2003) Lactate and oxygen constitute a fundamental regulatory mechanism in wound healing. Wound Repair Regen 11(6):504–509

Urbich C, Kaluza D, Fromel T, Knau A, Bennewitz K, Boon RA, Bonauer A, Doebele C, Boeckel JN, Hergenreider E, Zeiher AM, Kroll J, Fleming I, Dimmeler S (2012) MicroRNA-27a/b controls endothelial cell repulsion and angiogenesis by targeting semaphorin 6A. Blood 119(6):1607–1616. doi:10.1182/blood-2011-08-373886

Ushio-Fukai M, Nakamura Y (2008) Reactive oxygen species and angiogenesis: NADPH oxidase as target for cancer therapy. Cancer Lett 266(1):37–52. doi:10.1016/j.canlet.2008.02.044

Ushio-Fukai M, Tang Y, Fukai T, Dikalov SI, Ma Y, Fujimoto M, Quinn MT, Pagano PJ, Johnson C, Alexander RW (2002) Novel role of gp91phox-containing NAD(P)H oxidase in vascular endothelial growth factor-induced signaling and angiogenesis. Circ Res 91(12):1160–1167. doi:10.1161/01.res.0000046227.65158.f8

van Hinsbergh VW, Collen A, Koolwijk P (2001) Role of fibrin matrix in angiogenesis. Ann N Y Acad Sci 936:426–437

Varadi K, Rosing J, Tans G, Pabinger I, Keil B, Schwarz HP (1996) Factor V enhances the cofactor function of protein S in the APC-mediated inactivation of factor VIII: influence of the factor VR506Q mutation. Thromb Haemost 76(2):208–214

Vos HL (2006) Inherited defects of coagulation Factor V: the thrombotic side. J Thromb Haemost 4(1):35–40. doi:10.1111/j.1538-7836.2005.01572.x

Waltenberger J, Pardali E (2012) Oiling vascular growth: adipokines can induce (pathological) angiogenesis by using the VEGF/VEGFR system: experts' perspective. Cardiovasc Res 96(2):220–222. doi:10.1093/cvr/cvs180

Wang X, Zhao T, Huang W, Wang T, Qian J, Xu M, Kranias EG, Wang Y, Fan GC (2009) Hsp20-engineered mesenchymal stem cells are resistant to oxidative stress via enhanced activation of Akt and increased secretion of growth factors. Stem Cells (Dayton, Ohio) 27(12):3021–3031. doi:10.1002/stem.230

Wang X, Huang W, Liu G, Cai W, Millard RW, Wang Y, Chang J, Peng T, Fan GC (2014) Cardiomyocytes mediate anti-angiogenesis in type 2 diabetic rats through the exosomal transfer of miR-320 into endothelial cells. J Mol Cell Cardiol 74:139–150. doi:10.1016/j.yjmcc.2014.05.001

Weigand M, Hantel P, Kreienberg R, Waltenberger J (2005) Autocrine vascular endothelial growth factor signalling in breast cancer. Evidence from cell lines and primary breast cancer cultures in vitro. Angiogenesis 8(3):197–204. doi:10.1007/s10456-005-9010-0

Wurdinger T, Tannous BA, Saydam O, Skog J, Grau S, Soutschek J, Weissleder R, Breakefield XO, Krichevsky AM (2008) miR-296 regulates growth factor receptor overexpression in angiogenic endothelial cells. Cancer Cell 14(5):382–393. doi:10.1016/j.ccr.2008.10.005

Xie X, Yu H, Wang Y, Zhou Y, Li G, Ruan Z, Li F, Wang X, Liu H, Zhang J (2014) Nicotinamide N-methyltransferase enhances the capacity of tumorigenesis associated with the promotion of cell cycle progression in human colorectal cancer cells. Arch Biochem Biophys 564:52–66. doi:10.1016/j.abb.2014.08.017

Zhang Y, Daaka Y (2011) PGE2 promotes angiogenesis through EP4 and PKA Cgamma pathway. Blood 118(19):5355–5364. doi:10.1182/blood-2011-04-350587

Zhang X, Wang X, Zhu H, Kranias EG, Tang Y, Peng T, Chang J, Fan GC (2012) Hsp20 functions as a novel cardiokine in promoting angiogenesis via activation of VEGFR2. PLoS One 7(3):e32765. doi:10.1371/journal.pone.0032765

Ziemann E, Grzywacz T, Luszczyk M, Laskowski R, Olek RA, Gibson AL (2011) Aerobic and anaerobic changes with high-intensity interval training in active college-aged men. J Strength Cond Res 25(4):1104–1112. doi:10.1519/JSC.1100b1013e3181d1109ec1109

FSC
www.fsc.org
MIX
Papier aus verantwortungsvollen Quellen
Paper from responsible sources
FSC® C105338

If you have any concerns about our products,
you can contact us on
ProductSafety@springernature.com

In case Publisher is established outside the EU,
the EU authorized representative is:
Springer Nature Customer Service Center GmbH
Europaplatz 3, 69115 Heidelberg, Germany

Printed by Libri Plureos GmbH
in Hamburg, Germany